GENETICS AND EVOLUTION OF AGING

Contemporary Issues in Genetics and Evolution

VOLUME 3

Genetics and Evolution of Aging

Edited by
MICHAEL R. ROSE and CALEB E. FINCH

Contributions with an asterisk in the table of contents were first published in *Genetica*, Volume 91 (1993)

Kluwer Academic Publishers
DORDRECHT / BOSTON / LONDON

Library of Congress Cataloging-in-Publication Data

Genetics and evolution of aging / edited by Michael R. Rose and Caleb
   E. Finch.
       p.   cm. -- (Contemporary issues in genetics and evolution ; v.
   3)
    "In part a re-publication of a special issue of Genetica, vol. 91,
   no. 1-3" -- Pref.
    ISBN 0-7923-2902-3
    1. Aging--Genetic aspects. 2. Evolution.   I. Rose, Michael R.
   (Michael Robertson), 1955-   . II. Finch, Caleb Ellicott.
   III. Series.
   QP86.G397 1994
   574.3'72--dc20   94-16569

ISBN 0-7923-2902-3

Published by Kluwer Academic Publishers,
P.O. Box 17, 3300 AA Dordrecht, The Netherlands.

Kluwer Academic Publishers incorporates
the publishing programmes of
D. Reidel, Martinus Nijhoff, Dr W. Junk and MTP Press.

Sold and distributed in the U.S.A. and Canada
by Kluwer Academic Publishers,
101 Philip Drive, Norwell, MA 02061, U.S.A.

In all other countries, sold and distributed
by Kluwer Academic Publishers Group,
P.O. Box 322, 3300 AH Dordrecht, The Netherlands.

Cover illustration
There are two reasons for choosing this combination of
figures, beyond their simplicity. The first is that they
concern two of the most important species in the volume,
Caenorhabditis elegans and *Drosophila melanogaster.* The
second is that they illustrate both genetic methods, with the
mapping diagram of *C. elegans*, and evolutionary methods,
with the cladogram from *D. melanogaster.* Thus the
combination embodies the two elements of the title. (*C.
elegans* from the article by T.E. Johnson *et al.*;
D. melanogaster from the article by J.E. Fleming *et al.*).

Printed on acid-free paper

Printed in the Netherlands

Contents

* Contributions indicated with an asterisk were first published in *Genetica*, Volume 91 (1993).

M.R. Rose and C.E. Finch (eds.), Genetics and Evolution of Aging, 1, 1994.
© 1994 *Kluwer Academic Publishers. Printed in the Netherlands.*

Preface

This book is, in part, a re-publication of a special issue of *Genetica*, vol. 91, No. 1-3, pp. 1-290, also edited by us. That special issue in turn arose because of the interest that we felt in combining genetic, evolutionary, and other perspectives on the problem of aging. Our conviction is that the field of aging research turned a corner in the 1980's, and is now in a period of rapid progress. Genetic and evolutionary theories and methods have been central, in our opinion, for this renaissance of gerontology. Let this book stand as evidence that research on aging is no longer stagnant.

Most of the articles were reviewed anonymously by at least two referees. All authors were invited to contribute, but most submissions were substantially revised before publication. We have been gratified by the large proportion who sent us articles, and we regret those cases where difficulties of timing and coordination have prevented the inclusion of some of the proposed articles. The editors divided up the handling of manuscripts so as to prevent conflicts of interests and personalities, as well as to maximize the editorial expertise available in each review process.

Some articles were essentially unreviewed. These include the articles by Rose and Finch, Clark, Promislow and Tatar, Holliday, and Richardson and Pahlavani. These articles were prepared after all the refereed articles were in their final form for publication in *Genetica*. All but one of these did *not* appear in *Genetica*. This second group of authors was asked to provide their personal perspective on some issue, or issues, that the editors felt was either not adequately discussed in our first collection of articles or inadequately synthesized. In other words, we wanted some additional mortar to hold together the bricks of our edifice. In a few passages, these articles may be speculative or idiosyncratic, and we do not exclude ourselves from this stricture. But we wanted the book to 'breathe' somewhat more than the special issue for *Genetica* could.

We are grateful to many people for help with this book. But the Editor-in-Chief of *Genetica*, John McDonald, is particularly responsible for persuading us to undertake this task, as well as helping us along the way. We have also been pleased by the interest of the publishers, specifically Adrian Plaizier. We also thank Linda Mitchell and Amy Weinstein for their help with tracking down authors and referees. Theodore J. Nusbaum read the introductory article by the editors. We are also very grateful to our authors and referees, without whom this book would have been as the sound of one hand clapping.

Michael R. Rose
University of California, Irvine

Caleb E. Finch
University of Southern California, Los Angeles

To *Della* and *Doris*,
in appreciation of their patience

PART ONE
General perspectives on aging

Introduction

Aging is one of those subjects that many biologists feel is largely unknown. Therefore, they often feel comfortable offering extremely facile generalizations that are either unsupported or directly refuted in the experimental literature. Despite this unfortunate precedent, aging is a very broad phenomenon that calls out for integration beyond the mere collecting together of results from disparate laboratory organisms.

With this in mind, Part One offers several different synthetic perspectives. The editors, Rose and Finch, provide a verbal synthesis of the field that deliberately attempts to look at aging from both sides, the evolutionary and the molecular. The articles by Charlesworth and Clark both provide population-genetic perspectives on aging, the former more mathematical, the latter more experimental. Bell takes a completely different approach, arguing that aging may not be the result of evolutionary forces. Bell's model instead proposes that aging could arise from the progressive deterioration of chronic host-pathogen interactions. This is the first detailed publication of this model. It marks something of a return to the type of aging theories that predominated in the 1950's and 1960's, theories like the somatic mutation and error catastrophe theories. We hope that the reader will be interested by the contrast in views between the articles based on evolutionary theory and that of Bell.

M.R. Rose and C.E. Finch (eds.), Genetics and Evolution of Aging, 5–12, 1994.
© 1994 *Kluwer Academic Publishers. Printed in the Netherlands.*

The Janiform genetics of aging

Michael R. Rose[1] & Caleb E. Finch[2]
[1] *Department of Ecology and Evolutionary Biology, University of California, Irvine, CA 92717, USA*
[2] *Andrus Gerontology Center and Department of Biological Sciences, University of Southern California, Los Angeles, CA 90089, USA*

Received and accepted 9 July 1993

Introduction

The genetics of aging is Janus-faced, looking in one direction to gene function, molecular biology, and cell biology, looking in the other direction to fitness, population genetics, and evolution. The field will not be successful if its practitioners only look in one direction or the other. On one face, the molecular genetics of aging often makes discoveries in which the biochemical effect is more apparent than the functional significance. On the other face, the population genetics of aging has difficulties with mechanistic content, such that its theories and their tests are often put forward in a manner that resists biological interpretation. If we are to achieve a remotely satisfying scientific understanding of aging, we must bear both faces of the problem in mind.

To that end, the present collection of articles contains contributions chosen to balance the two views of aging. But we would go further. The problems of the genetics of aging are not merely a list of issues from molecular genetics and a list from population genetics, stapled together. A deeper understanding of ancient Janus, the Roman god, reveals that the two faces of Janus contemplate the simultaneous union of opposites, the bringing together of becoming and ending. In this manner, the specific problems of the genetics of aging simultaneously involve and implicate both molecular and population genetics. Indeed, their severance into two domains of discourse may occlude our senses due to the imposition of partial vision. But Janus looked both ways, and folded both views together, achieving a depth of understanding and control that ac-counted for this god's particular hold over Roman culture. Our suggestion would be that the future development of the genetics of aging must feature the integration of molecular and population genetic perspectives.

There are some qualifications that need to be appended to this bald programmatic statement. The two major components of the genetics of aging, molecular and population, must continue to cultivate their own particular strengths. Specialization has considerable value in its concentration of effort, resources, and attention. In addition, the admonition to integrate the study of aging is by no means specific to genetic contexts; it is salient for cognate issues in cell biology, neurobiology, demography, comparative biology, and related fields. Finally, we admit that any assertion of this kind is valueless unless it can be directly related to the outstanding problems of a field.

Therefore, in what follows we outline scientific problems of aging that, we propose, integrally require research from both population and molecular genetic perspectives. Some of these problems arise in the papers that we have collected here, but they are by no means exhaustively covered by them. We are not supposing that our list of problems is definitive, merely illustrative.

Genetic foundations of aging

How many loci determine patterns of aging?
There is perhaps no more elementary problem for the genetics of aging than the number of loci that influence the phenomenon. For some characters,

like eye color, this may not be a fundamental problem; it will be expected in such cases that a physiologically delimitable set of loci defines a pathway of pigment synthesis and postsynthetic modification. The accumulation of genetic complementation data could therefore proceed apace. For aging, there is no warrant for assuming automatically that such a well-defined situation exists. At least this warrant does not obtain unless it is supposed that aging is the result of some mechanism that has evolved specifically to kill off the chronologically older individuals (Weismann, 1889; Denckla, 1975). But such theories have long been exploded relics (Rose, 1991).

From an *evolutionary* perspective, there are good reasons for supposing that aging will be influenced by numerous loci, although not all of these need be polymorphic at any one time. The gravamen of the evolutionary analysis is that aging is defined by all those physiological processes that determine the long-term deterioration of survival and reproduction of the adult organism. But all loci that remain functional through evolutionary time, and thus do not degenerate through evolutionary time, must make some positive contribution to either survival or reproduction, leaving aside DNAs that derive from parasitic sequences, like transposable elements. Otherwise mutation pressure would make the loci degenerate in sequence, and become untranslatable. Given that almost all, nonparasitic, transcribed loci inherently affect survival or reproduction, and that aging is merely a last phase of deterioration in survival and reproduction, then any locus having an allele with a late-age effect is a locus contributing to the genetics of aging. The main complication to this line of argument is its neglect of environmental variation. An allele may have a large effect upon aging in some environments, but not others. This can greatly complicate the interpretation of the functional significance of an allele.

From a *molecular or cellular* perspective, there are also good reasons for supposing that aging will be influenced by numerous gene loci. Consider, as an example, that set of genes that allows cell replacement. This function varies widely between cell types in different organisms. At one extreme are many vertebrate cell types that maintain the capacity for proliferation and replacement. Erythrocytes, for example, show no loss of production from bone marrow stem cells during the lifespan. However, most neurons are irreplaceable. Nucleated postmitotic cells, whether in the organism or in cell culture systems, are thought to have the same sets of genes as any other cell, according to the doctrine of genomic totipotency. Any absence of replication and regeneration in certain cells or tissues of adults is considered part of the program of vertebrate development and differentiation, but also has profound consequences for aging. Drosophila and Caenorhabditis are at the other extreme, and as adults completely lack somatic cell replacement. Other sets of genes that differ between cells confer resistance to damage from depolarizing stimulae. The major differences between neuron types in the mammalian brain in vulnerability to excitatory amino acid neurotransmitters are attributed to the amount of calcium binding proteins: neurons differ in their expression of these genes. We suggest that, for each organism, the patterns of cell differentiation that arise during development have major influences on aging that may arise through myriad combinatorics of gene expression. Virtually any gene may thus contribute to adult aging, depending on the environmental circumstances, as a translated gene is likely to influence one or more of growth, differentiation, regeneration, proliferation, and repair, which in turn will shape aging.

Some of the articles published here are directly relevant to the problem of locus number. Fleming *et al.* (this issue) present evidence for the involvement of a hundred or so loci in the differentiation of lines of D. melanogaster with different rates of aging. Tyler *et al.* (this issue) demonstrate that the effect upon lifespan of genetic variation at the superoxide dismutase locus in Drosophila is relatively small. Both of these points suggest that aging phenotypes are determined by many loci, each of individually small effect. That is, aging appears to be a character better suited to quantitative genetics (cf. Falconer, 1981) than Mendelian genetics. This finding, if generally true, constitutes a profound constraint on the genetic analysis of aging. In particular, the same Mendelian approach that might be successful with eye color or sex determination might have little prospect for success with aging, if only because of the number of loci involved. On the other hand, the research of Jazwinski and Johnson (both this issue) suggests that much progress can be made with sufficiently powerful single-gene approaches, particularly when molecular methods are used.

Are aging loci polymorphic or monomorphic?
A common stereotype of genetic analysis is the expectation that loci coding for important functions will be monomorphic because of intense natural selection. For most characters, this stereotype is not likely to be universally valid, if only because of balancing selection arising intermittently. But in the case of aging, it is not likely to hold for still other reasons.

The first phenomenon of relevance is mutation accumulation of alleles with strictly age-specific effects (Edney & Gill, 1968; Charlesworth, 1980, this issue; Rose, 1991). Alleles that cause material differences in rates or mechanisms of aging may nonetheless be effectively neutral, compared to each other. This arises not because these alleles have no biologically detectable effects, but because they do not have any differential effects on fitness, providing only that their effects are expressed at sufficiently great adult ages. Thus, when such alleles arise, we expect them to accumulate by mutation pressure and genetic drift alone. In so doing, they will generate a great deal of genetic polymorphism.

A completely contrasting scenario arises when alleles have pleiotropic effects over multiple ages, particularly over both early and late ages. Such alleles could have large effects on the evolution of aging, because alleles with beneficial effects at early ages could be favored by natural selection notwithstanding later deleterious effects, which is referred to as 'antagonistic pleiotropy', since the strength of natural selection is much greater at earlier ages (Williams, 1957; Charlesworth, 1980, this issue). An additional effect of this type of selection is that it can lead to the establishment and maintenance of genetic polymorphism, because it can induce balancing selection (Rose, 1985). Thus two major population genetic mechanisms of aging can establish genetic polymorphisms, at least some of the time.

An alternative type of evolutionary theory is optimization of life-history (Partridge & Barton, 1993, this issue). Optimization mechanisms are not well-defined genetically, nor have the evolutionary trade-off functions built into these theories ever

been measured. In any case, it is clearly assumed that the optimization mechanism is supposed to lead to the monomorphic fixation of alleles at loci affecting aging. To the extent that this is a valid model of the evolution of aging, we expect little genetic variance for aging-related characters.

The molecular findings most pertinent to polymorphism are studies of the molecular genetics of Mhc (the main histocompatibility complex) loci (see Crew, this issue, inter alia). These studies reveal high levels of allelic variability among loci determining disease resistance. These high levels of genetic variability presumably arise from the function of these loci in antigen response. This Mhc system has been strongly implicated as a factor controlling aging (Crew, Yunis & Salazar, this issue).

At the whole-organism level, the general pattern found in outbreeding species is one of genetic polymorphism for aging-related characters (Rose, 1991). This genetic polymorphism may not be high in terms of heritability values (vid. Falconer, 1981), partly because of environmental effects on the characters studied. But it is clear from quantitative-genetic and selection studies that there is usually selectable genetic variation for aging in well-studied organisms that do not normally self or reproduce parthenogenetically. [See the Drosophila studies in this issue for specific examples.] Two mechanisms can explain these findings: mutation accumulation and antagonistic pleiotropy, as discussed above. For optimization theories, these results are anomalous, since evolutionary optimization is supposed to lead to monomorphism. Perhaps for this reason, optimization models for the population genetics of aging are not in widespread use.

Physiological genetics of aging

How many distinct physiological mechanisms determine aging?
If many, frequently polymorphic, loci determine aging, how many distinct physiological pathways do so? That is, is aging determined wholly or largely by a single mechanism, such as free-radical damage? Or are there a number of distinct physiological mechanisms that give rise to progressive deterioration in the older organism?

Evolutionary biology is only a partial guide to

this question, because it usually avoids questions that pertain to the physiological integration of the organism. A central population genetic issue here is patterns of pleiotropy (Rose, 1991). To what extent do loci act via independent causal pathways, versus loci having pleiotropic effects over a range of different physiological sub-systems? Physiological systems that are pleiotropically interconnected are not appropriately thought of, or studied, as independent. Some evolutionary genetic experiments indicate that loci involved in the postponement of Drosophila aging may be at least somewhat specific in the aging mechanisms that they affect (e.g. Service *et al.*, 1988), with some genetic mechanisms apparently determining distinct physiological mechanisms controlling aging.

The kinds of arguments developed for the developmental and cell biology of aging where the number of loci are concerned are also applicable here. It is evident from the biology of aging (Finch, 1990) that a number of different pathways of pathogenesis are involved, and these pathways in turn depend on a complex matrix of mechanisms of differentiation.

While it is apparent that some of the physiological mechanisms of aging may be independent from each other, and others not, we still have no idea concerning the scale at which such interdependence arises. Are the classical physiological systems of respiration, digestion, etc. independent from each other, with respect to the determination of aging? Or is the level at which disaggregation occurs quite different, where the genetics of aging are concerned?

To what extent is aging hierarchically integrated?
Perhaps the most radical form that a physiological interdependence hypothesis can take is the notion that aging is hierarchically integrated, as in Comfort's (1979) hierarchy of clocks. This theory supposes that there are master regulatory processes that set upper limits to survival or the continuance of reproduction, as a function of age. The main problem with this theory from an evolutionary standpoint is that it seems to require selection for something of no advantage in terms of Darwinian fitness.

For many iteroparous organisms, such selection may never occur. But in semelparous organisms, death arises shortly after reproduction. In terms of evolutionary logic, semelparity is classic big-bang reproduction, in which all the resources of the organism are mobilized for one, large-scale, suicidal burst of reproduction. There are evolutionary situations in which this is expected to be the favored life-history pattern (Charlesworth, 1980). Thus, in such organisms, the hormonal controls over maturation and reproduction are simultaneously the master controls for the processes that will bring about death, particularly because of the deleterious pleiotropic effects upon survival of alleles fostering more fecund semelparous reproductive episodes. At the organismal level, endocrine and other humoral systems generally are the natural candidates for master controls of the physiological hierarchy. We note that humoral systems control reproduction in innumerable species of plants as well as animals (cf. Roach, this issue), and therefore are likely to be key regulators of aging and life history in both kingdoms.

Aging and other features of biology

Until recently, the study of aging had a poor reputation compared to other branches of biology. Talks and papers on the subject were generally held to be poor relative to other biological research, in part because there were few widely identified paradigms in the biology of aging. Besides manipulations of lifespan in organisms or cell cultures, most changes of aging have been studied as isolated phenomena by biomedical subfields with little connectivity to larger issues in the biology of aging. However, we hope that the present issue of *Genetica* makes it clear that there are now a number of different lines of investigation into the genetics of aging that have attained such a degree of maturity that they can contribute substantially to biological research. In particular, some of the more important results of aging research shed light on the mechanisms by which important component processes of physiology operate.

Energy metabolism
Energy metabolism is involved in the determination of aging in at least two respects. Firstly, aging appears to be markedly affected by the allocation of energy from somatic functions to reproduction. This was already well-known in the case of semelparous organisms, like Pacific salmon and marsupial mice. In the case of these organisms, castra-

tion greatly increases life span (Robertson, 1961; Bradley, McDonald & Lee, 1980), suggesting that castration forestalls the transfer of resources from soma to reproduction.

Genetic and related studies have greatly increased the clarity with which this connection between aging and energetic metabolism can be made. Studies of the physiology of genetically-postponed aging in Drosophila reveal that flies with postponed aging have increased lipid and glycogen reserves, along with some alterations in metabolic rate (Service, 1987; Graves *et al.*, 1992). The functional consequences of this increased storage, or retention, is an increase in a number of performance characters, from stress resistance to flight duration. While it is not clear that these characters play any direct role in prolonging adult survival, the general enhancement of a number of somatic functions is apparent.

A second possible role of energetic metabolism in the determination of aging is alterations in the scavenging of the free-radical by-products of metabolism. As shown in the paper of Tyler *et al.* (this issue), a more active superoxide dismutase allele is increased in frequency in some genetically longer-lived Drosophila populations. This enzyme is thought to be of major importance in the scavenging of superoxide radicals. While this result is not free from problems of linkage disequilibrium when it comes to inferring a causal involvement for free-radical biochemistry in the postponement of aging, it is at least corroborative of such a hypothesis.

A variety of other circumstantial evidence also supports the involvement of free-radical scavenging in the control of aging. For example, free radical mechanisms are strongly implicated in atherogenesis, which is promoted by lipid peroxidation. The formation of glycated proteins, which slowly accumulate in basement membranes and other sites of very slow protein turnover, is also dependent on free radical mechanisms. Most recently, mitochondrial DNA deletions were found to accumulate during normal aging in the brain, but to a 50-fold greater degree in dopamine-rich regions in which free-radical production is thought to be higher (Soong *et al.*, 1992).

The general principle that is illustrated by this discussion of energy metabolism is that the control of aging may involve basic aspects of organismal or cell metabolism. Aging does not appear to depend on its own unique genetic or biochemical pathways. Rather, it is an outcome of generally important mechanisms.

Neurobiology and behavior
The reproductive schedule, including sexual maturation, is tightly controlled by hormonal mechanisms that coordinate behavior and physiology. The genetics of these intricate hierarchical mechanisms are poorly understood. It is apparent that these physiological systems allow diverse trade-offs, given the many examples of morbidity and mortality as a consequence of reproductive behaviors (Finch, 1990). At one extreme are the species of salmon and marsupial mice that die off as a cohort after reproduction. At the other extreme are the delayed onsets of reproductive tumors and bone loss in women given steroidal hormone supplements. The many mammals and birds with prolonged reproductive schedules that last one or more decades would appear to be selected for brains with low, rather than rapid, neurodegenerative trends. By contrast, rhesus monkeys begin to develop neuritic plaques and cognitive declines at about 20 years, whereas these are very rare in humans before the age of 60, even in familial Alzheimer disease (see Tanzi *et al.*, this issue). At the greatest extreme so far observed in primates is a new model for brain aging, the lemur *Microcebus murinus*, which shows extensive neuropathology by ten years (Maestre & Bons, 1993).

Immunology
The immune system in tetrapods has evolved a remarkable suite of mechanisms that protect against a vast range of invading organisms. Of particular importance are the antigen-presenting receptors that are encoded in genes of the Mhc. Some Mhc genes show extraordinary levels of polymorphisms. In inbred mice with different Mhc haplotypes, Crew (this issue) describes how some Mhc haplotypes are associated with lifespan in mice. While no example of this has come to light in humans so far, various Mhc genes are candidates for influences on rates of aging. Other genes not in the Mhc influence susceptibility to retroviruses that cause mid-life onset of moto-neurone degeneration in wild mice (Gardner, this issue). It thus seems likely that the immunology of aging will depend on numerous loci, with effects on a wide range of

physiological parameters. Many of these points underscore more general conclusions concerning polymorphism already discussed. The Mhc is also a genetic system that bears some hallmarks of physiological integration.

Diversity of aging

How phylogenetically diverse is aging?

Somewhat analogous to the problem of how many different loci are involved in the control of aging in any one organism is the question of the diversity of aging among disparate species. Moreover, it could be argued on first principles that many of the same principles are likely to apply to the question of phylogenetic diversity as apply to the diversity of genetic mechanisms within species.

One extreme scenario would be a universal mechanism of aging, or perhaps one as wide as a phylum, which would delimit survival and reproduction in the chronology of adult life in essentially all species within the taxon. Many aspects of the physiology of different taxa, be they phyla or classes, have such qualities, as witness the physiologies of respiration and excretion in mammals or insects. Even chordates and arthropods have widespread mechanisms in common at the level of organismal functions. At the level of molecular biology, there are even more universal features of life. Much of the research in the aging field in the 1950s and 1960s was based on the premise that such universal features would be found to determine aging, as reviewed critically in Finch (1990) and Rose (1991). Less explicit hopes of this kind may still motivate suggestions that the biochemical controls of life among insects or arthropods, even yeast, are likely to be the same as those of humans. [This will be discussed further below.]

A population-genetic argument that might support fairly universal mechanisms is that antagonistic pleiotropy, in which alleles with early beneficial effects are selectively favored despite later, pleiotropic, deleterious effects, will often involve early reproduction being enhanced at the expense of later survival or reproduction. Since the physiology of reproduction must have hormonal and other controls that precisely entrain it, these are also likely to be important, phylogenetically widespread, controls. But these are the same control mechanisms that are likely to give rise to antagonistic pleiotropy and the evolution of aging. Thus crucial mechanisms of aging could be phylogenetically widespread, at least to the extent that the physiological genetics of reproduction are phylogenetically uniform.

The alternative scenario is that the genetics of aging are as phylogenetically diverse as most other features of life, if not more so. One argument in favor of considerable diversity comes from mutation-accumulation at the population-genetic level, in which alleles with distinct effects only at late ages are not discriminated by natural selection, and therefore can accumulate by mutation pressure alone (Charlesworth, 1980, this issue). Each species is likely to accumulate a distinct set of such alleles both because it would have a unique range of possibilities for such alleles and because the specific alleles that occur by mutation will differ by chance alone.

For now it seems clear that theoretical arguments can be marshalled on both sides of the question of phylogenetic diversity in aging. This particular problem awaits a concerted attack on the problem from an evolutionary perspective. To the extent to which we can determine the genetic controls of aging in multiple species placed on a known phylogeny, we will be able to address the question of the phylogenetic diversity of aging with more than contending speculations.

What are the best model systems for the study of aging?

Even though we don't understand the phylogenetics of aging to any serious extent, there is still the question of which species are most worthwhile to study, genetically and otherwise. An obvious issue in addressing this question is whether or not medical considerations are the main motive for research. If they are not, then the best systems for the study of aging will usually be those research organisms which are best suited to rapid scientific progress, and that also undergo aging within a relatively short period of time. Yeast, nematodes, and fruit flies might then be the obvious choices for research, with numerous other developing research systems also of interest, like small fish that can be grown in aquaria. This is discussed further in Reznick's contribution (this issue).

Complicating this issue is that of medical relevance. As we do not have much definitive knowledge concerning the phylogenetics of aging, it would seem unwise to found hopes for medical breakthroughs on arthropods and pseudocoelomates. Therefore, medical research would seem to require work with mammals. In this context, the standard laboratory mouse and rat would seem the only promising systems, particularly the former, given its more extensive genetics. The papers of Gardner, Mobbs, Yunis and Salazar, and Crew, in this issue, stand as examples of the present state of research in this field. It has also been proposed that mice with genetically postponed aging be created by selection (e.g. Rose, 1990). Such organisms could be used in much the same way as the fruit fly research discussed in several of the articles of this number. This model system might provide the surest basis for medical intervention in human aging, as will be discussed below.

Medical consequences of aging genetics

Aging in human populations

As the article of Albin (this issue) reveals, somewhat abstruse population-genetic concepts like antagonistic pleiotropy and mutation accumulation might nonetheless be of material importance for contemporary human medical genetics. However, the extent to which older humans are particularly subject to genetic pathologies is an open question. (See also Martin, this issue.) The inferential problem is that, in the absence of molecular genetic information that strongly implicates a particular genetic mechanism, we are dependent on pedigrees to infer genetic etiology. But genotype determination of siblings, to say nothing of parents and grandparents, will be particularly difficult for the elderly, whose family members will often have died or moved away. The offspring of the elderly may be more to hand, but they may not yet be old enough for phenotype ascertainment.

How can human aging be postponed?

Ultimately, the most exciting thing about genetic research on aging is that it offers the potential of postponing human aging. As the work of Johnson with nematodes (this issue) and that of various Drosophila workers (this issue) illustrates, it is now eminently feasible to postpone aging genetically in

laboratory animals by a number of different means, from mutagenesis to selection. As with so many other areas of biological research in this century, the prospect of going from basic research with these simple organisms to medical applications is extremely tantalizing.

But, as discussed above, it is an open question whether or not gene-enzyme systems that can be used to postpone aging in nematodes or Drosophila can also be used to postpone the process in humans. It is possible. But any attempt to go from such genetic systems directly to the human, for now, must be considered fraught with uncertainty.

The more conservative route would be research with other mammalian systems. We cannot yet be sure that the mouse-to-man transition will be an easy one, though it seems likely to be easier than any transition from invertebrate to man, given homology. A caveat might be added here that potentially any genetic discovery, in any organism, might prove to be serendipitously important for aging research. An example of this is that findings in bacterial genetics led to the working out of the molecular genetics of xeroderma pigmentosum, a human aging-associated skin disease. However, in most cases, it should be easier to proceed from mouse genetics to human genetics than from any non-vertebrate system to humans.

The problem with advocacy of mammalian research is that it remains unclear that there are any appropriate model systems for genetic research. Inbred mice might die at particular ages from inbreeding depression, not normal mechanisms of mouse, to say nothing of human, aging (see Rose, 1991, Ch. 2; cf. Crew, this issue). In fact, inbred rodents are as subject to criticism as potential models as insects: there is no reason to be confident that their 'aging' processes will be homologous to those in humans, who are almost always outbred.

The great hope in this situation, at least for some, has been to select mice for postponed aging (Rose, 1990), analogously to the Drosophila stocks studied by Arking et al., Fleming et al., and Tyler et al. (all this issue). However, there are a number of problems with this proposal as well. Breeding mice for later reproduction would require a great many years of work, since 'later' in the mice might mean 10-12 months later, each generation. A twenty-generation experiment might take almost two decades to complete. Moreover, to be useful, the stocks resulting from selection must again be free of inbreeding depression. This would require the selection of hundreds of mice starting from thousands of mice, per line. Finally, as the studies of Fleming *et al.* and Tyler *et al.* (both this issue) illustrate, considerable replication of lines is required to make molecular genetic inferences from selectively-differentiated populations. Otherwise studies of such stocks are plagued by falsely positive inferences. Thus there must be multiple control and selected lines, say 4-8 for each treatment. This brings the total size for the experiment well into the tens of thousands of mice, far beyond the capacities of a normal research laboratory, or even a dedicated institute.

How soon all these constraints of time and scale can be overcome is difficult to say. To do so will require the allocation of considerable resources to the project. However, no other approach to the postponement of human aging has ever worked. Now may be the time for policy-makers to consider whether the recent achievements in the genetics of aging warrant a full-scale, genetically-oriented approach to the problem of human aging.

Acknowledgement

For a number of years, the work of the authors on the biology of aging has been supported by the National Institute on Aging, whose support is gratefully acknowledged.

References

Bradley, A. J., I. R. McDonald & A. K. Lee, 1980. Stress and mortality in a small marsupial (*Antechinus stuartii*, Macleay). Gen. Comp. Endocrinol. 40: 188-200.

Charlesworth, B., 1980. Evolution in Age-Structured Populations. Cambridge University Press, Cambridge.

Comfort, A., 1979. The Biology of Senescence, Third Ed. Churchill Livingstone, Edinburgh and London.

Denckla, W. D., 1975. A time to die. Life Sci. 16: 31-44.

Edney, E. B. & R. W. Gill, 1968. Evolution of senescence and specific longevity. Nature 220: 281-282.

Falconer, D. S., 1981. Introduction to Quantitative Genetics, Sec. Ed. Longman Press, London.

Finch, C. E., 1990. Longevity, Senescence, and the Genome. University of Chicago Press, Chicago.

Graves, J. L., E. C. Toolson, C. Jeong, L. N. Vu & M. R. Rose, 1992. Desiccation, flight, glycogen, and postponed senescence in *Drosophila melanogaster*. Phys. Zool. 65: 268-286.

Mestre, N. & N. Bons, 1993. Age-related cytological changes and neuronal loss in basal forebrain cholinergic neurons in *Microcebus murinus* (Lemuian, primate). Neurodegeneration 2: 25-32.

Partridge, L. & N. H. Barton, 1993. Optimality, mutation and the evolution of ageing. Nature 362: 305-311.

Robertson, O. H., 1961. Prolongation of life span of kokanee salmon (*Oncorhynchus nerka kennerlyi*) by castration before beginning of gonad development. Proc. Natl. Acad. Sci. USA 47: 609-621.

Rose, M. R., 1985. Life history evolution with antagonistic pleiotropy and overlapping generations. Theor. Pop. Biol. 28: 342-358.

Rose, M. R., 1990. Should mice be selected for postponed aging? A workshop summary. Growth, Dev. & Aging 54: 7-17.

Rose, M. R., 1991. Evolutionary Biology of Aging. Oxford University Press, New York.

Service, P. M., 1987. Physiological mechanisms of increased stress resistance in *Drosophila melanogaster* selected for postponed senescence. Physiol. Zool. 60: 321-326.

Service, P. M., E. W. Hutchinson & M. R. Rose, 1988. Multiple genetic mechanisms for the evolution of senescence in *Drosophila melanogaster*. Evolution 42: 708-716.

Soong, N. W., D. Hinton, G. A. Cortopassi & N. Arnheim, 1992. The adult brain is mosaic for a specific mtDNA deletion. Nature genetics 2: 318-323.

Weismann, A. J., 1889. Essays upon Heredity and Kindred Biological Problems. Clarendon Press, Oxford.

M.R. Rose and C.E. Finch (eds.), Genetics and Evolution of Aging, 13–21, 1994.

Evolutionary mechanisms of senescence

Brian Charlesworth
Department of Ecology and Evolution, The University of Chicago, 1101, E.57th St., Chicago, IL 60637, USA

Received and accepted 22 June 1993

Key words: senescence, life-history, population genetics, quantitative genetics, mutation

Abstract

This paper reviews theories of the evolution of senescence. The population genetic basis for the decline with age in sensitivity of fitness to changes in survival and fecundity is discussed. It is shown that this creates a presure of selection that disproportionately favors performance early in life. The extent of this bias is greater when there is a high level of extrinsic mortality; this accounts for much the diversity in life-history patterns among different taxa. The implications of quantitative genetic theory for experimental tests of alternative population genetic models of senescence are discussed. In particular, the negative genetic correlations between traits predicted by the antagonistic pleiotropy model may be obscured by positive correlations that are inevitable in a multivariate system, or by the effects of variation due to deleterious mutations. The status of the genetic evidence relevant to these theories is discussed.

Introduction

Characteristics such as the mortality rate of individuals as a function of age, the timing of the age of first reproduction, the spacing of reproductive episodes during life, and the average reproductive success of individuals at different ages, constitute the description of the life-history of a species. Life-histories are astonishingly variable, even within a relatively homogeneous group of organisms such as eutherian mammals, where maximum life-span in captivity ranges from a low of a few months for a small insectivore to a high of 80 years or more for humans (Comfort, 1979; Finch, 1991; Stearns, 1992). Reproductive characters, such as the age at first reproduction and the number of offspring per litter, show similar diversity within mammals (Harvey & Read, 1988; Promislow, 1991). The problem of explaining this diversity of life-history characteristics has attracted considerable attention from ecologists and evolutionists (Cole, 1954; Lack, 1954; Williams, 1957; Hamilton, 1966; Charlesworth, 1980; Dingle & Hegmann, 1982; Lande, 1982; Sibly & Calow, 1986; Partridge & Harvey, 1988; Rose, 1991; Finch, 1991; Stearns, 1992).

Despite this diversity, we can observe an almost universal life-history pattern: that of senescent decline in survival and reproductive performance with increased age after reproductive maturity (Medawar, 1952; Comfort, 1979; Rose, 1991; Finch, 1991; Stearns, 1992). This decline results from the deterioration of a multiplicity of different biological parameters with advancing age. No unitary proximate cause of senescence can apparently be found: almost everything gets worse with age. At first sight, this seems to constitute a severe challenge for conventional evolutionary theory, since natural selection, at least on a naive view, supposedly promotes increased fitness.

The answer to this challenge is provided by the standard evolutionary theories of aging, which state that senescence is a consequence of the greater selective premium on genes with favorable effects on survival or fecundity early in the life-history (Haldane, 1941; Medawar, 1946, 1952; Williams, 1957; Hamilton, 1966; Kirkwood & Holliday, 1979; Charlesworth 1980; Rose, 1991). Whatever the ways in which genes affecting survivorship or fecundity express their effects, this universal factor will lead to the evolution of a life-history in which mortality increases and fecundity decreases with advancing age, provided that genetic variability with the appropriately age-specific effects is available for use by natural selection.

The intensity of the selective premium on high early performance varies considerably with ecological circumstances, and this leads to variation in the rate at which senescent decline is expected to occur. But in species in which there is a clear distinction between reproductive and somatic structures (including all multicellular, sexually reproducing species and species where asexual eggs are produced, but excluding unicellular species and multicellular species reproducing exclusively by budding), senescence is an evolutionary consequence of the fact that somatic structures are simply packaging material for the genome, so that maximization of the individual's genetic contribution to future generations is not equivalent to maximization of the length of life.

In nature, extrinsic causes of mortality may be so high that it is difficult to detect increases in mortality rates with age, and a decline in fecundity may only occur at ages so great that few individuals are present in the population. Nevertheless, the study of captive populations in which individuals are protected from extrinsic causes of death as far as possible usually yields evidence for senescent decline, even in groups such as fish where senescence is often postponed (Comfort, 1979; Rose, 1991; Finch, 1991). Senescence can also be detected in the wild in groups such as large mammals that are especially suitable for demographic study (Caughley, 1977; Finch, 1991; Promislow, 1991).

Natural selection with age-structure

A proper understanding of the evolutionary origin of senescence requires a theory of the dynamics of natural selection on age-specific survival and fecundity rates, based on the principles of population genetics. Research over the past twenty years has contributed to the establishment of a solid body of theory of this kind (Charlesworth, 1980, 1990a; Lande, 1982; Orzack & Tuljapurkar, 1989; Rose, 1991). The theory now covers both changes in frequencies of single genes, and changes in the mean values of quantitatively varying characters controlled by many genes of relatively small effect. The main results are as follows.

Fitness with age-structure

It is most convenient to describe life-histories in terms of discrete time-intervals, such that individuals of a given age x have age-dependent probabilities, $P(x)$, of survival to the next time-interval, when they will be aged $x + 1$. (Sex differences will be ignored for these purposes.) Fecundity at age x, $m(x)$, can be described by the expected number of daughters that a female aged x will produce. Their chance of survival to the next time interval, when they are of age 1 and their surviving parents are age $x + 1$, is $P(0)$, so that the effective fertility of females aged x is $P(0)m(x)$. The environment is assumed to be constant, so that these demographic parameters can be treated as independent of time. A genotype i at a given locus will in general have its own set of life-history characteristics, i.e. a set of values of the probabilities of survival to age x ($x = 1, 2...$), $l_i(x) = P_i(0) P_i(1)...P_i(x-1)$, and fecundities at age x, $m_i(x)$. Summary statistics of the life-history are provided by quantities such as the *net reproduction rate*, $R = \Sigma\, l(x)m(x)$, which measures the expectation at birth of total lifetime reproductive output.

This representation is exact for species that reproduce at discrete time-intervals, such as annually breeding species of birds or mammals. It provides an approximate representation of the demography of continuously reproducing species such as man or *Drosophila*. The approximation can be made arbitrarily good by increasing the number of age-classes into which the life-history is divided.

With this formulation, and a number of restrictive assumptions (Charlesworth, 1980, Chap. 3), for many purposes the Darwinian fitness of a genotype can be equated to the *intrinsic rate of increase* of the life-history, r_i, given by the real root of the Euler-Lotka equation (Charlesworth, 1980, Chaps. 3-4)

$$\sum_x e^{-r_i x} l_i(x) m_i(x) = 1 \tag{1}$$

where the summation is taken over all ages x.

This is the rate of population growth per unit time which would be attained by a population that consists entirely of individuals with the life-history characteristics of the genotype in question. In the case of a heterozygous genotype, which cannot form a true-breeding population, r_i is the hypothetical rate of population growth that would be attained by a population, all of whose members have the life-history characteristics of the genotype in question.

Another quantity of importance is the *generation time* of a genotype, which is most conveniently measured by the quantity

$$T_i = \sum_x x\, e^{-r_i x} l_i(x) m_i(x) \tag{2}$$

This is the mean age of parents in a population growing at rate r_i.

If selection is fairly weak, r_i can be used to predict changes in the genetic composition of populations. In particular, the rate of change in frequency per unit time of a rare mutant gene that modifies the life-history is proportional to the difference in r between carriers of the mutant and wild-type. The rate of change of gene frequency per generation, which is analogous to the standard formulation for selection with discrete generations, can be obtained by multiplying the rate per unit time by the value of T for a standard genotype (Charlesworth, 1980, Chap. 4). The quantity Tr_i can be used as a non-dimensional measure of fitness with weak selection; genotypic differences in the Tr_i play a similar role to differences in the standard discrete-generation fitness measure w_i (Charlesworth, 1980, Chap. 4).

The theory is easily extended to situations in which the overall population growth rate is held to approximately zero by density-dependent governing factors; the rate of increase of a mutant in this case is determined by the value of r for the mutant, holding the density-dependent factors constant at the level that confers a zero rate of population growth on the wild-type population (Charlesworth, 1980, Chap. 4). If population size is close to being stationary, then the rate of change of gene frequency per generation is determined to a good approximation by the differences in net reproduction rate among genotypes (Charlesworth, 1980, Chap. 4). Since most natural populations are near to zero population growth on average (Lack, 1954), R thus provides a fitness measure that is useful for many practical purposes. For a given genotype i, $R_i - 1 \approx Tr_i$ with weak selection and stationary population size (Charlesworth, 1980, Chap. 4).

Extensions of the theory to the important case when the demographic parameters vary over time are considered by Charlesworth (1980, Chap. 4) and Orzack and Tuljapurkar (1989).

Sensitivity of fitness to age-specific changes in the life-history

In this section, I will show that natural selection places greater weight on early-acting genetic effects on fecundity or survival than on effects expressed later in the life-history, and that this effect is strongly dependent on the demographic environment imposed by extrinsic ecological factors, such as the death rate due to predation. The intuitive interpretation of this result is that deferring the effect of a gene on survival or fecundity to late life incurs the risk that the individuals who carry the gene may have died before its effects are expressed; this risk is higher the later in life it is expressed, and the higher the rate of mortality.

The effect on fitness of a small change in the probability of survival from age x to $x + 1$ is conveniently measured by the partial derivative of r with respect to the logarithm of $P(x)$, $\partial r/\partial \ln P(x)$. By multiplying by the generation time T, we can rescale to units of the non-dimensional per-generation fitness measure, Tr. $T\, \partial r/\partial \ln P(x)$ is thus a measure of the *sensitivity* of the per-generation measure of fitness to a change in survival at age x. It is given by the following equation, derived by Hamilton (1966):

$$S(x) = T\frac{\partial r}{\partial \ln P(x)} = \sum_{y=x+1} e^{-ry} l(y) m(y) \tag{3}$$

The corresponding sensitivity of R with respect to a change in fecundity at age x is

$$S'(x) = T\frac{\partial r}{\partial m(x)} = e^{-rx} l(x) \tag{4}$$

From the form of equation (3), it is easily seen that an increase in survival of a given magnitude that is confined to a single age has a greater impact on fitness if it occurs early rather than late in reproductive life (the timing of changes in pre-reproductive survival is irrelevant). Similarly, equation (4) implies that, if the rate of population growth is non-negative, a change in fecundity at a given age always has a greater impact the earlier it occurs. The magnitude of the effect of age is increased by a high rate of mortality, which causes the right-hand sides of these equations to fall off more rapidly with age. Increased population growth has an effect similar to that of increased mortality on the relation between the sensitivity functions and age. Further implications of these equations are discussed by Hamilton (1966), Charlesworth (1980, Chap. 5), and Rose (1991, Chap. 1).

Evolutionary explanations of senescence

This body of genetically-based theory has shed new light on the various evolutionary explanations of senescence, originally proposed by Haldane (1941), Medawar (1946, 1952), Williams (1957) and Hamilton (1966). These theories may be divided into three categories (Charlesworth, 1980, p. 217; Rose, 1991, Chap. 1).

Differential rates of gene substitution
According to this theory, there is a higher rate of incorporation into the population of favorable genes that increase survival at earlier ages within the reproductive period, compared with genes that act later (Hamilton, 1966). This would have the effect of gradually raising survival for young individuals relative to older ones. An initially non-senescent population, in which survivorship is independent of age, would gradually become converted into one which exhibits senescence, in the sense of exhibiting a decline in survivorship with advancing age. It seems, however, somewhat unlikely that this process could account for the more obviously pathological aspects of aging, which contribute largely to decline in survival in later life (Finch, 1991).

Mutation-accumulation
Deleterious alleles are maintained in populations at a large number of loci as a result of mutation pressure, and must contribute to a significant reduction in the average fitness of individuals in the population, compared with that of mutation-free individuals. Studies of *Drosophila*, for example, suggest that a typical gamete may have a probability near one of carrying a new mutation with a small, detrimental effect on fitness (Crow & Simmons, 1977; Houle *et al.*, 1992); the probability must be considerably higher in mammals, with their much larger

genomes (Kondrashov, 1988). Medawar (1952) and Edney & Gill (1968) suggested that deleterious mutations with effects limited to later ages would equilibrate at higher frequencies than mutations with early-acting effects. Taking into account the contributions of all loci, a lower mean survivorship would thus be expected for individuals of more advanced age. This theory has been placed on a quantitative basis by Charlesworth (1980, pp. 140-142, p. 218; 1990a).

Many rare genes causing hereditary diseases with delayed ages of onset are well-known in human genetics, and their fitness effects can be calculated from equation (1) and related equations (Charlesworth & Charlesworth, 1973). The effects of these genes on fitness are, as expected from equations (3) and (4), highly dependent on the demographic environment; a gene with a deleterious effect on survival late in life will have a smaller selective disadvantage in a population with a high rate of mortality from other causes. For example, individuals with Huntington's chorea suffered an estimated loss in fitness relative to normal individuals of about 15% in the US population of 1939-41, compared with a loss of 9% in the Taiwan population of 1906 (Charlesworth, 1980, p. 151). Over the long-term, this increased net selection pressure would cause the frequencies of genes with delayed age of onset to equilibrate at lower frequencies in low-mortality populations, leading to an improvement in survivorship late in life compared with high-mortality populations with the same genetic make-up. This process is thus capable in principle of explaining species differences in longevities referred to in the introduction, on the assumption that species differ in their demographic environment (e.g. decreased extrinsic mortality). Given that mutation rates per genome are probably high for life-history traits (Kondrashov, 1988; Charlesworth, 1990a), even though mutation rates per locus are extremely low, the process of adjustment of the life-history to a shift in the demographic environment need not take more than a few tens of generations under this model.

A process related to mutation-accumulation was proposed by Haldane (1941, pp. 192-194), who suggested that there might be significant selection for modifier genes which delay the age of onset of hereditary diseases. Medawar (1952) endorsed this as a probable major factor in the evolution of senescence. But Charlesworth (1980, pp. 218-219) showed that the magnitude of the selective advantage of such a modifier is on the order of the rate of mutation at the locus it is modifying. This process is thus unlikely to be of much evolutionary significance.

Antagonistic pleiotropy

Williams (1957) argued that it is likely that genes that increase survivorship or fecundity at one age or set of adjacent ages will have deleterious effects at other ages, since an increase in the efficiency of a physiological process that improves survival or fecundity will usually place a demand on resources that would otherwise be utilized in a different way or at a different time. From the principle that changes early in life have a larger effect on fitness than later changes, Williams proposed that selection will thus tend to cause the fixation of genes with positive early effects and negative later ones, rather than genes with the opposite pattern. An explicit physiological model of this kind of trade-off has been formulated in terms of repair and maintenance versus reproduction (Kirkwood, 1990). Genetic models of this *antagonistic pleiotropy* have been elaborated by Charlesworth (1980, pp. 208-209), Templeton (1980), and Rose (1982, 1985), confirming Williams' insight. As with the mutation-accumulation theory, the evolutionary pressure in favor of senescence will be greater in demographic environments where there is a high overall rate of mortality during the adult life-history.

Quantitative genetic models of multivariate systems

It is likely that genetic variation in life-history traits in natural populations will normally be under the control of many loci, each with a relatively small effect in comparison with the total range of variability in the trait. Many of the attempts to discriminate experimentally between the evolutionary theories described above have relied on quantitative genetic approaches to determine whether or not genetic correlations between the same trait at different ages are usually strongly negative (as might be expected under the antagonistic pleiotropy theory) or near zero (as might be expected under

the mutation-accumulation theory) (Bell & Koufopanou, 1986, Rose, 1991; Stearns, 1992). In order to evaluate the results of these experiments, it is essential to express the predictions of the models in quantitative genetic terms.

The selection equation with quantitative inheritance

Lande (1982), building on the work of Hill (1974, 1977), constructed such a theory for the case of weak, frequency-independent selection in density-independent environments. (It is not difficult to relax the assumption of density-independence [Charlesworth, 1990a]). He assumed that a given life-history phenotype is described by a vector z, such that the components of z completely determine the values of $l(x)$ and $m(x)$. For example, under the standard reproductive effort model, in which $P(x)$ is a decreasing function of $m(x)$ alone (Schaffer, 1974), the values of $m(x)$ for $x = 1,2$ etc., would form the components of z. Within a population, the vector z is assumed to follow a multivariate normal distribution within a population, whose variance-covariance matrix is the sum of genetic and environmental components. The genetic component of this matrix can be further partitioned into a matrix G of variances and covariances of *breeding values* or *additive genetic values* (Falconer, 1989, p. 128) of the components of z, such that g_{jk} is the genetic covariance in breeding value between components z_j and z_k of z (g_{jj} is the additive genetic variance in z_j). The matrix G can be estimated from data on resemblances between relatives for the traits in question (Kempthorne, 1957; Becker, 1984).

In general, the effect of natural selection on z is determined by the *selection gradient vector* $\nabla \bar{w}$, whose components are the partial derivatives of population mean fitness with respect to the respective components of the mean value of z, $\partial \bar{w}/\partial \bar{z}_j$ (Lande, 1982). In the present context, ∇w can be approximated with sufficient accuracy by the vector of derivatives of r with respect to the z_j, evaluated at the population mean of z (Charlesworth, 1990a), ∇r. The change per unit time in z_j is then given by

$$\Delta \bar{z}_j \approx \sum_k g_{jk} \left(\frac{\partial r}{\partial z_k} \right)_{\bar{z}} . \tag{5}$$

Biological implications

At equilibrium under selection, the changes in the z_j are zero for all j. This implies that either all the components of ∇r are equal to zero or that the determinant of G is zero. In the present context, the latter condition is biologically most likely; if the components of z are chosen to be life-history variables such as fecundity or survival, the derivatives of r must be positive (see equations [3] and [4] above). If there is additive genetic variance in at least some of the z_j, so that some of the g_{jj} are positive, the equilibrium version of equation (5) requires that there be some negative additive genetic covariances between the z_j, in order for the terms on the right-hand side to sum to zero (Charlesworth, 1990a).

Under rather general conditions, selection theory thus predicts that life history traits in equilibrium populations exhibiting additive genetic variance which is maintained by selection will be expected to show negative genetic covariances and correlations with some other additively variable traits. This has been known to specialists in animal improvement for a long time (Dickerson, 1955; Robertson, 1955). It is a consequence of the fact that selection exhausts additive genetic variance for net fitness (Fisher, 1930), so that any remaining additive genetic variation in fitness-related traits must reflect the properties of genes whose beneficial effects on some traits are counter-balanced by deleterious effects on other characters (antagonistic pleiotropy: see above). Rose (1982, 1985) has provided some explicit models of how this can be achieved.

This predicts that the most genetically variable components of the life-history are often likely to show some evidence of negative genetic correlations with other traits, reflecting the physiological trade-offs between traits envisaged with antagonistic pleiotropy. Unfortunately, there are two complications, which render this prediction harder to test than one would like. First, as suggested by Pease and Bull (1988) and proved mathematically by Charlesworth (1990a), some pairs of traits in a multivariate system must show non-negative genetic covariances in an equilibrium population. Thus, if genetic covariances or correlations are determined for pairs of characters taken from a multivariate set of unknown dimension (as is normally the case in experimental studies), findings of positive genetic correlations are to be expected in some instances, and do not shed any light on the validity or otherwise of the antagonistic pleiotropy model. Only if the multivariate system can be dissected into a number of sub-systems, in which pairs of variables are subject to trade-offs (as in the reproductive effort model) and are not related to any other variables, can positive genetic correlations be excluded (Charnov, 1989; Charlesworth, 1990a). Thus, the fact that positive genetic correlations between pairs of life-history traits are frequently found (Bell & Koufopanou, 1986) does not necessarily falsify the antagonistic pleiotropy theory. Findings of negative genetic correlations do, of course, suggest trade-offs, but the underlying functional relations among the variables in a multivariate system have an exceedingly remote connection with the patterns of pairwise genetic correlations that they generate (Charlesworth, 1990a).

The second difficulty is the possible confounding effect on the pattern of genetic covariances of variation arising from deleterious alleles maintained by mutation (Charlesworth, 1990a; Houle, 1991). Even if antagonistic pleiotropy is the rule for a substantial fraction of loci maintained in the population, it is likely that many mutations will have deleterious effects on multiple traits, rather than affecting trade-offs between different traits by lowering performance with respect to the ability to harvest or utilize resources. This will add positive genetic covariance terms to the overall G matrix, which may obscure the evidence for underlying trade-offs. Given that most deleterious alleles are recessive or partially recessive (Crow & Simmons, 1983), their effects on G are magnified in inbred populations (Charlesworth, 1990a; Houle, 1991). This suggests that such effects of deleterious mutations may be tested for by experimentally inbreeding populations, and determining whether there is a tendency for G to contain more positive covariance terms. Such an effect has been detected by Rose (1984).

Empirical tests of the theories

Tests of theories of life-history evolution or senescence by means of comparisons of different populations or taxa are reviewed by Partridge and Harvey (1988), Rose (1991), Finch (1991), Promislow (1991), and Stearns (1992). I will not discuss the details here. The general conclusion seems to be that many of the patterns expected on the theories presented above are borne out by the data. In particular, it seems that ecological circumstances that put a premium on early reproduction are indeed associated with increased relative values of early life-history characteristics. Where the expected correlations are not found, as in the failure to detect a relation between high extrinsic mortality and the rates of senescence in different species of mammals (Promislow, 1991), it is conceivable that inadequacies of the data are responsible (e.g. the difficulty of detecting senescence under natural conditions, when mortality rates are high). Further research will presumably clarify these ambiguities.

Evidence for the existence of genetic variability of the kind postulated by the rival theories of the evolution of senescence has been reviewed by Charlesworth (1990b), Rose (1991) and Finch (1991). It is clear that, at least in *Drosophila*, there is ample additive genetic variance for many life-history characteristics. Artificial selection on laboratory outbred stocks for greater length of life and reproduction late in life has been remarkably successful in increasing longevity and late female fecundity, usually (perhaps not always – see Partridge & Fowler, 1992) at the expense of early female fecundity. These experiments provide a model of how a life-history may evolve in response to a shift in the demographic environment of the population, in this case one which favors late performance over early performance. The fall in early fecundity that accompanies the rise in late fecundity and longevity in several of these experiments suggests that there may be negative genetic correlations of the type postulated in the antagonistic pleiotropy theory. Direct estimates of genetic parameters in the base stock used in Rose's experiments support this interpretation (Rose & Charlesworth, 1981; Service & Rose, 1985).

The results of the *Drosophila* selection experiments do not necessarily rule out a role for mutation-accumulation, as discussd by Rose (1991, p. 75). But apart from the fact that many examples of deleterious mutations with delayed age of onset, of the type postulated in the mutation-accumulation theory, are known in human genetics (Finch, 1991, Chap. 6), there is little empirical evidence in its favor at present. One of the few attempts to test it directly led to negative results (Rose & Charlesworth, 1980, 1981). This test was based on the fact

that, with mutation-accumulation, the higher equilibrium frequencies of late-acting deleterious alleles lead to an increase in the additive genetic variance of age-specific life-history traits with advancing age (see Charlesworth [1990a] for a specific model). No such effect was detected by Rose and Charlesworth. Kosuda (1985) reported an increase in variance in male mating success between chromosome lines in *D. melanogaster*, but did not carry out an analysis of variance components, so that the interpretation of the results is unclear.

A different approach has been taken by Houle *et al.* (in preparation). They accumulated spontaneous mutations for 44 generations on sets of second chromosomes of *D. melanogaster* maintained with a minimum of selection against deleterious mutations. The genetic variance-covariance matrix for a suite of life-history characters (early and late female fecundity, female longevity, and male longevity) generated by mutation was characterized. Correlations between pairs of traits are generally positive. The genetic correlations between early and late fecundity, male and female longevity, and between female longevity and late fecundity are all close to one, although the confidence intervals are wide. The estimate of per genome mutation rate for a measure of net fitness is very similar to that for individual components, such as fecundity and viability, consistent with the findings of high genetic correlations (Houle *et al.*, 1992). This provides little support for the mutation-accumulation theory, which requires decoupling of the effects of mutations at different ages. Other data on mutational covariances between life-history characters, suggesting generally but universally positive values, are reviewed by Charlesworth (1990a). Further information on the properties of mutational variation is badly needed.

Conclusions

We now have a well-established body of theory on the evolution of senescence, based firmly on populations genetic principles. The theory implies that senescence is an evolutionary response to the fact that the sensitivity of Darwinian fitness to changes in survival or fecundity declines with age, at a rate that is determined by the ecology of the population. Genetic studies of natural variation in life-history

20

characteristics within species and comparisons of different populations and taxa have yielded results that are broadly consistent with the general theoretical framework of life-history evolution.

Many of the conclusions from comparative studies do not depend strongly on the details of the physiological or genetic bases of life-history characters. They simply reflect the effect of the demographic environment of the species on the age-sensitivity of fitness. For this reason, we are presently far from being clear about the genetic mechanisms underlying many evolutionary patterns, although trade-offs between different traits almost certainly play an important role. Only further experimental work on the genetics of intraspecific variation is likely to settle such questions as the importance of mutation-accumulation versus antagonistic pleiotropy.

References

Becker, W. A., 1984. Manual of Quantitative Genetics. 4th ed. Academic Enterprises, Pullman, WA.

Bell, G. & V. Koufopanou, 1986. The cost of reproduction. Oxf. Surv. Ev. Bio. 3: 83-131.

Caughley, G., 1977. Analysis of Vertebrate Populations. Wiley Interscience, New York, N.Y.

Charlesworth, B., 1980. Evolution in Age-Structured Populations. Cambridge University Press, Cambridge, U.K.

Charlesworth, B., 1990a. Optimization models, quantitative genetics, and mutation. Evolution 44: 520-538.

Charlesworth, B., 1990b. Natural selection and life history patterns, pp. 21-40 in Genetic Effects on Aging, edited by D. E. Harrison. Telford Press, Caldwell, N.J.

Charlesworth, B. & D. Charlesworth, 1973. The measurement of fitness and mutation rate in human populations. Ann. Hum. Genet. 37: 175-187.

Charnov, E. L., 1989. Phenotypic evolution under Fisher's Fundamental Theorem of natural selection. Heredity 62: 97-106.

Cole, L. C., 1954. The population consequences of life history phenomena. Quart. Rev. Biol. 29: 103-137.

Comfort, A., 1979. The Biology of Senescence. 3rd. ed. Churchill Livingstone, London, U.K.

Crow, J. F. & M. J. Simmons, 1983. The mutation load in Drosophila, pp. 1-35 in The Genetics and Biology of Drosophila, Vol. 3c., edited by H. L. Carson, M. Ashburner and J. N. Thomson, Academic Press, London, U.K.

Dickerson, G. E., 1955. Genetic slippage in response to selection for multiple objectives. Cold Spring Harb. Symp. Quant. Biol. 20: 213-224.

Dingle, H. & J. P. Hegmann, 1982. Evolutionary Genetics of Life Histories. Springer-Verlag, New York, N.Y.

Edney, E. B. & R. W. Gill, 1968. Evolution of senescence and specific longevity. Nature 220: 281-282.

Falconer, D. S., 1989. An Introduction to Quantitative Genetics. 3rd. ed. Longman, London, U.K.

Finch, C. E., 1991. Longevity, Senescence, and the Genome. University of Chicago Press, Chicago, IL.

Fisher, R. A., 1930. The Genetical Theory of Natural Selection. Oxford University Press, Oxford U.K.

Haldane, J. B. S., 1941. New Paths in Genetics. Allen and Unwin, London.

Hamilton, W. D., 1966. The moulding of senescence by natural selection. J. Theor. Biol. 12: 12-45.

Harvey, P. H. & A. F. Read, 1988. How and why do mammalian life histories vary? pp. 213-231 in Evolution of Life Histories of Mammals: Theory and Pattern, edited by M. P. Boyce. Yale University Press, New Haven, C.T.

Houle, D., 1991. Genetic covariance of fitness correlates: what genetic correlations are made of and why it matters. Evolution 45: 630-648.

Houle, D., D. K. Hoffmaster, S. Assimacopoulos & B. Charlesworth, 1992. The genomic mutation rate for fitness in Drosophila. Nature 359: 58-60.

Kempthorne, O., 1957. An Introduction to Genetic Statistics. John Wiley, New York, N.Y.

Kirkwood, T. B. L., 1990. The disposable soma theory of aging, pp. 9-10 in Genetic Effects on Aging, edited by D. E. Harrison. Telford Press, Caldwell, N.J.

Kirkwood, T. B. L. & R. Holliday, 1979. The evolution of ageing and longevity. Proc. Roy. Soc. Lond. B. 205: 531-546.

Kondrashov, A. S., 1988. Deleterious mutations and the evolution of sexual reproduction. Nature 336: 435-440.

Kosuda, K., 1985. The aging effect on male mating activity in Drosophila melanogaster. Behav. Genet. 15: 297-303.

Lack, D. L., 1954. The Natural Regulation of Animal Numbers. Oxford University Press, Oxford, U.K.

Lande, R., 1982. A quantitative genetic theory of life history evolution. Ecology 63: 607-615.

Medawar, P. B., 1946. Old age and natural death. Modern Quarterly 1: 30-56.

Medawar, P. B., 1952. An Unsolved Problem of Biology. H. K. Lewis, London, U.K.

Orzack, S. H. & S. Tuljapurkar, 1989. Population dynamics in variable environments. Amer. Nat. 133: 901-923.

Partridge, L. & K. Fowler, 1992. Direct and correlated responses to selection on age at reproduction in Drosophila. Evolution 46: 76-91.

Partridge, L. & P. H. Harvey, 1988. The ecological context of life history evolution. Science 214: 1449-1455.

Pease, C. M. & J. J. Bull, 1988. A critique of methods for measuring life-history trade-offs. J. Evol. Biol. 1: 293-303.

Promislow, D. E. L., 1991. Senescence in natural populations of mammals: a comparative study. Evolution 45: 1869-1887.

Reznick, D., 1985. Costs of reproduction: an evaluation of the empirical evidence. Oikos 44: 257-267.

Robertson, A., 1955. Selection in animals: synthesis. Cold Spring Harb. Symp. Quant. Biol. 20: 225-229.

Rose, M. R., 1982. Antagonistic pleiotropy, dominance and genetic variation. Heredity 48: 63-78.

Rose, M. R., 1984. Genetic covariation in Drosophila life history: untangling the data. Amer. Nat. 123: 565-569.

Rose, M. R., 1985. Life history evolution with antagonistic pleiotropy and overlapping generations. Theor. Pop. Biol. 28: 342-358.

Rose, M. R., 1991. The Evolutionary Biology of Aging. Oxford University Press, Oxford, U.K.

Rose, M. R. & B. Charlesworth, 1980. A test of evolutionary theories of senescence. Nature 287: 141-142.

Rose, M. R. & B. Charlesworth, 1981. Genetics of life history in *Drosophila melanogaster*. I. Sib analysis of adult females. Genetics 97: 173-186.

Service, P. M. & M. R. Rose, 1985. Genetic covariation among life history components: the effect of novel environments. Evolution 39: 943-945.

Sibly, R. M. & P. Calow, 1986. Physiological Ecology of Animals: An Evolutionary Approach. Blackwell, Oxford, U.K.

Stearns, S. C., 1992. The evolution of life histories. Oxford University Press, Oxford, U.K.

Templeton, A. R., 1980. The evolution of life histories under pleiotropic constraints and *r*-selection. Theor. Pop. Biol. 18: 279-289.

Williams, G. C., 1957. Pleiotropy, natural selection and the evolution of senescence. Evolution 11: 398-411.

M. R. Rose and C. E. Finch (eds), Genetics and Evolution of Aging, 22–28, 1994.

Mutation-selection balance and the evolution of senescence

Andrew G. Clark
Department of Biology, Pennsylvania State University, University Park, PA 16802, USA

Key words: mutation-accumulation, *Drosophila*, antagonistic pleiotropy

Abstract

Populations of all organisms are faced with a continuous bombardment of mutations. These mutations have a range of effects from lethality to advantageous, and they have a range of pleiotropic effects as well. Population genetics theory tells us that a population in mutation-selection balance will have late-acting deleterious alleles in higher frequency than early-acting deleterious alleles. The mutation-accumulation model explains the evolution of senescence by the accumulation of late-acting deleterious alleles, and no form of pleiotropic effects needs to be specified. Alleles that cause earlier senescence can be favored by natural selection provided pleiotropic effects give those mutations a net reproductive advantage. Provided such mutations with antagonistically pleiotropic effects occur, this model may explain the evolution of senescence. To some extent, the relative importance of these two models can be inferred from the difference between the distribution of mutational effects and the equilibrium distribution of allelic effects of polymorphic loci. There are several experimental designs that, to varying degrees of success, give a quantitative assessment of the distribution of mutational effects on age-specific survival and fecundity. Thorough analysis of physiological mechanisms for senescence is highly relevant to this problem, because the mechanistic explanation for patterns of pleiotropy may be revealed by the physiology.

Population genetic theories of senescence

All organisms suffer from a continuous barrage of mutations. Advantageous mutations are most favored if they improve juvenile survival and early fecundity. Deleterious mutations are removed by natural selection, but mutations with effects only late in life are removed only slowly because selection against them is relatively weak. In a finite population of size N, deleterious mutations whose selection coefficient is less than $1/N$ will have dynamics dominated by drift, so they can rise to appreciable frequency. Even in a deterministic setting, the mutation-selection equilibrium will have more late-acting deleterious alleles at higher frequency. This statement of the mutation-accumulation model emphasizes that deleterious mutations are inescapable, and the degree to which we see senescent decline as a result of mutation accumulation depends on the age-specific distribution and pleiotropic effects of new mutations. If a substantial portion of mutations

have pleiotropic effects, such that they are beneficial at young ages but deleterious later on (antagonistic pleiotropy), then natural selection may result in an increase in frequency of these alleles, and the mean longevity (or late fecundity) of the population may actually decrease. Because the life history is driven by natural selection in this scenario, Partridge and Barton (1993) call this the optimality model.

Reznick (this volume) makes the point that the mutation accumulation and antagonistic pleiotropy models are not mutually exclusive. Both models can be valid at the same time, such that there is an accumulation of late-acting deleterious alleles and invasion of alleles with pleiotropic effects that decrease longevity. The fact that the two models are not exclusive can be bothersome to the experimentalist. The idea of Partridge and Barton (1993), to consider what effect an increase in mutation rate will be, illustrates the difficulty. If senescence is explained purely by antagonistic pleiotropy, and all mutations have antagonistic effects, then an increase in the mutation rate may have no effect on longevity. On the other hand, if senescence is explained entirely by mutation accumulation, an increase in the mutation rate will increase the equilibrium frequency of deleterious alleles at mutation-selection balance, and must reduce longevity. Even if 100 late-acting alleles are segregating in mutation-selection balance, all it takes is a single allele of large effect, whose effects happen to exhibit antagonistic pleiotropy, to have natural selection reduce the lifespan.

Attempts to test models of the evolution of senescence have come in many forms, and only a few will be mentioned here. The most influential design has been to select for delayed fecundity and examine correlated responses. Rose and Charlesworth (1981) found that selecting on late fecundity reduced early fecundity, consistent with antagonistic pleiotropy. By directly estimating additive, dominance and environmental components of variance in mortality in *D. melanogaster*, Hughes and Charlesworth (1994) found that all three components of variance increased dramatically with age. Furthermore, mortality had no significant heritability at young ages, but heritability differed significantly from zero among older flies. These results clearly support predictions of the mutation accumulation model, but raise the question of whether the magnitude of additive variance in mortality is consistent with a mutation-selection balance.

Because quantitative characters are influenced by combinations of genes and environments, the importance of environmental control in experiments to quantify effects on life history cannot be overemphasized. The simplest design for measuring survival distributions is to place a cohort of subjects in a cage and to follow the numbers that survive over time. In this design, the density is continually declining, and to the extent that mortality might be accelerated by crowding, estimates of rates of mortality might be in error (see in Graves & Mueller, this volume). Carey *et al.* (1992) obtained the surprising result that mortality rates levelled off among old medflies, and this pattern was observed both in cages (with the above potential problem) and in individually-reared flies. It may be simply that the starting population was highly heterogeneous, and that only the flies with the lowest mortality are left at the end. Consistent

with this idea, Hughes and Charlesworth (1994) observed highly significant variation in rates of mortality in *Drosophila*. We are nearing the time when the appropriate question will not simply be whether there is evidence for mutation accumulation, but to ask whether the magnitude of variation in rates of mortality is as great as we would expect. Sadly, the enormous sample sizes of Carey *et al.* (1992) may be necessary to have any statistical confidence in mortality rates of very old individuals.

To summarize, it would appear that the most progress could be made by accepting that senescence is a complex, quantitative trait and to employ the methods of theoretical quantitative genetics to make predictions about the relationship between mutational and equilibrium distributions of allelic effects. The empirical methods outlined in the next section may afford means for estimating the relevant parameters.

Mutation-accumulation experiments

One bit of information needed to determine where natural populations fall in the continuum between the mutation-accumulation and antagonistic pleiotropy models is the distribution of effects of spontaneous mutations. The power of this approach was best demonstrated by the important series of papers of Terumi Mukai and his students (Mukai, 1964, 1969; Mukai *et al.*, 1972). Mukai's mutation-accumulation lines showed a decline in mean fitness and an increase in variance among lines. By fitting the data to a model having a Poisson distribution of mutations, Mukai was able to estimate the rate of polygene mutation per generation. Departures from a linear trend over generations in the mean fitness allowed estimation of epistatic interaction. In a similar manner, mutational covariance for pairs of characters should be estimable (Houle *et al.*, 1992; Clark *et al.*, submitted).

The importance of the rate of polygenic mutation was made clear by Lynch (1988). Evolutionary quantitative genetic models have an annoying compounding of number of loci and magnitude of effect per allelic substitution (Turelli, 1984). Empirical estimates of mutation rates help determine where in this continuum reality lies. Synthesizing data from a variety of studies, Lynch (1988) found that the ratio of mutational variance to the environmental variance (V_m/V_e) was on the order of 0.001. This figure is high enough that many artificial selection experiments may be realizing a response that is at least partially due to the variation introduced to the selected population by mutation after the beginning of the experiment. Mutation accumulation experiments can be done in several ways, but the key feature is to arrest the action of natural selection. Mackay *et al.* (1992) estimated the mutational variance for bristle number in sib-mated lines of *Drosophila* by quantifying the divergence among lines, but their estimates ($V_m/V_e = 1.5$–3.3×10^{-3}) should be considered a lower bound, because stabilizing selection would reduce the variance among lines. To date, the only mutation accumulation experiment that quantified age-specific effects of new mutations is that of Houle *et al.* (1992). Using a design like that of Mukai (1964), Houle *et al.* constructed about 200 lines of *Drosophila* that had recessive mutations accumulating on the second chromosome. There was an overwhelming positive correlation among fitness components in this study, so that mutations that decreased early fecundity also tended to decrease late fecundity, and viability as well. Experiments of this sort allow fairly direct tests of equivalence of the distributions of mutational effects to the distributions of allelic effects extant in a population.

Longevity genes

Most mutations result in some loss of function, so it comes as no surprise that most mutations have deleterious effects that shorten the lifespan. The study of these mutations is unlikely to reveal much about the natural causes of senescence, because they result in pathological death at a younger age than the average for the population. According to the mutation accumulation model, the mean longevity in a population would be increased if the introduction of new mutations could be ceased, so lifespan is in a sense limited by presence of deleterious loss-of-function mutations. Some of these loss-of-function mutations may have gone to fixation by Muller's ratchet. In such a population, a mutation that

extends life probably represents a reversion of a fixed loss-of-function mutation, so the reversion is really a gain-of-function mutation. It would be a gain-of-function in a physiological mechanism that is responsible for most of the senescence. Such a mutation would be a repair in the weakest link in the population. Mutations that cause a dramatic increase in longevity with weak or undetectable pleiotropic effects, like the *age-1* mutation in *C. elegans* (Johnson, this volume), hold a special fascination. Surprisingly, deficiency analysis shows that *age-1* is a loss-of-function mutation, and yet it appears to be beneficial. Why would an organism express a gene that shortens the lifespan? The population geneticist would expect that such a gene must have an untested disadvantageous pleiotropic effect or that assessment of fitness effects is incomplete, perhaps due to gene by environment interaction. Otherwise, the *age-1* mutation should increase in frequency in a population.

Another approach is to apply molecular and genetic methods to manipulate genes that are thought *a priori* to be involved in a cellular mechanism of aging. Stearns and Kaiser (this volume) examined longevity and age-specific fecundity in *Drosophila* lines transformed with *P*-elements bearing the gene for the elongation factor EF-1α. EF-1α is responsible for docking of the aminoacyl-tRNA to the ribosome, and is required for protein synthesis. There is no particular reason to believe that this function decays with age and is causal to senescence in flies. Stearns and Kaiser found no effect of extra doses of EF-1α on male longevity and only inbred females reared at 25° lived longer. There was a highly significant treatment by temperature interaction. These results illustrate the complexity of cause of senescence and the importance of gene by environment interactions. They also demonstrate the difficulty of trying to manipulate a single gene and ascribing changes in broad phenotypic performance to the single targeted change. On the other hand, characterization of a set of random *P*-element inserts provides another approach to quantify the distribution of mutational effects. Such lines differ markedly in both longevity and age-specific fecundity (Clark and Guadalupe, *in prep.*), and, consistent with Houle *et al.* (1992), they exhibit primarily positive correlations between early and late age fitness components.

The response to artificial selection on late female fecundity will be achieved by changes in allele frequency at loci that cause the natural lifespan to be shorter than what it could be. If one could identify these loci, one might be able to assess whether they are highly mutable, or that they have pronounced pleiotropic effects. Loci responsible for quantitative differences between a pair of lines can be identified by mapping or by directly testing those thought to be physiologically relevant. Tyler *et al.* (this volume) present some of the latest results from Ayala's lab on superoxide dismutase (*Sod*) in *Drosophila*, and its role in mediating the response to selection for delayed senescence. Free radicals are known to cause damage to cells, and genes, like *Sod*, whose products help scavenge free radicals became targeted for physiological research. When Tyler *et al.* compared allele frequencies at *Sod* between selected and control lines, they found that selected lines had an elevated frequency of the more active *Sod* allele (consistent with the proposed role of *Sod* in preventing cell damage). Attempts to control the genetic background and to extract the *Sod* gene into a novel background however failed to detect an effect of *Sod* on longevity. These results need to be put in the context of the tantalizing molecular population genetic results, which show near monomorphism of one allelic form of *Sod*, consistent with natural selection 'sweeping' this allele to high frequency (Hudson *et al.*, 1994). Although interesting pleiotropic properties of genes of large effect can be obtained through these studies, the idea that senescence is caused by defects in one or two longevity genes is not very plausible. Although the pursuit of longevity genes is unlikely to yield much insight into general principles about the evolution of senescence, these studies have much merit in attempting to span the gap from the population genetic to the molecular-mechanistic causes of senescence.

The relevance of physiological mechanisms

A lesson from the evolutionary theory of senescence is that the pleiotropic effects of mutations play a critical role. To the extent that pleiotropic effects are antagonistic, alleles that cause earlier senescence can be made to increase in frequency by natural selection. Rose and Finch (this volume) argue that progress in understanding physiological mechanisms will shed light on the evolution of senescence by providing clues to the nature of pleiotropic effects. Several investigators have gotten good response in selecting lines of *Drosophila* for late female fecundity (Rose & Charlesworth, 1981; Rose, 1984; Luckinbill & Clare, 1985; Partridge & Fowler, 1992; Roper *et al.*, 1993). These selected lines and appropriate controls have been tested and compared in a myriad of physiological characters. Arking *et al.* (this volume) tested differences in resistance to six different stresses among controls and lines selected over 22 generations for delayed female reproduction. They observed that the age at 50% peak resistance differed between the control and selected lines, but that the temporal sequence of the six peak performance ages was the same. This suggests that the response manifested by the selected lines involved a general increase in stress tolerance as opposed to a stress-specific response. Some idea of the biochemical complexity of the difference between the selected lines is illustrated by Fleming *et al.* (this volume). Two-dimensional protein electrophoresis shows over 100 proteins that quantitatively differ between control and selected lines, and seven of them are consistent across replicates. Demonstration that the band differences are *causal* to the relevant senescence phenotypes will require more work, but consistency across replicates is very encouraging.

The importance of senescent changes in glucose metabolism and the possible cause of senescence through glucose hysteresis, is discussed by Mobbs (this volume). One of the most general observations regarding the physiological basis of longevity is that diet restriction lengthens life, implicating failures in energy metabolism as a general cause of senescence. To the extent that failure of normal glucose metabolism is a primary cause of senescence, there is no reason to assume that antagonistic pleiotropy would be common. While all of these physiological aspects are of interest in their own right, they leave an evolutionary biologist wondering whether senescence has a primary cause that is distinct from all these mechanisms, and that the general decline in many physiological processes may just be epiphenomena. This makes the physiology no less interesting or important, it simply raises the question of causality. The above mechanisms generally lead to the prediction of positive correlation between survival and fecundity, or between early and late age effects. One would like to know whether physiological mechanisms of senescence generally result in a tradeoff consistent with antagonistic pleiotropy. In this sense, the work of Partridge and coworkers (Partridge & Farquar, 1981; Partridge & Harvey, 1985; Fowler & Partridge, 1989; Partridge & Fowler, 1992), showing that reproduction itself has a cost in shortened lifespan, seems particularly relevant.

Unless some means of resistance is actively maintained, most organisms would succumb to attach by pathogens. Bell (this volume) explores the idea that senescence is hastened by pathogens, or more specifically that evolution of increasing virulence of pathogens within each host results in senescence even though the level of resistance of the host remains constant. This idea gets some indirect support through evidence of associations between senescence and variation in the major histocompatibility loci in the H2 region of the mouse (Crews & Yunis; Salazar, both in this volume). Gardner (this volume) did an extensive survey of feral mice, and found one population in which the murine leukemia virus was responsible for a substantial fraction of mortality and morbidity. These studies support Bell's hypothesis to the extent that they demonstrate that pathogens are important causes of mortality and that there is genetic variation in the ability to fight infection. But Bell argues that the evolution of the pathogens within each host may result in senescence even in the absence of either mutation accumulation of antagonistic pleiotropy mechanisms. On the other hand, pathogen-free mice still grow old and die, so the pathogen model cannot be an ultimate explanation for the evolution of senescence. Barring some form of antagonistic pleiotropy, the pathogen model can be thought of as a particular mechanism for the mutation accumulation model. Rather than accumulating deleterious mutations, however, the organisms face an increasing threat from pathogens as they age. Natural selection should favor alleles that strengthen the immune system at later ages, but the forces of selection favoring late-acting advantageous alleles is weak. In summary, the search for a single physiological basis for senescence is akin to the search for a single major longevity gene, and is unlikely to be successful. On the other hand, understanding the physiological basis of aging allows one to limit consideration of the kinds of pleiotropic effects that mutations might have.

Senescence and human genetic disorders

One advantage of considering humans in the study of senescence is that we are so acutely aware of the nature of physiological changes associated with senescence, and mutations that affect the rate of senescent decline come under scrutiny by the medical community. Martin (this volume) makes it abundantly clear that the genetic disorders that cause the most premature deaths in humans are highly polygenic. Although progress in mapping and cloning the genes responsible for conditions like cystic fibrosis and Duchenne muscular dystrophy is exciting and provides critical inroads to ameliorating these devastating genetic disorders, the fact remains that only a small fraction of the population die from single gene defects. Instead, most of us will die from conditions like heart disease, which have a significant, but polygenic, component of familial tendency. Even familial Alzheimers disease is genetically heterogeneous (Tanzi et al., this volume).

This bias in detecting mutations in humans results in seeing those which have large effects, and they are much more likely to be deleterious and to have pleiotropic effects that are positively correlated (i.e. to be deleterious in both longevity and fecundity). It is not surprising that Albin (this volume) finds no unambiguous evidence for antagonistic pleiotropy among human genetic disorders, but this does not constitute evidence against the model. It is likely that a large number of genes with subtle allelic effects are responsible for senescence. And we cannot say a priori what pattern of pleiotropic effects they will show. Strongly deleterious alleles have little effect on the longevity of a population, because they are kept at a low frequency in mutation-selection balance except in the case where the mutations have a very late age of onset. Late-onset mutations can achieve substantial allele frequency because, compared to early-onset mutations, natural selection is much less effective at removing late-onset disorders. To summarize, studies of human genetic disorders can tell us a great deal about physiological mechanisms, but the patterns of pleiotropic effects caused by these major defects are unlikely to tell us much about the evolution of senescence.

Another approach is to consider human longevity as a quantitative trait and apply quantitative genetic methods. Records of the Danish twin registry were used by James Vaupel and coworkers (Odense & Duke) to estimate the broad-sense heritability of lifespan among this population. Contrary to the anecdotes about familial tendencies for lifespan, they found that the heritability was not significantly different from zero (Vaupel, pers. comm.).

The challenge

As for most problems in population biology, the challenges that remain have both empirical and theoretical components. There is a need for a more complete theory of mutation-selection balance for quantitative genetic variation in early and late age fitness components. One of the simplest ways to formulate this model is to use the two-age class model of Partridge and Barton (1993), and suppose there is a one-locus continuum of alleles model with a bivariate distribution of mutational effects on juvenile and adult survival. How clearly can we derive predicted relations between the distribution of fitnesses, the distribution of mutational effects, and the equilibrium distribution of allelic effects? Under what circumstances do we expect negative genetic correlation between early- and late-age effects in a population that is in mutation-selection equilibrium? More theoretical work of this sort is needed before correct interpretation of empirical observations can be made.

The empirical data gathered to date have yet to cause a major re-thinking of the population genetic theory of senescence. Most of the data

have addressed the nature of extant variation and response to artificial selection. Houle et al. (1992) represents the first of what should be several attempts to quantify the distribution of age-specific mutational effects. What is the pleiotropic nature of new mutations? (Houle et al. found mostly positive correlations in fitness components). What is the distribution of pleiotropic effects of existing variation in the population? (There are disparate results from many labs). Do the distributions of mutational effects and equilibrium allelic effects differ significantly? If so, do the differences support one or another model for the evolution of senescence? An intriguing question raised by Partridge and Barton (1993) was what would happen if the mutation rate increased? Experiments like this are particularly well motivated. For the evolutionary biologist, the most informative experiments will be targeted toward a determination of where in the continuum from the mutation accumulation to the antagonistic pleiotropy models reality lies. Of course the ultimate dream is to unify the population genetic theory with the molecular and quantitative genetic data to provide an understanding of why such a melancholy thing as aging should be so universal to living things.

Acknowledgements

Supported by grant BSR-9007436 from the U.S. National Science Foundation.

References

Carey, J.R., P. Liedo, D. Orozco & J.W. Vaupel, 1992. Slowing of mortality at older ages in large medfly cohorts. Science 258: 457–461.

Clark, A.G. & R. Guadalupe, 1994. Probing the evolution of senescence in Drosophila melanogaster with P-element tagging (in preparation).

Clark, A.G., L. Wang & T. Hulleberg, 1994. Spontaneous mutation rate of modifiers of metabolism in Drosophila. Genetics (submitted).

Fowler, K. & L. Partridge, 1989. A cost of mating in female fruitflies. Nature 338: 760–761.

Houle, D., D.K. Hoffmaster, S. Assimacopoulos & B. Charlesworth, 1992. The genomic mutation rate for fitness in Drosophila. Nature 359: 58–60.

Hudson, R.R., K. Bailey, D. Skarecky, J. Kwiatowski & F.J. Ayala, 1994. Evidence for positive selection in the superoxide dismutase (*Sod*) region of *Drosophila melanogaster*. Genetics 136: 1329–1340..

Hughes, D. & A.G. Clark, 1988. Analysis of the genetic structure of life history of *Drosophila melanogaster* using recombinant extracted lines. Evolution 42: 1309–1320.

Hughes, K.A. & B. Charlesworth, 1994. A genetic analysis of senescence in *Drosophila*. Nature 367: 64–66.

Luckinbill, L.S. & M.J. Clare, 1985. Selection for life span in *Drosophila melanogaster*. Heredity 55: 9–18.

Lynch, M., 1988. The rate of polygenic mutation. Genet. Res. 51: 137–148.

Mackay, T.F.C., R.F. Lyman, M.S. Jackson, C. Terzian & W.G. Hill, 1992. Polygenic mutation in *Drosophila melanogaster*: Estimates from divergences among inbred strains. Evolution 46: 300–316.

Mukai, T., 1964. The genetic structure of natural populations of *Drosophila melanogaster*. I. Spontaneous mutation rate of polygenes controlling viability. Genetics 50: 1–19.

Mukai, T., 1969. The genetic structure of natural populations of *Drosophila melanogaster*. VII. Synergistic interaction of spontaneous mutant polygenes controlling viability. Genetics 61: 749–761.

Mukai, T., S.I. Chigusa, L.E. Mettler & J.F. Crow, 1972. Mutation rate and the dominance of genes affecting viability in *Drosophila melanogaster*. Genetics 72: 335–355.

Partridge, L. & N.H. Barton, 1993. Optimality, mutation and the evolution of ageing. Nature 362: 305–311.

Partridge, L. & M. Farquar, 1981. Sexual activity reduces lifespan of male fruitflies. Nature 294: 580–582.

Partridge, L. & P.H. Harvey, 1985. The costs of reproduction. Nature 316: 20–21.

Partridge, L. & K. Fowler, 1992 Direct and correlated responses to selection on age at reproduction in *Drosophila melanogaster*. Evolution 46: 76–91.

Partridge, L. & N.H. Barton, 1993. Optimality, mutation, and the evolution of ageing. Nature 362: 305–311.

Roper, C., P. Pignatelli & L. Partridge, 1993. Evolutionary effects of selection on age at reproduction in larval and adult *Drosophila melanogaster*. Evolution 47: 445–455.

Rose, M.R., 1984. Laboratory evolution of postponed senescence in *Drosophila melanogaster*. Evolution 38: 1004–1010.

Rose, M.R. & B. Charlesworth, 1981. Genetics of life history of *Drosophila melanogaster*. II. Exploratory selection experiments. Genetics 97: 187–196.

Service, P.M. & M.R. Rose, 1985. Genetic covariation among life history components: the effect of novel environments. Evolution 39: 943–945.

Turelli, M., 1984. Heritable genetic variation via mutation-selection balance: Lerch's zeta meets the abdominal bristle. Theor. Pop. Biol. 25: 138–193.

Vaupel, J.W. & J.R. Carey. 1993. Compositional interpretations of medfly mortality. Science 260: 1666–1667.

M.R. Rose and C.E. Finch (eds.), Genetics and Evolution of Aging, 29–42, 1994.
© 1994 *Kluwer Academic Publishers. Printed in the Netherlands.*

Pathogen evolution within host individuals as a primary cause of senescence

Graham Bell
Biology Department, McGill University, 1205 Ave Dr. Penfield, Montreal, Quebec, Canada H3A lBl

Received and accepted 22 June 1993

Key words: virus, selection, aging, senescence, HIV, AIDS

Abstract

This paper discusses a novel theory of senescence: the community of pathogens within each host individual evolves during the life-time of the host, and in doing so progressively reduces host vigour. I marshal evidence that asymptomatic host individuals maintain persistent populations of viral pathogens; that these pathogens replicate; that they are often extremely variable; that selection within hosts causes the evolution of pathogens better able to exploit the host; that selection is host-specific; and that such evolving infections cause appreciable and progressive deterioration. Experimental approaches to testing the theory are discussed.

The senescence of clones

The inescapable aging of multicellular organisms might arise in either of two fundamentally different ways. First, the internal capacity to resist and repair damage might be less in older individuals. Because the force of selection declines with age, germ-line mutations which impair function late in life will accumulate more rapidly than those with comparable effects early in life, and mutations which increase vigour early in life may spread even if they have severely deleterious effects later in life. This is the conventional evolutionary explanation of aging, which has received thorough theoretical exploration and substantial experimental support, and which is reviewed at length by other authors in this issue. The second possibility is that the external environment may systematically deteriorate during the lifetime of every individual, so that individuals will senesce even if they are able to maintain a constant level of defence. This paper is concerned exclusively with this possibility, which does not seem to have been investigated previously.

It is obviously out of the question that the physical environment should continually deteriorate during each individual's lifetime, since if this were true the earth would no longer be capable of supporting life. This is not necessarily true of the biotic environment. Antagonists such as pathogens and their hosts are involved in a continuous process of coevolution, in which the evolution of resistance by the host population is eventually overcome by the evolution of new infective abilities by the pathogen populations, which in turn creates selection for new sources of host resistance, and so on, endlessly. On different time-scales, viruses and bacteria, fungi and helminths will all play this continual game of tag with their hosts. Each new advance made by the pathogen population is conditional: it impairs the vigour of some host genotypes, but not of all, and the host population persists because of the increased proliferation of host genotypes which are not susceptible to the novel pathogen adaptation. Consequently, it is conceivable that the average individual host, at any time, may perceive the environment as continually deteriorating, because of genetic changes in pathogen populations which increase its susceptibility to disease, while the host population is nevertheless able to maintain itself through geological time.

To clarify this idea, consider a thought experiment which contrasts the fate of individuals with the fate of clones each consisting of many genetically identical individuals. Suppose that we constructed many replicate clones, all of the same genotype, and measured the life table of these replicates, each of which might comprise many individuals, in the same way that we might measure the life table of individuals within any of these clones. If we did measure the life table of individuals belonging to a given clone, we would almost certainly find that the rate of survival (or, in principle, some other component of fitness) decreased with age, indicating that individuals senesce. There is, however, no obvious reason to suspect that this would apply to replicate clones able to propagate themselves through time. It is conceivable that reproduction earlier in time is more important than later reproduction, perhaps because an early-reproducing clone would occupy available space or opportunities more quickly. But it is difficult to see how the reproductive activity of the founders of the clone could systematically reduce the vigour of their descendants, a quite separate set of individuals, many generations later. It seems reasonable to conclude that clones will not senesce, even though each individual member of the clone does. On this basis, we should expect to find that each clone has a fixed probability of becoming extinct in any interval of time, so that the number of surviving clones decreases exponentially through time. Clones with different genotypes might have different rates of extinction, of course; the important point is that for any given genotype this rate would be a constant that did not increase through time. This conclusion will fail if one of two additional processes is operating: if either the genotype itself, or the environment in which it is expressed, tends continually to deteriorate through time.

A clone descending from a single founder will not remain genetically uniform, even if its reproduction is strictly mitotic. Mutation will drive an increase in genetic diversity, until the input of new genetic diversity through mutation is just balanced by the erosion of diversity by selection, at which point the clonally propagated population is in equilibrium, and will maintain the same level of diversity from generation to generation. This argument is not quite true for finite populations, in which the occasional stochastic loss of the class of individuals with the fewest mutations will drive an irreversible increase in the mean number of mutations per individual. Because most mutations are somewhat deleterious, the vigour of a finite clone will tend to decrease through time (Muller, 1964; Haigh, 1978). The process can actually be observed in very small clones of asexual protists (Bell, 1988). This paper is not concerned with genetic deterioration of this sort.

A set of replicate clones growing in a natural environment represents a large stationary target for resident populations of parasites and pathogens, which will adapt so as to be able to exploit it more effectively. I assume for simplicity that each clone is completely uniform, removing this restriction later. At first, each clone will survive well, because there has not yet been sufficient time for local pathogens to evolve efficient methods of infection. As time goes on, selection will favour pathogen variants able to exploit this new resource effectively, and as the frequency of such variants increases the survival rate of *all* the replicate clones will fall. Evolution of the local pathogen community will therefore cause a decrease in clonal survival through time; the set of clones will senesce.

It is not known whether such a process of clonal aging occurs in natural populations. There are three possible sources of information. The first is the life table of clones, for organisms in which reproduction produces distinct individuals rather than physically connected ramets or zooids. Claims for the extreme longevity of some clones in plants which reproduce vegetatively, such as creosote bush (Vasek, 1980) and bracken (Olinonen, 1967), are commonplace, but beside the point: large organisms are usually long-lived, but the crucial information on the rate of mortality in relation to age is lacking. Secondly, deliberate experiments in which the survival rate of a clone which is continually replanted into a natural community is compared with that of the same clone maintained in the greenhouse would provide important information (Bell, 1992), but so far as I know none have ever been reported. Finally, accidental experiments are done every time agronomists develop a new clone or inbred line of crop plant and release it for large-scale cultivation. The result is very often a dramatic process of senescence in which the failure of the new variety is quickly brought about by the evolution of local pathogen populations (Barrett, 1981; Wolfe, 1985).

The theory of within-host selection

The relevance of clonal senescence to individual senescence is that multicellular individuals are mitotic clones of cells. The cells may vary dramatically in phenotype, but all are genetically identical, or nearly so. When a pathogen infects a host individual, it may proliferate so successfully that the host is killed quickly by the disease it causes. More often, the growth of the pathogen population is suppressed by host defences, and disease symptoms cease, or never appear. The absence of symptoms is normally taken to imply elimination of the pathogen. Suppose, however, that a small population of the pathogen, made up of somewhat more resistant forms, persists in host tissues. This population continues to reproduce, but is prevented from increasing in numbers, and thereby causing overt symptoms, by host defences. As time goes on, however, variants which are able to replicate somewhat more successfully will be selected in the pathogen population resident in the host. These variants are those pathogen genotypes which happen to be best able to exploit the genotype of the particular host in which they are evolving, and may differ from individual to individual. However, although the pathogen population within the host can evolve adaptation to the host genotype, the host itself cannot evolve, being an individual and not a population. As the pathogen population continues to evolve more effective ways of circumventing host defences, the rate of replication and the abundance of the pathogen will tend to increase, causing a progressive increase in the level of damage sustained by the host, which is macroscopically apparent as a senescent decline in survival or fecundity. Eventually, the host dies, either because the pathogen finally eludes host defences entirely and breaks out to cause acute disease, or because the indirect effects of increased levels of damage caused by evolving pathogens result in the host starving to death or being eaten by a predator.

There is one major complication in this otherwise simple theory that should be addressed before proceeding. Hosts which possess an immune system can evolve, in a sense, to a changing population of pathogens, by producing lymphocytes with altered specificities. The way in which they do this is strikingly similar to the way in which sexual lineages can maintain resistance to external populations of pathogens – the creation of variation by genetic recombination, followed by the increase in frequency under selection of variants which happen by chance to be resistant. Nevertheless, the range of pathogen genotypes to which a given host genotype can respond effectively is limited, and selection among pathogens can eventually cause the emergence of types able to evade immune surveillance. Immune systems are likely to retard but not to prevent the evolution of increasingly virulent resident populations of pathogens.

The theory that senescence is generally caused by the within-host selection of persistent pathogens requires that at least seven conditions be fulfilled before it can be taken seriously as a rival to the antagonistic-pleiotropy or mutation-accumulation theories.

(1) It must be demonstrable that all individuals maintain persistent populations of pathogens; ideally, most hosts should be infected with any given common pathogen.

(2) The pathogen must be capable of horizontal as well as (or rather than) vertical transmission. Parasites which are exclusively vertically transmitted, through the germ line, will be selected for benignity, and will evolve into mutualists such as plastids. Moreover, the pathogen should be horizontally transmitted among cells within the body, rather than being transmitted only by the reproduction of infected cells. This is not an absolute requirement, but greatly enlarges the possibilities of continuing evolution.

(3) The pathogen must replicate, otherwise selection cannot long continue.

(4) The pathogen must be variable, to sustain continued selection.

(5) It must be shown that selection actually occurs, and that it results in populations which are better adapted to exploit the host.

(6) Selection must be host-specific, since otherwise selection in one host would produce adaptation to many other hosts, and most infections would be well-adapted without the necessity for within-host evolution. Host-specific selection will lead to genetic differences among the pathogen populations in different hosts, even if these hosts were initially infected from the same source.

(7) Persistent infections must cause appreciable and progressive deterioration.

This onerous list of conditions excludes most pathogens immediately as candidates for producing senescent deterioration. Most metazoan parasites do not replicate within the host. The endogenous bacterial and fungal populations of large perennial plants, and the gut community of animals, are more relevant, but the most promising candidates are mycoplasmas, viruses and retroviruses. I shall now summarize the evidence that viruses and retroviruses, at least, satisfy all of the conditions outlined above. This is in no sense a review of all the relevant literature, which would be virtually a review of the literature of disease. All that I have done is to collate some recent papers, from which the earlier literature can be obtained.

The evolution of viruses within hosts

Viruses can persist in cell cultures for long periods of time. Human parainfluenza virus persisted through 147 passages over 29 months in culture (Murphy, Dimock & Kang, 1990). Polio virus maintained high titres in neuroblastoma cell lines for nine months (Colbere-Garapin *et al.*, 1989). Reovirus readily becomes persistent in epithelial cell lines, causing morphological and physiological changes in the process (Montgomery *et al.*, 1991). Bovine coronavirus continued to replicate in cell culture over a 120-day period of observation (Hofmann, Sethna & Brian, 1990).

Healthy asymptomatic adults are often cryptically infected by many viruses and other pathogens. Jarret *et al.* (1990) found that 18/20 saliva samples from asymptomatic adults contained herpes virus 6; the occasional cell from peripheral blood samples was also infected. A search using polymerase chain reaction of 23 individuals not known to have recently had chicken pox or shingles found that herpes virus was present in nervous tissue from a majority of them (Mahalingam *et al.*, 1990). A survey of over 1000 asymptomatic adults showed that nearly all carried persistent populations of Epstein-Barr virus (Imai, 1990; see also Niedobiteck *et al.*, 1991). Maedi and visna viruses often become persistent in sheep (Staskus *et al.*, 1991). Woodchuck often carry persistent infections of a hepatitis virus, and persistent infections can be established by experimental inoculation of intact

virus (Miller *et al.*, 1990). Avian leukosis virus is a transformation-defective variant of Rous sarcoma virus which becomes persistent in ducks despite the continued expression of neutralizing antibodies (Nehyba *et al.*, 1990). Strawberry cultivars thought to be free of virus turned out to carry persistent populations of pallidosis virus (Yoshikawa & Converse, 1990). Prokaryotes can also establish persistent infections with no clinical symptoms. Most mycoplasma infections appear to be symptomless, and may be almost universal. Faden and Dryja (1989) isolated an unusual slow-growing bacterium that would not be detected by conventional tests from human middle-ear fluid. Sunaina, Kishore and Shekhauat (1989) found that healthy potato tubers bore latent populations of *Pseudomonas*, perhaps maintained as symptomless infections of wild plants.

Persistent viruses can replicate. Niedobitek *et al.* (1991) showed that the persistence of Epstein-Barr virus requires continued viral replication, rather than the maturation of previously infected cells. Righthand (1991) showed that persistent non-lytic variants of echovirus 6 replicate. Lemon *et al.* (1991) measured the rate of replication of hepatitis A virus in cultured cells.

Persistent infections can cause continued and progressive deterioration. Persistent infection of the central nervous system by measles virus causes chronic and usually fatal disease (Carrigan & Knox, 1990). Theiler's virus, persistent in the central nervous system of mice, causes a gradual and progressive demyelination (Tangy *et al.*, 1991; Lipton *et al.*, 1991). Persistent viruses of the nervous system have been reviewed by Kristensson and Norrby (1986). Avian leukosis virus and the various hepatitis viruses cause liver necrosis and neoplastic diseases (Nehyba *et al.*, 1990). Continued damage is normally caused only by replicating virus; the amount of damage is proportional to the rate of replication, and persistent viruses, usually with low rates of replication, cause a low level of damage that cumulates through time. Persistent infections by hepatitis B virus are likely to lead to liver disease if the virus is replicating (Fattovich *et al.*, 1990). Coxsackie virus B3 persists in myocardium; the development of myocardial lesions is correlated with the rate of virus replication (Klingel *et al.*, 1992). Some endogenous retroviruses of mice are initially benign, as expected, but recombinant virus appearing during the lifetime of the host causes thymic lymphoma (Stoye & Coffin, 1985).

Persistent, rather than acute, infection can be caused by simple changes in intact viral genomes. Persistent populations of Theiler's virus in mice are characterized by mutational change in a capsid protein (McAllister *et al.*, 1990; Tangy *et al.*, 1991); a single nucleotide change causing a single amino-acid substitution may direct a shift from acute to persistent infection (Zurbriggen *et al.*, 1991). Similarly, a single amino-acid substitution in the glycoprotein encoded by choriomeningitis virus shifts its behaviour from acute infection followed by clearance to long-term persistence with immune suppression (Matloubian *et al.*, 1990; Salvato *et al.*, 1991). Persistent hepatitis B virus differed from the form present in the initial acute infection by two mutations, one of which prevented antigen expression (Raimondo *et al.*, 1990a).

The altered viruses of persistent infections escape immune surveillance through reduced transcription or reduced replication. Lytic gene expression is reduced or abolished early in the establishment of persistent infections by herpes simplex virus (Kosz-Vnechak, Coen & Knipe, 1990). The persistent forms of pseudorabies virus in swine have lower levels of transcription than those present in acute infections (Lokensgard, Thawley & Molitor, 1990). Persistent strains of Theiler's virus are less abundant and have lower rates of replication than virulent strains (Lipton *et al.*, 1991). Variant forms of hepatitis A virus which appear during persistent infections and which have altered antigenicity, including neutralization resistance to antibody, have lower rates of replication than forms with normal antigenic characteristics (Lemon *et al.*, 1991). A causal relationship between persistence and reduced replication was established by J. S. Li *et al.* (1991), who showed that a mutant of duck hepatitis B virus with a reduced rate of replication established a persistent infection when inoculated experimentally.

Persistence often follows an acute infection. Hepatitis B virus persists in the liver after serological recovery from an acute infection, and is capable of subsequent reactivation (Ling, Blum & Wands, 1990; see Raimondo *et al.*, 1990b). Mouse hepatitis virus persists in the spinal cord following acute infection, and may subsequently be amplified and spread to cause clinical disease (Perlman *et al.*, 1990). The infection of cultured cells by Theiler's virus kills most of the cells; the virus population in the surviving cells becomes persistent after repeated passage (Patrick *et al.*, 1990).

Virus populations of persistent infections are genetically variable. This has been repeatedly observed since the development of techniques for sequencing nucleic acids; recent examples include parainfluenza virus 3 (Murphy, Dimock & Kang, 1991), hepatitis B virus (Groetzinger & Will, 1992), equine infectious anaemia virus (Salinovich *et al.*, 1986; Alexandersen & Carpenter, 1991), visna virus of sheep (Clements *et al.*, 1980) and caprine arthritis encephalitis virus (Ellis *et al.*, 1987). RNA viruses as a whole are so variable, in large part because the infidelity of RNA polymerases yields a high mutation rate, that their populations should not be thought of as comprising a basic type around which some variation occurs, but rather as a mixture of indefinite and constantly changing composition (Domingo *et al.*, 1985). The literature on retroviral diversity, which may likewise be generated by imprecise replication, has been reviewed by Katz and Skalka (1990).

Viral genomes of persistent infections vary in their ability to attack different cell types. Murine lymphocytic choriomeningitis virus has two sites of infection: the central nervous sytem, where populations similar to the initial infection cause acute disease, and lymphoid tissue, where altered viruses cause a persistent infection. This difference is associated with a single amino-acid substitution in the viral glycoprotein (Ahmed *et al.*, 1991). Macrophage isolates of this virus tended to grow well in cultured macrophages but poorly in lymphocytes, and vice versa (King *et al.*, 1990), so the occurrence of tissue-specific selection can be modelled in culture. A single passage through the host causes the appearance of tissue-specific variants of bovine herpes virus 1 (Whetstone *et al.*, 1989). The divergence of viral populations in different tissues provides evidence for the occurrence of adaptive evolution within the host.

Some variants in heterogeneous viral populations can escape host defences and are cytopathic. Some variants of measles virus in persistent infections are resistant to interferon (Carrigan & Knox, 1990). Variants of hepatitis A virus isolated from persistently-infected monkey cells were cytophatic (Lemon *et al.*, 1991). The appearance of variants which evade immune surveillance has been reported for visna view of sheep (Stanley *et al.*, 1988) and for equine infectious anaemia virus (Salinovich *et al.*, 1986).

Selection can cause rapid and substantial change in the genetic composition of viral populations within cell cultures or hosts. Genetic changes can be detected in bovine herpes virus 1 after a single passage through the host (Whetstone *et al.*, 1989). Such changes cumulate through time. Diez *et al.* (1990) report the structure of part of the genome of foot-and-mouth disease virus after 100 passages in cell culture. Only 5/15 mutations were silent. There were nine amino-acid substitutions in the capsid protein, including some at sites which in other picornaviruses are highly conserved; thus, even highly-conserved sites can evolve when viral populations are exposed to selection in a novel environment. The evidence for selection in retroviral populations has been reviewed by Katz and Skalka (1990).

Selection may favour variants with reduced virulence and cytotoxicity in cell culture. This is often associated with the appearance of incomplete viral genomes in the form of so-called defective interfering particles, for example in human parainfluenza virus (Moscona, 1991), Murray Valley encephalitis virus (Poidinger, Coelen & MacKenzie, 1991), and tomato bushy stunt virus (Knorr *et al.*, 1991). Since these are parasites of intact viruses, their effect on the host is difficult to predict. However, in many cases selection alters the population of intact virus. Rubella virus maintained with antibody in cell culture evolved reduced replication and cytotoxicity; these changes took place in intact virus, though defective interfering RNAs also appeared (Abernathy, Wang & Frey, 1990). The passage of lymphocytic choriomeningitis virus through cultured mouse cells results in the evolution of non-pathogenic populations with altered genomes (Bruns *et al.*, 1990). Pelletier *et al.* (1991) selected persistent

mutants of polio virus in cultured cells, obtaining forms that could persist in non-neural tissue and had altered cell specificity. Measles virus populations may contain two forms resistant to interferon, one of which replicates normally and destroys cells, while the other has reduced replication and cytopathogenicity. The passage of populations containing both forms through nervous tissue of hamsters selects strongly for the less damaging form, causing chronic persistence of the virus. The outcome of such processes depends on the balance between vertical transmission (through the reproduction of infected cells) and horizontal transmission (through the movement of virus from infected to uninfected cells). Reducing the importance of horizontal transmission, for example by the presence of soluble antibody, will shift the balance towards vertical transmission and favour the evolution of benignity. The study of hepatitis A virus by Lemon *et al.* (1991) is a particularly clear description of the mechanism involved: altered forms with reduced antigenicity have greater rates of replication during persistent infection, while normal forms have greater rates during serial passage.

Selection also occurs in the living host. Groetzinger and Will (1992) comment that: 'There is increasing evidence for selection of hepatitis B virus mutants during the natural course of infection, after immunization, and during immune therapy. These mutants can prolong viral persistence, escape immune-mediated elimination, and influence the clinical sequelae of infection.' The appearance of genetically distinct persistent forms in living hosts is, of course, strong evidence for selection.

The outcome of selection may depend on host genotype and on cell type. The ability of herpes virus to establish persistent populations in mammalian hosts varies with the species and strain of the host (Gordon *et al.*, 1990). Borneo disease virus causes encephalitis in most strains of rat, but in one strain no clinical disease is apparent, despite persistent replication of the virus in the central nervous system of the host (Herzog, Frese & Rott, 1991). The mutagenesis of mouse fibroblasts readily produced cell lines resistant to mouse hepatitis virus, through reduced susceptibility to membrane fusion encoded by a viral glycoprotein (Daya *et al.*, 1990). It is probable, therefore, that the final genetic composition of viral populations, after selection, will vary among host individuals, each host presenting a different genetic environment for viral evolution. Kato *et al.* (1992) found extensive variation among hepatitis C virus isolated from different patients with respect to envelope protein genes.

These results suggest that a wide range of viruses, and perhaps other pathogens, develop highly variable populations within their hosts, and in consequence undergo selection for persistent types which through continued evolution produce cumulative and eventually substantial damage. The most detailed and extensive studies of variation and within-host selection, however, have concerned the human immunodeficiency virus HIV and its relatives, slow viruses which may persist asymptomatically for many years after an initial acute phase.

The special case of immunodeficiency viruses

Populations of HIV within single host individuals are extremely heterogeneous. (Hahn *et al.*, 1986; Saag *et al.*, 1988). There is extensive sequence variation in the *env* gene, which encodes envelope glycoproteins (see Howell *et al.*, 1991), with frequent amino-acid substitutions at sites critical in antibody and T-cell recognition (Simmonds *et al.*, 1990; Callahan *et al.*, 1990). Hypervariable sites in *env* of HIV-1 correspond to open regions of RNA secondary structure with little intramolecular base-pairing, which are mostly extracellular portions of the molecule (Le *et al.*, 1989). The substitution of a single amino-acid in such regions can reduce antibody binding specificity (Looney *et al.*, 1989; McKeating *et al.*, 1989). Sites which direct antagonistic interactions with the host are thus particularly variable and liable to change readily under selection (see Simmonds *et al.*, 1990; Balfe *et al.*, 1990).

High levels of variation have also been reported for *nef* (Delassus, Chenier & Wain-Hobson, 1991) and *rev* (Martins *et al.*, 1991) in cell culture, and for *env* of the related simian immunodeficiency virus (Burns & Desrosiers, 1991). The very general operation of selection in viral populations itself casts into doubt the relevance of culture studies, in which populations are exposed to a novel environment. However, the extreme heterogeneity of HIV has been confirmed *in vivo*. Y. Li *et al.* (1991) characterized ten genomes obtained directly from uncul-

tured brain tissue of an AIDS patient and found four full-length genomes with one or two LTRs, three defective genomes with internal deletions, two rearranged genomes with inverted LTRs, and an inverted proviral half flanked by host sequences. Transfection of the four full-length genomes showed that one was fully able to replicate and was comparable with cultured virus. Thus, HIV populations within the host include both fully replicating forms, which themselves show sequence variation, and complex mixtures of other viral entities.

HIV populations in different tissues are genetically different. The frequency of different variant sequence differs among brain, spleen, lymph and blood isolates (Epstein *et al.*, 1991; Haggerty & Stevenson, 1991), which differ from the populations found in cultured cells (Pang *et al.*, 1991). Terwilliger *et al.* (1991) found that allelic variation in *nef* differentially affected the ability of HIV-1 to replicate in different cell lines.

Selection occurs in HIV populations during serial passage in cell culture. Types which are initially rare may increase greatly in frequency (Zweig *et al.*, 1990; Vartanian *et al.*, 1991). Specific variants resistant to drugs are readily selected (Gao *et al.*, 1992). More importantly, passage in the presence of soluble antibody causes the evolution of resistant populations (Robert-Guroff *et al.*, 1986; Masuda *et al.*, 1990; McKeating *et al.*, 1991; Ashkenazi *et al.*, 1991; Rey *et al.*, 1991), and may occur even in highly conserved regions of the viral genome. Hoxie *et al.* (1991) found that the passage of attenuated HIV-1 with low binding affinity for CDA cells eventually resulted in the appearance of cytopathic variants whose binding affinities were comparable with those of normal cytopathic HIV-2 isolates. It is likely that there are several different routes by which resistance can be acquired, since it may or may not be associated with decreased binding affinity to soluble antibody (compare McKeating *et al.*, 1991, with Ashkenazi *et al.*, 1991). L'Age-Stehr *et al.* (1990) observed the evolution of a shorter latent period and increased virulence towards T-cells when HIV was maintained in cell culture.

Damage caused to the host is related to levels of viral replication. Rapidly-replicating isolates which induce syncitium formation characterize individuals who progress rapidly to AIDS and tend to die soon; individuals who remain asymptomatic or survive longer yield more slowly-replicating isolates (Tersmette *et al.*, 1990). The replication rate of HIV is correlated with the rate of loss of CDA+ lymphocytes (Tersmette *et al.*, 1989).

Sequence variation among tissues seems to be caused by tissue-specific selection. Peden, Emerman and Montagnier (1991) cultured isolates from three individuals on six different cell lines and found that replicative ability and tropism changed after repeated passage, causing divergence from a single ancestral sequence. The variety of tissues that can be infected by HIV increases during the lifetime of the host (Cheng-Mayer *et al.*, 1988).

Different loci can evolve independently as the result of recombination. Masuda and Harada (1990) obtained recombinant virus after co-culturing two distinct strains of HIV, and recombination seems to be a major source of heterogeneity of HIV populations within single hosts (Vartanian *et al.*, 1991; Delassus, Chenier & Wain-Hobson, 1991; Howell *et al.*, 1991). The three loci scored over four years from the same patient by Martins *et al.* (1991) all behaved differently.

Longitudinal studies provide convincing evidence for selection and evolution within single host individuals. Phillips *et al.* (1991) found that fluctuations in the specificity of cytotoxic T-cells for virus were matched by variability in proviral *gag* DNA epitope sequences. Some of these variants were not recognised by T-cells, and the accumulation of such variants would provide a basis for immune escape. Wolfs *et al.* (1991) studied *env* sequences from two individuals over five years. At the time of antibody seroconversion, genomic RNA levels peaked and the virus population was highly homogeneous. During the course of infection, heterogeneity increased and the consensus sequence shifted. A single amino-acid substitution was fixed in both patients; this substitution affected antibody binding specificity. After a subsequent increase of genomic RNA and progression to AIDS in one patient, a new variant with major changes in the principal neutralization domain of the viral genome emerged, again forming a homogeneous population.

Host-specific selection leads to variation among host individuals. The most detailed studies concern the *env* gene, and especially the hypervariable V3 domain of the envelope glycoprotein gp120 which it encodes. Extensive variation among individuals has been documented by Simmonds *et al.* (1990) and Epstein *et al.* (1991). The virus populations of six children infected from a single source diverged rapidly from the initial inoculum in different directions (Wolfs *et al.*, 1990).

Within-host selection in relation to other theories

The material that I have reviewed above shows that all the preconditions required by the hypothesis of within-host selection are met, at least in some instances. Whether or not they are *generally* true must await the accumulation of evidence. However, the availability of methods for amplifying viral genomes present at very low abundance in host tissues will make it possible to decide very soon whether the single most important and most surprising observation which the theory requires, the universality of persistent viral infection, is really the case.

Should all the conditions be shown to hold generally, it will still remain to be shown that they operate so as to produce senescence. An obvious experimental approach is to rear rigorously pathogen-free stocks by isolating sterile eggs and raising them in enclosed or filtered environments. The suggestion that senescence is caused by the diseases associated with internal populations of microbes was first made by Metchnikoff in the early 1900s. He emphasized the deleterious nature of toxins produced by the bacterial population of the large bowel, but also described the pathological behaviour of lymphocytes (whose phagocytic properties he discovered) and thus autoimmune diseases. His theory was quite different from that being advanced here: he attributed senescence to the long-continued action of bacterial toxins without reference to the possibility of pathogen evolution within the host. It was thought to be adequately refuted by the observation that animals reared in bacteria-free conditions were not immortal, and indeed usually died without notably exceeding their normal span. The current neglect of his ideas may not be completely justified. For example, Greenberg and

Burkman (1963) compared bacteria-free with normal adult flies, recording average lifespan for both sexes in two experiments. There was no difference between the two groups in two comparisons, but in the remaining two (concerning females in one case and males in the other) the bacteria-free flies lived much longer. However, I have been unable to find observations of the longevity of defined strains of animal under bacteria-free conditions which would support the crucial test of whether rates of mortality increase with age when cellular pathogens are excluded. A detailed quantitative life table for a colony of bacteria-free mice would be of great interest.

In this paper I have emphasized the role of viruses rather than bacteria in forming persistent populations within the host. It is possible, though very difficult, to exclude exogenous viruses, and maintain stocks of animals in completely axenic conditions. Some virus will be transmitted in the egg, of course, but because elements which are exclusively vertically transmitted are not expected to be pathogenic (mouse mammary tumour virus seems to be an exception), this is not necessarily a problem in principle. If such stocks could be shown to be virus-free and did not senesce, the theory would be supported. In practice, the possibility that some retroviruses are partly horizontally and partly vertically transmitted, and that purely endogenous viruses might in time evolve exogenous variants, might cloud the interpretation of a negative result. I have been unable to find any information on the lifespan of animals reared in virus-free conditions.

A more ingenious and more practicable experiment might test the prediction that within-host selection causes the evolution of more pathogenic strains, using genetically uniform host individuals produced asexually. Inoculation of a young host with viral isolates from an old member of the same clone should cause more rapid senescence. Moreover, this should not be the case when the inoculum is obtained from an unrelated individual. I do not know of any experiments of this kind.

It has been reported that individuals do not senesce appreciably in worms which proliferate vegetatively by a nearly equal binary fission (Bell, 1985). This has been taken to support the antagonistic-pleiotropy or mutation-accumulation hypotheses, because it will not be possible to select for the postponement of deleterious gene expression when parent and offspring are indistinguishable and therefore of the same age. At the same time, it seems to negate the hypothesis of within-host selection, which will occur in any piece of tissue, regardless of its status as parent or offspring. It seems desirable to repeat this observation, which has been criticized by Martinez and Levinton (1992), who report the occurrence of senescence in an annelid reproducing by fission. If it turns out to be generally true, it offers strong support for the conventional view and strong evidence against the contrary view being advanced in this paper.

The hypothesis of within-host selection, on the one hand, and related hypotheses of antagonistic pleiotropy and mutation accumulation, on the other hand, are quite different, but they are not mutually exclusive. Rather, they might complement one another in either of two ways. First, the reduced levels of maintenance and repair in older individuals that evolve as the consequence of the decrease in the force of natural selection with increasing age will make it easier for persistent populations of pathogens to evolve resistance to host defences. Secondly, the prospect that persistent pathogen populations will become steadily better adapted during the lifetime of the host further devalues the maintenance of vigour in old individuals, and therefore advances the schedules of antagonistic pleiotropy and mutation accumulation.

A very similar theory has been developed for the particular case of HIV infection by Nowak (1992; Nowak & May, 1991). This virus represents a special case because it attacks the very cells that are responsible for its suppression, and in consequence the dynamics of its evolution in the host are more complex. Nowak's theory differs from the simpler theory advanced in this paper in that the heterogeneity of the viral population plays a central role in disease progression, rather than merely providing a basis for selection of more effective variants. The crucial concept is that the immune cells whose production is specifically induced by a given viral genotype are active only against that genotype, whereas they can themselves be killed by any intact virus. It follows that more heterogenous viral populations will suppress the immune response more effectively, and that there exists a threshold value of viral diversity beyond which the immune system is incapable of restraining the virus and preventing progression to AIDS. It is likely that the immune response is not completely specific, so that new clones of immune cells are active not only against the particular viral genotype responsible for their induction, but also against a range of other genotypes. If viral genotypes exist which can survive a non-specific response (when this is not true, the initial infection is quickly cleared) but not the combined effect of the specific and non-specific responses, then there will be an initial viremia, perhaps accompanied by clinical symptoms, followed by a reduction in virus titre and a long and variable period of persistence before the viral diversity threshold is exceeded and a final phase of acute disease ensues. Differential-equation models of this process are strikingly successful in capturing the major features of the time-course of HIV infection. Whether they are preferable to the simpler theory of the sequential selection of more effective pathogen genotypes from a population whose diversity is maintained by high mutation rates is not clear. One possible approach might be to construct the phylogeny of the evolving viral population, because simple within-host selection implies the sequential replacement of currently abundant variants by slightly modified descendents, whereas Nowak's theory implies that there is no single successful line of descent. Alternatively, simple within-host selection would lead us to expect that the final HIV population causing AIDS would be more homogeneous than the persistent population during the asymptomatic phase, since the single viral strain able to overcome host defences would spread to fixation; this seems to correspond with the situation described by Wolfs *et al.* (1991). Nowak's theory leads us to expect, to the contrary, that the final HIV population will be maximally heterogeneous. However, he suggests that the breakdown of immune function which is eventually caused by increasing viral diversity itself permits rapidly-replicating viral genotypes to be selected, creating genetic uniformity as a consequence rather than a cause of the development of acute disease. This may well be the case; the coupled dynamics of HIV infection and the immune system may necessitate a more complicated explanation than is required for other diseases. The more general significance of the experimental and theoretical work on HIV is that is provides a dramatic illustration of how evolutionary processes driven by within-host selection cause a cumulative and eventually catastrophic collapse of host vigour.

References

Abernathy, E. S., C. Y. Wang & T. K. Frey, 1990. Effect of antiviral antibody on maintenance of long-term rubella virus persistent infection in Vero cells. Journal of Virology 64: 5183-5187.

Ahmed, R., C. S. Hahn, T. Somasundaram, L. Villarete, M. Matloubian & J. H. Strauss, 1991. Molecular basis of organ-specific selection of viral variants during chronic infection. Journal of Virology 65: 4242-4247.

Alexandersen, S. & S. Carpenter, 1991. Characterization of variable regions in the envelope and S3 open reading frame of equine infectious anemia virus. Journal of Virology 65: 4255-4262.

Ashkenazi, A., D. H. Smith, S. A. Marsters, L. Riddle, T. J. Gregory, D. D. Ho & D. J. Capon, 1991. Resistance of primary isolates of human immunodeficiency virus type 1 to soluble CD4 is independent of CD4-rgp120 binding affinity. Proceedings of the National Academy of Sciences of the United States of America 88: 7056-7060.

Balfe, P., P. Simmonds, C. A. Ludlam, J. O. Bishop & A. J. Leigh-Brown, 1990. Concurrent evolution of HIV-1 in patients infected from the same source. Journal of Virology 64: 6221-6231.

Barrett, J. A., 1981. The evolutionary consequences of monoculture, pp. 209-248 in Genetic Consequences of Man-made Change, edited by J. A. Bishop and L. M. Cook. Academic Press, London.

Bell, G., 1985. Evolutionary and non-evolutionary theories of senescence. The American Naturalist 124: 600-603.

Bell, G., 1988. Sex and Death in Protozoa: the History of an Obsession. Cambridge University Press.

Bell, G., 1992. Five properties of environments, pp. 33-56 in Molds, Molecules and Metazoa, edited by P. R. Grant and H. S. Horn. Princeton University Press.

Bruns, M., T. Krazberg, W. Zeller & F. Lehmann-Grube, 1990. Mode of replication of lymphocytic choriomeningitis virus in persistently infected cultivated mouse L cells. Virology 177: 615-624.

Burns, D. P. W. & R. C. Desrosiers, 1991. Selection of genetic variants of simian immunodeficiency virus in persistently infected rhesus monkeys. Journal of Virology 65: 1843-1854.

Callahan, K. M., M. M. Fort, E. A. Obah, E. L. Reinherz & R. F. Siliciano, 1990. Genetic variability in HIV-1 gp120 affects interactions with HLA molecules and T cell receptor. Journal of Immunology 144: 3341-3346.

Carrigan, D. R. & K. K. Knox, 1990. Identification of interferon-resistant subpopulations in several strains of measles virus: Positive selection by growth of the virus in brain tissue. Journal of Virology 64: 1606-1615.

Cheng-Mayer, C., D. Seto, M. Tateno & J. A. Levy, 1988. Biological features of HIV-1 that correlate with variation in the host. Science 240: 80-82.

Clements, J. E., F. S. Pedersen, O. Narayan & W. A. Haseltine, 1980. Genomic changes associated with antigenic variation of visna virus during persistent infection. Proceedings of the National Academy of Sciences of the United States of America 77: 4454-4458.

Colbere-Garapin, F., C. Christodoulou, R. Crainic & I. Pelletier,

1989. Persistent poliovirus infection of human neuroblastoma cells. Proceedings of the National Academy of Sciences of the United States of America 86: 7590-7594.

Daya, M., F. Wong, M. Cervin, G. Evans, H. Vennema, W. Spaan & R. Anderson, 1989. Mutation of host cell determinants which discriminate between lytic and persistent mouse hepatitis virus infection results in a fusion-resistant phenotype. Journal of General Virology 70: 3335-3346.

Delassus, S., R. Cheynier & S. Wain-Hobson, 1991. Evolution of human immunodeficiency virus type 1 nef and long terminal repeat sequences over 4 years in vivo and in vitro. Journal of Virology 65: 225-231.

Diez, J., M. Davila, C. Escarmis, M. G. Mateu, J. Dominguez, J. J. Perez, E. Giralt, J. A. Melero & E. Domingo, 1990. Unique amino acid substitutions in the capsid proteins of foot-and-mouth disease virus from a persistent infection in cell culture. Journal of Virology 64: 5519-5528.

Domingo, E., E. Martinez-Salas, F. Sobrino, J. C. de la Torre, A. Portela et al., 1985. The quasispecies (extremely heterogeneous) nature of viral RNA genome populations: biological relevance – a review. Gene 40: 1-8.

Ellis, T. M., G. E. Wilcox & W. F. Robinson, 1987. Antigenic variation of CAEV during persistent infection of goats. Journal of General Virology 68: 3145-3152.

Epstein, L. G., C. Kuiken, B. M. Blumberg, S. Hartman, L. R. Sharer, M. Clement & J. Goudsmit, 1991. HIV-1 V3 domain variation in brain and spleen of children with AIDS: Tissue-specific evolution within host-determined quasispecies. Virology 180: 583-590.

Faden, H. & D. Dryja, 1989. Recovery of a unique bacterial organism in human middle ear fluid and its possible role in chronic otitis media. Journal of Clinical Microbiology 27: 2488-2491.

Fattovich, G., L. Brollo, A. Alberti, G. Giustina, P. Pontisso, G. Realdi & A. Ruol, 1990. Chronic persistent hepatitis type B can be a progressive disease when associated with sustained virus replication. Journal of Hepatology (Amsterdam) 11: 29-33.

Gao, Q., Z. Gu, M. A. Parniak, X. Li & M. A. Wainberg, 1992. In vitro selection of variants of human immunodeficiency virus type 1 resistant to 3'-azido-3'-deoxythymidine and 2',3'-dideoxyinosine. Journal of Virology 66: 12-19.

Gordon, J. J., J. L. C. McKnight, J. M. Ostrove, E. Romanowski & T. Araullo-Cruz, 1990. Host species and strain differences affect the ability of an HSV-1 ICPO deletion mutant to establish latency and spontaneously reactivate in vivo. Virology 178: 469-477.

Greenberg, B. & A. M. Burkman, 1963. Effect of B-vitamins and a mixed flora on the longevity of germ-free adult houseflies, Musca domestica L. Journal of cellular and comparative Physiology 62: 17-22.

Groetzinger, T. & H. Will, 1992. Sensitive method for identification of minor hepatitis B mutant viruses. Virology 187: 383-387.

Haggerty, S. & M. Stevenson, 1991. Predominance of distinct viral genotypes in brain and lymph node compartments of HIV-infected individuals. Viral Immunology 4: 123-132.

Hahn, B. H., G. M. Shaw, M. E. Taylor, R. R. Redfield, P. D. Markham, S. Z. Salahuddin, F. Wong-Staal, R. C. Gallo, E. S. Parks & W. P. Parks, 1986. Genetic variation in HTLVIII/

LAV over time in patients with AIDS or at risk for AIDS. Science 232: 1548-1553.

Haigh, J., 1978. The accumulation of deleterious genes in a population-Muller's Ratchet. Theoretical Population Biology 14: 251-267.

Herzog, S., K. Frese & R. Rott, 1991. Studies on the genetic control of resistance of black hooded rats to Borna disease. Journal of General Virology 72: 535-540.

Hofmann, M. A., P. B. Sethna & D. A. Brian, 1990. Bovine coronavirus messenger RNA replication continues throughout persistent infection in cell culture. Journal of Virology 64: 4108-4114.

Howell, R. M., J. E. Fitzgibbon, M. Noe, Z. Ren, D. J. Gocke, T. A. Schwartzer & D. T. Dubin, 1991. In vivo sequence variation of the human immunodeficiency virus type 1 env gene: Evidence for recombination among variants found in a single individual. AIDS Research and Human Retroviruses 7: 869-876.

Hoxie, J. A., L. F. Brass, C. H. Pletcher, B. S. Haggarty & B. H. Hahn, 1991. Cytopathic variants of an attenuated isolate of human immunodeficiency virus type 2 exhibit increased affinity for CD4. Journal of Virology 65: 5096-5101.

Imai, S., 1990. Virological and immunological studies on inapparent Epstein-Barr virus infection in healthy individuals: In comparison to immunosuppressed patients and patients with infectious mononucleosis. Hokkaido Journal of Medical Science 65: 481-492.

Jarrett, R. F., D. A. Clark, S. F. Josephs & D. E. Onions, 1990. Detection of human herpesvirus-6 DNA in peripheral blood and saliva. Journal of Medical Virology 32: 73-76.

Kato, N., Y. Ootsuyama, T. Tanaka, M. Nakagawa, T. Nakazawa, K. Muraiso, S. Ohkoshi, M. Hijikata & K. Shimotohno, 1992. Marked sequence diversity in the putative envelope proteins of hepatitis C viruses. Virus Research 22: 107-123.

Katz, R. A. & A. M. Skalka, 1990. Generation of diversity in retroviruses. Annual Review of Genetics 24: 409-445.

King, C. C., R. De-Fries, S. R. Kolhekar & R. Ahmed, 1990. In vivo selection of lymphocyte-trophic and macrophage-tropic variants of lymphocytic choriomeningitis virus during persistent infection. Journal of Virology 64: 5611-5616.

Klingel, K., C. Hohenadl, A. Canu, M. Albrecht, M. Seemann, G. Mall & R. Kandolf, 1992. Ongoing enterovirus-induced myocarditis is associated with persistent heart muscle infection: Quantitative analysis of virus replication, tissue damage, and inflammation. Proceedings of the National Academy of Sciences of the United States of America 89: 314-318.

Knorr, D. A., R. H. Mullin, P. Q. Hearne & T. J. Morris, 1991. De novo generation of defective interfering RNAs of tomato bushy stunt virus by high multiplicity passage. Virology 181: 193-202.

Kosz-Vnechak, M., D. N. Coen & D. M. Knipe, 1990. Restricted expression of herpes simplex virus lytic genes during establishment of latent infection by thymidine kinase-negative mutant viruses. Journal of Virology 64: 5396-5402.

Kristensson, K. & E. Norrby, 1986. Persistence of RNA viruses in the central nervous system. Annual Reviews of Microbiology 40: 159-184.

L'Age-Stehr, J., M. Niedrig, H. R. Glederblom, J. W. Sim-

Brandenburg, M. Urban-Schriefer, E. P. Rieber, J. G. Haas, G. Riethmueller & H. W. L. Ziegler-Heitbrock, 1990. Infection of the human monocytic cell line Mono Mac6 with human immunodeficiency virus types 1 and 2 results in long-term production of virus variants with increased cytopathogenicity for CD4-positive T cells. Journal of Virology 64: 3982-3987.

Le, S. Y., J. H. Chen, D. Chatterjee & J. V. Maizel, 1989. Sequence divergence and open regions of RNA secondary structures in the envelope regions of the 17 human immunodeficiency virus isolates. Nucleic Acids Research 17: 3275-3288.

Lemon, S. M., P. C. Murphy, P. A. Shields, L. H. Ping, S. M. Feinstone, T. Cromeans & R. W. Jansen, 1991. Antigenic and genetic variation in cytopathic hepatitis A virus variants arising during persistent infection: Evidence for genetic recombination. Journal of Virology 65: 2056-2065.

Li, J. S., I. Fourel, C. Jaquet & C. Trepo, 1991. Decreased replication capacity of a duck hepatitis B virus mutant with altered distal pre-S region. Virus Research 20: 11-22.

Li, Y., J. C. Kappes, J. A. Conway, R. W. Price, G. M. Shaw & B. H. Hahn, 1991. Molecular characterization of human immunodeficiency virus type 1 cloned directly from uncultured human brain tissue: Identification of replication-competent and replication-defective viral genomes. Journal of Virology 65: 3973-3985.

Lipton, H. L., M. Calenoff, P. Bandyopadhyay, S. D. Miller, M. C. Dal-Canto, S. Gerety & K. Jensen, 1991. The 5' noncoding sequences from a less virulent Theiler's virus dramatically attenuate GDVII neurovirulence. Journal of Virology 65: 4370-4377.

Lokensgard, J. R., D. G. Thawley & T. W. Molitor, 1990. Pseudorabies virus latency: Restricted transcription. Archives of Virology 110: 129-136.

Looney, D. J., A. G. Fisher & S. D. Putney, 1988. Type-restricted neutralization of molecular clones of HIV. Science 241: 357-359.

Mahalingam, R., M. Wellish, W. Wolf, A. N. Dueland, R. Cohrs, A. Vafai & D. Gilden, 1990. Latent varicella-zoster viral DNA in human trigeminal and thoracic ganglia. New England Journal of Medicine 323: 627-631.

Martinez, D. E. & J. S. Levinton, 1992. Asexual metazoans undergo senescence. Proceedings of the National Academy of the USA 89: 9920-9923.

Martins, L. P., N. Chenciner, B. Asjo, A. Meyerhans & S. Wain-Hobson, 1991. Independent fluctuation of human immunodeficiency virus type 1 rev and gp41 quasispecies in vivo. Journal of Virology 65: 4502-4507.

Masuda, T., S. Matsushita, M. J. Kuroda, M. Kannagi, K. Takatsuki & S. Harada, 1990. Generation of neutralization-resistant HIV-1 in vitro due to amino acid interchanges of third hypervariable env region. Journal of Immunology 145: 3240-3246.

Matloubian, M., T. Somasundaram, S. R. Kolhekar, R. Selvakumar & R. Ahmed, 1990. Genetic basis of viral persistence: Single amino acid change in the viral glycoprotein affects ability of lymphocytic choriomeningitis virus to persist in adult mice. Journal of Experimental Medicine 172: 1043-1048.

McAllister, A., F. Tangy, C. Aubert & M. Brahic, 1990. Genetic mapping of the ability of Theiler's virus to persist and demyelinate. Journal of Virology 64: 4252-4257.

McKeating, J., P. Balfe, P. Clapham & R. A. Weiss, 1991. Recombinant CD4-selected human immunodeficiency virus type 1 variants with reduced gp120 affinity for CD4 and increased cell fusion capacity. Journal of Virology 65: 4777-4785.

Metchnikoff, E., 1904. The Nature of Man. [I have read the 1938 translation of Sir P. Chalmers Mitchell, published by Watts & Co, London.]

Miller, R. H., R. Girones, P. J. Cote, W. E. Hornbuckle, T. Chestnut, B. H. Baldwin, B. E. Korba, B. C. Tennant, J. L. Gerin & R. H. Purcell, 1990. Evidence against a requisite role for defective virus in the establishment of persistent hepadnavirus infections. Proceedings of the National Academy of Sciences of the United States of America 87: 9329-9332.

Montgomery, L. B., C. Y. Y. Kao, E. Verdin, C. Cahill & E. Maratos-Flier, 1991. Infection of a polarized epithelial cell line with wild-type reovirus leads to virus persistence and altered cellular functions. Journal of General Virology 72: 2939-2946.

Moscona, A., 1991. Defective interfering particles of human parainfluenza virus type 3 are associated with persistent infection in cell culture. Virology 183: 821-824.

Muller, H. J., 1964. The relation of recombination to mutational advance. Mutation Research 1: 2-9.

Murphy, D. G., K. Dimock & C. Y. Kang, 1990. Viral RNA and protein synthesis in two LLC-MK-2 cell lines persistently infected with human parainfluenza virus 3. Virus Research 16: 1-16.

Murphy, D. G., K. Dimock & C. Y. Kang, 1991. Numerous transitions in human parainfluenza virus 3 RNA recovered from persistently infected cells. Virology 181: 760-763.

Nehyba, J., J. Svoboda, I. Karakoz, J. Geryk & J. Hejnar, 1990. Ducks: A new experimental host system for studying persistent infection with avian leukemia retroviruses. Journal of General Virology 71: 1937-1946.

Niedobitek, G., L. S. Young, R. Lau, L. Brooks, D. Greenspan, J. S. Greenspan & A. B. Rickinson, 1991. Epstein-Barr virus infection in oral hairy leukoplakia: Virus replication in the absence of a detectable latent phase. Journal of General Virology 72: 3035-3046.

Nowak, M. A., 1992. Variability of HIV infections. Journal of theoretical Biology 155: 1-20.

Nowak, M. A. & R. M. May, 1991. Mathematical biology of HIV infections: antigenic variation and diversity threshold. Mathematical Biosciences 106: 1-21.

Oinonen, E., 1967. Sporal regeneration of bracken in Finland in the light of the dimension and age of its clones. Acta Forest Fennica 83: 1-96.

Pang, S., H. V. Vinters, T. Akashi, W. A. O'Brien & I. S. Y. Chen, 1991. HIV-1 env sequence variation in brain tissue of patients with AIDS-related neurologic disease. Journal of Acquired Immune Deficiency Syndromes 4: 1082-1092.

Patrick, A. K., E. L. Oleszak, J. L. Leibowitz & M. Rodriguez, 1990. Persistent infection of a glioma cell line generates a Theiler's virus variant which fails to induce demyelinating disease in SJL/J mice. Journal of General Virology 71: 2123-2132.

Peden, K., M. Emerman & L. Montagnier, 1991. Changes in growth properties on passage in tissue culture of viruses derived from infectious molecular clones of HIV-1-LAI, HIV-1-MAL, and HIV-1-ELI. Virology 185: 661-672.

Pelletier, I., T. Couderc, S. Borzakian, E. Wyckoff, R. Crainic, E. Ehrenfeld & F. Colbere-Garapin, 1991. Characterization of persistent poliovirus mutants selected in human neuroblastoma cells. Virology 180: 729-737.

Perlman, S., G. Jacobsen, A. L. Olson & A. Afifi, 1990. Identification of the spinal cord as a major site of persistence during chronic infection with a murine coronavirus. Virology 175: 418-426.

Phillips, R. E., S. Rowland-Jones, D. F. Nixon, F. M. Gotch, J. P. Edwards, A. O. Ogunlesi, J. G. Elvin, J. A. Rothbard & C. R. M. Bangham et al., 1991. Human immunodeficiency virus genetic variation that can escape cytotoxic T cell recognition. Nature (London) 354: 453-459.

Poidinger, M., R. J. Coelen & J. S. Mackenzie, 1991. Persistent infection of Vero cells by the flavivirus Murray Valley encephalitis virus. Journal of General Virology 72: 573-578.

Raimondo, G., R. Schneider, M. Stemler, V. Smedile, G. Rodino & H. Will, 1990a. A new hepatitis B virus variant in a chronic carrier with multiple episodes of viral reactivation and acute hepatitis. Virology 179: 64-68.

Raimondo, R., M. Stemler, R. Schneider, G. Wildner, G. Squadrito & H. Will, 1990b. Latency and reactivation of a precore mutant hepatitis B virus in a chronically infected patient. Journal of Hepatology (Amsterdam) 11: 374-380.

Rey, F., G. Donker, B. Spire, C. Devaux & J. C. Chermann, 1991. Selection of an HIV-1 variant following anti-CD4 treatment of infected cells. Aids-Forschung 6: 243-246.

Righthand, V. F., 1991. Transmission of viral persistence by transfection of human cultured cells with RNA of a persistent strain of echovirus 6. Microbial Pathogenesis 11: 57-66.

Robert-Guroff, M., M. S. Reitz, W. G. Robey & R. C. Gallo, 1986. In vitro generation of an HLTV III variant by neutralizing antibody. Journal of Immunology 137: 3306-3309.

Saag, M. S., B. H. Hahn, J. Gibbons, Y. Li, E. S. Parks, W. P. Parks & G. M. Shaw, 1988. Extensive variation of HIV-1 in vivo. Nature (London) 34: 440-444.

Salinovich, O., S. L. Payne, R. C. Montelaro, K. L. Hussain, C. J. Issel et al., 1986. Rapid emergence of novel antigenic and genetic variants of equine infectious anaemia virus during persistent infection. Journal of Virology 57: 71-80.

Salvato, M., P. Borrow, E. Shimomaye & M. B. A. Oldstone, 1991. Molecular basis of viral persistence: A single amino acid change in the glycoprotein of lymphocytic choriomeningitis virus is associated with suppression of the antiviral cytotoxic T-lymphocyte response and establishment of persistence. Journal of Virology 65: 1863-1869.

Simmonds, P., P. Balfe, C. A. Ludlam, J. O. Bishop & A. J. L. Brown, 1990. Analysis of sequence diversity of hypervariable regions of the external glycoprotein of human immunodeficiency virus type 1. Journal of Virology 64: 5840-5850.

Stanley, J., L. M. Bhaduri, U. Narayan & J. E. Clements, 1987. Topographical rearrangements of visna virus envelope glycoprotein during antigenic drift. Journal of Virology 61: 1019-1028.

Staskus, K. A., L. Couch, P. Bitterman, E. F. Retzel, M. Zupan-

cic, J. List & A. T. Haase, 1991. In situ amplification of visna virus DNA in tissue sections reveals a reservoir of latently infected cells. Microbial Pathogenesis 11: 67-76.

Stoye, J. P. & J. M. Coffin, 1985. Endogenous retroviruses, volume 2 in RNA Tumor Viruses, edited by R. Weiss, N. Teich and J. Coffin. Cold Spring Harbor Laboratory, New York.

Sunaina, V., V. Kishore & G. S. Shekhawat, 1989. Latent survival of Pseudomonas solanacearum in potato tubers and weeds. Zeitschrift fur Pflanzenkrankheiten und Pflanzenschutz 96: 361-364.

Tangy, F., A. McAllister, C. Aubert & M. Brahic, 1991. Determinants of persistence and demyelination of the DA strain of Theiler's virus are found only in the VP1 gene. Journal of Virology 65: 1616-1618.

Tersmette, M., R. E. Y. De-Goede, J. K. M. Eeftink-Schattenkerk, P. T. A. Schellekens, R. A. Coutinho, J. Goudsmit, J. M. A. Lange, F. De-Wolf, J. G. Huisman & F. Miedema, 1989. Association between biological properties of human immunodeficiency virus variants and risk for AIDS and AIDS mortality. Lancet 1: 983-985.

Tersmette, M., R. A. Gruters, F. De-Wolf, R. E. Y. De-Goede, J. M. A. Lange, P. T. A. Schellekens, J. Goudsmit, H. G. Huisman & F. Miedema, 1989. Evidence for a role of virulent human immunodeficiency virus (HIV) variants in the pathogenesis of acquired immunodeficiency syndrome: Studies on sequential HIV isolates. Journal of Virology 63: 2118-2125.

Terwilliger, E. F., E. Langhoff, D. Gabzuda, E. Zazopoulos & W. A. Haseltine, 1991. Allelic variation in the effects of the nef gene on replication of human immunodeficiency virus type 1. Proceedings of the National Academy of Sciences of the United States of America 88: 10971-10975.

Vartanian, J. P., A. Meyerhans, B. Asjo & S. Wain-Hobson, 1991. Selection, recombination, and G to A hypermutation of human immunodeficiency virus type 1 genomes. Journal of Virology 65: 1779-1788.

Vasek, F. C., 1980. Creosote bush: long-lived clones in the Mohave desert. American Journal of Botany 67: 246-255.

Whetstone, C., J. Miller, D. Bortner & M. Van-Der-Maaten, 1989. Changes in the restriction endonuclease patterns of four modified-live infectious bovine rhinotracheitis virus (IBRV) vaccines after one passage in host animal. Vaccine 7: 527-532.

Wolfe, M. S., 1985. The current status and prospects of multiline cultivars and variety mixtures for disease resistance. Annual Review of Phythopathology 23: 251-273.

Wolfs, T. F. W., J. J. De-Jong, H. Van-Den-Berg, J. M. G. H. Tijnagel & W. J. A. Krone, 1990. Evolution of sequences encoding the principal neutralization epitope of human immunodeficiency virus 1 is host dependent, rapid, and continuous. Proceedings of the National Academy of Sciences of the United States of America 87: 9938-9942.

Wolfs, T. F. W., G. Zwart, M. Bakker, M. Valk, C. L. Kuiken & J. Goudsmit, 1991. Naturally occurring mutations within HIV-1 V3 genomic RNA lead to antigenic variation dependent on a single amino acid substitution. Virology 185: 195-205.

Yoshikawa, N. & R. H. Converse, 1990. Strawberry pallidosis disease: Distinctive double stranded RNA species associated with latent infections in indicators and in diseased strawberry cultivars. Phytopathology 80: 543-548.

Zurbriggen, A., C. Thomas, M. Yamada, R. P. Roos & R. S. Fujinami, 1991. Direct evidence of a role for amino acid 101 of VP-1 in central nervous system disease in Theiler's murine encephalomyelitis virus infection. Journal of Virology 65: 1929-1937.

Zweig, M., K. P. Samuel, S. D. Showalter, S. V. Bladen, G. C. Dubois, J. A. Lautenberger, D. R. Hodge & T. S. Papas, 1990. Heterogeneity of Nef proteins in cells infected with human immunodeficiency virus type 1. Virology 179: 504-507.

PART TWO
Diversity of aging

Introduction

In contrast to the widespread existence of aging, particularly among multicellular animals, is the diversity of its manifestations. Different species seem to become decrepit with adult age in markedly different ways. Aging theories in the 1960's, like the somatic mutation theory, asserted that this diversity was merely the superficial heterogeneity generated by the action of powerful, universal, molecular mechanisms on different life-forms.

Evolutionary theories of aging have a similar cast, except that they precisely invert this causal pyramid, placing the unifying force that generates aging at the level of population genetics: the decline in the force of natural selection with adult age (vid. Charlesworth, this volume). This evolutionary account then leads to the explanation of the diversity of aging in terms of the specific effects of general evolutionary mechanisms, a familiar issue within evolutionary biology.

Whatever explanation is offered for the diversity of aging processes, that diversity itself is an important subject for investigation. Unfortunately, most of our knowledge of aging comes from just two taxonomic Classes: Mammalia and Insecta. And even our knowledge of Insecta is tiny beyond the genus *Drosophila*. Thus Parts Three and Four of this volume treat *Drosophila* and mammals, respectively. In Part Two, we have some illustrative, or rather exemplary, treatments of aging in other taxonomic groups. Two of the genera discussed, *Sacchromyces* and *Caenorhabditis*, are among the best known organisms from the standpoint of the experimental genetics of aging. The other groups, plants and crustaceans, are little known from a genetic point of view. However, the study of heterogeneous groups of organisms can proceed by methods other than the genetic. Promislow and Tatar present the state-of-the-art with respect to the comparative study of aging, revealing how modern phylogenetic methods can considerably enhance the power of inference from comparative data. All told, these studies suggest that the study of aging in diverse species will be essential to future progress in the field.

M. R. Rose and C. E. Finch (eds), Genetics and Evolution of Aging, 45–53, 1994.
© 1994 *Kluwer Academic Publishers. Printed in the Netherlands.*

Comparative approaches to the study of senescence: bridging genetics and phylogenetics

Daniel E.L. Promislow[1] & Marc Tatar[2]
[1] *Department of Biology, Queen's University, Kingston, Ontario, K7L 3N6 Canada*
[2] *Graduate Group in Ecology, University of California, Davis, CA 95616-8584, USA*

Introduction

Experimental genetics provides perhaps the most powerful set of tools one can use to understand biological phenomena. The studies presented in this volume elucidate the genetic structure underlying variation in longevity, particularly the role of genetic covariation, and the potential for future evolutionary change. What, then, can biologists who study senescence do with the comparative method – an approach that dates back to Buffon, Locke, Pliny and even Aristotle (Ridley, 1983) – that they cannot do with modern genetics? Modern genetic studies based on allelic variation can only provide a slice through time, and so cannot tell us what has happened in the past, nor why. Comparative studies provide this historical perspective. And where genetic studies provide depth of information on a single species, comparative studies provide taxonomic breadth, enabling us to generalize specific results across a broad array of species.

George Sacher introduced comparative techniques to understand the diversity of life span among animals (Sacher, 1959). Sacher (1959) found that a significant amount of the variation in life span in mammals could be explained by variation in body size. Since then, many studies have used a similar approach to address the cause of aging, comparing interspecific variation in life span with DNA repair rates (Hart & Setlow, 1974), brain size (Sacher & Staffeldt, 1974; Sacher, 1978), titres of superoxide dismutase (Cutler, 1984), and so forth. However, until recently, comparative gerontological studies like these had not addressed genetic theories for the evolution of senescence. Furthermore, these studies all suffered from conceptual and methodological problems that obscured and

weakened their validity (discussed in Promislow, 1993).

In recent years, statistically robust comparative methods (e.g. Felsenstein, 1985; Grafen, 1989; Harvey & Pagel, 1991; Martins & Garland, 1991) have begun to be applied to questions about the evolution of senescence. In the first part of this chapter, we examine ways in which these studies have furthered our understanding. We also outline the sorts of comparative studies that would allow us to generalize findings from genetics on single-species to explain broad patterns among diverse taxa. In each case, we find a number of unresolved problems in the comparative biology of aging. Thus, in the second part of our discussion, we examine the various challenges that lie ahead for a comparative biology of senescence, and perhaps for evolutionary studies of aging in general.

Insights from the comparative method

Testing evolutionary hypotheses

While there are many comparative reviews of aging in both natural and laboratory populations (Comfort, 1979; Nesse, 1988; Newton, 1989; Finch, 1990; Finch *et al.*, 1990; Gavrilov & Gavrilova, 1991; Promislow, 1991; Rose, 1991; Austad, 1992; Austad, 1993), few have actually used these data to test genetic, evolutionary theories of aging. A recent study of senescence in natural populations of mammals considered several evolutionary theories (Medawar, 1952; Williams, 1957; Kirkwood, 1977) but found only equivocal support (Promislow, 1991; Gaillard *et al.*, 1994). Similarly, Schnebel and Grossfield (1988) used data from laboratory populations of

12 species of *Drosophila* to test Williams's (1957) concept of antagonistic pleiotropy (AP, reviewed in this volume by Rose & Finch; Partridge & Barton; Charlesworth), and here, too, the results were inconclusive (Schnebel & Grossfield, 1988; Promislow, 1991; Gaillard *et al.*, 1994). (Comparative approaches have also been used to examine the role of DNA repair rates in studies of aging, but Promislow (1990) has questioned the validity of these studies). The quality of currently available data on mortality rates in natural populations may preclude comparative tests of evolutionary theories of aging (Gaillard *et al.*, 1994). But the quality of even relatively noisy laboratory data may be sufficient to address these evolutionary questions. Here we reconsider Schnebel and Grossfield's (1988) data on 12 species of laboratory-reared *Drosophila* to illustrate how appropriate comparative methods can test evolutionary hypotheses even with noisy data.

Schnebel and Grossfield tried to determine whether Williams's (1957) prediction of negative correlations between early-life and late-life fitness traits holds across taxa. To test the theory, they measured age-specific clutch size, mating success at three different ages, and average maximum life span among replicate vials, among all twelve species, and then examined the correlations between life span and both clutch size and mating success. They posited that the AP theory for the evolution of senescence would be supported by significant negative correlations of either early-age fecundity or mating success with life span, and positive correlations between survival and either late-age fecundity or mating success. Schnebel and Grossfield failed to find the expected correlations, and so concluded that the AP theory could not explain interspecific life history variation.

However, there are two major problems with this analysis (Promislow, in review). First, absolute measures of fecundity may be a misleading indicator of age-specific investment in reproduction. Schnebel and Grossfield (1988) compared fecundity at each of three ages. Within a species, if total lifetime fecundity remains constant, then changes to early fecundity will correlate with changes in *relative* investment in fecundity early in life. But among different species, fecundity may vary widely, so measures

of early fecundity may not indicate relative investment in early fecundity. In the 12 species of *Drosophila* that Schnebel and Grossfield (1988) examined, total fecundity ranged from 54 eggs in *D. montana* to 200 eggs in *D. melanogaster*. But that does not necessarily mean that *D. melanogaster* made a greater *relative* investment early in life compared to *D. Montana*.

Second, Schnebel and Grossfield (1988) compared the twelve species using analysis of variance, assuming that each species is a statistically independent point. But each species is more closely related to other species within its three-species taxon than to species in other taxa. Thus, the species cannot be considered independent (see Harvey & Pagel, 1991). In this case, among each of the four taxonomic sub-groups, ecological or physiological variation may obscure the patterns we expect to find. Among populations within a species, or even among closely related, ecological similar species, there may be a trade-off of the sort expected between reproductive investment and survival. Among different ecotypes, however, this trade-off may not be apparent.

Fortunately, we can resolve both of these difficulties using the original data. To provide measures of age-specific fecundity that were independent of total fecundity, Promislow (in review) calculated early relative fecundity as the ratio of early fecundity to total fecundity measured at all three ages, and late relative fecundity as late fecundity divided by total fecundity. In addition, mating success is defined as the slope of the least-squares regression line for percentage of individuals mating versus age. And to correct for phylogenetic effects, Promislow performed separate analyses within each of the four groups of closely related species, and used a randomization test (see Manly, 1991) to increase statistical power.

On the basis of this reanalysis, Promislow (in review) reached dramatically different conclusions. The original data did, indeed, support the prediction from Williams's hypothesis. Promislow found that, contrary to the claims of the original study, early fecundity was significantly, negatively correlated with late survival and late fecundity, and that species with relatively high early mating success also had lower survival.

We can draw four lessons from this reanalysis. First, correct results from comparative studies of senescence depend on using the appropriate metric to measure senescence and fitness-related traits. Second, comparative studies need to consider the non-independence of species as data points. In some cases, this makes little difference to the outcome of the study. In this case, the correlation among species may have been confounded by ecological or physiological effects on fecundity or survival. Third, randomization methods are useful when small sample sizes limit statistical power using traditional methods. This may be particularly important in studies on senescence, where good data may be available for only a few species. Finally, Promislow's reanalysis supports predictions derived from Williams's (1957) antagonistic pleiotropy theory for the evolution of senescence.

We can use the comparative method to do more than just evaluate existing evolutionary hypotheses. We can also derive novel hypotheses and predictions from previously unrecognized phylogenetic patterns, and test the taxonomic generality of results from single-species genetic studies (Harvey & Pagel, 1991; Austad, 1993).

Generating novel patterns and hypotheses

Just as the natural historian can identify novel biological phenomena that the evolutionary theorist then tries to explain (Grafen, 1987), so the comparative biologist may act as a natural historian of phylogeny, identifying patterns in need of explanation (e.g. McLennan, 1991).

Austad (1993) discusses a classic example of this phenomenon, where observations made early in the twentieth century on the allometry of life span and metabolic rate led to the 'rate-of-living' hypothesis (Pearl, 1928; Sacher, 1959). Even in recent comparative studies of senescence, new patterns continue to be identified and novel theory developed. In Promislow's (1991) comparative study of senescence in mammals, he found that rates of senescence did not increase with increasing extrinsic sources of mortality, contrary to expectations based on the model of Hamilton (1966). And along similar lines, Abrams (1993) developed a theory that gave explicit predictions for the relationship between extrinsic mortality

rates and senescence. In his model, he found that the prediction of a positive correlation between extrinsic mortality and rates of senescence holds only under rather restrictive conditions.

As our knowledge of the natural history of aging grows (Finch, 1990), newly discovered patterns of aging should stimulate new theoretical developments. For example, recent work has found a slowing of the rate of change in age-specific mortality at advanced ages in the Mediterranean fruit fly (Carey *et al.*, 1992) and *Drosophila* (Curtsinger *et al.*, 1992). In these flies, age-specific increases in mortality rate begin to decelerate at the most advanced ages, and mortality may even become age-independent at a surprisingly low level. If we find clear taxonomic patterns for this trait, we may be able to generate explanatory hypotheses or to reformulate the current mechanistic explanations. For instance, if late-age mortality only decelerates in holometabolous insects (insects with complete metamorphosis), this might suggest a causative role for complex life histories with post-mitotic adult somas. Or perhaps we may find that mortality does not decelerate in species with non-continuous population dynamics (as occurs in species with seasonally synchronized life cycles). Such a finding might lead us to examine the dependency of the age-specific force of natural selection on the assumptions of the Euler-Lotka formalization, which serves as the basis of demographic models of senescence (Charlesworth, this volume). These hypothetical ideas are examples of how we might generate novel *post hoc* hypotheses when patterns are described with comparative techniques. Such hypotheses could then provide the impetus for further comparative tests or for population genetic analyses.

Generalizing specific findings

The success of genetic approaches to the study of senescence is limited to explaining patterns of aging in a small number of organisms: primarily the fruit fly, mouse, yeast and nematode (see chapters throughout this volume). But more critically, genetic studies are constrained by the extant and inducible genetic variability one finds within populations. In some cases, the comparative method may provide a potent tool to generalize specific genetic findings.

Experimental genetic techniques (e.g. selection, inbreeding and transmission analyses) depend on allelic variation, and are thereby limited by the available genetic variation in the study population. But genes that affect life history traits (e.g. life span, age at reproduction, fecundity) may be under strong selection and thus fixed within populations, even if these same genes have contributed to the evolutionary differentiation of life history strategies among taxa. The EF-1α gene provides an illustrative example. The EF-1α gene is thought to be homozygously fixed in *Drosophila* (Shepherd *et al.*, 1989). Because of this lack of polymorphism, its potential role in the evolution of senescence in *Drosophila* could not be inferred using standard population genetics. But by inserting extra copies of the gene EF-1α into the *Drosophila* genome, Shepherd *et al.* (1989) found that life span was extended significantly (but see Stearns and Kaiser, this volume).

If there were variation among species either in the presence or activity of EF-1α, we could use a comparative approach to examine its role in determining patterns of aging. For example, one could compare the activity levels of the EF-1α gene in diverse species (perhaps by quantitative PCR of cDNA), to test whether variation in these levels correlates with interspecific variation in age-specific patterns of mortality or reproductive investment.

These studies would still require time-consuming and costly genetic analysis for each of the species we included in our comparative analysis. As the following hypothetical comparative study illustrates, in some cases we may be able to gain insights without incurring these costs. From a genetic study of the nematode, *Caenorhabditis elegans*, Van Voorhies (1992) reported that a gene for reduced sperm production (*spe-26*) led to a 65% increase in mean lifespan. Van Voorhies was careful not to speculate on the importance of these patterns outside the lab, nor on the impact of spermatogenesis in species other than *Caenorhabditis*. However, from the safety of our armchairs, we are led to wonder whether this study might tell us something more general about sperm production and life span among invertebrates, or whether it is a phenomenon limited in scope to the genus *Caenorhabditis*, or even to a particular laboratory population of nematodes.

As a working hypothesis, assume that spermatogenesis is costly, so greater male investment in spermatogenic machinery should lead to shorter male lifespan (*ceteris paribus*). A simple measure of male investment in 'spermatogenic machinery' could be the fraction of an organism, by weight, devoted to testes and accessory structures. If spermatogenesis affects the rate of senescence, then taxa with relatively high investment by weight should have relatively short lifespan. A statistically significant relationship between relative investment in spermatogenesis and relative lifespan would support the possibility that Van Voorhies's (1992) observations are a more taxonomically widespread phenomenon.

Our final example comes from work we discussed in the preceding section on the deceleration of age-specific mortality in the Mediterranean fruit fly (Carey *et al.*, 1992) and *Drosophila* (Curtsinger *et al.*, 1992). Can we extend the observations on flies to other taxa? The original finding is strengthened by the fact that a relative of the medfly (the Mexican fruit fly *Anastrepha*, J.R. Carey, pers. comm.) and a second laboratory population of *Drosophila* (A. Clark, pers. comm.) also show this deceleration. But these studies do not extend the generality beyond the order Diptera. We need to examine life tables based on sufficiently large sample size, and across many taxa, to determine the distribution and generality of Methuselah-like cohorts. This information may then give us sufficient insight to understand the causes of these unexpected patterns.

Challenges for the comparative biology of aging

Our examples above illustrate how a comparative approach can be used to explore the evolution of senescence. But we are left with a serious difficulty – most comparative studies cannot infer the direction of causality from statistical correlation – which raises two important questions. First, what comparative approaches will permit strong inferences about the relationships among traits? And second, can we generate specific, unique and testable comparative predictions from evolutionary theories of aging?

Predictions and tests of genetic hypotheses

As with all scientific approaches (Platt, 1964), comparative studies must examine how inferences are drawn between causation and correlation. Consider our earlier discussion of Van Voorhies's (1992) study on sperm production and aging in *Caenorhabditis*. Let us assume that across a variety of invertebrate taxa, we find a negative relationship between investment in spermatogenesis and life span. This correlation might suggest that selection for increased spermatogenic investment could have resulted in a corresponding increase in the rate of aging, but the causal arrows may point in the other direction. That is, short life span in males could select for increased investment in sperm production to maximize the probability that a male fertilizes a female before he dies. Work presented in this volume (Service and Fales) shows that the causal arrow can, indeed, be drawn in the other direction. They found that flies selected for delayed senescence had greater sperm competitive ability than control (rapid senescence) lines. Note, however, that while the causal arrow moves in the opposite direction, the correlation is also opposite to that we would have predicted from Van Voorhies's (1992) result.

Even when we try to test specific evolutionary predictions, the causal relationships we seek from our experimental observations may be ambiguous. In Promislow's (in review) reanalysis of patterns of aging in *Drosophila* species, we cannot distinguish whether over evolutionary time, an increase in early fecundity has led to a decrease in survival, a decrease in survival has led to an increase in early fecundity, or whether some unexamined third factor has affected both fecundity and survival independently.

To some extent, these causal ambiguities can be resolved. Given sufficient data and a detailed enough phylogeny, we can deduce the phylogenetic order in which evolutionary events occur (e.g. Huey & Bennett, 1987; Sillén-Tullberg, 1988; Donoghue, 1989; Losos, 1990; Richman & Price, 1992). We can then use this information to test hypotheses for the direction of causal arrows. But as Lauder *et al.* (1993) point out, these types of analysis are not without difficulty. Let us hypothesize a causal relationship between two traits, *A* and *B*, where a high value of trait *A* selects for an increase in trait *B*. Imagine that we find a statistically significant correlation between *A* and *B* across many independent taxa. It is possible that the relationship between the two traits is confounded by some unexamined, third trait, *C*. Perhaps an increase in *A* selects for an increase not in *B*, but in *C*. But because *B* and *C* are positively, genetically correlated, there is a concomitant increase in *B* as *A* changes. If we are analyzing a sufficiently large number of independent evolutionary events, we can use partial regression analysis to test for changes in *C*, controlling for *B*, and changes in *B*, controlling for *C*. If *A* has selected on *C*, not *B*, then the first test should be significant and the second non-significant. Statistical methods now exist that provide for this type of detailed comparative analyses (Bell, 1989; Grafen, 1989; Pagel, 1992).

Perhaps even more critical is the need to derive testable, falsifiable comparative hypotheses for the evolution of senescence. Predictions from genetic theories of senescence are more ambiguous than one might have first thought. Medawar (1952) and Williams (1957) addressed potential mechanisms underlying senescence that could arise from the age-specific force of natural selection. They considered genetic models where the expression of deleterious effects are age-specific. In the case of mutation accumulation, the alleles have only late acting deleterious effects, while for antagonistic pleiotropy, the same alleles have beneficial consequences early in life. Williams set the stage for testing the model of antagonistic pleiotropy by positing nine specific predictions of how factors like extrinsic mortality, gender and demographic selection will influence the rate of senescence, and subsequent models have extended these predictions further (e.g. Charlesworth, 1980). Ironically, though, these predictions may be equally compatible with mutation accumulation as the underlying mechanism.

For instance, Williams (1957) expected the likelihood of survival to be lower in males relative to females because of the prevalence of male-male competition. In that case, the force of natural selection among males will decrease at a faster rate than for females and would lead to a greater accumulation of genes with antagonistic effects in males. Males are predicted to have a more rapid rate of increase in morbidity and mortality and a more negative correlation between early and late fecundity. But the precipitous decline of the force of natural selection among males will also permit greater accumulation of late expressed mutations that reduce late life survival and fecundity. So compared to females, males expend a greater proportion of their reproductive effort early in life and have relatively accelerated somatic degeneration and mortality. The consequence for comparative studies is that we may not be able to use sex-specific patterns of aging as an exclusive test of the model of antagonistic pleiotropy. These sorts of ambiguities make it difficult to interpret a correlation like that observed in the *Drosophila* data of Schnebel and Grossfield (1988): does an increase in early relative reproductive output give rise to a decrease in lifespan (antagonistic pleiotropy) or are both affected by the accumulation of late acting mutations (and thus correlated indirectly).

To apply comparative biology to the genetics of senescence, we need to develop a new set of predictions that permit discrimination between our current genetic models. Rose and Finch offer such a prediction in their introductory chapter to this volume. They suggest that we examine the extent of phylogenetic diversity in causes of aging. Antagonistic pleiotropy is thought to give rise to increasing adult mortality through selection on increased early reproduction. Since the physiology of reproduction is relatively phylogenetically uniform, we might expect phylogenetic diversity in mechanisms of aging to match that for reproductive physiology among a broad taxonomic spectrum. By contrast, if senescence arises due to mutation accumulation, each species will be expected to accumulate a unique set of mutations, and consequently, mechanisms of senescence should be phylogenetically diverse and independent of the relatively invariant patterns of diversity for reproductive physiology.

Similarly, we suggest that under antagonistic pleiotropy, phylogenetic increases in early-age fecundity should correlate with increases in the rate of increase in adult mortality. In contrast, mutation accumulation will lead to decreases in late-age fecundity and late-age survival over evolutionary time, but early-age fecundity should not be systematically affected. In effect, we can ask whether an increase in mortality over evolutionary time is associated with an increase in early reproductive effort or a decrease in late reproductive effort. Under antagonistic pleiotropy, we see reproductive senescence as constrained or even counteracted by changes in mortality. Under mutation accumulation, reproductive senescence should be concordant with but causally independent of changes in mortality rates. Work with dynamic programming models by Peter Abrams and Donald Ludwig (unpublished manuscript) on the disposable soma theory of senescence underscores the nature of our predictions. Abrams and Ludwig find that antagonistic pleiotropy will lead to a variety of age-specific mortality patterns that are constrained by the form of the fecundity schedule. They also emphasize, however, that comparable optimization analyses have not been explored for mutation accumulation. Thus, predicting how the relationship between fecundity and mortality will differ between the alternative genetic mechanisms is still an open question.

We have illustrated our predictions in Figure 1. Imagine that we start with an ancestral population (solid lines) in which fertility declines and mortality increases with age. With antagonistic pleiotropy, an increase in early fecundity (dashed line, Fig. 1a) leads to an increase in the rate of increase of age-specific mortality (dashed line, Fig. 1c). Fecundity late in life either remains the same, or decreases below that of the ancestral population. Under mutation accumulation, the acceleration in mortality rates (dashed line, Fig. 1c) is matched by a steeper decline in fecundity (dashed line, Fig. 1b), but with no change early in life. This comparison is complicated by the fact that in a population of constant size, fertility and survival rates must balance each other, so for a given mortality schedule, the fertility schedule is somewhat constrained, and the two are likely to be negatively correlated (Sutherland *et al.*, 1986). If

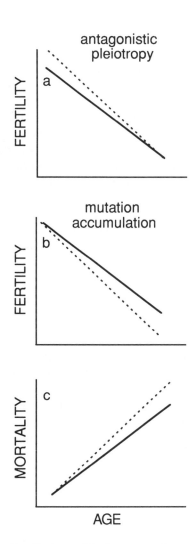

Fig. 1. Graphical model for comparing two models (antagonistic pleiotropy, or AP, and mutation accumulation, or MA) for the evolution of senescence. Solid line is the original population. Dashed lines represent trajectory after increase in senescence. In both scenarios, early mortality does not change while late mortality increases (*c*). Under AP, changes in late-age mortality are correlated with changes in early-age fecundity (*a*). Under MA, late-age mortality and late-age fecundity are both influenced by late-acting deleterious mutations. (See text for further details).

we begin with a population that is stationary in size (intrinsic rate of increase, *r*, equal to zero), and if mutation accumulation then raises the rate of increase in mortality rates and the rate of decrease in fertility rates throughout the life span, the population would rapidly go extinct. So we might expect the line for fertility (Fig. 1b) to be shifted upwards, and we need to devise a way

to normalize fertility rates among taxa to facilitate the comparison of relative changes in age-specific fecundity. At this point, the problem remains unresolved.

Finally, as our knowledge of the genetic basis of aging grows (Crew; Dixon, this volume), we may be able to compare actual genes or gene families among species. For example, Crew's work on the MHC locus (this volume) points to the value of comparing the risk of pathogen-related mortality in natural populations (see also Bell, this volume) with the degree of MHC variability across a diverse range of species.

Measuring senescence

One of the most difficult aspects of comparative studies of aging is knowing just what to compare. Previous studies (e.g. Promislow, 1993) argued that maximum or mean life span and survivorship are not useful measures of aging, either within or among species. While these measures are all informative, they lack any indication of progressiveness with age, which is critical in definitions of senescence. Rose, for instance, (1991, p. 20) defines senescence as '. . . a persistent decline in the age-specific fitness components of an organism due to internal physiological deterioration' and later emphasizes that aging is a 'protracted decline' (p. 220). Without measures of progression with age, it is not possible to quantify the persistence and the degree of deterioration. Furthermore, a measure such as maximum life span is confounded by the sample size of the population in which it is measured, and is unlikely to have been shaped by natural selection. In light of this, Promislow (1993) argued that the best measure of aging for comparative studies may be the rate of increase in age-specific mortality (see, for example, Finch *et al.*, 1990; Promislow, 1991; Tatar & Carey, in press; Tatar *et al.*, in press), and others have argued that evolutionarily meaningful measures of aging should incorporate age-specific changes in fertility (Abrams, 1991; Partridge & Barton, this volume).

Measures of age-specific survival and fertility are not without their own shortcomings. Age-related changes in survival rate do not indicate longitudinal change in individual risk, and changes in survival within a cohort can be affected by variation in quality among individuals (Vaupel & Yashin, 1985; Partridge & Barton, this volume). Changes in fertility may arise from age-related changes in energy allocation that have nothing to do with any 'internal physiological deterioration'. The most appropriate measure of aging may be some value that integrates individual deterioration in physiological and reproductive performance with cohort-specific changes in mortality rates. We should emphasize that these issues are just as relevant to intraspecific studies as they are to comparative ones.

Conclusion

The examples, both real and hypothetical, that we present here illustrate the strengths of a comparative approach to aging. First, initial comparative studies on aging in natural populations (Nesse, 1988; Newton, 1989; Promislow, 1991) have shown that senescence is, in fact, a general phenomenon in the wild. But while rates of senescence differ among species, we have not yet been able to explain this variation (Gaillard *et al.*, 1993).

Second, we have been able to generate novel, testable hypotheses from previously unrecognized phylogenetic patterns. We referred earlier to Abrams (1993) theoretical work, which corresponds with observations from Promislow's (1991) study of aging in mammals. Other examples may present themselves as the field continues to grow.

Finally, we know of no comparative gerontological example in which the authors have taken an intraspecific result and tried to extend it to interspecific patterns. However, we hope that our discussion of the EF-1α gene, mortality deceleration, and Van Voorhies's (1992) work on sperm production and aging, will not only make explicit the ways to generalize specific findings, but will even spur others on to tackle these sorts of problems with comparative data.

Have there been any valid comparative tests of genetic theories for the evolutionary of senescence? As a general test of how natural selection affects age-specific life history, we suggest that Promislow's (in review) reanalysis of Schnebel and Grossfield's data is at least consistent with theoretical prediction. But the correlation detected in the study does not permit an exclusive test of the model of antagonistic pleiotropy. The existing predictions arising from William's (1957) theory are simply too ambiguous.

In considering the potential for a comparative biology of the genetics of aging, we have been led to evaluate not only a variety of strengths to this approach, but also some difficult shortcomings. These problems are not, however, unique to the comparative biologist. They are issues that need to be addressed at every level of genetic analysis, from the 'micro' genetic to the phylogenetic, if we are to develop a general understanding of aging in the context of evolution.

Acknowledgements

Discussions with J. Vaupel, J. Curtsinger, and J. Carey during the early stages of this manuscript were most helpful. R.D. Montgomerie provided helpful comments on previous drafts of this manuscript. D. Promislow was funded in part by NSERC grants to R.D. Montgomerie and P.D. Taylor. M. Tatar was funded by the National Institute of Aging program project, 'Oldest Old-Mortality – Demographic Models and Analysis.'

References

Abrams, P., 1991. Fitness costs of senescence: the evolutionary importance of events in early adult life. Evolutionary Ecology 5: 343–360.

Abrams, P.A., 1993. Does increased mortality favor the evolution of more rapid senescence? Evolution 47: 877–887.

Austad, S. N., 1992. Primate longevity: its place in the mammalian scheme. American Journal of Primatology 28: 251–261.

Austad, S.N., 1993. The comparative perspective and choice of animal models in aging research. Aging Clin. Exp. Res. 5: 259–267.

Bell, G., 1989. A comparative method. Amer. Natur. 133: 553–71.

Carey, J.R., P. Leido, D. Orzco & J.W. Vaupel, 1992. Slowing of mortality rates at older ages in large medfly cohorts. Science 258: 457–461.

Charlesworth, B., 1980. Evolution in Age-Structured Populations. Cambridge University Press, Cambridge, U.K.

Comfort, A., 1979. The Biology of Senescence, 3rd ed. Churchill Livingstone, Edinburgh, U.K.

Curtsinger, J.W., H.H. Fukui, D.R. Townsend & J.W. Vaupel, 1992. Demography of genotypes: Failure of the limited lifespan paradigm in *Drosophila melanogaster*. Science 258: 461–463.

Cutler, R.G., 1984. Evolutionary biology of aging and longevity in mammalian species, pp. 1–147 in Aging and cell function, edited by J.E. Johnson. Plenum Press, New York.

Donoghue, M.J., 1989. Phylogenies and the analysis of evolutionary sequences, with examples from seed plants. Evolution 43: 1137–1156.

Felsenstein, J., 1985. Phylogenies and the comparative method. Amer. Natur. 125: 1–15.

Finch, C.E., 1990. Longevity, Senescence and the Genome. University of Chicago Press, Chicago, IL.

Finch, C.E., M.C. Pike & M. Witten, 1990. Slow mortality rate accelerations during aging in some animals approximate that of humans. Science 249: 902–906.

Gaillard, J.-M., D. Allaine, D. Pontier & N.G. Yoccoz, 1994. Senescence in natural populations of mammals: A comment. Evolution (April issue).

Gavrilov, L.A. & N.S. Gavrilova, 1991. The Biology of Life Span: A Quantitative Approach. Harwood Academic Publishers, New York.

Grafen, A., 1987. Measuring sexual selection: why bother?, pp. 221–233 in Sexual Selection: Testing the Alternatives, edited by J.W. Bradbury & M.B. Andersson. John Wiley & Sons, Chichester, U.K.

Grafen, A., 1989. The phylogenetic regression. Phil. Trans. Roy. Soc. (Lond.) 326: 119–157.

Hamilton, W.D., 1966. The moulding of senescence by natural selection. J. Theor. Biol. 12: 12–45.

Hart, R.N. & R.B. Setlow, 1974. Correlation between deoxyribonucleic acid excision-repair and lifespan in a number of mammalian species. P.N.A.S., USA. 71: 2169–2173.

Harvey, P.H. & M.D. Pagel, 1991. The Comparative Method in Evolutionary Biology. Oxford University Press, Oxford, U.K.

Huey, R.B. & A.F. Bennett, 1987. Phylogenetic studies of co-adaptation: preferred temperatures versus optimal performance temperatures of lizards. Evolution 41: 1098–1115.

Kirkwood, T.B.L., 1977. Evolution and ageing. Nature (Lond) 270: 301–304.

Lauder, G.V., A.M. Leroi & M.R. Rose, 1993. Adaptations and history. Trends Ecol. Evol. 8; 294–297.

Losos, J.B., 1990. Concordant evolution of locomotor behaviour, display rate and morphology in *Anolis* lizards. Animal Behaviour 39: 879–890.

Manly, B.F.J., 1991. Randomization and Monte Carlo Methods in Biology. Chapman and Hall, New York.

Martins, E.P. & T.J. Garland, 1991. Phylogenetic analyses of the correlated evolution of continuous characters: A simulation study. Evolution 45: 534–557.

McLennan, D.A., 1991. Integrating phylogeny and experimental ethology: from pattern to process. Evolution 45: 1773–89.

Medawar, P.B., 1952. An Unsolved Problem in Biology. H.K. Lewis, London.

Nesse, R.M., 1988. Life table sets of evolutionary theories of senescence. Experimental Gerontology 23: 445–453.

Newton, I., 1989. Lifetime Reproductive Success in Birds. Academic Press, London.

Pagel, M.D., 1992. A method for the analysis of comparative data. J. Theoret. Biol. 156: 431–442.

Pearl, R., 1928. The Rate of Living. Knopf, New York.

Platt, J.R., 1964. Strong inference. Science 146: 347–353.

Promislow, D.E.L., 1990. Mortality patterns in natural populations of mammals and their consequences. D.Phil. University of Oxford, Oxford, U.K.

Promislow, D.E.L., 1991. Senescence in natural populations of mammals: A comparative study. Evolution 45: 1869–1887.

Promislow, D.E.L., 1993. On size and life: Progress and pitfalls in the allometry of life span. J. Gerontology: Biological Science 48: B115–B123.

Promislow, D.E.L. Antagonistic pleiotropy: A new perspective on interspecific comparisons. Evolution (in review).

Richman, A.D. & T. Price, 1992. Evolution of ecological differences in the Old World leaf warblers. Nature 355: 817–821.

Ridley, M., 1983. The Explanation of Organic Diversity: The Comparative Method and Adaptations for Mating. Oxford University Press, Oxford.

Rose, M.R., 1991. Evolutionary Biology of Aging. Oxford University Press, Oxford.

Sacher, G.A., 1959. Relationship of lifespan to brain weight and body weight in mammals, pp. 115–133 in C.I.B.A. Foundation Symposium on the Lifespan of Animals, edited by G.E.W. Wolstenholme & M. O'Connor. Boston, Ma. Little, Brown and Co.

Sacher, G.A., 1978. Longevity and ageing in vertebrate evolution. Bioscience 28: 297–301.

Sacher, G.A. & E.F. Staffeldt, 1974. Relationship of gestation time to brain weight for placental mammals. Amer. Natur. 108: 593–616.

Schnebel, E.M. & J. Grossfield, 1988. Antagonistic pleiotropy: An interspecific *Drosophila* comparison. Evolution 42: 306–311.

Shepherd, J.C.W., U. Walldorf, P. Hug & W.J. Gehring, 1989. Fruit flies with additional expression of the elongation factor EF-1α live longer. Proc. Natl. Acad. Sci. USA. 86: 7520–7521.

Sillén-Tullberg, B., 1988. Evolution of gregariousness in aposematic butterfly larvae: a phylogenetic analysis. Evolution 42: 293–305.

Sutherland, W.J., A. Grafen & P.H. Harvey, 1986. Life history correlations and demography. Nature 320: 88.

Tatar, M. & J.R. Carey. Sex mortality differential in the bean beetle: reframing the question. American Naturalist (in press).

Tatar, M., J.R. Carey & J.W. Vaupel. Long term cost of reproduction without accelerated senescence in *Callosobruchus maculatus*. Evolution (in press).

Van Voorhies, W.A., 1992. Production of sperm reduces nematode lifespan. Nature 360: 456–458.

Vaupel, J.W. & A.I. Yashin, 1985. Heterogeneity's ruses: Some surprising effects of selection on population dynamics. American Statistician 39: 176–195.

Williams, G.C., 1957. Pleiotropy, natural selection, and the evolution of senescence. Evolution 11: 398–411.

M.R. Rose and C.E. Finch (eds.), Genetics and Evolution of Aging, 54–70, 1994.
© 1994 *Kluwer Academic Publishers. Printed in the Netherlands.*

The genetics of aging in the yeast *Saccharomyces cerevisiae*

S. Michal Jazwinski
Department of Biochemistry and Molecular Biology, and LSU Center on Aging, Louisiana State University Medical Center, New Orleans, LA 70112, USA

Received and accepted 22 June 1993

Key words: longevity-assurance genes, *RAS*, senescence, asymmetric reproduction, development

Abstract

The yeast *Saccharomyces cerevisiae* possesses a finite life span similar in many attributes and implications to that of higher eukaryotes. Here, the measure of the life span is the number of generations or divisions the yeast cell has undergone. The yeast cell is the organism, simplifying many aspects of aging research. Most importantly, the genetics of yeast is highly-developed and readily applicable to the dissection of longevity. Two candidate longevity genes have already been identified and are being characterized. Others will follow through the utilization of both the primary phenotype and the secondary phenotypes associated with aging in yeast. An ontogenetic theory of longevity that follows from the evolutionary biology of aging is put forward in this article. This theory has at its foundation the asymmetric reproduction of cells and organisms, and it makes specific predictions regarding the genetics, molecular mechanisms, and phenotypic features of longevity and senescence, including these: GTP-binding proteins will frequently be involved in determining longevity, asymmetric cell division will be often encountered during embryogenesis while binary fission will be more characteristic of somatic cell division, tumor cells of somatic origin will not be totipotent, and organisms that reproduce symmetrically will not have intrinsic limits to their longevity.

Introduction

The yeast *Saccharomyces cerevisiae* is a microbial eukaryote that undergoes an asymmetric form of cell division referred to as budding (Strathern, Jones & Broach 1981, 1982). Although it exists in a unicellular state, yeast can develop along four alternate pathways: haploid and diploid cells can reproduce vegetatively by dividing mitotically, haploid cells can mate in a sexual conjugation process, while diploid cells reverse this in meiosis (sporulation), and diploid cells can undergo a dimorphic transition to a pseudo-filamentous form. The last of these pathways has been discovered recently. Because both haploids and diploids can multiply vegetatively, the haplophase and the diplophase can be maintained stably, simplifying a variety of genetic manipulations. Indeed, this organism is a genetic powerhouse, fulfilling the role of a eukaryotic *Escherichia coli*. The entire genome, some 15 Mb and sixteen chromosomes strong, has been physically mapped (Link & Olson, 1991). The DNA sequence of chromosome III has been determined (Oliver *et al.*, 1992), and the remaining chromosomes will soon be sequenced. Yeast can be cultured in a defined medium containing only a carbon source, salts, trace elements and vitamins. In a rich medium with good aeration, the cell doubling time of the culture is as short as 80 minutes, providing ample material for study. The organism is, however, famous for anaerobic growth, during which it produces ethanol.

In this article, I will summarize some of the advantages of yeast in aging research. I will then discuss briefly the phenomenology of aging in this organism. Evidence for a 'senescence factor' will be discussed; this forms a springboard for a molecular genetic analysis of yeast longevity. This will be followed by an analysis of some of the possible molecular mechanisms of aging that follow from the basic characteristics of yeast aging. Two candidate longevity-determining genes in yeast will be presented. The question of why yeasts age will be addressed, and an ontogenetic theory of aging will be elaborated. Finally, some comments on the place of model systems in aging research will be put forward.

Advantages of yeast for aging studies

Several features of yeasts make them eminently suitable for a genetic and molecular analysis of aging:

1. The cell and the organism are one, allowing the analysis of aging at the cellular level without recourse to arguments concerning the relevance of such studies to the situation *in vivo*.
2. Features that are intrinsic to aging can be examined without confounding effects of extracellular factors, such as hormones and interactions with other cells.
3. The powerful tools of yeast genetics (Guthrie & Fink, 1991), classical and molecular, can be applied to the study of aging.
4. The fundamental phenomena associated with aging in yeast have been described (Jazwinski, 1990, 1990a).
5. Yeast undergoes single cell aging, such that an individual cell senesces randomly and independently of other yeast cells. The individual yeast is mortal, while the population is immortal. Therefore, this organism ages in a fashion similar to many higher organisms rather than undergoing clonal senescence, as do individual cells from multicellular organisms. The individual aging cell can be identified and followed unambiguously, because yeasts divide asymmetrically by budding.
6. Biomarkers for both nominal and functional age of yeasts are available.
7. Bulk quantities of age-synchronized yeasts can be prepared readily (Egilmez, Chen & Jazwinski, 1990).
8. Similar to higher eukaryotes, yeast longevity is a polygenic trait (C. Pinswasdi & S. M. Jazwinski, unpublished results), as evidenced by the segregation pattern of life span in the meiotic progeny of a diploid strain.
9. Several genes that determine yeast longevity have been identified and cloned (Jazwinski, 1993).

Basic phenomenology of yeast aging

Age changes

Barton (1950) was the first to point to the possibility that changes occur with age in the reproductive properties of yeast cells. However, it was Mortimer and Johnston (1959) who demonstrated that the budding or reproductive (replicative) capacity of individual yeasts is finite. They showed that yeast cells divide a certain number of times, then assume a granular appearance and frequently lyse. Whether or not the cells that persist are alive and metabolizing after cessation of cell division is an open question. Recently, it has been shown that protein synthesis declines with replicative age due to a decrease in the activity and recruitment of ribosomes (Motizuki & Tsurugi, 1992). Although ribosomal RNA levels increase in cells with replicative age, this increase does not keep pace with the increase in cell size (S. P. Kale & S. M. Jazwinski, unpublished results; Motizuki & Tsurugi, 1992). The effective decrease in ribosomal RNA concentration would also contribute to a decline in protein synthesis with age in yeast. Unfortunately, one must extrapolate these decrements to suggest that yeasts that cease dividing at the end of their life span either die outright, cease metabolizing or at least greatly reduce their biochemical activity. Although it remains open, the possibility of a post-mitotic life span is difficult to associate with any terminal differentiation process.

The measure of the yeast life span is the number of cell divisions (cell cycles) or generations. Chronological age does not appear to play a part in yeast aging (Muller *et al.*, 1980). Over a long period of time, however, yeast cells lose viability when maintained in a non-dividing stationary phase (Choder, 1991) or G_0 (Drebot, Johnston & Singer, 1987), to utilize cell cycle terminology. Thus, metabolic time may still be running, albeit slowly. This issue has not as yet been addressed satisfactorily in the context of yeast longevity, notwithstanding the studies of Muller *et al.* (1980). Aging and senescence is a stochastic phenomenon in yeast, to which Gompertz analysis can be readily applied. The mortality rate of yeast cells increases exponentially with

Table 1. Morphological and physiological changes during yeast aging.

Characteristic	Change	References
Cell size	Increase	Bartholomew & Mittwer, 1953; Mortimer & Johnston, 1959; Johnson & Lu, 1975; Egilmez, Chen & Jazwinski, 1990
Cell shape	Altered	Chen & Jazwinski, unpublished results
Granular appearance		Mortimer & Johnston, 1959
Surface wrinkles		Mortimer & Johnston, 1959; Muller, 1971
Loss of turgor		Muller, 1971
Cell fragility (prior to death)	None	Egilmez, Chen & Jazwinski, 1990
Cell lysis		Mortimer & Johnston, 1959
Bud scar number	Increase	Barton, 1950; Bartholomew & Mittwer, 1953; Beran *et al.*, 1967; Egilmez, Chen & Jazwinski, 1990
Cell wall chitin	Increase	Egilmez, Chen & Jazwinski, 1990
Vacuole size	Increase	Egilmez, Chen & Jazwinski, 1990
Generation (cell cycle) time	Increase	Mortimer & Johnston, 1959; Egilmez & Jazwinski, 1989
Response to pheromones (haploids)	None	Muller, 1985
Mating ability (haploids)	Decrease	Muller, 1985
Sporulation ability (diploids)	Increase	Sando *et al.*, 1973
Cessation of division at G_1/S boundary of cell cycle (putative)		Egilmez & Jazwinski, 1989
Senescence factor		Egilmez & Jazwinski, 1989
Mutability of mtDNA	Decrease	James *et al.*, 1975
Telomere length	None	D'mello & Jazwinski, 1991
Specific gene expression	Altered	Egilmez, Chen & Jazwinski, 1989
rRNA levels	Increase	Kale & Jazwinski, unpublished results; Motizuki & Tsurugi, 1992
Cellular rRNA concentration	Decrease	Kale & Jazwinski, unpublished results; Motizuki & Tsurugi, 1992
Protein synthesis	Decrease	Motizuki & Tsurugi, 1992
Ribosome activity, polysome recruitment	Decrease	Motizuki & Tsurugi, 1992

replicative age (Pohley, 1987; Jazwinski, Egilmez & Chen, 1989). Thus, yeast ages like other species that produce offspring repeatedly. The mean replicative life span of a yeast strain usually falls between 20 and 30 cell divisions, while the maximum may approach as many as 50. The daughter cells produced by the individual mother cell during its life span start from scratch. Their clocks are reset, as it were, and they have before them the capacity for a full life span. This conclusion is based on the analysis of Johnston (1966) with relatively limited data. Some indication can be found in that study suggesting that this statement may not be completely correct. In our studies, we have found that there is no decline in the life span of daughters produced through at least the first half of the mother's life span (C. Pinswasdi & S. M. Jazwinski, unpublished results). However, it is important to note that daughters produced late in the life span frequently exhibit reduced longevity, and this effect is amplified within a lineage on continuous selection of such daughters (A. Hogel & I. Muller, personal communication). Replicative life span is finite not only in laboratory strains, for *S. cerevisiae* captured in the wild also display this attribute (D. S. Franklin & S. M. Jazwinski, unpublished results).

Yeasts undergo a variety of morphological and physiological changes as they progress through their life span (Jazwinski, 1990, 1990a). Thus, it is reasonable to speak of an aging process. Many of these age-related changes have been discussed in detail before and are only tabulated here (Table 1).

Bud scars

Each time a yeast divides the mother cell is marked in the process by a chitin-containing structure called the bud scar (Bartholomew & Mittwer, 1953; Cabib, Ulane & Bowers, 1974). These bud scars accumulate on the surface of the mother cell, and the possibility was raised that yeasts age because they exhaust available budding sites (Mortimer &

Johnston, 1959). A refinement of the proposal that bud scars limit yeast life span was also advanced by Mortimer and Johnston (1959). They suggested that the accumulation of bud scars limits the ratio of metabolically active surface to volume. The arguments against the proposals that bud scars limit yeast life span have been reviewed (Jazwinski, 1990, 1990a), and they are summarized in Table 2. Some of these arguments have been expanded more recently. Examination of the life spans of a series of isogenic strains of increasing ploidy showed that no difference in life span was observed between a diploid and a tetraploid (Muller, 1971). This experiment has been refined by examining a completely isogenic haploid/diploid pair (D. S. Franklin & S. M. Jazwinski, unpublished results). A yeast strain of the *a* mating type was induced to change its mating type to α. This was achieved by transforming the strain with a plasmid carrying the *HO* gene behind an inducible promoter. After induction of

the mating type switch, the strain was cured of the plasmid. The *a* and α cells mated to form the *a*/α diploid, which was sporulated to provide *a* and α cells isogenic with the diploid. The life spans of the three isogenic strains were identical. This experiment indicates that there is no difference in life span between the two mating types. Furthermore, it shows that surface-to-volume ratio is not a major determinant of yeast life span, for the greatest decrease in this ratio occurs going from a haploid to a diploid. It is also worth noting that this result argues against mutation as the cause of aging.

The life span of daughter cells of young mothers as compared to that of daughters of older mothers does not differ (C. Pinswasdi & S. M. Jazwinski, unpublished results). Yet, in this experiment, the daughters of the young mothers were approximately five-fold less voluminous than those of the older mothers and had a correspondingly greater surface-to-volume ratio. A similar conclusion can

Table 2. Evidence against bud scars as the determinant of yeast longevity.

Evidence/argument	Source
<50% of possible sites for scars is ever occupied.	Bartholomew & Mittwer, 1953; Mortimer & Johnston, 1959
Cell wall expands more than necessary to oblige a bud scar at each cell division.	Beran *et al.*, 1967; Johnson & Lu, 1975
Bud scars can overlap.	Bartholomew & Mittwer, 1953
Life spans of individual cells of a given strain vary, mean and maximum life spans of different strains also vary.	Jazwinski, 1990, 1990a
Overexpression of certain genes extends longevity without affecting bud scars.	Chen, Sun & Jazwinski, 1990; Jazwinski, 1993
Adaptive increases in surface-to-volume ratio are not the rule during evolution over about 300 generations in a chemostat.	Adams *et al.*, 1985
Isogenic strains of increasing ploidy (and size) up to tetraploid do not differ in life span.	Muller, 1971; Franklin & Jazwinski, unpublished results
The large daughters of older mothers do not differ in life expectancy from the small daughters of younger mothers.	Pinswasdi & Jazwinski, unpublished results; Egilmez & Jazwinski, 1989; Johnston, 1966
Ethanol increases yeast life span without affecting bud scars.	Muller *et al.*, 1980
Induced deposition of chitin, the major component of the bud scar, does not curtail life span.	Egilmez & Jazwinski, 1989
Yeasts display the Lansing effect[1], which cannot be mediated by the bud scar.	Hogel & Muller, personal communication
Transmission of the mother cell effect on generation time within a lineage cannot be mediated by the bud scar.	Egilmez & Jazwinski, 1989
Life span of zygote reflects that of the older parent, even though the number of bud scars per unit of cell surface is smaller than in the parents.	Muller, 1985
Overexpression of v-Ha-*ras* at moderate levels increases life span irrespective of the increase in cell size.	Chen, Sun & Jazwinski, 1990

[1] In the case of yeasts, refers to the extinction of a lineage through the continuous selection of the final daughters produced during the life span.

be drawn from the work of Johnston (1966) referred to earlier. Finally, it has been shown that the deposition of chitin, the major component of the bud scar, in the cell wall does not reduce the life span (Egilmez & Jazwinski, 1989). This latter study was carried out with a cell division cycle (*cdc*) mutant that accumulates this polysaccharide in the cell wall while it is arrested in cell cycle traversal at nonpermissive temperature as an unbudded cell. Despite all of this evidence, the occurrence of cell wall and membrane changes that lead to senescence is a possibility that has not been completely ruled out. It seems likely that cell wall alterations may be the result of aging rather than its cause.

Genetic determinants

Yeast longevity is genetically determined. This follows from the fact that the mean and the maximum life span are characteristic features of a strain and that they vary from one yeast strain to another. Furthermore, the mean and maximum life spans of a strain are correlated (Egilmez & Jazwinski, 1989). Yeast populations evolving in a chemostat acquire adaptive changes (mutations) in a limited number of loci that display rather complex epistatic or functional interactions (Paquin & Adams, 1983, 1983a; Adams *et al.*, 1985). It is possible that yeast longevity does not involve an extremely large number of genes either. About 70% of the yeast genome is not essential for cell growth and division, even though the yeast genome is almost entirely unique (Goebl & Petes, 1986), leaving ample room for genes that may be involved in longevity.

These considerations do not mean that genetic mutations are the normal cause of aging in yeast. The qualifier 'normal' is stressed because it is expected that many defects would reduce cell viability and therefore shorten life span dramatically. One such mutation is *rad*3 that confers UV sensitivity (Muller, 1985). During aging, auxotrophic mutations or petites (mutations impairing mitochondrial respiration) do not accumulate. Also, there is no correlation between the age of a mother cell at the birth of a daughter and the growth rate of cells cloned from this daughter (Muller, 1971). The life spans of spontaneous and induced petites are not significantly different from that of the parent strain (Muller & Wolf, 1978). In this context, it is interesting that the mutability of ρ^+ cells to ρ^- (petite) decreases with age (James *et al.*, 1975). The

genetic segregation of longevity is not characteristic of a mitochondrially-determined trait (C. Pinswasdi & S. M. Jazwinksi, unpublished results). As Mortimer and Johnston (1959) pointed out, nondisjunction or recessive lethal genetic changes, other deleterious nuclear events or random depletion of essential autonomous cytoplasmic constituents are not likely to explain aging in yeast, because the original population is already at equilibrium for such events. Thus, genetic damage theories are not attractive as explanations for yeast aging (Jazwinski, 1990). Furthermore, the longevity of a population of daughter cells is similar to that of the original population of mothers (Johnston, 1966; C. Pinswasdi & S. M. Jazwinski, unpublished results). On the basis of these properties, it was pointed out that aging in yeast does not appear to be the result of accumulation of amplified sequences from the mitochrondrial genome (Egilmez & Jazwinski, 1989) as occurs in the fungus *Podospora anserina*, for example (Munkres, 1985). On the other hand, in the *nib*1 mutant an elevation in copy number of extrachromosomal DNA elements and abnormal cell enlargement occur, and this is associated with a severe reduction in division potential (Holm, 1982). However, a mutation that severely curtails life span or leads to clonal lethality is difficult to interpret, because the attenuation of virtually any vital function should ultimately result in loss of viability. Obviously, mutation frequency is much too low for a mutation of this sort to be the cause of cell death in a wild-type population. In zygotes produced from matings of young and old yeast cells, the life span reflects that which would have remained to the older mating partner (Muller, 1985). This excludes the exhaustion of a substance or the loss of an organelle or function as the cause of aging, because the young mating partner should provide these (Muller, 1985). The immediate causes of death at the end of the yeast life span may be many, given the variety in morphology of cells that have ceased dividing and the lysis of some of these.

Biomarkers of aging in yeast

Yeast cells increase in size as they progress through their replicative life span (Mortimer & Johnston, 1959; Johnson & Lu, 1975; Egilmez & Jazwinski,

1989; Egilmez, Chen & Jazwinski, 1990). It is likely that this increase is simply related to the growth that occurs on passage through consecutive cell cycles, particularly because it appears that growth rate does not play an important role in determining life span (Tyson, Lord & Wheals, 1979; Muller *et al.*, 1980; Egilmez, Chen & Jazwinski, 1990). Nevertheless, it constitutes an excellent marker for the nominal age of a cell (in cell divisions) or of a cohort of cells. The increases in the number of bud scars and in cell wall chitin constitute additional biomarkers for the nominal age of a yeast cell, and accurate ones indeed (Egilmez, Chen & Jazwinski, 1990). Similar comments pertain to many of the other age changes listed in Table 1.

In contrast to these markers of nominal age, the generation time, the elapsed time between production of consecutive buds, constitutes a biomarker for functional or physiological age. The generation time of individual yeast cells increases with replicative age (Egilmez & Jazwinski, 1989; Egilmez, Chen & Jazwinski, 1990). As cells enter the phase of exponential increase in mortality, the increase in generation time accelerates. Finally, this increase becomes acute two to three generations prior to cessation of cell division (Mortimer & Johnston, 1959). In one instance, we have found the final divisions to approach 50 h (J. Sun & S. M. Jazwinski, unpublished results). The increase in generation time with age is the most adequate indicator of functional age of yeasts, and it is the primary biomarker for the aging process from which others, such as increase in cell size, are likely derived. For individual cells, the generation time defines the position of the cell in its individual life span and provides an accurate predictor of the time of cessation of division (Egilmez & Jazwinski, 1989; Egilmez, Chen & Jazwinski, 1990). The status of this parameter as a biomarker for aging is further bolstered by the fact that a longer-lived cohort of cells displays a shorter generation time than a shorter-lived cohort at any given age (Egilmez & Jazwinski, 1989). Finally, strains with greater longevity display a delayed increase in generation time with nominal age as compared to strains of shorter life span (Egilmez & Jazwinski, 1989).

The increase in generation time with replicative age is apparently confined to the unbudded portion of the cell cycle or G_1 (Egilmez & Jazwinski, 1989;

Motizuki & Tsurugi, 1992a). The addition of 17β-estradiol reduces this increase in old cells but has no effect on young cells (Motizuki & Tsurugi, 1992a). This action is only apparent in medium containing glucose but not glycerol as a carbon source, suggesting an effect on glycolysis. In either medium, the hormone elevates cAMP levels close to those found in young cells. However, the estradiol elevates ATP levels in old cells close to those of young cells, when the cells are grown on glucose but not on glycerol. Finally, exogenous cAMP shortens the unbudded portion of the cell cycle of old cells grown in glucose. These results suggest that a deficit in energy metabolism may lead to the increase in generation time with replicative age in yeast.

Senescence factor

Most interestingly, the generation times of daughter cells mimic those of the mothers from which they were derived (Egilmez & Jazwinski, 1989) (Fig. 1). Above the age of about ten generations, the mother and its daughter appear to be synchronized in the cell cycle at the point at which the cell buds. This point is normally associated temporally with entry into the DNA-synthetic S phase of the cell cycle (Pringle & Hartwell, 1981). In the case of old mother cells, the resulting synchrony in budding leads to the growth of the daughters well beyond the usual size prior to their first budding. Significantly, daughter cells recover from the mother cell effect on their generation time within three cell divisions, and they begin dividing at the rate characteristic of their own age, as young cells (Egilmez & Jazwinski, 1989). Importantly, this recovery takes place whether the cells are left in contact with their mothers or whether they are placed at a distance from them. These results suggest that the senescent phenotype, manifested by the mother cell effect on generation time, is a dominant feature in yeast. They also suggest that this senescent phenotype is governed by a diffusible, cytoplasmic factor(s) (senescence factor) that is produced by old cells and transmitted to their daughters, where it is diluted, degraded or inactivated. A formal analysis of this phenomenon indicates that the senescence factor accumulates in the mother cells (Hirsch & Witten, 1991). The senescence factor can be transmitted from a mother to its daughter, from this daughter to the grand-daughter, and so on within a

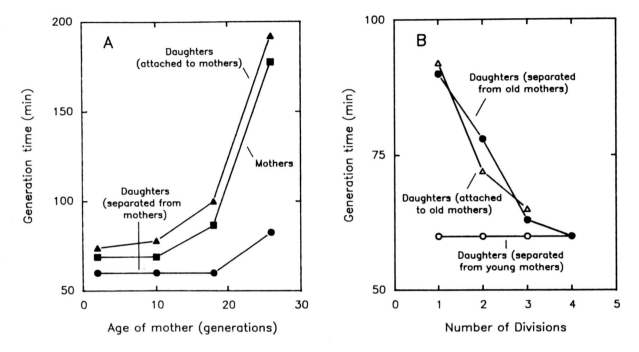

Fig. 1. The mother cell's effect on the daughter's generation time. (A) The generation times of mother cells were measured at various points during their life spans. The generation times of the daughters they produced at different points in their life span were also recorded, both during the daughters' first cell cycles while they remained attached to the mothers and after the daughters had been separated for two to three cell divisions. (B) The generation times of daughter cells produced by young (2-generation old) and old (24-generation old) mothers were determined during the first few divisions after birth. The daughters were either separated from the mothers or left attached (in the case of the daughters of old mother cells, only). (Reproduced from: Egilmez, N. K. & S. M. Jazwinski, 1989. Evidence for the involvement of a cytoplasmic factor in the aging of the yeast *Saccharomyces cerevisiae*. J. Bacteriol. 171: 37-42, with permission from the American Society for Microbiology.)

pedigree (Egilmez & Jazwinski, 1989). Thus, continued cell-cell contact is not necessary for transmission. However, the senescence factor is ultimately depleted in this pedigree. Studies in which cells of various ages were mated with young yeast cells (Muller, 1985) can also be interpreted to indicate the operation of a senescence factor. The young cell does not rescue the old yeast cell. The life span of the zygote reflects that remaining to the older partner of the mating pair. The results suggest that aging is not due to recessive changes in the old cell.

The depletion of the senescence factor in a pedigree, described above, does not appear to be gradual. Instead, it appears to display an all-or-none pattern, perhaps reflecting a threshold phenomenon (Egilmez & Jazwinski, 1989). From this perspective, the drastic increase in generation time during the last few divisions in a cell's life span (Mortimer & Johnston, 1959), which appears to signal impending cessation of division, may indicate the tra-

versal of such a threshold. The accumulation of the senescence factor late in the life span may in fact overcome the cell, leading to cessation of division. The cells in a yeast population that die early often fail to release their daughters in a viable form, something long-lived cells infrequently fail to do (Johnston, 1966). Perhaps a more rapid accumulation of senescence factor than average or an increased susceptibility to it leads to the demise of the daughter cell. This result, in addition to the lysis of some cells, provides the only clear evidence available that cells actually die at the end of their life spans. The gradual increase in generation time during the life span probably is indicative of the aging process, while the drastic increase signals imminent cell demise. It will be of interest to determine how the rapid accumulation of senescence factor in the last few generations of the mother cell affects its last few daughters, given the greater amounts these daughters undoubtedly receive from their mothers. In this regard, it is worth reiterating that daughters produced late in the life span frequently exhibit reduced longevity, and this effect is amplified within a lineage on continuous selection of such daughters (A. Hogel & I. Muller, personal communication). A complete description of the senescence factor ultimately must include a determination of whether it acts catalytically or stoichiometrically in the cell.

Molecular mechanisms of aging

Given the phenomenology of yeast aging, it seems intuitively obvious that the cell cycle and replicative life span are intimately related. In fact, there is some evidence for coordination of successive cell cycles in yeast (Pringle & Hartwell, 1981). The way we like to view yeast aging is depicted in the cell spiral model (Jazwinski, Egilmez & Chen, 1989; Jazwinski, 1990; Jazwinski, Chen & Jeansonne, 1990). As shown in Fig. 2, a yeast cell enters the cell spiral at the top and proceeds through several consecutive cell cycles, producing a daughter at each turn of the spiral. These daughters can enter the cell spiral at the top. However, the mother cell can only continue down, and ultimately dies. At each successive turn there exists a certain, decreasing probability that the mother will continue, imparting both a stochasticity to the aging process and the characteristic increase in mortality rate. Thus, the individual cell cycles that constitute the spiral are connected. The cell spiral implies the operation of a 'molecular memory'. Many of the *CDC* gene products are synthesized in amounts sufficient to sustain more than one cell cycle (Byers & Sowder, 1980; Sclafani *et al.*, 1988). Thus, yeast cells possess a molecular memory of previous cell cycles. A prototypic example of a molecular mechanism that coordinates gene expression with events in prior cell cycles is the control of *HO* gene transcription in yeast (Nasmyth & Shore, 1987). In this example, an inducer of the *HO* gene is synthesized at a later point in the cell cycle than the *HO* gene product, which itself is synthesized late in the pre-DNA-synthetic G$_1$ phase of the cell cycle. The net effect is the expression of *HO* in mother cells that have divided at least once, and its lack in daughters prior to their first cell division. Another recent example is the preferential utilization of cyclins in mother and daughter yeasts (Lew, Marini & Reed,

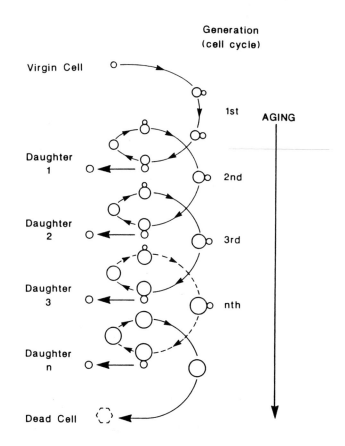

Fig. 2. The cell spiral model. See text for details. The progressively increasing length of the unbudded (G$_1$) portion of successive cell cycles is not rendered, nor is the decreasing probability of further advance down the spiral. (Reproduced from: Jazwinski, S. M., N. K. Egilmez & J. B. Chen, 1989. Replication control and cellular life span. Exp. Gerontol. 24: 423-436, with permission from Pergamon Press plc.)

1992). Cln1 and Cln2 are utilized preferentially in daughters, while Cln3 is preferentially used by mothers. The consequences of the activity of certain genes can be long-lasting, encompassing several generations (Park, Finley & Szostak, 1992).

Molecular memory can be embodied in transcriptional states of chromatin. The epigenetic inheritance of different regulatory states of chromatin at a certain genetic locus in yeast (Pillus & Rine, 1989) points to this. The establishment of these states may be related to transcriptional silencers associated with origins of DNA replication (Rivier & Rine, 1992). These origins need not be utilized frequently. In any case, the inheritance of these states can be maintained for at least ten rounds of division (Pillus & Rine, 1989). Post-translational mechanisms could also be involved in molecular memory. A prime candidate is protein phosphorylation. This mechanism is especially relevant considering the importance of cyclin and *CDC*28 protein kinase phosphorylation for cell cycling (Lewin, 1990). Multiple phosphorylation sites may be involved. Certain phosphorylations may open up other sites for phosphorylation, until finally a site can be phosphorylated that plays a direct role in activity. This mechanism is operative in glycogen synthetase (Cohen, 1985). This discussion of molecular mechanisms underlying the finite replicative life span of yeasts is not exhaustive. It is meant to indicate that mechanisms exist that can readily accommodate the major features of yeast aging. It is the goal of genetic analysis of the aging phenomenon to identify the specific mechanisms involved.

It should be stressed, however, that a common pathway or genetic program related to final arrest and demise of yeasts does not necessarily denote a strict one-to-one dependence of events on the function of specific genes. It is possible that given a final terminal state the sequence of events from that point on can only follow a certain molecular and cellular logic without the need for recourse to a highly-ordered, genetically-programmed scenario. In essence, the response might be passive rather than active.

Candidate longevity-assurance genes in yeast

LAG1
The senescent phenotype in yeast, which is manifested by an increase in generation time with age, and in particular the evidence for a senescence factor produced by old mother cells (Egilmez & Jazwinski, 1989), suggest that there are changes in gene activity during the yeast life span. Among the genes that are differentially expressed, we would expect to find genes that play a role in determining longevity. Thus, an initial strategy in the molecular genetic approach for identification of such genes was to clone genes that are differentially expressed during the yeast life span.

In order to effect this strategy, a procedure for the preparation of age-synchronized cells in bulk quantities was necessary. Such a procedure was devised (Egilmez, Chen & Jazwinski, 1990). The asymmetric cell division of yeast which is so useful in examining the aging parameters of individual cells renders preparation of large quantities of age-synchronized cells non-trivial, because with each cell division the population always contains approximately 50% newborn cells. Consequently, the number of older cells decreases exponentially with nominal age. To circumvent this difficulty, virgin cells that have never divided are prepared from a stationary culture and allowed to complete two cell divisions in a synchronous fashion. As described earlier, yeast cells become larger with each cell division. Thus, the virgin cells, now mothers, can be separated from their two daughters by velocity sedimentation. The two-generation-old mothers obtained in this way are returned to culture for three generations. The five-generation-old mothers are then prepared by sedimentation. This process is reiterated until cells of the desired age are obtained. The nominal and functional age of the cells in the preparations was verified in several ways (Egilmez, Chen & Jazwinski, 1990). Most importantly, the remaining life spans of the cells in the preparations are entirely consistent with the nominal age (in cell divisions) of the preparations, and these cells age normally.

With the procedure for aged-cell preparations in hand, a scheme was devised to clone genes that are differentially expressed during the yeast life span. mRNA was prepared from young and from old cells and used to synthesize cDNA probes. These probes were used to screen a yeast genomic DNA library to identify genes that are preferentially expressed in young cells and those preferentially expressed in old cells (Egilmez, Chen & Jazwinski, 1989). Six distinct genes were cloned initially in this fashion. The total is now fourteen, although not all of these have been characterized to the same extent. The genes display specific patterns of expression during the life span, and they can be characterized as young-specific, old-specific and even middle-age-specific (Egilmez, Chen & Jazwinski, 1989). For at least two of the genes, no dependence of the levels of expression on the growth state of the cells was found, indicating that expression is specifically age-dependent. It has not yet been determined whether the differences in mRNA levels observed are due to transcriptional regulation or changes in mRNA stability. Importantly, the alterations in transcript prevalence are not random but are reproducible from one aging cohort to the next.

At least partial DNA sequence information has been obtained on five of the original six differentially-expressed genes that were cloned (J. B. Chen, A. M. Childress, N. P. D'mello, J. Sun & S. M. Jazwinski, unpublished results). These sequences do not identify similar genes in any of the sequence databases. Thus, these are novel genes. One of these genes, *LAG*1, has been characterized in significant detail (N. P. D'mello, A. M. Childress, D. S. Franklin, S. P. Kale, N. E. Jeansonne, C. Pinswasdi & S. M. Jazwinski, unpublished results; Jazwinski, 1993). From its sequence, it appears to code for a protein with two transmembrane domains. There are clusters of putative phosphorylation sites at the N- and C-termini. There is also a PEST (proline, glutamic acid, serine, threonine) region at the C-terminus. This sequence motif is often found in proteins that are turned over rapidly. *LAG*1 is present in single copy in the yeast genome, and no other sequences with significant similarity have been found in yeast DNA. It maps on the right arm of chromosome VIII. The expression of *LAG*1 decreases precipitously with replicative age. There are two transcription start sites that are utilized by the promoter of this gene, and these sites are used at about five-fold different levels.

*LAG*1 is not essential for vegetative growth and division, because it can be deleted or disrupted in haploid cells. These mutations also do not affect mating ability. However, one interesting phenotype is displayed by the gene disruption. The loss of function of the gene results in a 40% reduction in both the mean and the maximum life span. The mean life span drops from about 21 to about 12 cell divisions. In a simple organism like yeast, such a moderate reduction in the number of divisions available to the cell suggests that this is a longevity-assurance gene, because the number of divisions is the measure of the life span; that is, it defines longevity. Furthermore, the characteristic asymmetry between the mother and the daughter in terms of life expectancy is preserved in the *LAG*1-disruption strain. The strain does not undergo clonal senescence, but individuals in the population exhibit a finite life span, as usual. On the basis of these results and because expression of the gene declines with age, we are calling this gene *LAG*1 for *l*ongevity-*a*ssurance *g*ene.

We have also been examining the effects of overexpression of *LAG*1 on yeast longevity (N. P.

D'mello & S. M. Jazwinski, unpublished results; Jazwinski, 1993). Massive overexpression of the gene from the yeast *GAL*1 promoter results in a marked attenuation of viability; however, some of the cells that survive can live longer. We have achieved more moderate levels of expression using a binary expression system in which the glucocorticoid receptor is expressed from a constitutive yeast promoter on one plasmid and *LAG*1 from the glucocorticoid response element on another. Graded levels of expression are obtained by altering glucocorticoid concentration in the growth medium. These studies are complicated by deleterious effects of the hormone at higher concentrations; however, up to 60% increase in mean and maximum life span can be obtained. Although these studies are still in progress, the picture that is emerging is that overexpression early in the life span, when endogenous levels of *LAG*1 are high, is detrimental, while expression late in the life span, when endogenous levels are low, can extend longevity. These results support the conclusion that *LAG*1 is a longevity-assurance gene in yeast.

Importantly, sequences similar to yeast *LAG*1 have been detected in human DNA on Southern genomic blots (N. P. D'mello, N. E. Jeansonne & S. M. Jazwinski, unpublished results; Jazwinski, 1993). Recently, mRNA has been detected in two human cell types in culture using yeast *LAG*1 as a hybridization probe on Northern blots (N. E. Jeansonne, J. Sun, S. P. Kale & S. M. Jazwinski, unpublished results). These hybridizations were carried out under moderately-high stringency, indicating a good deal of sequence relatedness. The putative human homolog is now being cloned. If a functional similarity of the human homolog of yeast *LAG*1 can be demonstrated in human cells, this will augment the utility of yeast as a model system for the study of aging and longevity.

RAS

The cloning of *LAG*1 provides an example of the 'virtuous' approach to the identification of genes that determine longevity using the molecular genetic strategy. The other approach involves shortcuts, and I call it the 'lucky guess' approach. It is the way we have implicated *RAS* in yeast longevity. Yeast possesses two homologs of the mammalian proto-oncogene c-*ras*. They are called *RAS*1 and *RAS*2 (Broach & Deschenes, 1990). The simple question that we chose to address was whether a change in the activity of a single gene could extend the yeast life span. In particular, we first wanted to examine whether the transforming version or oncogene would prolong yeast proliferation (Chen, Sun & Jazwinski, 1990). The Harvey murine sarcoma virus transforming gene v-Ha-*ras* was placed under the control of the galactose-inducible yeast promoter *GAL*10 on a plasmid that is maintained at single copy in yeast cells. Several constructs were actually prepared differing five-fold in the expression levels of v-Ha-*ras* mRNA (and protein). A near doubling in the mean and maximum number of cell division cycles was observed at moderate levels of v-Ha-*ras* expression (Chen, Sun & Jazwinski, 1990), demonstrating that the activation of a single gene can extend life span. At five-fold higher levels of expression of v-Ha-*ras*, no life span prolongation was observed. This lack of life span extension by the higher levels of v-Ha-ras is not due to a decrease in cell viability (Chen, Sun & Jazwinski, 1990). Instead, it appears to be due to the down-regulation of the signal generated by the higher levels of the oncoprotein similar to what has been observed in mammalian cells (Hirakawa & Ruley, 1988).

RAS in yeast exerts an effect on both cell growth or metabolism and on cell division (Broach & Deschenes, 1990). *RAS*2, in particular, is involved in cell size modulation (Baroni *et al.*, 1989). We have been able to uncouple the growth (cell size)-stimulatory effect of v-Ha-*ras* from its cell cycle-stimulatory effect that leads to extended longevity in yeast (Chen, Sun & Jazwinski, 1990). Both moderate and high levels of expression of v-Ha-*ras* resulted in a rapid initial increase in cell size during the life span. However, only moderate expression extended life span. These results indicate that there is no obligatory coupling between cell size or increase in cell size and life expectancy in yeast. The yeast genes *RAS*1 and *RAS*2 are highly pleiotropic, reflecting their central role in growth control (Broach & Deschenes, 1990). This pleiotropy is consistent with the pleiotropic nature of the aging process. However, it should be kept in mind that not all of the pleiotropic effects of *RAS* need be specific aspects of the aging process, and certainly not all, like increase in cell size for example, need play a causal role. The question can be raised regarding the lack of immortalization of yeast cells

by v-Ha-*ras*. Several explanations have been suggested (Chen, Sun & Jazwinski, 1990). These include the requirement for a second genetic event, an imbalance in growth control that ultimately leads to cell demise, and the down-regulation of the signal generated by v-Ha-*ras*. Perhaps a more fundamental explanation is based on the fact that yeasts are intact organisms and not simply individual cells from a multicellular organism. There may well exist intrinsic limits to the longevity of intact organisms that are not as readily surpassed as those to the immortality of their individual cells.

With the results with the *ras* oncogene pointing the way, it was necessary to examine the possibility that one or both of the endogenous yeast *RAS* genes might play a role in determining yeast longevity. Thus far, the focus has been on *RAS*2 (J. Sun, S. P. Kale & S. M. Jazwinski, unpublished results; Jazwinski, 1993). It has been found that the expression of *RAS*2 decreases with replicative age. Importantly, overexpression of *RAS*2 from the *GAL*10 promoter results in a 30% increase in both the mean and the maximum life span of yeasts. In these ways, *RAS*2 very much resembles *LAG*1. Therefore, we suggest that *RAS*2 is a longevity-assurance gene in yeast. *RAS*2 not only extends yeast life span, but it also appears to postpone the aging of yeast cells. We have found that overexpression of *RAS*2 delays the increase in generation time of yeast cells with replicative age, to an extent commensurate with its life span-extending effect. It also appears to limit the magnitude of the increase in generation time. Thus, yeasts overexpressing *RAS*2 not only live longer but remain fit and young longer too.

The genetics of the *RAS*1/*RAS*2 system is one of the most highly developed areas in yeast molecular and cell biology (Broach & Deschenes, 1990). Many of the genes that function in this system have been marked by mutation and cloned. *RAS* is a central integrator of cell growth and cell division in yeast. It is also part of the nutritional sensor of the cell, monitoring the nutrient status of the environment (Broach & Deschenes, 1990). *RAS* functions along at least two pathways in yeast, although others may well exist. These are the cAMP pathway, in which *RAS* stimulates the activity of adenyl cyclase, and the pathway stimulating inositol phospholipid turnover, which may be more akin to the major *ras* pathway in mammalian cells. A wild-type strain that takes up cAMP does not display an extended life span in the presence of this nucleotide. We have not obtained any evidence clearly favoring the involvement of the cAMP pathway in life span-extension, despite examining this pathway genetically at several levels (J. Sun & S. M. Jazwinski, unpublished results). In this context, it is interesting that Motizuki and Tsurugi (1992a) found a shortening of generation time in old cells upon administration of exogenous cAMP to a wild-type strain. In fact, however, these investigators were measuring the change in the time it took the cells to emerge from stationary phase and not simply generation time. In addition, their studies with estradiol discussed earlier were correlative and not causal.

RAS in yeast 'pushes' the cell to continue growing and dividing. The putative senescence factor alluded to earlier would serve to 'pull' it back and arrest it in the cell cycle. There may exist a reciprocal relationship between Ras and the senescence factor. The relative levels of these gene products during the life span may determine whether or not an individual cell continues dividing or senesces at any point along the cell spiral.

Recently, something akin to a developmental process has been described in yeast. This is the dimorphic transition from a yeast phase to a pseudo-filamentous phase in *S. cerevisiae* (Gimeno *et al.*, 1992). This transition, not unlike that in other dimorphic yeasts, occurs on nutrient limitation of diploid cells. It manifests itself as a change in cell morphology from ellipsoid to elongated or rod-like. The elongated yeast cells change their budding pattern from bipolar to unipolar, and they then grow in strings of cells with periodic branches. After a time of outgrowth from the edges of colonies which has been termed foraging for nutrients, the filamentous outgrowths cease, and they become covered with small round cells called blastospores. The role of these small round cells is not clear, and it is certainly not known whether they function as true spores. I believe that this dimorphic transition and pseudo-filamentous growth just discussed is closely related to yeast longevity. In essence, some of the yeast cells, rather than permanently arresting on nutrient limitation, divide an additional but finite number of times. Haploid yeasts appear simply to arrest division in an orderly fashion on nutrient limitation just as they do when they enter stationary phase (Pringle & Hartwell, 1981). It should be reiterated that both haploid and diploid yeasts display a limited replicative life span, and that in an isogenic haploid/diploid pair it is statistically identical (D. S. Franklin & S. M. Jazwinski, unpublished results).

The *RAS*2 gene, which is part of the nutritional sensor of the yeast cell and integrates cell metabolism and cell division (Broach & Deschenes, 1990), plays a role in filamentous growth; overexpression of this gene augments the phenomenon in diploids (Gimeno *et al.*, 1992). As discussed above, overexpression of *RAS*2 prolongs the replicative life span of haploids, similarly to the life span extension by v-Ha-*ras* (Chen, Sun & Jazwinski, 1990). Thus, the genetic control of longevity in both haploids and diploids involves *RAS*. In haploid cultures that are starved for nutrients, we have observed budding patterns resembling pseudo-filamentous growth (N. E. Jeansonne, D. S. Franklin & S. M. Jazwinski, unpublished results). The cells remain ellipsoid but tend to grow in strings with branches, due to a shift from an axial to a bipolar budding pattern. In axial budding, a new bud appears in the vicinity of the previous bud on the mother cell, while in bipolar budding new buds alternate between the two poles of the ellipsoid mother cell. Once the genetic dissection of filamentous growth and yeast longevity are sufficiently advanced, it will be possible to answer the question of the extent of overlap of the two processes. In any case, the involvement of *RAS*2 in these two processes warrants the reminder that *ras* is not only a proto-oncogene in higher eukaryotes, but it also is important in development (Barbacid, 1987).

'Why' do yeasts age?

The question of why yeasts age can be phrased better by asking what the selective pressure is for the evolution of yeast longevity. In order to address this question, it is necessary to consider yeast clones in the wild. Yeasts are largely HO^+ diploids in nature, however, ho^- haploids have a finite replicative life span as well. Elsewhere, I have argued that aging, senescence and cell death have evolved to limit polyploidization of yeasts (Jazwinski, 1990). These arguments focus around the significance of mating-type control in the yeast life cycle, and the consequences of errors in this control. In this example, the selective pressure would be to limit life span. On the other hand, the need to forage for nutrients might serve to increase longevity. There may well exist a balance between these two opposite modes of selection, and depending upon shifting ecological conditions one or the other may predominate in a dynamic equilibrium.

It is also necessary to consider the well-founded tenets of the evolutionary biology of aging (Rose, 1991). There is good reason to believe on the basis of the evolutionary theory of aging that, in fact, there is no selection on longevity. Instead, there is selection for an appropriate level of reproductive fitness, and post-reproductive life span is simply a derivative of this that is free to drift with no direct selection. From this standpoint, an individual yeast cell need divide only twice to generate an exponentially-expanding population. In the next section, I will try to generalize from what we know of yeast longevity, taking the evolutionary theory of aging as a point of departure. However, I will attempt to formulate a mechanistically pleasing proposal.

Asymmetric reproduction: soma and germ line

The asymmetric reproduction (division) of yeast cells lies at the foundation of yeast aging and yeast longevity. This is intuitively obvious given the basic phenomenology of yeast aging, in which the daughter cell is produced by asymmetric division (budding) and generally has the same probability to lead a full life span as its mother did at the beginning of hers. However, it is not simply the case that the difference in size between mother and daughter explains the difference in reproductive capacity between the two, as discussed earlier. Where is the essential asymmetry? We do not know the answer to this question. I suspect that this answer might provide a key to the genetics of yeast longevity and possibly to that of other organisms as well. In any case, the mother cell appears to give up some of its resources to its daughters with each cell division. The yeast mother cell is the equivalent of the soma, and the nascent daughter cell is the equivalent of the germ line, because in yeast the cell and the organism are one. The germ line is immortal, while the soma is mortal.

There are several examples of asymmetry at the molecular and cellular levels in yeast that could serve as paradigms for the asymmetric distribution of reproductive capacity between mother and daughter. One example of this asymmetry is the genetic machinery regulating *HO* expression and mating-type switch (Nasmyth & Shore, 1987), as suggested earlier (Jazwinski, 1990). Regulation in this case is achieved by the apparent asymmetric segregation of the transcription factor Swi5 into the mother cell. Another example is the preferential usage of different cyclins in mother and daughter cells (Lew, Marini & Reed, 1992). The cyclins are proteins that form complexes with protein kinases that are important in regulating cell cycle traversal. It is postulated that mother cells, which need not grow significantly in size before budding again, utilize the constitutively-expressed Cln3. The daughters must await the induced expression of Cln1 and Cln2. The bases for the cyclin preferences are not clear, however. Yet a third example is the asymmetric mitotic segregation of components of the spindle pole body preferentially into the bud (Vallen *et al.*, 1992). The spindle pole body is embedded in the nuclear membrane and organizes both nuclear and cytoplasmic microtubules. Duplication of the spindle pole body is a very early event in the cell cycle. One of the proteins required for this process becomes associated with only one of the daughter spindle pole bodies, and this is usually the one that enters the bud. Budding pattern and cell polarity (Hartwell, 1991) are established in yeast in a series of events that may be related to this asymmetric segregation of daughter spindle pole bodies. Budding pattern and cell polarity constitute a fourth example of genetically-determined asymmetry in the yeast cell. The determination of cell polarity involves many genes, including ones that code for *RAS*-related GTP-binding proteins (Pringle, 1993). All four of the examples of asymmetry listed are likely to be related through the genes that lie high in the hierarchy of regulation of mating-type switch and cell cycle control, the *SWI* genes (Nasmyth & Shore, 1987; Andrews, 1992).

In contrast to the paradigms for asymmetric distribution of reproductive capacity listed above, it is clear that the shortening of telomeres with each cell division is not the natural cause of aging in yeast as it might be in human fibroblasts. No shortening of telomeres is detected in yeast mother cells with replicative age or the daughters of old mothers (D'mello & Jazwinski, 1991). There is also no preferential co-segregation of chromatids of similar replicative age to mother or daughter cells (Neff & Burke, 1991; D'mello & Jazwinski, 1991). The distinction between yeasts and individual fibroblasts from the human organism is significant. Fibroblasts, unlike yeasts, undergo clonal senescence. Interestingly, yeasts do as well when the *EST*1 gene is mutated resulting in telomere attrition (Lundblad & Szostak, 1989). Thus, the major causal factors limiting replicative life span may be distinct in the yeast cell and an individual cell from a multicellular organism.

Cell polarity and asymmetric (physically and/or functionally) cell division are fundamental features of eukaryotic development. They can encompass partitioning of components of the egg cytoplasm or perhaps asymmetric segregation of products (including transcriptional states of chromatin) generated in the embryo, and they might be caused by signaling from other cells. Recent studies point to mechanisms by which information is differentially localized in the oocyte cytoplasm (Mowry & Melton, 1992). Much of *Caenorhabditis elegans* embryonic development involves the generation of cell lineages through asymmetric cell divisions (Kemphues *et al.*, 1988). Interestingly, positional memory established in embryos can be retained in adults in a heritable fashion in cells (Donoghue *et al.*, 1992).

On the basis of these observations, I propose an ontogenetic theory of longevity: (1) Limited life span is an evolutionary vestige of a primitive set of related processes in eukaryotes: asymmetric cell division, cell polarity, and pattern formation. In many instances, the original connection between life span limitation and these processes is now obscure: limited life span remains, and asymmetric cell division, cell polarity and pattern formation remains, seemingly unrelated. For metazoans to evolve, the molecular and cellular pre-requisites for asymmetric cell division (physical and functional), cell polarity and pattern formation had to be established. (These three processes are basic to the progressive specialization that comprises development. A fundamental aspect of these related processes is the unequal distribution of resources, as elaborated below.) It is the establishment of these molecular and cellular pre-requisites in evolution, at the genetic level, that is the source of finite life span. Today, only limited examples are available of organisms in which there exists a necessity for limited longevity as a direct consequence of asymmetric division, cell polarity and pattern formation. (One such example are yeasts.) However, this necessity can be traced back in evolution for other organisms. In these latter organisms, this evolutionary rationale will have predictable genetic consequences. Some of these genetic consequences will be common for a broad spectrum of organisms, as their roots lie close to the primitive connection between longevity and asymmetric division, cell polarity and pattern formation. Other genetic consequences will be as different from one organism to the next, as the evolved developmental specialization processes that distinguish them. (2) Although the fertilized egg is totipotent, the somatic cells and tissues that appear as embryogenesis and development progress become specialized (differentiation) and limited in potential. (This, of course, is exclusive of the privileged germ line.) The primordial basis for this loss of potential in somatic cells is asymmetric cell division (cell polarity) which at the level of the individual cell effects partitioning of cell resources or potential. (Initially, this may be partitioning of the determinants in the egg cytoplasm.) (3) The metazoan organism is an extension of the fertilized egg through the process of development. (This is especially obvious in organisms that possess mosaic eggs.) In this way, the above statements at the cellular level translate to the organismal level; that is, the limited potential of individual somatic cells inexorably results in the limited life span of the individual organism.

The predictions are: (1) Those genetic determinants and molecular mechanisms, such as *RAS*-like genes, that play a role in asymmetric cell division, cell polarity and pattern formation will also be important in determining longevity, although which ones do so may differ from species to species and not all of them will be necessarily involved in any given species. The corollary is that *RAS*-like genes, and GTPase genes in general, are many and conserved throughout eukaryotes. This statement is derived from our understanding of the genetics of cell polarity in yeasts. (2) True asymmetric divisions occur in embryogenesis. Later, in the adult, binary fission is the major mode of cell division, except in specialized stem cells. If asymmetric cell divisions were prevalent in the adult, the partitioning of potential would be continuous and hinder the generation of populations of equipotent, functional somatic cells. (3) Somatic cells can divide by binary fission because their potential is limited (as defined above). Such cells will undergo clonal senescence. Symmetric division implies immortality; however, somatic cells have already experienced reduction in potential during the asymmetric divisions that occur in development. (4) Tumor cells of somatic cell origin will not be totipotent. Such tumor cells will possess a finite number of genetic changes that distinguish them from clonally-senescing somatic cells. This follows from the progressive loss of potential in development, during asymmetric cell division. (5) Cells of the germ line will not experience telomere shortening. Otherwise, the germ cells would not be totipotent. (6) Organisms that reproduce symmetrically will not possess intrinsic limits to their longevity. Asymmetry is fundamental to finite life span according to the theory.

Place of model systems in aging research

Aging and senescence occur because the organism deteriorates and does not repair intrinsic and extrinsic damage. There are probably no aging genes as such. However, the capacity to carry out maintenance and repair is determined by genes. Genes not only determine this in a general way by establishing the overall constitution of the organism, they are also directly involved in crucial maintenance and repair processes. These crucial processes may vary from one species to the next both because of the organism's overall constitution and because of the external ecological factors within which a given species operates. The genes that specify those crucial processes are the major longevity-determining genes in that species. For another species, those major longevity genes may be different. All of these major longevity genes constitute the entire spectrum of the genetics of aging. For any given species, the genes from this spectrum that are not the major determinants will frequently in fact be minor determinants or modifiers. The action of these modifiers may be difficult to discern, as compared to that of the major determinants. However, they can be studied as major determinants in another species. This argues for a comparative approach to aging research. It also argues for the use of several, genetically malleable model systems in

which the major determinants can readily be dissected.

Many of the specific aspects or manifestations of aging (phenotype) may lead more or less directly to longevity-determining genes (genotype). However, it is also probable that many will simply identify the 'gray hair genes'. After all, longevity is the primary phenotype and aging the secondary phenotype. If the ontogenetic theory of longevity enunciated above is taken seriously, many of the specific aspects of aging may indeed be very far removed from the longevity-determining genes.

Acknowledgements

The work from my laboratory was supported by grants from the National Institute on Aging of the National Institutes of Health and by the American Federation for Aging Research (AFAR), Inc.

References

Adams, J., C. Paquin, P. W. Oeller & L. Lee, 1985. Physiological characterization of adaptive clones in evolving populations of the yeast, Saccharomyces cerevisiae. Genetics 110: 173-185.

Andrews, B. J., 1992. Dialogue with the cell cycle. Nature 355: 393-394.

Barbacid, M., 1987. ras genes. Annu. Rev. Biochem. 56: 779-827.

Baroni, M. D., E. Martegani, P. Monti & L. Alberghina, 1989. Cell size modulation by CDC25 and RAS2 genes in Saccharomyces cerevisiae. Mol. Cell. Biol. 9: 2715-2723.

Bartholomew, J. W. & T. Mittwer, 1953. Demonstration of yeast bud scars with the electron microscope. J. Bacteriol. 65: 272-275.

Barton, A. A., 1950. Some aspects of cell division in Saccharomyces cerevisiae. J. Gen. Microbiol. 4: 84-87.

Beran, K., I. Malek, E. Streiblova & J. Lieblova, 1967. The distribution of the relative age of cells in yeast populations, pp. 57-67 in Microbial Physiology and Continuous Culture, edited by E. O. Powell, C. G. T. Evans, R. E. Strange and D. W. Tempest. Her Majesty's Stationery Office, London, England.

Broach, J. R. & R. J. Deschenes, 1990. The function of RAS genes in Saccharomyces cerevisiae. Adv. Cancer Res. 54: 79-139.

Byers, B. & L. Sowder, 1980. Gene expression in the yeast cell cycle. J. Cell Biol. 87: 6a.

Cabib, E., R. Ulane & B. Bowers, 1974. A molecular model for morphogenesis: The primary septum of yeast. Curr. Top. Cell. Regul. 8: 1-32.

Chen, J. B., J. Sun & S. M. Jazwinski, 1990. Prolongation of the yeast life span by the v-Ha-*RAS* oncogene. Molec. Microbiol. 4: 2081-2086.

Choder, M., 1991. A general topoisomerase I-dependent transcriptional repression in the stationary phase in yeast. Genes & Dev. 5: 2315-2326.

Cohen, P., 1985. The role of protein phosphorylation in the hormonal control of enzyme activity. Eur. J. Biochem. 151: 439-448.

D'mello, N. P. & S. M. Jazwinski, 1991. Telomere length constancy during aging of *Saccharomyces cerevisiae*. J. Bacteriol. 173: 6709-6713.

Donoghue, M. J., R. Morris-Valero, Y. R. Johnson, J. P. Merlie & J. R. Sanes, 1992. Mammalian muscle cells bear a cell-autonomous, heritable memory of their rostrocaudal position. Cell 69: 67-77.

Drebot, M., G. C. Johnson & R. A. Singer, 1987. A yeast mutant conditionally defective only for reentry into the mitotic cell cycle from stationary phase. Proc. Natl. Acad. Sci. USA 84: 7984-7952.

Egilmez, N. K. & S. M. Jazwinski, 1989. Evidence for the involvement of a cytoplasmic factor in the aging of the yeast *Saccharomyces cerevisiae*. J. Bacteriol. 171: 37-42.

Egilmez, N. K., J. B. Chen & S. M. Jazwinski, 1989. Specific alterations in transcript prevalence during the yeast life span. J. Biol Chem. 264: 14312-14317.

Egilmez, N. K., J. B. Chen & S. M. Jazwinski, 1990. Preparation and partial characterization of old yeast cells. J. Gerontol. 45: B9-17.

Gimeno, C. J., P. O. Ljungdahl, C. A. Styles & G. R. Fink, 1992. Unipolar cell divisions in the yeast S. cerevisiae lead to filamentous growth: Regulation by starvation and *RAS*. Cell 68: 1077-1090.

Goebl, M. G. & T. D. Petes, 1986. Most of the yeast genomic sequences are not essential for cell growth and division. Cell 46: 983-992.

Guthrie, C. & G. R. Fink (editors), 1991. Guide to Yeast Genetics and Molecular Biology. Academic Press, San Diego, CA.

Hartwell, L., 1991. Pathways of morphogenesis. Nature 352: 663-664.

Hirakawa, T. & H. E. Ruley, 1988. Rescue of cells from *ras* oncogene-induced growth arrest by a second, complementing, oncogene. Proc. Natl. Acad. Sci. USA 85: 1519-1523.

Hirsch, H. R. & M. Witten, 1991. Dilution theory applied to a senescence factor in the ageing of yeast cells. FASEB J. 5: A1475.

Holm, C., 1982. Clonal lethality caused by the yeast plasmid 2μ DNA. Cell 29: 585-594.

James, A. P., B. F. Johnson, E. R. Inhaber & N. T. Gridgeman, 1975. A kinetic analysis of spontaneous ρ^- mutations in yeast. Mutat. Res. 30: 199-208.

Jazwinski, S. M., 1990. Aging and senescence of the budding yeast *Saccharomyces cerevisiae*. Molec. Microbiol. 4: 337-343.

Jazwinski, S. M., 1990a. An experimental system for the molecular analysis of the aging process: The budding yeast *Saccharomyces cerevisiae*. J. Gerontol. 45: B68-74.

Jazwinski, S. M., N. K. Egilmez & J. B. Chen, 1989. Replication control and cellular life span. Exp. Gerontol. 24: 423-436.

Jazwinski, S. M., J. B. Chen & N. E. Jeansonne, 1990. Replication control and differential gene expression in aging yeast,

pp. 189-203 in The Molecular Biology of Aging, edited by C. E. Finch and T. E. Johnson. Alan R. Liss, New York, NY.

Jazwinski, S. M., 1993. Genes of youth: Genetics of aging in baker's yeast. ASM News 59: 172-178.

Johnson, B. F. & C. Lu, 1975. Morphometric analysis of yeast cells IV. Increase of the cylindrical diameter of *Schizosaccharomyces pombe* during the cell cycle. Exp. Cell Res. 95: 154-158.

Johnston, J. R., 1966. Reproductive capacity and mode of death of yeast cells. Antonie van Leeuwenhoek J. Microbiol. Serol. 32: 94-98.

Kemphues, K. J., J. R. Priess, D. G. Morton & N. Cheng, 1988. Identification of genes required for cytoplasmic localization in early C. elegans embryos. Cell 52: 311-320.

Lew, D. J., N. J. Marini & S. I. Reed, 1992. Different G1 cyclins control the timing of cell cycle commitment in mother and daughter cells of the budding yeast S. cerevisiae. Cell 69: 317-327.

Lewin, B., 1990. Driving the cell cycle: M phase kinase, its partners, and substrates. Cell 61: 743-752.

Link, A. J. & M. V. Olson, 1991. Physical map of the *Saccharomyces cerevisiae* genome at 110-kilobase resolution. Genetics 127: 681-698.

Lundblad, V. & J. W. Szostak, 1989. A mutant with a defect in telomere elongation leads to senescence in yeast. Cell 57: 633-643.

Mortimer, R. K. & J. R. Johnston, 1959. Life span of individual yeast cells. Nature 183: 1751-1752.

Motizuki, M. & K. Tsurugi, 1992. The effect of aging on protein synthesis in the yeast *Saccharomyces cerevisiae*. Mech. Ageing Dev. 64: 235-245.

Motizuki, M. & K. Tsurugi, 1992a. Effect of 17β-estradiol on the generation time of old cells of the yeast *Saccharomyces cerevisiae*. Biochem. Biophys. Res. Commun. 183: 1191-1196.

Mowry, K. L. & D. A. Melton, 1992. Vegetal messenger RNA localization directed by a 340-nt RNA sequence element in *Xenopus* oocytes. Science 255: 991-994.

Muller, I., M. Zimmermann, D. Becker & M. Flomer, 1980. Calendar life span versus budding life span of *Saccharomyces cerevisiae*. Mech. Ageing Dev. 12: 47-52.

Muller, I., 1971. Experiments on ageing in single cells of *Saccharomyces cerevisiae*. Arch. Mikrobiol. 77: 20-25.

Muller, I., 1985. Parental age and the life-span of zygotes of *Saccharomyces cerevisiae*. Antonie van Leeuwenhoek J. Microbiol. Serol. 51: 1-10.

Muller, I. & F. Wolf, 1978. A correlation between shortened life span and UV-sensitivity in some strains of *Saccharomyces cerevisiae*. Molec. Gen. Genet. 160: 231-234.

Munkres, K. D., 1985. Aging in fungi, pp. 29-43 in Review of Biological Research in Aging, Vol. 2, edited by M. Rothstein. Alan R. Liss, New York, NY.

Nasmyth, K. & D. Shore, 1987. Transcriptional regulation in the yeast life cycle. Science 237: 1162-1170.

Neff, M. W. & D. J. Burke, 1991. Random segregation of chromatids at mitosis in *Saccharomyces cerevisiae*. Genetics 127: 463-473.

Oliver, S. G. *et al.* (147 authors), 1992. The complete DNA sequence of yeast chromosome III. Nature 357: 38-46.

Paquin, C. & J. Adams, 1983. Frequency of fixation of adaptive mutations is higher in evolving diploid than haploid yeast populations. Nature 302: 495-500.

Paquin, C. E. & J. Adams, 1983a. Relative fitness can decrease in evolving asexual populations of *S. cerevisiae*. Nature 306: 368-371.

Park, E.-C., D. Finley & J. W. Szostak, 1992. A strategy for the generation of conditional mutations by protein destabilization. Proc. Natl. Acad. Sci. USA 89: 1249-1252.

Pillus, L. & J. Rine, 1989. Epigenetic inheritance of transcriptional states in S. cerevisiae. Cell 59: 637-647.

Pohley, H.-J., 1987. A formal mortality analysis for populations of unicellular organisms (*Saccharomyces cerevisiae*). Mech. Ageing Dev. 38: 231-243.

Pringle, J. R., 1993. Cell polarity in yeast. Annu. Rev. Cell. Biol. 9 (in press).

Pringle, J. R. & L. H. Hartwell, 1981. The *Saccharomyces cerevisiae* cell cycle, pp. 97-142 in The Molecular Biology of the Yeast *Saccharomyces*: Life Cycle and Inheritance, edited by J. N. Strathern, E. W. Jones & J. R. Broach. Cold Spring Harbor Press, Cold Spring Harbor, NY.

Rivier, D. H. & J. Rine, 1992. An origin of DNA replication and a transcription silencer require a common element. Science 256: 659-663.

Rose, M. R., 1991. Evolutionary Biology of Aging. Oxford University Press, New York, NY.

Sando, N., M. Maeda, T. Endo, R. Oka & M. Hayashibe, 1973. Induction of meiosis and sporulation in differently aged cells of *Saccharomyces cerevisiae*. J. Gen. Appl. Microbiol. 19: 359-373.

Sclafani, R. A., M. Patterson, J. Rosamond & W. L. Fangman, 1988. Differential regulation of the yeast *CDC* gene during mitosis and meiosis. Mol. Cell. Biol. 8: 293-300.

Strathern, J. N., E. W. Jones & J. R. Broach (editors), 1981. The Molecular Biology of the Yeast *Saccharomyces*: Life Cycle and Inheritance. Cold Spring Harbor Press, Cold Spring Harbor, NY.

Strathern J. N., E. W. Jones & J. R. Broach (editors), 1982. The Molecular Biology of the Yeast *Saccharomyces*. Metabolism and Gene Expression. Cold Spring Harbor Press, Cold Spring Harbor, NY.

Tyson, C. B., P. G. Lord & A. E. Wheals, 1979. Dependency of size of *Saccharomyces cerevisiae* cells on growth rate. J. Bacteriol. 138: 92-98.

Vallen, E. A., T. Y. Scherson, T. Roberts, K. van Zee & M. D. Rose, 1992. Asymmetric mitotic segregation of the yeast spindle pole body. Cell 69: 505-515.

M. R. Rose and C. E. Finch (eds.), Genetics and Evolution of Aging, 71–82, 1994.
© 1994 *Kluwer Academic Publishers. Printed in the Netherlands.*

Evolutionary senescence in plants

Deborah Ann Roach
Department of Zoology, Duke University, Durham, NC 27708-0325, USA

Received and accepted 22 June 1993

Key words: plant senescence, plant demography, life history

Abstract

Senescence is a decline in age-specific survival and reproduction with advancing age. Studies of evolutionary plant senescence are designed to explain this decline in life history components within the context of natural selection. A review of studies of plant demography reveals senescent declines in both annual and perennial plants, but also suggests that there are some plant species which may not be expected to show senescence. Thus, future comparative studies of closely related species, with and without senescence, should be possible. The assumptions of the major evolutionary theories of senescence are evaluated for their validity with respect to plants. Different plant species violate one or more of the assumptions of the theories, yet the consequences of violating these assumptions have never been investigated. Whereas, to date, evolutionary senescence has been studied only indirectly in plants, it is concluded that plants provide good experimental systems for clarifying our understanding of senescence in natural populations.

Senescence is a decline in age-specific survival and reproduction with advancing age. Evolutionary studies of senescence are designed to explain the persistence of this deleterious phenomenon in populations within the context of natural selection. Experimental tests of the theories of senescence have traditionally used *Drosophila, Caenorhabditis,* or *Tribolium,* and they have notably not included any plant species. Yet, the wealth of information from the fields of plant demography and plant population biology suggests that plants may offer unique opportunities for comparative studies on senescence. This paper will review plant demographic studies for senescence patterns, and will consider the current theories of senescence as they may be applied to plants.

The natural history of age-specific mortality and fecundity

Definitions

In the botanical literature the term 'senescence' has been used in several different contexts, many of them very unrelated to the use of the term in evolutionary studies. Most commonly, it is used to describe physiological processes and hormonal changes involved in the abscission or deterioration of plant parts (cf. Thimann, 1980; Nooden & Leopold, 1988; Rodriquez, Sanchez-Tames & Durzam, 1989). The modular construction of plants has led physiologists and population biologists alike to consider the individual plant as a 'population' of ramets and other repeated structures (Harper, 1977; Buss, 1987). Leaves, for example, are considered to have their own life history, with a seasonal demography which may be completely independent of root demography or demography of the whole plant. Senescence in this context is considered to be the orderly degenerative process leading to death of plant parts. It can occur at every stage of the whole plant life cycle, and may include senescence of the cotyledons and endosperm, senescence during cell differentiation, sequential and seasonal leaf senescence, or even whole plant senescence. Senescence in this modular physiological sense is an active process, which requires energy and protein synthesis, and involves redistribution of nutrients and photosynthates. The relationships between these modular, physiological processes and the demographic changes in life history components, which occur during evolutionary senescence, are for the most part very remote or non-existent. Clearly, a decline in reproduction and survival with advancng age reflects a decline in the performance of many different physiological functions, and physiological studies of plant senescence have given valuable insights into these processes. However, these studies have failed to adequately explain *why* there should be a demographic decline with age. Senescence in the modular, physiological sense is generally considered a beneficial process which increases individual fitness by allowing a plant to get rid of old redundant structures or as a prelude to the onset of harsh environments (e.g. leaf fall). Senescence in the evolutionary sense is a deleterious phenomenon which decreases the fitness of the individual. Because of this potential confusion over the use of the term plant senescence, it is important to emphasize that in this review 'senescence' will be used only in the evolutionary sense. Another confusing term that often appears in the literature is 'aging'. Aging in both the plant and animal literature has been used to describe any age-related change regardless of whether it does or does not affect life-history components in a senescent manner (Medawar, 1952, for plants; Leopold, 1975, for exception; Rose, 1991).

Considerable information is available on plant longevity (for reviews see Molish, 1938; Wanger-

mann, 1965). Individuals of each plant species have a characteristic maximum longevity which may range from a few weeks, for some ephemeral annuals, to over 1,000 years for many conifer species. It has been suggested that the extreme longevity of many plant species implies that senescence does not occur (Leopold, 1981; Nooden, 1988; Watkinson & White, 1986). But senescence, in the evolutionary sense, is defined as a change in the shape of the survivorship curve late in the life cycle. To say that a species characteristically has a long life says nothing about whether it shows senescence or not; data on maximum longevity give the endpoints for survivorship curves, not the shape of the curve (Bell, 1992). In spite of a long life span, individuals may still begin to die at a higher rate at advanced ages. The data which are critical to senescence studies deal with age-specific survivorship and fecundity.

Determinate annuals

There are two types of growth forms for annual plants: determinate and indeterminate. Determinate annuals are monocarpic (semelparous), and die abruptly after seed set. The main meristem is used for inflorescence formation, and thus further vegetative growth is not possible. The life cycle in these species is synchronized with recurrent seasonal events, and the transitions to different life stages are cued to photoperiod (Wareing & Phillips, 1981). Many studies with determinate annuals have shown a sharp decrease in survival late in life. For example, in a study with *Phlox drummondii*, Leverich and Levin (1979) found a constant decrease in survival at the seed stage due to seed predation. After emergence, survival is high until the latest stages of the life cycle when it decreases sharply (Fig. 1).

Physiologists have proposed the term 'programmed senescence' to describe the collapse of monocarpic plants following reproduction. This decrease in survival following reproduction in monocarpic annuals is an intrinsic decline in late-life survival and as such is actually a good example of evolutionary senescence. The close link between death and reproduction has been established in a number of studies which have shown that the removal of the reproductive structures generally increases the life span of individual plants relative to reproducing controls (Molisch, 1938; Nooden, 1988; Leopold, 1961). Although not quite as ex-

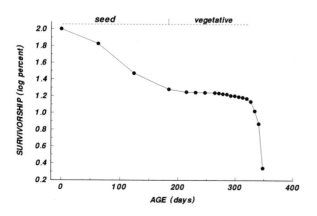

Fig. 1. Survivorship curve for a determinate annual *Phlox drummondii*. (Data from Leverich & Levin, 1979).

treme as in monocarpic plants, a close relationship between reproduction and survival has been found in *Drosophila* (Partridge, 1986). Physiologists have been interested in describing the trigger which eventually results in whole plant death (cf. Thimann, 1980), but this does not explain *why*, in the evolutionary sense, there should be such a close relationship between reproduction and survival.

Indeterminate annuals

Indeterminate annual plants flower when they are small, and continue to grow, flower, and set seed, until an extrinsic climatic event, for example frost or drought, causes death of any surviving individuals at the end of the growing season. When growing conditions deteriorate, the survivorship curve becomes abruptly vertical. For indeterminate annual species, death at the end of the growing season is usually caused by extrinsic causes. A precise definition of evolutionary senescence proposed by Rose (1991) is a decline in age-specific fitness components due to internal physiological deterioration. This definition thus excludes increased death due to deterioration of the environment. An implicit expectation is that if these plants were grown under protected conditions, away from the extreme climatic events, their total life span would be longer and the shape of the survivorship curve would be quite different. Unfortunately, no data are available on the shape of the survivorship curve of indeterminate annuals in the absence of climatic extremes, thus it is not known whether these species could in fact show intrinsic senescence. There is, however, limited evidence that some indeterminate annual species may show senescence, even prior to the loss of the population due to the deterioration of the environment. Bell, Lechowicz and Schoen (1991) found an increase in mortality rate in the last eight weeks of the season for *Impatiens pallida*. They suggest that this increase was due to an increasing burden of pathogens, but there were no data to support this. Increasing vulnerability to pathogens or disease with age could be a manifestation of senescence, as it leads to increasing mortality.

The shape of the survivorship curve from field studies, for any type of species, cannot be used as conclusive evidence for or against the presence of senescence. Environmentally imposed mortality rates may mask inherent survivorship patterns. In fact, Medawar (1952) suggested that senescence

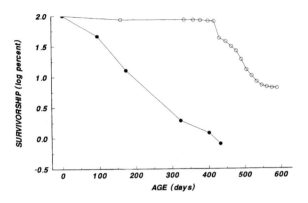

Fig. 2. Survivorship curves for *Rumex hastatulus* under field (closed circle) and greenhouse (open circle) conditions.

was only characteristic of populations in 'captive' environments, for which the level of random mortality had been reduced. His illustrative examples included man and *Drosophila*, which in captive or laboratory populations have increased numbers of individuals surviving the early life stages, and which then manifest a decreasing survival late in life. There has been no test of this hypothesis with animals, because comparative studies of populations in wild vs. captive environments are difficult, if not impossible to undertake; with plants, however, this contrast is possible. Plants can be grown under natural field conditions, or the idealized conditions of the greenhouse or growth chamber where the levels of random background mortality should be reduced. This study has been done with a plant species *Rumex hastatulus*, a species with a short but variable life span (D. A. Roach, unpubl.). Seed was germinated in the greenhouse and divided into two groups. 680 plants were planted back into a natural field, and 760 plants were transplanted into pots, with sterilized field soil, and kept in the greenhouse. The seedlings were transplanted in March and this species usually flowers in May. However, apparently none of the plants in either the field or the greenhouse were large enough to flower by this time. They instead flowered the following year. In the field, mortality was high, and relatively constant; less than 1% of the population survived to flower 14 months after planting (Fig. 2). In contrast, survival in the greenhouse was very high until after flowering, when there was a sharp increase in mortality. Half of the population survived this first episode of reproduction and was still alive six months later. These data suggest that even if senescence exists in a population, it may be difficult to observe unless early survival is increased, resulting in increased numbers of individuals in the later age classes. Moreover, it further suggests that even if senescence is not observed in wild populations, it does not mean that it does not exist.

Short-lived perennials

Perennials can also be either determinate or indeterminate. Determinate, or monocarpic perennials, flower only once, and the individual dies shortly thereafter. Monocarpic species may include typical 'biennials' such as *Digitalis purpurea* (foxglove) or *Daucus carota* (wild carrot), to longer lived monocarps such as *Phyllostachys bambusoides* (bamboo), which can live 120 years before flowering (Janzen, 1976). Most of these species have either a constant rate of mortality throughout their life cycle, or show a high rate of early mortality during establishment, followed by a high rate of survival for the few survivors (Fig. 3). Moreover, in these species it is not unusual to find less than one percent of the population surviving to reproduction, making it difficult to detect whether senescence actually occurs. There may be senescence for monocarpic perennials at the tail end of the survivorship curve, but given the extremely small number of individuals which live to this stage, any change in the slope of the survivorship curve is difficult to detect.

Indeterminate perennial plants are iteroparous. They often reproduce more than once, and reproduction and death are not closely coupled as in monocarpic species. General life history theory predicts that an iteroparous life history will be relatively advantageous under conditions in which the ratio of seedling to adult mortality is high (Schaffer & Gadgil, 1975). Most data on survival for perennials are obtained from 'depletion curves' in which individuals are marked at the beginning of a study irrespective of age. These individuals are then followed for several years, and their survival is recorded. The classic study of this type by Tamm on perennial herbs in a meadow in Sweden (Tamm, 1956, 1972a,b) spanned 14-30 years, and showed low constant survival for nearly all species (Harper, 1967). Similar linear curves have been found in other studies (Sarukhan & Harper, 1973; An-

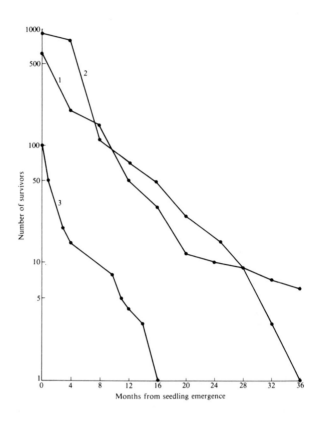

Fig. 3. Survivorship curves of three monocarpic perennials: (1) *Grindelia lanceolata*, (2) *Pastinaca sativa*, and (3) *Melilotus alba*. (Fig. 3.4 in Silvertown (1982)).

tonovics, 1972). However, it is not possible to construct survivorship curves from these studies. What is needed is the relative survival of known age classes, pooled over several cohorts to avoid cohort specific events (e.g. a poor season at the end of life). Unfortunately, there are almost no such data on the survivorship curves of non-monocarpic perennials. In the one study that documented the age-specific survival from emergence to death over 17 years, Canfield (1957) found evidence for decreased survival with increasing age in several tussock range grasses. Until more long-term detailed studies like this are done, however, no generalizations about the slope of survivorship curves in natural populations of perennial plant species can be made.

Long-lived perennials

Many perennial species can grow clonally; a genetic individual is referred to as a 'genet'; each clone is referred to as a ramet. While ramets may be born and old ones die, the genet can have an extremely long life span. There are many examples of exceptionally long-lived individuals in clonal species (for a review see Cook, 1983). The ages of clones are usually determined by the rates of radial spread and the diameter of the clone. For example, for the creosote bush (*Larrea tridentata*) the central stems die, but a clonal expansion of the ramets forms a ring of shrubs which advance radially. Extrapolating back at known rates of expansion, the largest clone has been estimated to be between 9,000-11,000 years old (Vasek, 1980). Given the remarkable age of individuals of many clonal species, many researchers have concluded that individual genets are potentially immortal and that there is no senescence in these species. Unfortunately, it is nearly impossible to test this hypothesis because the extremely long lives of many of these species make it impossible to construct survivorship curves. Furthermore, the clonal growth may make it impossible to follow individual genets for any period of time. It is unfortunate, but understandable, that there is at this date no information on the rates of late-life survival for clonal species, and therefore statements about senescence or even average longevity in such species must be viewed with skepticism.

Studies on the mortality rates in trees are in most cases obtained from vertical life tables in which individuals are aged using growth rings, leaf scars, or other approaches to obtain the age structure for a population. Alternatively, longer term studies estimate tree death rate from five- to ten-year remeasurement of tagged individuals. These data confound variation across age classes with variation in yearly recruitment conditions, recruitment success, successional changes, or variation in sampling techniques, yet some general patterns can be considered. Similar to other long-lived plant species, trees generally have high early mortality followed by a decline in mortality with increasing age or size (cf. Sarukhan, 1980; Hibbs, 1979; Harcombe & Marks, 1983). Long-term studies suggest that the death rate in many tree species may increase in the oldest age classes (Lorimer & Frelich, 1984; Harcombe, 1987). Decline in late survival may be due to size-dependent stress and a decrease in net assimilation rates with an increasing burden of respiratory tissue (Clark, 1983; Franklin, Shugart & Harmon, 1987), or a decline in vigor with age (cf.

Leopold, 1981). This decline is reflected in decreasing annual growth increments or in changes in the susceptibility of trees to insects, disease or viruses (Leopold, 1981). Clark (1983) notes that a number of insect pests of trees prefer old trees. For example, the susceptibility of trees to insects such as western pine beetle, spruce bud worm and bronze birch borer increases with age. The vulnerability of older-aged perennials to a physiological decline is often a function of extremely large size (Steenbergh & Lowe, 1983). Because there is likely to be a correlation between size and age there may be age-specificity and true senescence, although clearly in these cases size rather than age *per se* is the critical factor. It can be shown that the decline in vigor and survival of older-aged trees is a function of the tree as a whole and not necessarily of the component parts which make up the tree. The slowed growth rates of fruit trees, for example, can be restored by pruning, cutting or grafting (Leopold & Kriedeman, 1975; Leopold, 1981). Further, no significant differences were found between photosynthesis and respiration measurements on cuttings from 100-1000 year old yew trees (Zajaczkowska, Lotocki & Morteczka, 1984).

It has been suggested that in some species, even-aged, monospecific tree stands can show what is termed 'cohort senescence'. As the trees grow older, they become vulnerable to damage due to storms or snow. An accumulation of broken branches then causes an increase in the populations of wood-boring beetles and fungal pathogens which then triggers 'senescence' in the stand. This phenomenon has been found for *Abies* spp. (Sprugel, 1976; Kohyama & Fujita, 1981), *Nothofagus* (Ogden, 1988), and *Metrosideros* (Mueller-Dombois, 1983; Stewart & Veblen, 1983). The term senescence has been applied in these instances because younger-aged cohort stands are not vulnerable to this type of uniform die-back. The vulnerability may, however, not be directly age-related but result from all of the trees in the stand being at a similar physiological stage. A recent model of this phenomenon has shown that the 'wave' pattern, which is observed during the die-back of *Abies*, is not dependent on age (Sato & Iwasa, 1993). Moreover, die-back in general is more likely caused by ecological factors rather than the result of senescence inherent in the trees, because individuals growing in mixed-species or even mixed-aged stands are not vulnerable to this pattern of mortality, and generally have a longer average life-span (J. Odgen, pers. comm.).

Fecundity

Senescence may also be apparent through a decrease in fecundity with age, but unfortunately there are few data on the comparative reproductive output of plants during different stages of the life cycle. Several studies with grasses have shown a decline in fecundity with age (Canode & van Keuren, 1963; Stark, Hafenrichter & Klages, 1949). Evans and Canode (1971) found a per unit decline in seed production for 'Newport' Kentucky bluegrass (*Poa pratensis*) over a seven year period following seed sowing. Furthermore, they found that nitrogenous fertilizer increased the weight but not the number of seeds produced.

In trees, several different fecundity patterns with age have been observed. Some species show a linear increase with age (Bullock, 1980), some peak at intermediate ages (Enright & Odgen, 1979), and some species show an increase and then a constant level of reproductive output (Pinero, Martinez-Ramos & Sarukhan, 1984). If size is used, instead of age, many tree species show an exponential increase of fecundity with size (cf. Hubbell, 1980; Watkinson & White, 1986). Harper and White (1974) suggest that there is a decline in the fecundity of many tree species at the latest stages of the life cycle. In citrus species, for example, there has been a crude observation of a decline the 'boxes per tree' after 20 years (Savage, 1966). In some tree species, a decline in fecundity at very late ages appears to be a hazard of extremely large size, just as late survival often is. For example, in *Quercus* spp. (Goodrum, Reid & Boyd, 1971) and *Libocedrus bidwillii* (J. Odgen, pers. comm.), the decline in late seed production is a function of reduced crown as older trees lose their branches. How general this decline in fecundity in late aged trees is is not clear. In tropical trees (Sarukhan, 1980) no decline has been found.

There is also very little known about changes in the quality of the offspring produced by older individuals. In an annual plant *Geranium carolianum*, the size of the seeds produced declined over the growing season, probably as a function of a deteriorating environment but perhaps also as a result of the age of the maternal plant (Roach, 1986a). In perennials, there are almost no data on seedling viability from older parents; Bosch (1971) showed a decline in seed viability in older redwoods but this may have been due to the local environment of the tree as much as an influence of age (Finch, 1991).

Summary

In conclusion, in spite of a wealth of information on plant demography, data on survival and reproduction of plants at later stages of their life cycle are very limited. Despite the fact that the demography of plants is comparatively easy, there have been few studies which have followed a population of individuals for their entire life cylce, particularly for species which live more than a few months (Harper & White, 1974). The evidence to date suggests that a senescent decline in survival, and in reproduction, is found in some plant species. For some extremely long-lived species it will be difficult if not impossible to obtain good data on late-life performance, yet further studies on shorter lived species may show senescence to be a more common phenomenon of plant populations.

Plants and the evolutionary theories of senescence

There have been two major theories which have been proposed to explain the evolution of the deleterious phenomenon of senescence: pleiotropy and mutation accumulation. Both of these theories provide an explanation for a decline in reproduction and survival with age. The pleiotropic theory suggests that senescence will evolve if a mutation with a late-acting deleterious effect is favored by selection because that mutation also has a positive effect early in life (Williams, 1957). The mutation accumulation theory of senescence suggests that due to a post-maturation decline in the strength of selection on age-specific characters, mutations with a deleterious effect at the end of the life cycle will accumulate, resulting in senescence (Medawar, 1952). Such accumulation may result from ineffective selection against recurrent deleterious mutations, or from genetic drift leading to an increase of late age-specific deleterious genes in the population. Both theories suggest that senescence is the result of the failure of natural selection to act against genes with late age-specific deleterious effects (Charlesworth, 1980). There are several assumptions which are implied by either or both of these theories, including age-specific gene action, a decline in the force of selection with age, pleiotropic gene action, and a separation of the germ and soma cell lines. Each of these assumptions will be considered in turn as they may be applied to plants.

Age-specific gene action

A primary assumption of any evolutionary theory is that there is age-specific gene action that affects survival and fecundity; age-specific gene effects are a central assumption of general life-history theory. The life cycle of most organisms can be divided into different developmental stages. At a minimum, this will include a pre-reproductive and a reproductive stage. Depending on the organism and its environment, the relative duration of these two stages will vary. Age-specific survival and fecundity patterns are the best evidence for age-specific gene action. The ability to modify these patterns through selection experiments, for example in applied crop programs, demonstrates that these traits are under genetic control. Recent molecular studies have also demonstrated evidence for age-specific gene action in plants. For example, it has been shown that the 'agamous' gene in *Arabidopsis* is only active during the flowering period (Yanofsky *et al.*, 1990).

Age-specific gene action *per se* may not be found in all species of plants for several reasons. First, due to seed dormancy, it may be difficult to define 'age' for an individual. If age is defined, like in animals, as the time since zygote formation, then a species which has variable seed dormancy will at the time of germination have a population with a mixed age-structure. Because of this variability, it may be more appropriate to speak of stage- rather than age-specific gene action. In other words, there may be certain genes which have an effect at the time of germination and others which have an effect at time of flowering, etc. In fact even in *Drosophila*, which has a variable length to the larval period, it may be more appropriate to speak of stage-specific gene action. In other plant species, for example many 'biennials' which must reach a minimum size or growth rate before flowering, it may be more appropriate to consider size-specific

gene action. The evolution of senescence requires that gene action be isolated to discrete life stages.

Senescence may not be expected to evolve if age-specific gene action is absent, for example in species with modular growth. Perennial plants vary in the amount of their past which they carry with them (Harper, 1977), and the amount of senescence that they eventually manifest may be highly correlated with this. In species with clonal growth, the somatic tissues are constantly being renewed. It is thus difficult to determine the age of an individual, because the different modular parts of the individual are different ages. In the case where clonal offspring can become detached and completely independent of the aging parental growth form, immortality of the individual genet may result. There are currently no data on age-specific gene action in these species; however, there are many shorter-lived clonal growing plant species, for example grass species, which could be used for demographic studies, and to test whether age-specific gene action and senescence is found in species which have a turnover of somatic tissues.

Decline in the force of selection with age

Evolutionary theories of senescence are based on the assumption that the response to selection is related to the age or set of ages at which genes have an effect on survival or fecundity. As the life cycle progresses, the strength of selection declines in part because random mortality decreases the number of individuals who live to a late age. Furthermore, in many organisms there is a decrease in age-specific fecundity following a period of peak reproduction earlier in the life cycle. This decrease in late-life fecundity, coupled with the fact that fewer individuals live to a late age, results in a population in which older individuals are contributing only a relatively small proportion of the genes for the next generation. Selection on a trait which is only expressed late in the life cycle will be weak relative to selection on a trait which has an earlier age-specific effect. This assumption of a decline in the force of selection with age leads to the prediction that it is difficult to eliminate late-acting deleterious mutations from the population; thus senescence may develop.

In some species of plants, such as trees, there is an increase in reproductive output with increasing age and size (Harper & White, 1974). It has been shown that this increase in fecundity with age will retard, but not prevent, the decline in the force of selection with age (Hamilton, 1966; Charlesworth, 1980). Trees and other organisms with similar reproductive patterns are thus expected to show senescence, albeit at a slower rate.

There are, however, some types of plant species which do not show a decline in the force of selection with age, and thus which may not be expected to show senescence. As noted earlier, some species may show stage-specific instead of age-specific gene action. For species with a completely stage-structured life-cycle, Caswell (1982) has shown that gene effects at larger sizes may be more important than effects at smaller sizes. In age-structured populations, senescence is expected to evolve because gene effects earlier in the life cycle are more important than at later stages. Violation of this assumption will modify the selection pressures which favor senescence. If clonal plants, for example, can be shown to have stage-specific rather than age-specific life cycles, then the chances of finding senescence in these species are greatly reduced. Senescence will not evolve unless there is a correlation between age and stage (or size) and between stage (size) and death rate (Kirkpatrick, 1984).

Pleiotropic gene action
A further assumption, specifically of the pleiotropic theory of senescence, is that there is antagonistic pleiotropic gene action. The two ways to measure this type of gene action are with selection studies in which a negatively correlated response to selection is observed in a second trait not under selection, or in quantitative genetic studies when negative genetic correlations between two traits are measured. In plants, the most direct evidence for pleiotropic constraints between different stages of the life cycle is from an experiment with an annual plant *Geranium carolinianum* (Roach, 1986b). This species is a weedy winter annual found in fields and waste places throughout the United States. In North Carolina, it germinates in October, persists as a rosette close to the ground until April, then bolts to 25 cm tall, flowers, and dies. Three distinct life stages can be defined for this species: early juvenile with cotyledons and one leaf, late juvenile rosette, and adult bolted plant. Seeds from maternal half-sib families, from three populations, 15 km apart, were planted in a common garden. Plants were harvested at three times during the growing season, corresponding to the three life stages. At each harvest, the leaf area and dry weight of the plant parts was measured. There were positive phenotypic but negative genetic correlations between early juvenile and adult traits. Negative genetic correlations in plants can also be inferred from studies with closely related species in different habitats in which a negative relationship between seed size and seed number has been observed (cf. Werner & Platt, 1976). This work with *Geranium*, however, was the first to directly show negative genetic correlations between different life stages. It has been suggested that monocarpic senescence may be an extreme example of pleiotropic constraints between reproduction and survival (Rose, 1991). The existence of negative genetic correlations in this annual plant lends strong support to this explanation.

Additional support for the existence of antagonistic pleiotropic genes in plants can be inferred from a study with *Poa annua*, annual meadow grass (Law, Bradshaw & Putwain, 1977; Law, 1979). In this study, vegetative tillers were sampled from 28 populations, and grown in a common garden. Seeds were collected from the tillers (a 'family') and grown in an experimental field following germination. Families derived from frequently disturbed meadows had higher early reproduction but shorter life spans than families from less disturbed populations (Law, Bradshaw & Putwain, 1977). There was also a negative family correlation between early and late reproduction. Families with large numbers of inflorescences in the first season tended to have small numbers in the second. The mortality risks of families with high rates of reproduction early in life were also relatively higher (Law, 1979). This was not a rigorous quantitative genetic experiment, and there may be correlated environmental variables which would select for different life histories in the disturbed and undisturbed meadows, but the experiment was sufficiently large to suggest that the results may describe underlying genetic differences between the populations. The results of this study with *Poa*, together with the study with *Geranium*, suggest gene action which would be consistent with the pleiotropic theory of senescence.

Separation of germ and soma
It has been repeatedly stated in the literature that

plant senescence may be less clearly defined than animal senescence because plants, unlike animals, do not show a clear distinction between germ and soma plasm (cf. Kirkwood & Holliday, 1979; Rose, 1991). Williams (1957) suggested that aging should be found to be inevitable for any organism in which there is this distinction. Comparative data indicate that all species with a well defined soma separate from their germ line do senesce when observed under the proper conditions (Rose, 1991). Conversely, species which clearly lack separation of the soma, such as sea anemone, prokaryotes, and some protozoa, appear not to age (Finch, 1991). In mammals, insects, and many other animals, the germ cells are segregated early in development. In contrast, in plants the germ line does not segregate from the somatic cells until just before reproduction, and therefore it could be argued that plants should not senesce. Although the importance of the germ/soma distinction is often cited, the precise reason why this segregation is necessary for the evolution of senescence has never been explicitly stated. The basic theories of age-specific gene action should hold with or without an early separation of the germ line. What may be more important, yet correlated with germ and soma separation, is the distinction between generations (Bell, 1992). For clonal species, a sexually produced genet is the start of a new generation, yet the ability to produce ramets, which may become totally independent of the parent plant, means that age-specific traits and gene effects have no meaning at the level of the genet (Harper & Bell, 1979); the genet becomes a collection of ramets of different ages. Independence of clonally produced individuals means that age-specific gene action at the level of the clone (genet) is extremely unlikely. Furthermore, if the quality of ramets produced does not vary with the age of the genet then the distinction between generations becomes lost. Moreover, in those species in which individuals can continue to grow clonally, maintaining or increasing their size through the production of ramets, survival and reproduction may not decline with age. Clonal plant species may therefore offer an interesting contrast, in comparative studies, to other non-clonal plants; they provide a unique experimental tool in which genotypes can be easily replicated. However, I know of no comparative studies of senescence in closely related clonal and non-clonal species of plant or animal.

Conclusions

A review of the data from plant demography studies shows that there is a substantial number of species which show a decline in survival with age, or where such a decline is strongly suggested (see also Watkinson, 1992). Senescence, however, must include not only a decline in survival with age, but also internal physiological deterioration (Rose, 1991). Thus, despite the sharp increase in mortality at the end of the growing season in indeterminate annuals, these species may be said to senesce only if they show increased mortality during the growing season, before the environment deteriorates, or if they show a decline in survival in the absence of environmental changes. Moreover, it must be shown that the death of individuals in a population is not only age-dependent, but also independent of any changes in the biotic or abiotic environment. There are several other increases in mortality which are excluded from this definition of senescence, including many of the cases of 'cohort senescence' seen in fir-waves, when the phenotypic change is due to stand composition and any increases in mortality with increasing size, as in many trees and other species such as the saguaro cactus (Steenbergh & Lowe, 1983). Size dependent mortality, *per se*, cannot be termed senescence unless size differences are a consequence of age differences. If evidence can be found that trees and other species which show size dependent mortality also show age-specific physiological deterioration, for example changes in age-specific immunity to diseases, then these species could be said to senesce. There is no reason not to expect age-specific gene expression in trees, but this has not to date been investigated explicitly. A final group of plant species which may show no senescence are clonal plants. Age-specific gene action may not occur in species which have the ability to renew somatic tissues. In order for immortality of a genet to develop, the quality of ramets produced by different aged genets must be independent of the age of the genet, and newly produced ramets must become physiologically independent of parental tissues. If complete independence is not achieved, then age-dependent changes may occur in the older parental tissues, such as increased susceptibility to viral infection. These changes, because of their connectedness, may influence even the younger ramets and eventually bring about senescent declines. This may be the case in the tussock range grasses, where despite a 'modular' growth form, an increase in mortality with age was found (Canfield, 1957). If future studies show that senescence is common with tussock growth forms, it may be because the vegetative tillers remain connected to the parental plant (Harper, 1977).

Plants present excellent opportunities for comparative studies of species with and without senescence. Basic observations on gene action and age-specific selection should reveal fundamental differences in the evolution of the life histories of different species. The unique feature of plants is that a comparative analysis of species with different growth forms, and different expected levels of senescence, could be done with closely related species. A review of any local flora will reveal closely related species with annual vs. perennial life spans or clonal vs. non-clonal growth forms. Comparative work, particularly with plants, may also demonstrate the need for further expansion of the theories of senescence. Different plant species violate one or more of the assumptions of the theories, yet the consequences of violating these assumptions have never been investigated.

In addition to the opportunities for comparative study, there are several other reasons why plants are good organisms with which to test the evolutionary theories of senescence. One of the central controversies in the *Drosophila* senescence literature has revolved around experimental conditions which may present novel environments. Novel experimental conditions created by handling organisms, and even bringing them into the laboratory, will influence the expression of traits and the interrelationships between these traits (Service & Rose, 1985; Reznick, Perry & Travis, 1986; Clark, 1987). Field vertebrates must be captured and sampled, and there are additional problems of mark and recapture. Genetic studies with many animals are notoriously difficult. Plants, on the other hand, can be studied in their natural environments with minimal experimental disturbance. Furthermore, quantitative genetic studies can be done in the field in which the species are naturally found (cf. Roach, 1986b), thus many of the complications inherent with many animal populations can be avoided.

Through the use of minimal experimental manipulation, it should be possible to determine whether random mortality factors actually mask senescence in plant species which do not obviously show it. If this is the case, then there may be three categories of senescence with respect to plant species: first, those species which show senescence in wild populations; second, those species which have the potential to show senescence, but which for ecological reasons do not; and third, those species which do not senesce either in the wild or under protected conditions. The theoretical expectation is that the genetic constraints which may be found in the first and second species groupings will not be found in the third. Through the use of quantitative genetic studies in the field, it should be possible to do studies to test these expectations in comparative studies.

Future research on the evolution of senescence in plants should to be designed to answer several questions, as this review has pointed out. There is a need for good data on age-specific mortality patterns. Despite the large number of studies done on plant demography, there have in fact been very few studies which have followed individuals through to the end of their life cycles. Given the high level of random mortality in natural populations, large studies are required to detect subtle shifts in the slope of the survivorship curves. Beyond the basic demography, quantitative genetic approaches can also be used in natural populations to study senescence. There should be an attempt to obtain genotype-specific survivorship curves; these would indicate not only if senescence occurs in the populations, but also if there is variation for age-specific effects. With respect to the pleiotropic theory of senescence, is there evidence for antagonistic pleiotropic gene action between traits expressed at the different life stages? Except for the evidence from *Geranium* (Roach, 1986b), we have no direct evidence for this in plant species. In a comparative study, it would be useful to know how the genetic correlations between different life stages compares in senescent vs. apparently non-senescent plant species. An increasing level of additive genetic variance at the later life stages could be expected according to the mutation accumulation theory of senescence, although there may also be developmental reasons for different levels of variance at different life stages. Further information on age-specific gene action would also be useful, perhaps using joint molecular- and quantitative-genetic approaches.

The empirical work on plants to date has addressed assumptions of the senescence theories, but has not tested the theories themselves. The tractability of plants as experimental organisms suggests that it will be possible to directly test theories in the future. Senescence is an excellent illustrative example of non-adaptive life history evolution and the failure of natural selection to overcome genetic constraints. Future research on evolutionary plant senescence will help to explain not only senescence, but also the constraints on life history patterns in general.

Acknowledgements

I thank Janis Antonovics and David Reznick for valuable comments on earlier drafts of this manuscript.

References

Antonovics, J., 1972. Population dynamics of the grass *Anthoxanthum odoratum* on a zinc mine. J. Ecol. 60: 351-366.

Bell, G., 1992. Mid-life crisis. Evolution 46: 854-856.

Bell, G., M. J. Lechowicz & D. J. Schoen, 1991.

Bell, G., M. J. Lechowicz & D. J. Schoen, 1991. The ecology and genetics of fitness in forest plants III. Environmental variance in natural populations of Impatiens pallida. J. Ecol. 79: 697-713.

Bosch, C. A., 1971. Redwoods: A population model. Science 172: 345-349.

Bullock, S. H., 1980. Demography of an undergrowth palm in littoral Cameroon. Biotropica 12: 247-255.

Buss, L. W., 1987. The Evolution of Individuality. Princeton University Press, New York.

Canfield, R. H., 1957. Reproduction and life span of some perennial grasses of Southern Arizona. J. Range Management 10: 199-203.

Canode, C. L. & R. W. van Keuren, 1963. Seed production characteristics of selected grass species and varieties Wash. Agr. Exp. Sta. Bull. 647, 15 pp. Cited in Harper and White (1974).

Caswell, H., 1982. Stable structure and reproductive value for populations with complex life cycles. Ecology 63: 1223-1231.

Charlesworth, B., 1980. Evolution in Age-Structured Populations Cambridge University Press, Cambridge.

Clark, J. R., 1983. Age-related changes in trees. J. Aboric. 9: 201-205.

Clark, A. G., 1987. Senescence and the genetic-correlation hangup. Am. Nat. 129: 932-940.

Cook, R. E., 1983. Clonal plant populations. Am. Sci. 71: 244-253.

Enright, N. & J. Ogden, 1979. Applications of transition matrix models in forest dynamics: *Aravcaria* in Pupua, New Guinea and *Nothifagus* in New Zealand. Australian Journal of Ecology 4: 3-23.

Evans, D. W. & C. L. Canode, 1971. Influence of nitrogen fertilization and burning on seed production of Newport Kentucky Bluegrass. Agron. J. 63: 575-580.

Finch, C. E., 1991. Longevity, Senescence, and the Genome. University of Chicago Press, Chicago.

Franklin, J. F., H. H. Shugart & M. E. Harmon, 1987. Tree death as an ecological process. Bioscience 37: 550-556.

Goodrum, P. D., V. H. Reid & C. E. Boyd, 1971. Acorn yields, characteristics, and management criteria of oaks for wildlife. J. Wildl. Mgmt. 35: 520-532.

Hamilton, W. D., 1966. The moulding of senescence by natural selection. J. Theor. Biol. 12: 12-45.

Harcombe, P. A., 1987. Tree life tables. Bioscience 37: 557-568.

Harcombe, P. A. & P. L. Marks, 1983. Five years of tree death in a *Fagnus-Magnolia* forest, southeast Texas (USA). Oecologia 57: 49-54.

Harper, J. L., 1967. A Darwinian approach to plant ecology. J. Ecol. 55: 247-270.

Harper, J. L., 1977. Population Biology of Plants. Academic Press, London and New York.

Harper, J. L. & A. D. Bell, 1979. The population dynamics of growth form in organisms with modular construction, in Population Dynamics, edited by R. Anderson. Symposium of the British Ecological Society.

Harper, J. L. & J. White, 1974. The demography of plants. Annual Review of Ecology and Systematics 5: 419-463.

Hibbs, D. E., 1979. The age structure of a striped maple population. Can. 1980. J. For. Res. 9: 504-508.

Hubbell, S. P., 1980. Seed predation and the coexistence of tree species in tropical forests. Oikos 35: 214-229.

Janzen, D. H., 1976. Why bamboos wait so long to flower. Annual Review of Ecology and Systematics 7: 347-391.

Kirkwood, T. B. L. & R. Holliday, 1979. The evolution of aging and longevity. Proc. R. Soc. Lond. B205: 531-546.

Kirkpatrick, M., 1984. Demographic models based on size, not age, for organisms with indeterminate growth. Ecology 65: 1874-1884.

Kohyama, T. & N. Fujita, 1981. Studies on the *Abies* population of Mt. Shimagare. I. Survivorship curve. Bot. Mag. Tokyo. 94: 55-68.

Law, R., A. D. Bradshaw & P. D. Putwain, 1977. Life history variation in *Poa annua*. Evolution 31: 233-246.

Law, R., 1979. The cost of reproduction in annual meadow grass. Am. Nat. 113: 3-16.

Leopold, A. C., 1961. Senescence in plant development. Science 134: 1727-1732.

Leopold, A. C., 1975. Aging, senescence and turnover in plants. Bioscience 25: 659-662.

Leopold, A. C., 1981. Aging and senescence in plant development, pp. 1-12, in Senescence in Plants, edited by K. V. Thimann, CRC Press, Boca Raton, FL.

Leopold, A. C. & P. E. Kriedeman, 1975. Plant Growth and Development, 2d ed. McGraw-Hill, New York.

Leverich, W. J. & D. A. Levin, 1979. Age-specific survivorship and reproduction in *Phlox drummondii*. Am. Nat. 113: 881-903.

Lorimer, C. G. & L. E. Frelich, 1984. A simulation of equilibrium diameter distributions of sugar maple (*Acer sacchara-*

rum). Bull. Torrey Bot. Club 111: 193-199.

Medawar, P. B., 1952. An Unsolved Problem of Biology, H. K. Lewis, London.

Molish, H., 1938. The Longevity of Plants. Science Press, Lancaster, Pennsylvania.

Mueller-Dombois, D., 1983. Canopy dieback and successional processes in Pacific forests. Pacific Science 37: 317-325.

Nooden, L. D., 1988. Whole plant senescence, pp. 391-442 in Senescence and Aging in Plants, edited by L. D. Nooden and A. C. Leopold. Academic Press, San Diego.

Nooden, L. D. & A. C. Leopold, 1988. Senescence and Aging in Plants. Academic Press, San Diego.

Ogden, J., 1988. Forest dynamics and stand-level dieback in New Zealand's *Nothofagus* forests. Geojournal 17: 225-230.

Partridge, L., 1986. Sexual activity and life span, pp. 45-54 in Insect Aging, Strategies and Mechanisms, edited by K. G. Collatz and R. S. Sohal. Springer-Verlag, Berlin.

Pinero, D., M. Martinez-Ramos & J. Sarukhan, 1984. A population model of *Astrocaryum mexicanum* and a sensitivity analysis of its finite rate of increase. J. Ecol. 72: 977-991.

Reznick, D. N., E. Perry & J. Travis, 1986. Measuring the cost of reproduction: A comment on papers by Bell. Evolution 40: 1338-1344.

Roach, D. A., 1986a. Time of seed production and dispersal in *Geranianum*: Identifying the important components of fitness for an annual plant. Ecology 67: 572-576.

Roach, D. A., 1986b. Life history variation in *Geranium carolianum*. I. Covariation between characters at different stages of the life cycle. Am. Nat. 128: 47-57.

Rodriquez, R., R. Sanchez-Tames & D. J. Durzan, 1989. Plant Aging Basic and Applied Approaches. NATO ASI Series A: Life Sciences. Vol. 186.

Rose, M. R., 1991. Evolutionary Biology of Aging. Oxford University Press, New York.

Sarukhan, J., 1980. Demographic problems in tropical systems, pp. 161-188 in Demography and Evolution in Plant Populations, edited by O. T. Solbrig. Blackwell, Oxford.

Sarukhan, J. & J. L. Harper, 1973. Studies on plant demography: *Ranunculus repens* L., *R. bulbosus* L. and *R. acris* L.: I. Population flux and survivorship. J. Ecology 61: 675-716.

Sato, K. & Y. Iwasa, 1993. Modelling of wave regeneration (shimagare) in subalpine *Abies* forest: population dynamics with spatial structure. Ecology (in press).

Savage, Z., 1966. Citrus yield per tree by age. Fla. Agr. Exp. Sta. Agr. Exten. Serv. Econ Ser. 66-3, 9 pp. Cited in Harper and White (1974).

Schaffer, W. M. & M. D. Gadgil, 1975. Selection for optimal life histories in plants, pp. 142-156 in Ecology and Evolution of Communities, edited by M. L. Cody and J. M. Diamond. Belknap Press, Cambridge, Mass. and London England.

Service, P. M. & M. R. Rose, 1985. Genetic covariation among life-history components: The effect of novel environments. Evolution 39: 943-945.

Silvertown, J. W., 1982. Introduction to Plant Population Ecology. Longman, London.

Sprugel, D. G., 1976. Dynamic structure of wave-regenerated *Abies balsamea* forests in the north-eastern United States. J. Ecol. 64: 889-911.

Stark, R. H., A. L. Hafenrichter & K. H. Klages, 1949. The production of seed and forage by mountain brome as influenced by nitrogen and age of stand. Agron. J. 41: 508-512.

Steenbergh, W. F. & C. H. Lowe, 1983. Growth and demography. Ecology of the Saguaro: III. Monograph series, no. 8. Washington, DC: U.S. Government Printing Office. Cited in Finch (1991).

Stewart, G. H. & T. T. Veblen, 1983. Forest instability and canopy tree mortality in Westland, New Zealand. Pacific Science 37: 427-431.

Tamm, C. O., 1956. Further observations on the survival and flowering of some perennial herbs. I. Oikos 7: 273-292.

Tamm, C. O., 1972a. Survival and flowering of perennial herbs. II. The behaviour of some orchids on permanent plots. Oikos 23: 23-28.

Tamm, C. O., 1972b. Survival and flowering of perennial herbs. III. The behaviour of *Primula veris* on permanent plots. Oikos 23: 23-28.

Thimann, K. V., ed., 1980. Senescence in Plants. CRC Press, Boca Raton, Florida.

Vasek, F. C., 1980. Creosote bush: long-lived clones in the Mojave desert. Am. J. Bot. 67: 246-255.

Wangermann, E., 1965. Longevity and ageing in plants and plant organs. Encyclopedia Plant Physiol. 15: 1026-1057.

Wareing, P. F. & I. D. J. Phillips, 1981. Growth and Differentiation in Higher Plants. Pergamon.

Watkinson, A. R., 1992. Plant senescence. TREE 7: 417-420.

Watkinson, A. R. & J. White, 1986. Some life-history consequences of modular construction in plants. Phil. Trans. Roy. Soc. (London) Ser. B. 313: 31-51.

Werner, P. A. & W. J. Platt, 1976. Ecological relationships of co-occurring golden rods (*Solidago*: Compositae). Amer. Nat. 110: 959-971.

Williams, G. C., 1957. Pleiotropy, natural selection, and the evolution of senescence. Evolution 11: 398-411.

Yanofsky, M. F., H. Ma, J. L. Bowman, G. N. Drews, K. A. Feldmann & E. M. Meyerowitz, 1990. The protein encoded by the *Arabidopsis* homeotic gene agamous resembles transcription factors. Nature 346: 35-39.

Zajaczkowska, J., A. Lotocki & H. Morteczka, 1984. Gas exchange in yew cuttings from trees of different ages. Sylwan 128: 25-29.

M.R. Rose and C.E. Finch (eds.), Genetics and Evolution of Aging, 83–95, 1994.
© 1994 *Kluwer Academic Publishers. Printed in the Netherlands.*

Comparing mutants, selective breeding, and transgenics in the dissection of aging processes of *Caenorhabditis elegans*

Thomas E. Johnson, Patricia M. Tedesco & Gordon J. Lithgow
Institute for Behavioral Genetics, University of Colorado, Boulder, CO 80309, USA

Received and accepted 22 June 1993

Key words: longevity, nematodes, senescence, genetic transformation

Abstract

The genetic analysis of aging processes has matured in the last ten years with reports that long-lived strains of both fruit flies and nematodes have been developed. Several attempts to identify mutants in the fruit fly with increased longevity have failed and the reasons for these failures are analyzed. A major problem in obligate sexual species, such as the fruit fly, is the presence of inbreeding depression that makes the analysis of life-history traits in homozygotes very difficult. Nevertheless, several successful genetic analyses of aging in *Drosophila* suggest that with careful design, fruitful analysis of induced mutants affecting life span is possible. In the nematode *Caenorhabditis elegans*, mutations in the *age-1* gene result in a life extension of some 70%; thus *age-1* clearly specifies a process involved in organismic senescence. This gene maps to chromosome II, well separated from a locus (*fer-15*) which is responsible for a large fertility deficit in the original stocks. There is no trade-off between either rate of development or fertility versus life span associated with the *age-1* mutation. Transgenic analyses confirm that the fertility deficit can be corrected by a wild-type *fer-15* transformant (transgene); however, the life span of these transformed stocks is affected by the transgenic array in an unpredictable fashion. The molecular nature of the *age-1* gene remains unknown and we continue in our efforts to clone the gene.

Introduction

Three distinct strategies have been used to obtain genetic variants that have increased life span. These strategies are selective breeding, mutant induction, and the construction of transgenic animals. These genetic strategies, as well as providing long-lived animals for *in vivo* analysis, allow the study of the mechanistic processes involved in aging and senescence at the cellular or molecular level. These approaches, when joined with interspecies comparisons, constitute the range of genetic 'tricks' that can be brought to bear in the understanding of the mechanistic basis of aging and senescence. Of course the strongest approaches combine more than one method. For example, the best method of studying cellular function or development combines an unbiased, single-gene, mutational approach with

transgenic methods for studying disrupted gene function (Watson *et al.*, 1992).

Advantages of mutational analysis. The mutational approach has significant advantages over both selective breeding and transgenic analyses. A major advantage of mutagenesis is that the results are unbiased by the expectations of the investigator; any mutational event that results in viable offspring and leads to an increase in life span can be detected. Moreover, unlike selective breeding, mutational analysis is not limited to allelic variants already present in the population (Johnson, 1988); mutations can be induced in any gene, even those fixed in the species. Thus, the identification of these mutants is limited only by the ingenuity of the investigator and breadth of the screen. Also, unlike transgenic approaches, there is neither bias as to which genes are picked for analysis nor limitation based on the availability of cloned genes. The identification of single-gene mutants affecting some aspect of the aging processes and subsequent study of these processes, by means of coupled *in vitro* manipulation and *in vivo* analysis, thus represents the ideal strategy for finding genes that are important in aging.

Selective breeding. The selective breeding approach has been successfully applied in *Drosophila melanogaster* by two different laboratories (Rose, 1984; Luckinbill *et al.*, 1984; see also Fleming *et al.*, this volume). This strategy has been effective in producing strains with increased maximum life span and has uncovered a plethora of tantalizing associations between increased longevity and other physiological traits, including early ovary weight, early fecundity, total body weight, lipid content, desiccation resistance, ethanol resistance, starvation resistance, and flight duration (summarized in Hutchinson & Rose, 1990). All of these characteristics uniformly differentiated the five long-lived populations of Rose (1984). Increased early female fertility (Luckinbill *et al.*, 1984), cessation of female reproduction (Arking, 1988), flight duration (Luckinbill *et al.*, 1988), later age of reduced amino acid incorporation (Pretzlaff & Arking, 1989), prolonged phototactic and geotactic abilities (Arking & Wells, 1990), resistance to paraquat (Arking *et al.*, 1991), better survival after male mating (Luckinbill *et al.*, 1988), altered glucose-6-phosphate dehydrogenase isozymes (Luckinbill *et al.*, 1990), and altered glycogen levels (Graves *et al.*, 1992) have all been associated with increased longevity in the less highly replicated studies of selected lines of Luckinbill *et al.* (1984). Allelic variants at the superoxide dismutase (SOD) locus (Rose *et al.*, 1990) and in unknown proteins detected by 2D gel electrophoresis (Fleming *et al.*, this volume) have also been tentatively associated with increased life spans. Attempts to reconstruct increased-life strains by breeding for a high activity SOD allelic variant or by constructing transgenic flies using SOD have met with no or little success (Rose, 1993, and personal communication).

Thus the two successful selections for increased life span have not yet identified candidate genes that can explain much of this increase in longevity. For example, Rose estimates that only a few per-

cent of the increase in life span in his lines results from altered allele frequencies at the SOD locus (Rose, personal communication). Most unfortunately, the extended life span of these strains seems to be controlled by many, perhaps even hundreds, of genes (Hutchinson & Rose, personal communication) that may be extending life through general effects stemming mainly from heterosis. Thus, the identification of individual genes associated with even marginal increases in life span is quite difficult using selected lines of *Drosophila*. Because of the highly outbred nature of the strains, there seems to be little expectation that quantitative trait loci (QTLs) can be mapped using traditional approaches (Rose, personal communication), although Fleming *et al.* (this volume) propose a clever alternative method for identifying significant changes. Despite these drawbacks, both selection studies in *Drosophila* (Arking *et al.*, 1991; Rose *et al.*, 1990) suggest that genes coding for proteins involved in protection against free-radicals (e.g. SOD) have been differentially selected in long-lived lines.

Several problems potentially affecting the validity of these selective breeding studies in *Drosophila* were raised at a recent meeting on the molecular biology of aging at Cold Spring Harbor. It seems clear that selective breeding, while successful in producing stocks with longer life spans than the parents, may not have successfully produced stocks that are longer-lived than the best laboratory wild-type stocks or lines that were newly derived from nature. Careful studies comparing life spans of the selectively-bred lines and those of laboratory stocks displaying various longevities should be carried out under the same laboratory conditions to determine if the long-lived strains are longer lived than laboratory stocks under identical environmental conditions. It also seems possible that the increased life of these selected *Drosophila* strains could result from a self-imposed dietary restriction effect.

Long-life mutants in Drosophila. Unfortunately, all attempts to identify mutants in *Drosophila* with extended life spans have been unsuccessfull. Before going on to describe studies that have successfully identified mutants with extended longevity in the nematode, the failure to detect mutant genes that lead to longer life span (gerontogenes) in

Drosophila should be addressed in more detail. Roberts and Iredale (1985) reported the only (unsuccessful) attempt to identify *Drosophila* mutants with increased longevity. Their selection was indirect and was based on selection for delayed loss of climbing ability. The authors concluded that there may be no such mutants in *Drosophila*. Indeed, their lack of success could result from a lack of such loci in the fruit fly. However, several other explanations are also possible: (1) the selected traits and life span could have no genetic correlation, (2) multiple mutations with deleterious effects on life span could have been induced at the same time thus masking the desired mutant phenotypes, or (3) inbreeding depression resulting from producing homozygotes for these recessive mutant alleles could mask the long-life phenotype. Indeed it is clear that long-life mutants exist in *Drosophila*; for example, a female sterile mutant of *D. subobscura* has increased longevity (Maynard Smith, 1958). This suggests that similar mutations in *D. melanogaster* (e.g. St Johnson & Nusslein-Volhard, 1992) were not detected. Thus, the screen used by Roberts and Iredale (1985) leaves much to be desired.

The possibility that mutational analysis in a carefully inbred laboratory stock could identify mutant alleles leading to increased life span has not been carefully explored in *Drosophila* and is usually dismissed out of hand based on the short life spans of inbred laboratory strains. Nevertheless, the now classic studies of Gould and Clark (1977) and the more recent studies of Leffelaar and Grigliatti (1984) show that mutations leading to shorter than normal life can be studied in an inbred background. Rose (1991) has argued that such studies are worthless because of epistatic effects resulting from inbreeding of recessive deleterious mutant alleles, but similar stocks are being used for *Drosophila* transgenic analysis which may be even more prone to problems resulting from epistasis (Stearns, Kaiser & Hillesheim, 1993).

Drosophila as a model for mammalian aging. In the absence of any additional way to ascertain causality of increased life span in *Drosophila*, any attempt to use similar protocols for selective breeding to postpone age at death in selected mouse lines may be premature. Indeed, evidence collected here at IBG suggests that there is little or no correlation between age of individual females at loss of fertility

and the age at death of these animals; (Johnson, unpublished). Mice undergo 'menopause' midway through life, strongly contrasting with *Drosophila* where females remain reproductive until a few days before death (Rose, 1984; Arking, 1987). The mid-life menopause that is observed in most or all mammals (Finch, 1990) contrasts dramatically with the reproductive schedule of fruit flies (Rose, 1984; Arking, 1987). Thus, there is significant reason to doubt the validity of the fruit fly as a model for mammalian life-histories. Indeed, largely untested evolutionary theory suggests that there should be little cross-species conservation in genetic specification of senescence processes (Rose, 1991) and one might think that this would be especially true when there were dramatic differences in life-history. A discussion of possible approaches to identifying gerontogenes through the production of long-lived strains of mammals appeared as a series of essays (Charlesworth, 1988; Harrison, 1988; Johnson, 1988; Rose, 1988) and a subsequent editorial (Rose, 1990).

Review of research in C. elegans *on the genetics of aging*

Caenorhabditis elegans is a self-fertilizing, hermaphroditic nematode species. It can be grown in Petri plates on a simple medium and diet of *Escherichia coli*. It has been widely used as an experimental organism in analyses of development, muscle physiology, and behavior. The cell lineage is almost invariant, which has facilitated the description of the complete somatic cell lineages of the hermaphrodite and male. The small size of the animal (1.2 mm length), its 2-day period of development, and its optical transparency also facilitate the analysis of cell lineage. These and other areas pertaining to *C. elegans* biology are covered in a comprehensive review (Wood, 1988).

Genetic tools. Simple techniques for classical and molecular genetic analysis as well as more sophisticated molecular genetic tools have been developed (Wood, 1988). The isolation of mutants is straightforward in *C. elegans*, and a large number of mutants have been identified after treatment with ethyl methanesulphonate (EMS) or other mutagens (Herman, 1988). The recent beginnings of quantitative and population genetic analyses prove that unique

Table 1. Advantages of *C. elegans* for the genetic analysis of aging.

- Small (1.2 mm)
- Rapid life cycle (< 3 days)
- Short life span (20 days)
- Self-fertilizing hermaphrodite
- Easy isolation of recessive mutants
- Lack of inbreeding depression
- Spontaneous males (obligate outcrossers)
- Dauer larvae (> 90 day survival)
- Cryogenic preservation of strains
  - no loss of mutants
  - no accumulation of modifiers or suppressors
  - all stocks in same genetic background
- Genetic transformation methods available
- Transposon-mediated mutagenesis
- Physical map essentially complete: cosmid and YAC arrays
- Genome sequencing in progress
- Many laboratories
- Many types of mutants (developmental, behavioral, longevity, etc.)

---

applications are available in these areas as well (Brooks & Johnson, 1991; Van Voorhies, 1992; Johnson & Hutchinson, in press). Molecular genetic analysis of this species has progressed rapidly such that *C. elegans* is a principal model for the human genome project (Sulston *et al.*, 1992; Olson, 1992). A method for inducing mutations with the Tc1 transposable element has facilitated the cloning of genes identified only by mutations (Herman, 1988). Efficient techniques for genetic transformation (Mello *et al.*, 1991) and the availability of overlapping arrays of cosmids that cover as much as 95% of the genome (Coulson *et al.*, 1986; 1988) offer possibilities for genetic analysis almost unrivaled by any other metazoan species.

There are several additional advantages in the use of *C. elegans* for genetic analyses of aging and senescence as listed in Table 1. Two principal advantages, easy identification of animals carrying recessive mutant alleles and a lack of inbreeding depression, result from the fact that *C. elegans* is a self-fertilizing hermaphrodite. As a consequence, recessive mutations can be isolated with relative ease. This facet, perhaps more than any other, has been important in the identification of long-lived mutants (Klass, 1983) and in their subsequent analysis (Friedman & Johnson, 1988 a, b: Johnson, 1990).

Characterization of length of life. Interestingly, several different wild-type strains (Johnson, 1984; Johnson *et al.*, 1990; Johnson & Hutchinson, in press) have fairly similar life expectancies, maximum life spans, and survival curves, consistent with a functional significance for length of life or a tight relationship between length of life and some other life-history traits. A number of different types of mutants have been identified in *C. elegans*. These include morphological variants such as short, squat 'dumpies' (Dpy), longer than normal (Lon), etc.; behavioral variants such as uncoordinated (Unc), worms that roll instead of swimming smoothly (Rol), etc.; and more unusual mutants such as temperature-sensitive lethals which can affect almost any stage of life and transformers that cause strains of one sexual genotype to masquerade as the other sex. In a preliminary survey seeking strains that had normal life expectancy – and could therefore be used to map variants with longer life – we discovered that many mutants (30%) have life expectancies not significantly different from wild type (Johnson, 1984; Johnson *et al.*, 1990).

RI strains. Recombinant inbred (RI) strains were derived by crossing the N2 (18.2 day mean life span) and Bergerac (16.6 day mean life span) strains (Johnson & Wood, 1982). These RI lines have been assayed several times for length of life. Mean life spans of these lines vary three-fold, ranging from 13.8 days to 37.9 days (Johnson, 1987). More importantly, maximum life spans of the RI lines are also altered; maximum life spans both shorter (17 days) and up to 63% longer (63 days) than N2 (40 days) are observed among the RI lines. As expected, there are strong positive correlations between mean life span and the 90th percentile, the 95th percentile, or maximum life span (Johnson, 1987). These strong positive correlations are not trivial, because a higher mean life span can result either from decreased early life mortality or from increased maximum life span. Thus, in these lines, longer life results from an increase in life expectancy at all chronological ages.

The shape of the survival curve was examined in

detail in four selected RI lines and in the two parental genotypes (Johnson, 1987). The exact shape of the survival curves varies slightly between lines, but all are rectangular. We undertook a more detailed analysis of the kinetics of mortality to see if mortality increases exponentially with chronological age as modeled by the Gompertz equation. In both wild-type and RI strains, the age-dependent component increases exponentially with increasing chronological age. This is most clearly seen by plotting mortality rate against chronological age on a semilogarithmic scale (Johnson, 1987). Because increased mean life span could result from lower basal mortality rate or from a slower rate of increase in mortality with chronological age, we asked how these components varied in the RI lines. The age-dependent component varies between lines and explains most of the variance in length of life (Johnson, 1987). No significant change in the age-independent component was observed.

Co-inheritance of fertility and life span. Theoretical models for the evolution of senescence (Rose, 1991) predict the possible co-inheritance of fertility and life span in these recombinant inbred lines. Both life expectancy and hermaphrodite fertility showed significant heritability in each of three trials (Brooks & Johnson, 1991). However, no significant phenotypic or genetic covariance for life span and fertility was observed. Age-specific fertility was significantly correlated with fertility on consecutive days at the phenotypic level and also often significantly correlated at the genetic level. Correlations for fertility between more distant days were usually negative but not significant. No negative trade-offs between fertility and life span were observed. We are currently completing a project mapping QTLs for life span and fertility to chromosomal regions; negative associations between fertility levels and life span were, in general, not observed (Brooks & Johnson, in preparation).

Three to five independently segregating genes are estimated to be specifying these traits within the RI lines. Two single-gene markers were used to generate strain distribution patterns and were found to be associated with a statistically significant effect on life span and/or fertility. There was also evidence for a significant environmental component affecting fertility and length of life, which leads us to be cautious about generalizing the observed positive covariances to environments encountered in the wild.

Identification and mapping of age-1 mutations

The hx542 allele is a recessive mutation in the *age-1* gene. It was induced by EMS in a *C. elegans* strain already carrying the *fer-15* (*b26*) mutation which blocks fertility at 25°, thus facilitating the screen for mutants with increased life span (Klass, 1983). The *hx542* allele results in a 40% increase in life expectancy and a 60% increase in maximal life span at 20°C; at 25°C, *age-1* (*hx542*) averages a 65% increase in mean life span and a 110% increase in maximum life span (Friedman & Johnson, 1988a). Mutant males show extended life spans to a lesser degree (Johnson, 1990). The *age-1* mutant strains derived by Klass also displayed a 75% decrease in hermaphrodite self-fertility at 20° that was not observed in the DH26 parental strain, in which the *hx542* allele was originally identified (Klass, 1983; Friedman & Johnson, 1988b).

The formal possibility exists that the life-extension effects of *age-1* (*hx542*) result from 'self-imposed' dietary restriction. A series of studies on the behavior of these strains make it clear that the long-lived mutant strains do not ingest less food than wild type (Johnson *et al.*, 1990) nor is their behavior significantly different from wild type (Friedman & Johnson, 1988b; Johnson *et al.*, 1990; Duhon & Johnson, in preparation). Pharyngeal pump rates (an indirect assessment of rate of food ingestion) and rates of uptake of radiolabelled food were comparable for the long-lived stocks and for wild-type controls (Johnson *et al.*, 1990). Moreover, if dietary intake is restricted by lowering bacterial concentration, an additional and comparable extension of life span is added onto both *age-1* and wild-type strains (Johnson *et al.*, 1990; Johnson, unpublished). These findings show that the processes acting to increase life span in the *age-1* mutant strains, whatever their nature, function independently of, that is to say in addition to, the life-extension effects of dietary restriction.

Initial mapping of age-1. We followed the segregation of *age-1* in crosses back to wild type and found that the long-life (Age) and reduced-fertility at 20°C (here referred to as 'Brd' for <u>br</u>ood size) phenotypes cosegregated (Friedman & Johnson,

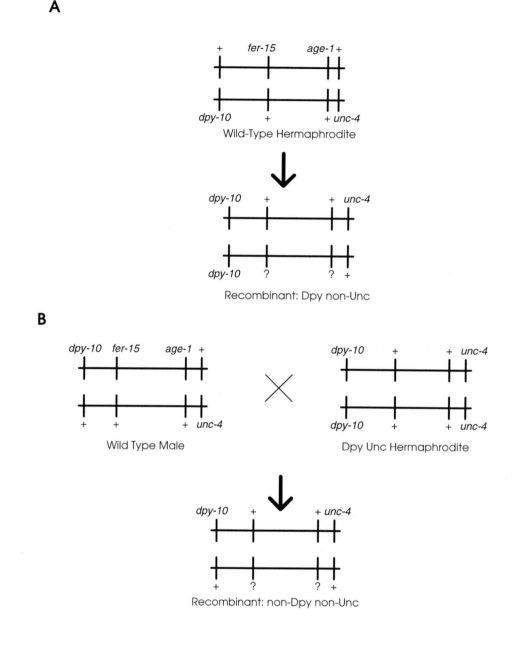

Fig. 1. Genetic strategies for detecting recombinants in *C. elegans*. A. Typical strategy. A phenotypically wild-type hermaphrodite, heterozygous for the gene(s) of interest and for selectable markers on either side, is generated by crossing appropriate parents and is allowed to self fertilize. Recombinant progeny are detected as being Unc non-Dpy or Dpy non-Unc and are cloned and allowed to self fertilize, producing some animals that are homozygous for the recombinant chromosome which are cloned and analyzed for the gene(s) of interest. B. Novel mapping scheme used herein to generate recombinants around *age-1*. Note that the markers (*dpy-10* and *unc-4*) are in trans and that recombination occurs in males. Recombinant progeny are identified by crossing heterozygous males with homozygous *dpy-10 unc-4* hermaphrodites and picking non-Unc non-Dpy F$_1$ progeny which represent one class of recombinants. These worms segregate the recombinant chromosome which is subsequently identified and analyzed in homozygous form among self progeny. The reciprocal class of recombinants is identified by using a configuration where *age-1* and *fer-15* are *cis* to *unc-4*. Thus, all recombinants are behaviorally and morphologically normal (non-Unc and non-Dpy).

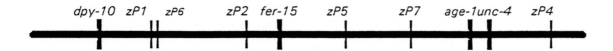

Fig. 2. Genetic map resulting from multi-point mapping scheme. Parental strains were crossed using a scheme essentially as shown in Fig. 1B except that one parent contained additional markers derived from another wild strain (Bergerac BO) that contains many additional markers detectable at the DNA level. Several additional markers are composed of Tc1 elements which are shown on the map as *zP1-zP7* (Johnson, Hutchinson, Lithgow & Lindsay, manuscript in preparation).

1988a). This is consistent with any of three models: (1) *age-1* may be a further mutation in *fer-15* that now affects both fertility and life span; (2) *age-1* may be tightly linked to *fer-15* and the life span and fertility characters are both determined by *age-1*; or (3) *age-1* and *fer-15* are different genes and *age-1* alone affects life span while *fer-15* affects fertility. These models were distinguished through additional mapping strategies.

Fine structure mapping of age-1. Two methods are available for fine-structure mapping in *C. elegans* (Herman, 1988): multi-point crosses and deficiency analysis. Recombinants generated using the *trans*-crossing scheme usual to *C. elegans* genetic crosses (Brenner, 1974) are easily identified as being either phenotypically Dpy but non-Unc, or Unc but non-Dpy (Fig. 1A). Using such a scheme, Brd cosegregated with *fer-15* (*b26*), but mapping for the Age

phenotype was ambiguous (Johnson *et al.*, 1990). This inconsistency in mapping could have resulted from the variability between assays, since they were completed over an 18-month period, or could have resulted from problems with marker effects of *dpy-10* (*e120*) and/or *unc-4* (*e124*).

To eliminate both problems, further assays were completed in a short period of time and a novel scheme for generating recombinants that are morphologically and behaviorally wild type was employed (Fig. 1B). This approach led to the assignment of *age-1* to a region well to the right of *fer-15*, nearer to *unc-4* (Johnson, Hutchinson, Lithgow & Lindsay, in preparation). This mapping assignment was unambiguous and was aided by the presence of multiple Tc1 elements that were segregating within the cross. Approximate positions of the genes and the relevant Tc1 elements are shown in Fig. 2.

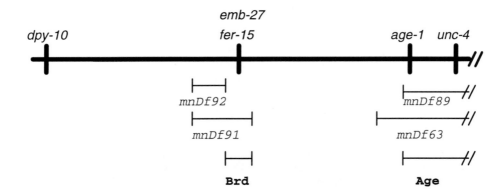

Fig. 3. Summary of results from deficiency mapping of *fer-15* and *age-1* (Johnson, Hutchinson & Tedesco, submitted). Some deficiences are shown below the genetic map. Two classes of deficiencies map *fer-15* to the center of linkage group II and are represented on the map by *mnDF91* and *mnDF92*. Heterozygotes for *mnDF91/fer-15* (*b26*) have lower fertility while heterozygotes for *mnDF92/fer-15* (*b26*) have fertility levels that are essentially wild type. Similarly *mnDF63/age-1* or *mnDF89/age-1* heterozygotes have life spans that are characteristic of *age-1* homozygotes.

Deficiency mapping. An alternate approach, deficiency mapping, takes advantage of a series of deficiencies that collectively cover most of the region between *dpy-10* and *unc-4* (Sigurdson, Sparier & Herman, 1984). One class of eight deficiencies (*mnDF91* and seven others) deletes the *fer-15* gene while a second class of three deficiencies (*mnDF92* and two others) does not (Fig. 3). Deficiency maps are constructed by making strains that are heterozygous for the deficiency and the mutation being studied and determining whether the heterozygote is mutant or normal (Johnson *et al.*, 1990). All eight of the deficiencies that fail to complement (i.e. delete) *fer-15* also fail to complement Brd and the three deficiencies that complement *fer-15* also complement Brd. Both deficiency classes complement *age-1* (Johnson, Hutchinson & Tedesco, manuscript submitted). Thus, these deficiencies define a genetic region containing *fer-15* and the Brd phenotype, but not *age-1*. Deficiency heterozygotes for either *mnDf63* or *mnDF89* are long-lived and map *age-1* to a position near *unc-4*. The *fer-15* gene was positioned to the physical map by identifying DNA polymorphisms associated with the end points of most of the first two classes of deficiencies (Johnson, Hutchinson & Tedesco, submitted). We have identified the right-hand end point of ten of eleven deficiencies and the left-hand end point of five of six deficiencies examined and have mapped the positions of these end points using cosmids from a contig which covers much of chromosome II. The cosmid T13F9 contains the *fer-15* gene and also reveals a DNA polymorphism that cosegregates with the *fer-15* (*b26*) mutant allele in Southern blots (Johnson, Hutchinson & Tedesco, submitted).

Transgenic lines

The method for constructing new strains of *C. elegans* by transformation with recombinant clones involves microinjection into the syncytial hermaphrodite gonad. The injected DNA forms a semi-stable, high-molecular-weight, extra-chromosomal tandem array that is packaged into a minichromosome during oogenesis (Stinchcomb *et al.*, 1985; Mello *et al.*, 1991). Current methodology involves detection of the transformation events by coinjection with pRF4, a plasmid carrying the dominant behavioral marker, *rol-6* (*su1006*), which causes transformed animals to exhibit a 'rolling' behavior (Rol), instead of normal sinusoidal movement (Mello *et al.*, 1991).

Over a period of some 12 months almost 1000 worms have been injected in our lab. About 270 of the injected animals produced one or more Rol transformants among their F_1 progeny and 64 independent lines produced Rol F_2 progeny that could be analyzed. To confirm the genetic separation of *fer-15* and *age-1*, we injected TJ411 [*age-1* (*hx542*) *fer-15* (*b26*)] hermaphrodites with the T13F9 cosmid predicted to complement *fer-15* (*b26*) but not *age-1* (*hx542*). Some of these lines were analyzed for complementation of the *b26* Ts and Brd phenotypes (Table 2) and for effects of these transformation events on life span (Table 3).

Although all transformed lines carrying T13F9 complemented the Ts character of *fer-15* (*b26*) at 25°, the fertility deficit was only weakly complemented, if at all, at 20°. Moreover, the life spans of the transgenic strains were in many cases dramatically shorter than life spans normal for *age-1* mutants (Table 3). Life spans ranged from 27 days to 35 days for non-integrated strains. These studies differed from earlier controls where pRF4 was injected alone; there smaller or non-significant effects on life span were detected (Fig. 4). Since the extragenic DNA arrays in transformed lines are somewhat unstable, these lines were also examined after generating 'integrated' lines where the extra-chromosomal array was integrated randomly into the genome using *v*-irradiation (Mello *et al.*, 1991). Several of these lines now contained completely stable (presumably integrated) arrays; stable lines still produced 10% non-Rol phenocopies which invariably threw Rol progeny when cloned. These integrated lines were much better in complementing the 20° fertility-deficit of *fer-15* (*b26*) (Table 2), but significantly shortened life span (Table 3). That these lines, both integrated and non-integrated, contained transforming DNA was confirmed by probing blots of total genomic DNA with the T13F9 cosmid originally used to inject the worms. This probe detects a polymorphism associated with the *b26* allele (Fig. 5).

The variability in life span and fertility of transgenic strains may result from the variable size of the tandem arrays that are formed after injection and from variability in the amount and form of the DNA in the array. Integrated transgenic strains

Table 2. Fertility of Transformants[a].

Strain (Genotype)	Number of progeny					
	20			25		
	Mean	SD	N	Mean	SD	N
N2	223.2	34.2	10	134.9	41.2	9
TJ1052 (*age-1*)	199.5	43.1	15	113.9	13.7	10
TJ411 (*age-1 fer-15*)	33.9	16.4	15	0.1	0.3	10
CH1035 (*rol-6*)	176.2	72.7	14	113.5	43.6	10
Non-integrated transgenic strains[b]						
TJ411 + pRF4	36.8	14.9	18	0.3	0.7	8
TJ411 + pRF4 + pJS303	20.8	12.3	14	0	0	10
TJ411 + pRF4 + pJS303 + T13F9 (4.2)	35.3	19.4	11	1.1	2.5	10
TJ411 + pRF4 + pJS303 + T13F9 (9.17)	38.1	30.7	12	1.0	2.0	9
TJ411 + pRF4 + pJS303 + T13F9 (16.3)	50.5	32.5	14	3.4	6.8	10
TJ411 + pRF4 + pJS303 + T13F9 (16.7)	42.9	31.2	15	3.4	7.3	7
TJ411 + pRF4 + pJS303 + T13F9 (116.5)	29.1	13.0	14	0	0	9
TJ411 + pRF4 + pJS303 + T13F9 (116.9)	37.9	25.7	14			
Integrated transgenic strains[c]						
TJ411 + pRF4 + pJS303 + T13F9 (2251)	170.4	86.6	11	31.0	40.5	5
TJ411 + pRF4 + pJS303 + T13F9 (2253)	122.1	39.4	10	10.8	18.8	10
TJ411 + pRF4 + pJS303 + T13F9 (2312)	36.9	48.2	14	6.9	9.3	8
TJ411 + pRF4 + pJS303 + T13F9 (2313)	28.9	30.5	8	5.0	10.4	10
TJ411 + pRF4 + pJS303 + T13F9 (2315)	44.5	38.5	10	7.0	13.0	10

[a] Transgenic strains carry either extrachromosomal or integrated arrays of one or more distinct plasmids and or cosmids.

[b] Various transformants carried pRF4, pJS303 and the T13F9 cosmid, as described in the table. pRF4 contains the *rol-6* (dominant roller) marker; pJS303 is a plasmid that may increase the stability of lines; T13F9 is a cosmid carrying *fer-15*. Four independent arrays are represented (classes 4, 9, 16, and 116) that result from distinct injections; strains 16.3 and 16.7 may not be independent and 116.5 and 116.9 may not be independent.

[c] Stains were identified after γ-ray-induced integration of the extragenic array in strain 16.7 and produced all genetically Rol progeny. Class 22 and class 23 are independent events; subclasses are not independent.

Fig. 4. Mean life span and standard errors of various strains involved in the transgenic analyses. The plasmid pRF4, which contains a dominant roller allele of *rol-6*, was injected into TJ411 [*fer-15(b26) age-1 (hx542)*]. Mean life spans and standard errors were 20.3 ± 2.4 days for N2 (+), 38.0 ± 2.4 days for uninjected TJ411 (*hx542*), 27.2 ± 2.9 days for TJ411 transformed with pRF4 but which are not phenotypically Rol anymore, and 30.0 ± 3.7 days, 35.7 ± 3.4 days, and 29.8 ± 3.5 days for three different transgenic lines transformed with pRF4 (*age-1 + pRF4*). All of the strains were significantly longer lived than N2 ($p < .05$) but none of the transgenics were significantly different from TJ411.

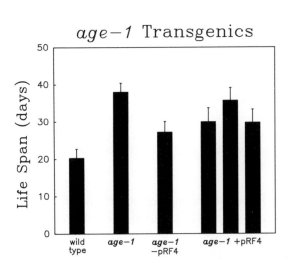

age−1 Transgenics

Table 3. Life Span of Transformants[a].

Strain (Genotype)	Life span (days)			P	
	Mean	SEM	N	N2	TJ1052
N2	24.8	1.9	38	–	–
BA713	27.0	1.3	41	.327	.022
CH1035 (*rol-6*)	13.7	2.0	16	<.001	<.001
TJ1052 (*age-1*)	33.2	2.0	26	.010	–
TJ411 (*age-1 fer-15*)	23.4	1.3	38	.735	<.001[b]
	Non-integrated transgenic strains[c]				
TJ411 + pRF4 + pJS303	35.2	2.5	26	.002	.368
TJ411 + pRF4 + pJS303 + T13F9 (4.2)	33.8	2.5	28	.005	.664
TJ411 + pRF4 + pJS303 + T13F9 (16.7)	27.6	4.7	11	.301	.019
TJ411 + pRF4 + pJS303 + T13F9 (116.5)	30.2	3.1	23	.142	.710
	Integrated transgenic strains from 16.7[c]				
2251[d]	17.1	3.1	13	.023	<.001
2253	12.1	1.7	19	<.001	<.001
2315	12.2	2.1	16	<.001	<.001

[a] All experiments were conducted at 20 ° using standard survival conditions (Johnson & Wood, 1982). Survivals were compared using the modified Kruskal-Wallis statistic as implemented in version 4.0 Advanced Statistics for SPSS/PC+ calculated according to the algorithm of Lee and Desu (1972).
[b] TJ411 failed to live longer in these tests and was recovered from frozen stocks subsequent to these assays. TJ1052 is therefore used as the long-lived control for statistical comparisons.
[c] All transformants carried pRF4, pJS303 and the T13F9 cosmid, as described in the text. Strain nomenclature follows that in Table 2.
[d] Both Rol and non-Rol individuals were identified and followed in separate populations until death. No statistically significant differences in survival were observed between Rols and non-Rols.

clearly can express genes at higher levels than are seen in the non-integrated lines, but this expression may be associated with effects on longevity which may result from inappropriate levels or patterns of expression. Another possibility is that inappropriate expression of other genes on the T13F9 cosmid lead to the reduction in life span. These possibilities remain to be distinguished by other analyses. One obvious way around these problems is to develop methodology for detecting rare homologous recombination events as is done in humans ES cells (Capecchi, 1989); several laboratories are currently experimenting with ways to generate and detect such recombinants in *C. elegans*.

Effects on other physiological and developmental traits

The length of developmental periods and the length of the reproductive period are unrelated to increased life span in the RI strains or in *age-1* mutants (Johnson, 1987; Friedman & Johnson; 1988a).

Lengthened life is due entirely to an increase in post-reproductive life span. Development, reproduction, and life span are each under independent genetic control. General motor activity decays linearly with chronological age in all RI genotypes examined (Johnson, 1987) and in *age-1* mutants (Duhon & Johnson, unpublished). The decay in general motor activity is both correlated with and a predictor of mean and maximum life span in the RIs, suggesting that both share at least one common rate-determining component (Johnson, 1987). These observations, when combined with earlier studies suggesting a model in which development is completed before the onset of aging (Johnson *et al.*, 1984), show that three distinct physiological processes occur simultaneously during senescence in *C. elegans*: (a) life span determining events, (b) events specifying length of egg lay, and (c) events specifying rate of movement loss.

age-1 (*hx542*) does not markedly affect the length of the reproductive period. All of the increase in life expectancy is due to an increase in the

TRANSGENICS

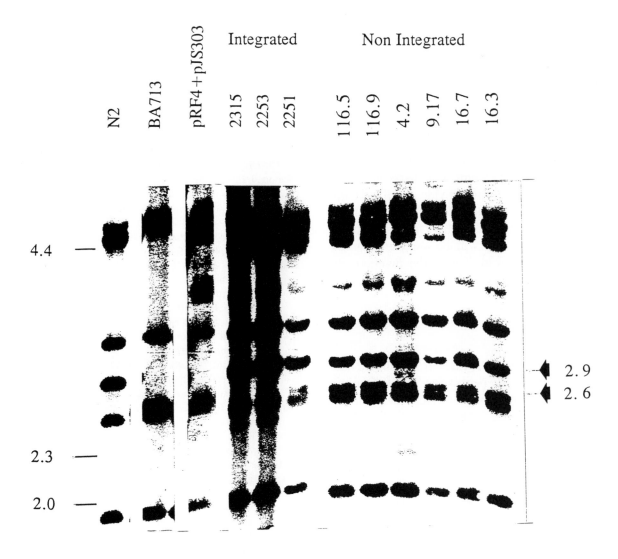

Fig. 5. Southern blots of strains transgenic for T13F9. Genomic DNA was digested with *HindIII*, followed by electrophoresis and blotting onto membranes. The blots were probed with the T13F9 cosmid containing the *fer-15* gene. A 2.6 kb restriction fragment (RFLP) is seen in genomic DNA from BA713 (lane 2) which is not seen in the N2 wild-type strain (lane 1). This RFLP cosegregates with the *fer-15* (*b26*) allele (Johnson *et al.*, in preparation). Both this 2.6 kb RFLP and the wild-type band at 2.9 kb are detected in genomic DNA extracted from the transgenic strains containing either integrated arrays (lanes 4-6) or extrachromosomal tandem arrays (lanes 7-12). In addition, novel higher-molecular-weight fragments can be seen in all transgenic lines.

length of postreproductive life. Two other measures of physiological age do not seem to be altered in *age-1* (lipofuscin content, Le Noir & Johnson, unpublished and lysosomal enzyme levels, Conley & Johnson, unpublished), suggesting that *age-1* changes one of the aspects of aging in *C. elegans* – that limiting length of life – without dramatically affecting other physiological systems.

Summary and perspectives

This mutation in *age-1* is the only instance of a genetic locus in which the mutant form results in lengthened life and has no effect on fertility. Consistent with the recessive nature of *age-1* (*hx542*) is the interpretation that longer life results from the elimination of normal gene function; deficiency analysis suggests that *age-1* (*hx542*) may be a null allele. The molecular cloning and characterization of this locus are likely to provide significant insights into the mechanistic basis of aging and senescence. We anticipate that many more mutations that lead to longer life will be uncovered in *C. elegans* and we hope to identify the functions of these new mutant genes as well.

Acknowledgements

We thank W. L. Conley, J. Le Noir, S. Duhon, and E. W. Hutchinson for permission to cite unpublished data. Supported by grants from the National Institutes of Health (K04-AG00369, R01-AG08322, and R01-AG10248) and from the Paul Glenn Foundation for Aging Research. Some stocks were supplied by and are available through the Caenorhabditis Genetics Center, which is supported by contract NO1-AG-9-2113 between the NIH and the curators of the University of Minnesota.

References

Arking, R., 1987. Evolution of longevity in animals, pp. 1-22 in Genetic and Environmental Determinants of Longevity in *Drosophila*, edited by A. D. Woodhead & K. H. Thompson. Plenum Publishing Corp., NY.

Arking, R., 1988. Genetic analysis of aging processes in *Drosophila*. Exp. Aging. Res. 14: 125-135.

Arking, R., S. Buck, A. Berrios, S. Dwyer & G. T. Baker III,

1991. Elevated paraquat resistance can be used as a bioassay for longevity in a genetically based long-lived strain of *Drosophila*. Dev. Genet. 12: 362-370.

Arking, R. & R. A. Wells, 1990. Genetic alteration of normal aging processes is responsible for extended longevity in *Drosophila*. Devel. Genet. 11: 141-148.

Brooks, A. & T. E. Johnson, 1991. Genetic specification of life span and self-fertility in recombinant-inbred strains of *Caenorhabditis elegans*. Heredity 67: 19-28.

Brenner, S., 1974. The Genetics of *Caenorhabditis elegans*. Genetics 77: 71-94.

Capecchi, M. R., 1989. Altering the genome by homologous recombination. Science 244: 1288-1292.

Charlesworth, B., 1988. Selection for longer-lived rodents. Growth Devel. Aging 52: 211.

Coulson, A., J. Sulston, S. Brenner & J. Karn, 1986. Toward a physical map of the genome of the nematode *Caenorhabditis elegans*. Proc. Natl. Acad. Sci. USA 83: 7821-7825.

Coulson, A., R. Waterston, J. Sulston & Y. Kohara, 1988. Genome linking with yeast artificial chromosomes. Nature 335: 184-186.

Finch, C. E., 1990. Longevity, Senescence, and the Genome. The University of Chicago Press, Chicago.

Fleming, J. E., G. S. Spicer, R. C. Garrison & M. R. Rose, 1993. Two-dimensional protein electrophoretic analysis of postponed aging in *Drosophila*. Genetica (In press).

Friedman, D. B. & T. E. Johnson, 1988a. A mutation in the age-1 gene in Caenorhabditis elegans lengthens life and reduces hermaphrodite fertility. Genetics 118: 75-86.

Friedman, D. B. & T. E. Johnson, 1988b. Three mutants that extend both mean and maximum life span of the nematode, *Caenorhabditis elegans*, define the age-1 gene. J. Gerontol. Biol. Sci. 43: B102-109.

Gould, A. B. & A. M. Clark, 1977. X-ray induced mutations causing adult life-shortening in *Drosophila melanogaster*. Exp. Gerontol. 12: 107-112.

Graves, J. L., E. C. Toolson, C. Jeong, L. N. Vu & M. R. Rose, 1992. Desiccation, flight, glycogen and postponed senescence in *Drosophila melanogaster*. Physiol. Zool. (In press).

Harrison, D. E., 1988. Mini-Editorial Introduction: Selection for longer-lived rodents. Growth Devel. Aging 52: 207.

Herman, R. K., 1988. Genetics, pp. 17-45 in The Nematode *Caenorhabditis elegans*, edited by W. B. Wood, Cold Spring Harbor Press, Cold Spring Harbor, NY.

Hutchinson, E. W., 1993. Genetica (In press).

Hutchinson, E. W. & M. R. Rose, 1990. Quantitative genetic analysis of postponed aging in *Drosophila melanogaster*, pp. 65-85 in Genetic Effects on Aging II, edited by D. E. Harrison, Telford Press, Caldwell, NJ.

Johnson, T. E., 1984. Analysis of the biological basis of aging in the nematode, with special emphasis on *Caenorhabditis elegans*, pp. 59-93 in Invertebrate Models in Aging Research, edited by D. H. Mitchell & T. E. Johnson, CRC Press, Boca Raton, FL.

Johnson, T. E., 1987. Aging can be genetically dissected into component processes using long-lived lines of *Caenorhabditis elegans*. Proc. Natl. Acad. Sci. USA. 84: 3777-3781.

Johnson, T. E., 1988. Thoughts on the selection of longer-lived rodents. Growth Devel. Aging 52: 207-209.

Johnson, T. E., 1990. The increased life span of age-1 mutants in

Caenorhabditis elegans results from lowering the Gompertz rate of aging. Science 249: 908-912.

Johnson, T. E., D. B. Friedman, N. Foltz, P. A. Fitzpatrick & J. E. Shoemaker, 1990. Genetic variants and mutations of *Caenorhabditis elegans* provide tools for dissecting the aging processes, pp. 101-127 in Genetic Effects on Aging. Volume II, edited by D. E. Harrison. Telford Press, NY.

Johnson, T. E. & E. W. Hutchinson, Absence of strong heterosis for life span and other life history traits in *Caenorhabditis elegans*. Genetics, in press.

Johnson, T. E., E. W. Hutchinson & P. M. Tedesco, Genetic and physical mapping of fer-15 and age-1 using chromosome deficiencies in *Caenorhabditis elegans* (Submitted).

Johnson, T. E., D. H. Mitchell, S. Kline, R. Kemal & J. Foy, 1984. Arresting development arrests aging in the nematode *Caenorhabditis elegans*. Mech. Ageing Dev. 28: 23-40.

Johnson, T. E. & W. B. Wood, 1982. Genetic analysis of lifespan in *Caenorhabditis elegans*. Proc. Natl. Acad. Sci. USA 79: 6603-6607.

Klass, M. R., 1983. A method for the isolation of longevity mutants in the nematode *Caenorhabditis elegans* and initial results. Mech. Ageing Dev. 22: 279-286.

Lee, E. & M. Desu, 1972. A computer program for comparing k samples with right-censored data. Comp. Progs. Biomed. 2: 315-321.

Leffelaar, D. & T. A. Grigliatti, 1984. A mutation in *Drosophila* that appears to accelerate aging, Develop. Genet. 4: 199-210.

Luckinbill, L. S., R. Arking, M. J. Clare, W. C. Cirocco & S. A. Muck, 1984. Selection for delayed senescence in *Drosophila melanogaster*. Evolution 38: 996-1003.

Luckinbill, L. S., J. L. Graves, A. Tomkiw & O. Sowirka, 1988. A qualitative analysis of some life-history correlates of longevity in *Drosophila melanogaster*. Evol. Eco. 2: 85-94.

Luckinbill, L. S., V. Riha, S. Rhine & T. A. Grudzien, 1990. The role of glucose-6-phosphate dehydrogenase in the evolution of longevity in *Drosophila melanogaster*. Heredity 65: 29-38.

Maynard Smith, J., 1958. The effects of temperature and egg laying on the longevity of *Drosophila subobscura*. Journal of Experimental Biology 35: 832-842.

Mello, C. C., J. M. Kramer, D. Stinchcomb & V. Ambros, 1991. Efficient gene transfer in *C. elegans*: extrachromosomal maintenance and integration of transforming sequences. The EMBO J. 10: 3959-3970.

Olson, M. V., 1992. The lessons from the nematode. Curr. Biol. 5: 221-223.

Pretzlaff, R. & R. Arking, 1989. Patterns of amino acid incorporation in long-lived genetic strains of *Drosophila melanogaster*. Exp. Geront. 24: 67-81.

Roberts, P. A. & R. B. Iredale, 1985. Can mutagenesis reveal major genes affecting senescence? Exp. Geront. 20: 119-121.

Rose, M. R., 1984. Laboratory evolution of postponed senescence in *Drosophila melanogaster*. Evolution 38: 1004-1010.

Rose, M. R., 1988. Response to 'Thoughts on the selection of longer-lived rodents'-Rejoinders. Growth Devel. Aging 52: 209-211.

Rose, M. R., 1991. Evolutionary Biology of Aging. Oxford Univ. Press, NY.

Rose, M. R., J. E. Fleming, G. Spicer, R. E. Tyler & F. J. Ayala, 1990. Molecular genetics of postponed aging in *Drosophila* (Abstract). The Gerontologist 30: 252A-253A.

Rose, M. R., 1990. A workshop summary: should mice be selected for postponed aging? Growth Devel. Aging 54: 7-15.

Rose, M. R. 1993. Genetica (In press).

Sigurdson, D. C., G. J. Spanier & R. K. Herman, 1984. *Caenorhabditis elegans* deficiency mapping. Genetics 108: 331-345.

St Johnston, D. & C. Nusslein-Volhard, 1992. The origin of pattern and polarity in the *Drosophila* embryo. Cell 68: 201-219.

Stinchcomb, D. T., J. E. Shaw, S. H. Carr & D. Hirsh, 1985. Extrachromosomal DNA transformation of *Caenorhabditis elegans*. Mol. Cell Biol 5: 3484-3496.

Stearns, S. C., M. Kaiser & E. Hillesheim, 1993. Effects of fitness components of enhanced expression of elongation factor EF-1α in *Drosophila melanogaster*: I. The contrasting approaches of molecular and population biologists. Genetica (In press).

Sulston, J., Z. Du, K. Thomas, R. Wilson, L. Hillier, R. Staden, N. Halloran, P. Green, J. Thierry-Mieg, L. Qiu, S. Dear, A. Coulson, M. Craxton, R. Durbin, M. Berks, M. Metzstein, T. Hawkins, R. Ainscough & R. Waterston, 1992. The *C. elegans* genome sequencing project: a beginning. Nature 356: 37-41.

Van Voorhies, W. A., 1992. Production of sperm reduces nematode lifespan. Nature 360: 456-458.

Watson, J. D., M. Gilman, J. Witkowski & M. Zoller, 1992. Recombinant DNA. Cold Spring Harbor Press, Cold Spring Harbor, NY.

Wood, W. B., 1988. The Nematode *Caenorhabditis elegans*. Cold Spring Harbor Press, Cold Spring Harbor, NY.

M. R. Rose and C. E. Finch (eds.), Genetics and Evolution of Aging, 96–105, 1994.
© 1994 *Kluwer Academic Publishers. Printed in the Netherlands.*

New model systems for studying the evolutionary biology of aging: crustacea

David Reznick
Department of Biology, University of California, Riverside, CA 92521, USA

Received and accepted 22 June 1993

Abstract

Progress in any area of biology has generally required work on a variety of organisms. This is true because particular species often have characteristics that make them especially useful for addressing specific questions. Recent progress in studying the evolutionary biology of senescence has been made through the use of new species, such as *Caenorhabditis elegans* and *Drosophila melanogaster*, because of the ease of working with them in the laboratory and because investigators have used theories for the evolution of aging as a basis for discovering the underlying mechanisms.

I describe ways of finding new model systems for studying the evolutionary mechanisms of aging by combining the predictions of theory with existing information about the natural history of organisms that are well-suited to laboratory studies. Properties that make organisms favorable for laboratory studies include having a short generation time, high fecundity, small body size, and being easily cultured in a laboratory environment. It is also desirable to begin with natural populations that differ in their rate of aging. I present three scenarios and four groups of organisms which fulfill these requirements. The first two scenarios apply to well-documented differences in age/size specific predation among populations of guppies and microcrustacea. The third is differences among populations of fairy shrimp (Anostraca) in habitat permanence. In all cases, there is an environmental factor that is likely to select for changes in the life history, including aging, plus a target organism which is well-suited for laboratory studies of aging.

Introduction

How do we know that genes are on chromosomes or that they are responsible for inheritance? This knowledge is so basic to our understanding of biology and has been accepted for so long that most of us have forgotten, or never knew, the answer. If we began the search for answers to these questions with the discovery of chromosomes, then resolving them took over 60 years and the efforts of dozens of investigators (Moore, 1985). It also involved work on diverse organisms including flatworms, sea urchins, nematodes, grasshoppers, various domesticated plants and animals, and finally the fruit fly. Of all the lessons that can be learned from this history, one is that scientific discovery is a combination of chance and design. A significant component of both chance and design is the choice of species for study; key discoveries often were de-

pendent upon special properties of the study organism, such as the number, size, or shapes of chromosomes (Moore, 1985). Because different qualities were important at different phases of answering these questions, the work ultimately involved the use of a large variety of organisms. These same elements of chance, design, and diversity underlie most of the accepted principles of biology.

Our efforts to understand the biology of aging already have followed, and will continue to follow, the same pattern. Much of our understanding to date has come from organisms such as mice and rats, chosen for their 'biological relevance' to humans, but significant progress has been made in the past decade with organisms as unlikely as fruit flies and nematodes (Rose, 1991). These two model systems illustrate the same lesson as above, which is that special properties of organisms often enable us to answer questions that were previously not answerable, or not even considered. My goal is to first illustrate why these organisms made it possible to address new questions, to propose criteria for finding additional species for studying the aging process, then to illustrate this point with two examples. The comparative biology of senescence suggests that it is a complex phenomenon which cannot be explained by a single cause; instead, it appears to have evolved independently several times and has characteristics that vary dramatically among even closely related species. Understanding this diversity will require work on a diversity of organisms (Finch, 1991).

The evolutionary biology of aging

One new approach to studying aging is via the evolution of the aging process. This new approach is dominated by work on *Drosophila melanogaster*. It began by artificially selecting for differences in age-specific reproductive success and achieved, as a correlated response, differences in the rate of aging (Rose & Charlesworth, 1980, 1981, a,b). Once such differences were obtained, it was possible to search for their correlates and, ultimately, their underlying causes. Examples of correlates of delayed aging in *Drosophila* include a reduction in fecundity early in life (Rose & Charlesworth, 1981a), a difference in how fat reserves were allocated to somatic versus reproductive tissues in newly eclosed adults (Service, 1987), and increased resistance to various forms of environmental stress (Service, *et al.*, 1985). These correlates are leading to additional work to unravel the causes of aging at different levels of biological organization.

Special properties of fruit flies that make this kind of experiment possible include the ease of mass-culturing them in the laboratory, their short generation time, and the accumulated knowledge of their life history and biology (Rose, 1991). By following these and other criteria, it should be possible to find other species that are likely to show similar genetic variation in the rate of aging, and hence provide additional models for studying the evolution of aging. Before developing this logic further, and proposing specific examples, I will briefly review the evolutionary approach to the study of aging, since this perspective motivated the *Drosophila* work and the examples that follow.

There are two models of the evolution of aging which can be applied to *Drosophila*, or to any other model system. One, proposed by Williams (1957), is referred to as 'antagonistic pleiotropy'. Williams proposed that genes associated with enhanced performance early in life (e.g. earlier maturity or more offspring) reduce performance later in life (e.g., shorter lifespan or fewer offspring late in life). In other words, the same genes can have a correlated influence on the early and late features of the life history. There are arguments coming from the field of quantitative genetics (e.g. Falconer, 1989) that such genetic variation is likely to persist as a feature of natural populations.

The second theory, called 'mutation accumulation', was proposed by Medawar (1952). Medawar's theory is based on the principle that, as an organism gets older, its reproductive contribution to the next generation declines in 'importance'. You can think of this as an application of the principle of compound interest. The young produced early in life of the parent potentially mature and begin to have young much sooner that young born late in life of the parent. This means that the early-born young will contribute to the fitness of the parent sooner and have a larger impact on the fitness of the parent than the late-born young, in the same way that money placed in an Individual Retirement Account early in your life will accumulate more interest and be worth more by retirement than money saved late in life. When deleterious mutations occur, meaning mutations that reduce the fitness of the parent in some way, it is predicted that natural selection will favor the deferral of their expression until later in life, when they will have a smaller influence on the individual's fitness. Also, deleterious mutations which naturally have their impact late in life will be less rapidly 'weeded out' by the process of natural selection.

It is important to bear in mind that Williams model views senescence as part of the normal life cycle of the organism; it is functionally related to the other components of the life history, including the age at maturity and pattern of investment of resources in reproduction. Medawar's model views senescence as, in part, a passive by-product of the declining influence of natural selection on older age classes (Rose, 1991). These models do not represent exclusive alternatives. Both could contribute to the observed patterns of senescence in a given species. It is also important to bear in mind that when these models are generalized to include factors like age-specific mortality or density dependence, their predictions are not as simple and clear cut as implied here (Charlesworth, 1980; Abrams, 1993). For the purposes of this discussion, however, I will deal with just the simple predictions summarized above.

To summarize, investigators have demonstrated that it is possible to select for lines that differ in the aging process, and hence to characterize the correlates and possible causes of these changes. Theory has presented two genetic mechanisms for the evolution of aging. The work to date on *Drosophila melanogaster* implies that both of these evolutionary mechanisms, antagonistic pleiotropy and mutation accumulation, account for different aspects of aging in this species (Mueller, 1987; Service, Hutchinson & Rose, 1988).

Identifying new study organisms

While directly selecting for lineages that differ in the rate of aging is a powerful tool for developing new systems for studying the evolutionary biology of aging, other means may be possible. The comparative method represents a second means for identifying populations with the desired biological properties. Applying this method can make a greater diversity of species available for the study of the evolutionary mechanisms of senescence. This method involves first identifying natural populations that differ in some aspect of their biology that is likely to influence the pattern of aging. The next step is to document that such differences in the rate of aging exist. Once such differences are established, we again have a system that can be used to probe the possible mechanisms that cause these differences.

Existing descriptions of the natural history of many species suggest that it will be possible to find populations within a species that differ either in the overall mortality rate or in age-specific mortality. One common cause of such patterns is differences in predation; e.g., populations may differ either in the overall presence or absence of predators or in the species of predators that are present. The types of predators in turn often differ in their overall rates of predation or in the size/age classes of prey

selected. Other possible causes of differences in mortality pattern might be disease, parasites, or the patterns of resource availability.

If differences among populations in overall mortality rate, age-specific mortality or age-specific reproductive success exist, then we expect the evolution of differences in the life history, such as age-specific reproductive investment and the rate of aging. The theory of life history evolution that makes such predictions (e.g., Gadgil & Bossert, 1970; Law, 1979; Charlesworth, 1980, plus many more) generally deals with variables expressed early in the life history, such as age at maturity and early life reproduction. In some cases they explicitly incorporate senescence (Charlesworth, 1980). Changes in the rate of aging can be inferred from the application of the antagonistic pleiotropy or mutation accumulation models. For example, a high adult mortality rate is predicted to select for earlier maturity and increased reproductive effort early in life (Gadgil & Bossert, 1970; Law, 1979; Michod, 1979; Charlesworth, 1980). William's theory predicts that these changes early in the life cycle will be correlated with decreased performance late in life via a reduction in fecundity and/or reduced lifespan. Medawar's theory predicts that such selection will allow the more rapid accumulation of deleterious mutations that act later in the life cycle, because fewer individuals survive to advanced ages.

A second criterion for choosing organisms is their ease of culture in the laboratory, accessibility for field studies, and life history. Important life history characteristics are a short generation time and high fecundity (Rose, 1991). A short generation time is a virtue because it results in shorter experiments. High fecundity will make it easier to design and execute experiments, because it is easier to generate a sufficient number of experimental subjects. It is also desirable for the organism to be small, so that it can be housed in large numbers in a small area.

One possible limitation of the use of short-lived organisms is that they may not be subject to certain time-dependent mechanisms of aging. Some molecular/cellular aspects of aging, such as somatic cell DNA mutations, may take decades to develop and may play an important role in the aging process of longer-lived organisms, but may not be important in those with short life cycles. Since we currently have little understanding of the relative importance of different mechanisms of aging, it is not possible to evaluate the importance of this limitation at this time.

Guppies as a model system

One organism that satisfies all of these criteria is the guppy (*Poecilia reticulata*). This species has already served well in studies of senescence (e.g. Comfort, 1960) and is described elsewhere as a potential model for studying the evolutionary biology of aging (Reznick, 1993). Briefly, Reznick (1993) summarizes fifteen years of work on life history evolution in guppies from the island of Trinidad. I compared the life histories of natural populations of guppies from high and low predation localities. 'High predation' localities are those where guppies co-occur with the pike cichlid *Crenicichla alta* and other predators that prey preferentially on large, mature size classes of guppies. 'Low predation' localities are those where guppies co-occur with just the killifish *Rivulus harti*. *Rivulus* preys less intensely on guppies and then preferentially on small, immature size classes. High and low predation localities are often found in the same stream. There are many streams with these two types of communities, providing many independent replicates for each type.

Guppies from high predation sites mature at an earlier age, devote more of their consumed resources to reproduction, reproduce more often, and produce more and smaller offspring per litter. All of these differences have a genetic basis. By transplanting guppies or predators to new localities, I have shown that these life history patterns can evolve in as little as eleven years, or 30 to 60 generations. These changes are consistent with the predictions of life history theory. Both theories for the evolution of aging predict that guppies from the high predation localities should display more rapid aging. Antagonistic pleiotropy predicts this as a consequence of earlier maturity and higher allocation to reproduction early in life. Mutation accumulation predicts this as a consequence of their shorter lifespan and hence weaker selection on performance late in life. Assaying the aging patterns of guppies from these different sites is obviously the next step.

Crustacea as model systems

Three groups of crustacea, the Copepoda (a class), the Cladocera (an order in the class Branchiopoda), and the Anostraca (a second order in the class Branchiopoda) can follow two different scenarios that are likely to select for life history variation, including variation in the rate of aging. The first scenario is interpopulation differences in the species of predators with which these organisms co-occur, and hence in predator mediated mortality. The second is interpopulation differences in habitat permanence. In this case, the target organisms all inhabit ephemeral habitats, such as temporary rain pools in deserts. These habitats vary considerably in their duration, and hence in their capacity to sustain the crustaceans.

Scenario 1: Interpopulation differences in predation

I will treat the Cladocera and Copepoda together here, even though they are distantly related, because they are commonly found together in nature and are subject to the same forms of natural selection. They are often referred to collectively as 'microcrustacea'. The predation scenario applies to populations inhabiting permanent bodies of water, where they may co-occur with a variety of vertebrate and invertebrate predators. Bodies of water often differ in the types of predators present and the predators differ in the size classes of prey that they feed upon. In these sorts of habitats, the target species of microcrustacea range in adult body size from 0.5 to 3 mm. They feed by filtering unicellular algae and small bits of organic debris from the water column.

The classic study by Brooks and Dodson (1965) demonstrated that the presence or absence of fish predators had a profound impact on the types of species of microcrustacea present in a community. They worked in freshwater lakes in Connecticut, where the fish predators were landlocked populations of *Alosa pseudoharengus*, a fish that feeds on small prey in the water column. In the presence of fish, the community was dominated by smaller species of microcrustacea, such as *Bosmina longirostris* (Cladocera) and *Cyclops bicuspidatus* (Copepoda). In lakes that do not have fish predators, the key predators on the microcrustacea are small invertebrates, such as *Chaoborus* larvae (the aquatic life stage of midges). Large-bodied zooplankton, such as members of the genus *Daphnia* (Cladocera) or *Mesocyclops* (Copepoda) tend to dominate such communities.

These differences in the structure of microcrustacea communities are the combined consequence of the size preferences of the predators and competition among the different species of prey. In the absence of fish, the larger species are able to outgrow the size classes which are preyed upon by *Chaoborus*. They also tend to be competitively dominant over the smaller crustacea, so they become numerically dominant in the body of water. Fish are capable of consuming all prey above a given size and prefer to prey upon the largest available microcrustacea. The consequence is that, when fish are present, the larger species of crustacea are sometimes eliminated and the smaller species become numerically dominant (Brooks & Dodson, 1965; Hall *et al.*, 1976).

These patterns of predation and associated differences in microcrustacea communities have been found in a variety of other circumstances and the literature published on this subject after Brooks and Dodson is immense (only a small portion is cited here). For example, Dodson (1970) reports a parallel example in ponds in Colorado where the key vertebrate predator is the salamander *Ambystoma tigrinum*, rather than fish. Hall, Cooper and Werner (1970) report a similar pattern for experimental ponds in New York, where the key vertebrate predators are sunfish (Centrarchidae). Finally, Hall *et al.* (1976) summarize investigations in a variety of aquatic ecosystems. Many of these studies incorporate manipulations, such as predator additions, which confirm the cause and effect relationship between predators and the microcrustacea communities.

While most investigators have concerned themselves with the ecological aspects of these interactions, some authors have argued that the predators are also selecting for life history differences among the different species of prey. For example, Lynch (1980) reports that the most successful species of crustacea, when exposed to invertebrate predation, are those with fast growth and delayed maturity. This enables them to quickly outgrow the size classes that are susceptible to predation. Here again, he was mostly concerned with the influences

of predation on differences among species; however, it is also true that the same species can be found in lakes with and without fish predation, so it is reasonable to predict that these differences in predation will also cause life history evolution within a species.

One prediction for differences in the rate of aging in these two types of communities is based on differences in adult mortality rates and hence follows the guppy example. Large bodied species co-occurring with just invertebrate predators will experience high mortality rates as juveniles, but low mortality rates once they outgrow their predators. When these same species are found with fish predators, their probability of survival as adults should be considerably lower because the bigger they are, the more attractive they are to their predators. Both models for the evolution of senescence predict more rapid aging in the populations that co-occur with fish predators. Under the Williams model (antagonistic pleiotropy), more rapid aging is a cost of earlier maturity and increased allocation of resources to reproduction early in life (changes that are predicted by life history theory as a consequence of high rates of adult mortality). Under the Medawar model (mutation accumulation), more rapid aging is a consequence of the low probability of adult survival, and hence the higher probability of accumulating late-acting, deleterious mutations. In the real world, these differences in predation are likely to be associated with other factors that could influence the evolution of the life history, such as resource availability. This means that predicting and evaluating differences among populations in aging may not be so straightforward; nevertheless, it is a start.

In spite of the strong argument for predicting within species variation in life histories, very little work has been done to address evolution within a species in response to predation. What is available at least suggests that predators can select for changes in the life history and other aspects of the biology of their prey.

Neill (1992) evaluated the influence of invertebrate versus vertebrate predators on the tendency of their microcrustacea prey to undergo daily vertical migrations. Such migrations have been widely studied as a response to predation. For example, in one fish-free lake studied by Neill, all age classes of the copepod *Diaptomus kenai* migrate to deeper water during the night to avoid nocturnal predation by *Chaoborus trivittatus*. This midge is sufficiently large to consume even adult copepods. The copepods return to the surface during the day to feed. Neill demonstrated that this vertical migration is not present in the absence of these predators, but is quickly induced when predators are added to enclosures. A neighboring lake contains cutthroat trout (*Oncorhyncus clarki*), but *Chaoborus flavicans*, rather than *C. trivitattus*. *C. flavicans* also preys nocturally on microcrustacea found near the surface of the lake, but it is too small to prey upon adult *D. kenai*. One can induce nighttime migrations into deep water in juvenile, but not adult, copepods from this lake by introducing either *Chaoborus* into an experimental enclosure. The adults will remain near the surface throughout the day and night. A consequence is that this copepod is eliminated when it is kept in the presence of *C. trivitattus*. Neill has demonstrated that there is likely to be a genetic basis for this difference in behavior. Even though this study does not consider features of the life history, it is important because it demonstrates that the predators can indeed select on specific age-classes of prey and can select for differences among neighboring populations of prey. It is this age-specificity of predator effects that is the source of predictions for evolved differences in aging.

Wyngaard (1986 a,b) compared the life histories of the copepod *Mesocyclops edax* from ponds in Michigan and Florida. The Florida populations co-occur with fish predators similar to those in the original Brooks and Dodson study, while no such fish predators were present in the Michigan lakes. In laboratory studies on the second generation of lab-reared stocks, she found that stocks from Michigan had delayed maturity (similar to Lynch's inter-specific comparisons), but smaller adult body sizes than their counterparts from Florida. They also produced far fewer eggs per litter. Because these were second-generation, lab-reared organisms that were reared singly at a common temperature and food availability, these differences are likely to be genetic. These lakes differ in a variety of other ways. For example, the Michigan ponds tended to have low nutrient levels and low productivity, while the Florida lakes had moderate to high productivity, so it is not possible to attribute these life history patterns to predation alone. The sorts of differences in life history documented here are also those that can

be associated with differences in the rate of aging. Tessier (1986) reports similar life history differences among populations of *Holopedium gibberum* (Cladocera).

Spitze (1991) successfully selected for changes in the life history of *Daphnia pulex* (a large bodied cladoceran) in a relatively short-term laboratory study. He used *Chaoborus* as predators, with the control being populations that were not exposed to predators. The *Chaoborus* larvae fed primarily on juvenile instars, so the effect of the predators was increased juvenile mortality rates relative to the control population. He began by mixing equal numbers of individuals from 45 parthenogenetic clones that were electrophoretically distinguishable. The response to selection was gauged by changes in the relative abundance of these clones and by the life history characteristics of the clones, as measured on individuals reared in isolation. Population density and food availability was controlled throughout the experiment. After four generations of selection, he found that *Chaoborus* predation had selected for larger body size, higher fecundity, and earlier maturity. These results also imply that *Chaoborus* predation selected for a higher rate of resource acquisition or higher conversion efficiency, plus a higher growth rate. They are only partially consistent with my expectations (I would have predicted delayed maturity), but the discrepancy may in part be due to differences between the life histories of the favored lines when measured in isolation, versus their manifestation in the experimental populations. Spitze considers other explanations in his discussion. The study demonstrates the ability of predators, or an investigator, to select for changes in the life history, as Rose and his colleagues did for *Drosophila*, and the feasibility of using these organisms for large-scale laboratory experiments.

The composite picture from these studies is that there can be substantial differences among populations in size-specific predation. This translates into differences in age-specific survival and can in turn result in the evolution of life history traits, including the age and size at maturity and fecundity. Although not very many examples have been developed, such differences in life history traits have been found within a species. These differences in field demography create the expectation of natural differences in longevity and the rate of aging, although such differences have not been sought.

Other aspects of the biology of these species and their prior history of use in laboratory research also suggest that they are good choices for studying senescence. There are well-established techniques for rearing them in the laboratory, as outlined in the methods sections of the above cited papers, or in publications devoted to describing the utility of microcrustacea for other types of investigations (e.g. Goulden *et al.*, 1982). The methods include means for precisely controlling food availability. There are even existing methods for constructing complete lifetables based on individual performance in their natural environment (Wyngaard, 1983).

The repeatability of lab rearing techniques and the feasibility for using these organisms for laboratory studies of aging is well illustrated by the large number of published studies, including some which deal specifically with aging (Ingle, Wood & Banta, 1937; Dunham, 1938; Lynch & Ennis, 1983). Their life history also makes them particularly suitable for laboratory studies. Their small body size means that they are easily accommodated in small containers; Lynch (1989) kept each individual *Daphnia pulex* (a large bodied species) in 40 ml of medium. They attain maturity rapidly; Allan and Goulden (1980) report ages at maturity for Cladocera which range from as little as three days for small bodied species like *Bosmina* and *Moina* up to eight days for large bodied species such as *Simocephalus* and *Daphnia*. The Copepoda tend to take longer to develop (Allan & Goulden, 1980). Maximum lifespans for *Daphnia pulex* ranged from 50 to 70 days (Lynch & Ennis, 1983). Smaller-bodied, more rapidly developing species are likely to have shorter lifespans. Fecundity tends to increase with the average size of the species and with age and size within a species, with the maximum clutch sizes of the larger species often exceeding 20 eggs (Allan & Goulden, 1980). Since a clutch is produced every few days, there is the potential to obtain a large number of offspring from each individual in an experiment.

One aspect of the reproductive biology of the Cladocera can be seen as either a cost or a benefit to their use in laboratory studies. All species have cyclical parthenogenesis; some clones within a species are obligately parthenogenetic. If conditions are favorable and constant, then parthenogenetic reproduction can be sustained indefinitely. If conditions deteriorate, then males will be produced and the next generation of eggs will be ephipia, or resting eggs. Thus far, all published laboratory studies have dealt solely with parthenogenetically reproducing clones because of the difficulties of initiating a new generation from the resting eggs (Fugate, 1992). This form of reproduction can be seen as a cost because it limits our abilities to do quantitative genetics experiments, though broad-sense heritabilities are readily estimable (e.g. Lynch, 1984). This form of reproduction can be seen as a benefit because it is possible to generate a large number of genetically identical individuals for the evaluation of environmental influences on the life history and for making comparisons among genotypes.

The Copepoda all have sexual reproduction, but they will produce eggs that either undergo direct development or resting eggs that require special treatment, such as drying or chilling, before they can be hatched. It is possible for them to produce directly developing eggs indefinitely if the laboratory conditions are favorable.

Scenario 2: Interpopulation differences in habitat permanence

A second type of aquatic community is one which is ephemeral in nature, either because it only holds water or because it is unsuitable for aquatic life for some other reason, such as freezing solid, for part of the year. Belk and Cole (1975), Broch (1965), and Hartland-Rowe (1972) present descriptions of the ecology of such temporary aquatic habitats. These communities differ in the mean and variance of their duration. Such variation can also select for differences in the life history of the organisms that comprise the community. Many such habitats are host to all three groups of crustacea, but most available information is about the fairy shrimp. There is far less information about the natural history of these communities than for Scenario 1, and far less precedence for using these organisms in laboratory studies. Nevertheless, they appear to have favorable properties for laboratory studies and there are good arguments for expecting them to have inter-population differences in life histories, including aging.

First, consider a comparison of the duration of rainpools in the eastern versus western Mojave desert. The pools of the eastern Mojave are filled by summer rains, usually driven by monsoon-like storms that develop over the deserts of Arizona. The pools of the western Mojave are filled by winter rains driven by the easterly winds blowing over the Pacific Ocean and the cooler temperatures on the land, versus the water. Because of the differences in season, the summer pools dry out much more quickly than the winter pools. For example, in 1991, five pools in the western Mojave (winter rains) lasted for an average of 59.2 days (range 37-89 days). In that same year, seven pools in the eastern Mojave (summer rains) lasted for an average of 8.4 days (range 5-11 days) (S. Morey, pers. commun.). In a similar survey conducted over a five year period in the desert near Big Bend National Park in Texas, Newman (1987) found that pools filled by summer rains lasted an average of 8.1 days (range 3-13 days, n = 82).

Organisms that occupy these pools must compress the aquatic phase of the lifecycle into the time period when water is available. These substantial differences in duration should therefore select for differences in the early life history. The best comparative studies done in these pools concern the development rate of the frogs that use them for breeding, such as species of spadefoot toads (genera *Scaphiopus* and *Spea*). Although not immediately relevant to my goals, these studies illustrate how pool duration can influence the life history. At one extreme, *Scaphiopus couchi* specializes in breeding in the short-lived, summer pools. It can develop from a fertilized egg to a newly metamorphosed froglet in as little as eight days (Newman, 1987). The mean development time is actually longer than the mean duration of pools. In most cases, the pools dried out before any of the tadpoles were able to complete development (Newman, 1987). Species in these same genera which develop in the longer-lived pools generally take in excess of 30 days before they metamorphose (S. Morey, pers. commun.). These differences can be replicated in the laboratory under controlled light, temperature, water quality, and food availability and are thus likely to have a genetic basis (S. Morey, pers. commun.).

The effects of such a temporary environment should be even more dramatic for crustacea, since the entire lifecycle must be compressed within the duration of the pool. Many species are specialized for life in a narrow range of ephemeral habitats; however, some are widespread and are likely to be exposed to considerable variation in pool duration. For example, *Branchinecta lindahli* is a likely inhabitant of the western Mojave pools, but its range extends eastward and south into Baja, where it is also likely to be found in pools filled by summer rains (Fugate, 1992). Pools of short duration should select for rapid development and perhaps high fecundity, since the probability of survival as an adult is low. This prediction follows from the same life history theory cited for the predation scenario and, by the same reasoning, I would also expect there to be differences among populations in the rate of aging.

Unfortunately, there has been very little consideration of the influence of the conspicuous factor of pool duration on the life history of fairy shrimp. The only intraspecific comparison that I have found is for egg size and egg number in *Streptocephalus seali*, a species that is found over much of the continental United States. Rain pool populations from Arizona and California are reported to produce more and smaller eggs than their counterparts from longer-lived pools in different parts of their range (Belk, 1977; Belk, Anderson & Hsu, 1990).

The life history of fairy shrimp begins with the production of an egg that initiates development when laid, but then enters a resting phase (Broch, 1965). The resting phase is broken by cues associated with recent pool filling, such as exposure to water with a low ionic concentration, similar to rainwater or snowmelt (Brown & Carpelan, 1971). The eggs must be dried and then rewetted for hatching (Bernice, 1972). They can be stored at low temperatures for long periods of time before initiating the next generation (Anderson & Hsu, 1990). There is variation in the response to rewetting, so that some eggs will hatch on the first exposure, while others require being soaked and rewetted two or more times before they will hatch (Al-Tikrity & Grainger, 1990). This variation is perhaps an adaptation to the uncertainty of pool duration.

All species of fairy shrimp have sexual reproduction and, with rare exceptions, produce only resting eggs. Reported ages at maturity in nature are as short as fourteen days for *Branchinecta mackini* from southern California (Brown & Carpelan, 1971), 17 days for *Streptocephalus vitreus* from Kenya (Hildrew, 1985), and 9 days for *Streptocephalus seali* from Louisiana (Moore, 1955). Prophet (1963) reports ages at maturity in the lab of

9 days for *Branchinecta lindahli* (20C), 11 days for *Streptocephalus texanus* (20C), and 13 days for *Eubranchipus serratus* (15C). Since adult-sized fairy shrimp were seen in the eastern Mojave rain-pools (Morey, pers. commun.) it is likely that some species are capable of more rapid development.

There is only fragmentary information available for the remainder of the life history. For example, Belk (1977) reports that females of *Streptocephalus seali* begin to reproduce when as small as 12.8 mm in length and produced from 38 to 315 eggs per clutch, with clutch size increasing rapidly with body size. Belk also reports single observations on seven other species, all ranging from 16.8 to 22.8 mm in length, with clutch sizes ranging from 52 to 846 eggs. The only more detailed description that I could find was for the brine shrimp (*Artemia salina*) by Browne *et al.* (1984). They report, as mean values for 11 to 20 females from each of 12 populations, interbrood intervals of three to six days, a total of 4 to 19 broods per lifetime, and longevities ranging from 58 to 127 days. While it is risky to generalize from *Artemia* to the fairy shrimp, these represent the best descriptions of the life history for any species in the order, plus they present a good description of standardized laboratory rearing procedures. It appears that this group has the desired properties of rapid development, small body size, and high fecundity. Further descriptions of the biology and zoogeography of this group can be found in Belk, Dumont and Munuswamy (1991). The fairy shrimp therefore have desireable properties as a laboratory organism because of their rapid development and high fecundity. The small number of available laboratory studies suggest that they are not too difficult to rear. Their requirement for an enforced resting stage would add to the time necessary for a multigeneration experiment, but the option of controlling and synchronizing hatch date would facilitate the execution of experiments.

In conclusion, existing work on a variety of organisms can be exploited to develop new model systems for studying the evolutionary biology of aging. Important aspects of natural history, such as predator-mediated mortality, habitat permanence, or anything else that will systematically influence lifespan or age specific reproductive success, are also factors that potentially influence the evolution of life history. Two theories for the evolution of senescence also suggest that these factors will se-lect for changes in the rate of aging. It may therefore be possible to use such information to identify different populations within a species that differ in their aging characteristics. These could then represent the necessary raw material to evaluate the mechanisms responsible for these differences in aging. One implicit message here is that there could be a fruitful partnership formed between those interested in senescence and scientists from other disciplines, such as ecology or population biology.

References

Abrams, P. A., 1993. Does increased mortality favor the evolution of more rapid senescence? Evolution, in press.

Allan, J. D. & C. E. Goulden, 1980. Some aspects of reproductive variation among fresh water zooplankton, pp. 388-410 in Evolution and Ecology of Zooplankton Communities, edited by W. C. Kerfoot. University Press of New England, Hanover, NH.

Al-Tikrity, M. R. & J. N. R. Grainger, 1990. The effect of temperature and other factors on the hatching of the resting eggs of *Tanymaztix staognatus* (L) (Crustacea, Anostraca). J. Therm. Biol. 15: 87-90.

Anderson, G. & S.-Yu Hsu, 1990. Growth and maturation of a North American fairy shrimp, *Streptocephalus* sedi (Crustacea: Anostraca): a laboratory study. Freshwater Biology 24: 429-442.

Belk, D., 1977. Evolution of egg size strategies in fairy shrimps. Southwest. Natur. 22: 99-105.

Belk, D., G. Anderson & S.-Yu. Hsu, 1990. Additional observations on variations in egg size among populations of *Streptocephalus seali* (Anostraca). J. Crust. Biol. 10: 128-133.

Belk, D. & G. A. Cole, 1975. Adaptational biology at desert temporary-pond inhabitants, pp. 207-226 in Environmental physiology of desert organisms, edited by N. F. Hadley. Dowden, Hutchinson and Ross, Inc., Stroudsburg, PA.

Belk, D., H. J. Dumont & N. Munuswamy, 1991. Studies on large branchiopod biology and aquaculture. Kluwer Academic Publishers, Dordrecht.

Bernice, M. R., 1972. Hatching and postembryonic development of *Streptocephalus dichotomus* Baird. Hydrobiologia 40: 251-278.

Broch, E.S., 1965. Mechanism of adaptation of the fairy shrimp *Chirocephalopsis bundyi* Forbes to the temporary pond. Cornell Experiment Station Memoir 392: 1-48.

Brooks, J. L. & S. I. Dodson, 1965. Predation, body size and composition of plankton. Science 150: 28-35.

Brown, L. & L. H. Carpelan, 1971. Egg hatching and life history of a fairy shrimp *Branchinecta mackini* Dexter (Crustacea: Anostraca) in a Mohave desert playa (Rabbit Dry Lake). Ecology 52: 41-54.

Browne, R. A., S. E. Sallee, D. S. Grosch, W. O. Segreti & S. M. Purser, 1984. Partitioning genetic and environmental components of reproduction and lifespan in *Artemia*. Ecology 65: 949-960.

Charlesworth, B., 1980. Evolution in age-structured populations. Cambridge University Press, Cambridge.

Comfort, A., 1960. Effect of delayed and resumed growth on the longevity of a fish (*Lebistes reticulatus*, Peters) in captivity. Gerontologia 8: 150-155.

Dodson, S. I., 1970. Complementary feeding niches sustained by size-selective predation. Limnol. and Oceanogr. 15: 131-137.

Dunham, H. H., 1938. Abundant feeding followed by restricted feeding and longevity in *Daphnia*. Physiol. Zool. 11: 399-407.

Falconer, D. S., 1989. Introduction to Quantitative Genetics, third edition. Longman, London.

Finch, C. E., 1991. New models for new perspectives in the biology of senescence. Neurobiology and Aging 12: 625-634.

Fugate, M., 1992. Speciation in the fairy shrimp genus *Branchinecta* from North America. Ph.D. thesis, University of California, Riverside.

Gadgil, M. & W. H. Bossert, 1970. Life historical consequences of natural selection. Amer. Natur. 104: 1-24.

Goulden, C. E., R. M. Comotto, J. A. Hendrickson, Jr., L. L. Hornig & K. L. Johnson, 1982. Procedures and recommendations for the culture and use of *Daphnia* in bioassay studies, pp. 139-160 in Aquatic toxicology and Hazard Assessment: Fifth Conference, ASTM STP 766, edited by J. G. Pearson, R. B. Foster and W. E. Bishop. American Society for Testing Materials.

Hall, D. J., W. E. Cooper & E. E. Werner, 1970. An experimental approach to the production dynamics and structure of fresh water animal communities. Limnol. Oceanogr. 15: 839-928.

Hall, D. J., S. T. Threlkeld, C. W. Burns & P. H. Crowley, 1976. The size efficiency hypothesis and the size structure of zooplankton communities. Ann. Rev. Ecol. Syst. 7: 177-208.

Hartland-Rowe, R., 1972. The limnology of temporary waters and the ecology of Euphyllopoda, pp. 15-31 in Essays in Hydrobiology, edited by R. B. Clarke and R. J. Wooton. University of Exeter, U.K.

Hildrew, A. G., 1985. A quantitative study of the life history of a fairy shrimp (Branchiopoda: Anostraca) in relation to the temporary nature of its habitat, a Kenyan rainpool. J. Anim. Ecol. 54: 99-110.

Ingle, L., T. R. Wood & A. M. Banta, 1937. A study of longevity, growth, reproduction and heart rate in *Daphnia longispina* as influenced by limitations in quantity of food. J. Exp. Zool. 76: 325-352.

Law, R., 1979. Optimal life histories under age-specific predation. Amer. Natur. 114: 399-417.

Lynch, M., 1980. The evolution of cladoceran life histories. Quart. Rev. Biol. 55: 23-42.

Lynch, M., 1984. The limits to life history evolution in *Daphnia*. Evolution 38: 465-482.

Lynch, M., 1989. The life history consequences of resource depression in *Daphnia pulex*. Ecology 70: 246-256.

Lynch, M. & R. Ennis, 1983. Resource availability, maternal effects and longevity. Exp. Geront. 18: 147-165.

Medawar, P. B., 1952. An unsolved problem of biology. H. K. Lewis, London.

Michod, R. E., 1979. Evolution of life histories in response to age-specific mortality factors. Amer. Natur. 113: 531-550.

Moore, J. A., 1985. Science as a way of knowing – Genetics. Amer. Zool. 25: 1-165.

Moore, W. G., 1955. The life-history of the spiny-tailed fairy shrimp in Louisiana. Ecology 36: 176-184.

Mueller, L. D., 1987. Evolution of accelerated senescence in laboratory populations of *Drosophila*. Proc. Natl. Acad. Sci. USA 84: 1974-1977.

Neill, W. E., 1992. Population variation in the ontogeny of predator-induced vertical migration of copepods. Nature 356: 54-57.

Newman, R. A., 1987. Effects of density and predation on *Scaphiopus couchi* tadpoles in desert ponds. Oecologia 71: 301-307.

Prophet, C. W., 1963. Physical-chemical characteristics of habitats and seasonal occurrence of some Anostraca in Oklahoma and Kansas. Ecology 44: 798-801.

Reznick, D. N., 1993. Life history evolution in guppies (Poecilia reticulata): guppies as a model for studying the evolutionary biology of aging. Experiment of Gerontology, in press.

Rose, M. R., 1991. Evolutionary biology of aging. Oxford University Press, Oxford.

Rose, M. R. & B. Charlesworth, 1980. A test of evolutionary theories of senescence. Nature 287: 141-142.

Rose, M. R. & B. Charlesworth, 1981a. Genetics of life history in *Drosophila melanogaster*. I. Sib analysis of adult females. Genetics 97: 173-186.

Rose, M. R. & B. Charlesworth, 1981b. Genetics of life history in *Drosophila melanogaster*. II. Exploratory selection experiments. Genetics 97: 187-196.

Service, P. M., 1987. Physiological mechanisms of increased stress resistance in *Drosophila melanogaster* selected for postponed senescence. Physiol. Zool. 60: 321-326.

Service, P. M., E. W. Hutchinson, M. D. MacKinley & M. R. Rose, 1985. Resistance to environmental stress in *Drosophila melanogaster* selected for postponed senescence. Physiol. Zool. 58: 380-389.

Service, P. M., E. W. Hutchinson & M. R. Rose, 1988. Multiple genetic mechanisms for the evolution of senescence in *Drosophila melanogaster*. Evolution 42: 708-716.

Spitze, K., 1991. *Chaoborus* predation and life history evolution in *Daphnia pulex*: temporal pattern of population diversity, fitness and mean life history. Evolution 45: 82-92.

Tessier, A. J., 1986. Life history and body size evolution in *Holopedium gibberum* Zaddach (Crustacea: Cladocera). Freshwater Biol. 16: 279-286.

Williams, G. C., 1957. Pleiotropy, natural selection and the evolution of senescence. Evolution 11: 398-411.

Wyngaard, G. A., 1983. In situ life table of a subtropical copepod. Freshwater Biology 13: 275-281.

Wyngaard, G. A., 1986a. Genetic differentiation of life history traits in populations of *Mesocyclops edax* (Crustacea: Copepoda). Biol. Bull. 170: 279-295.

Wyngaard, G. A., 1986b. Heritable life history variation in widely separated populations of *Mesocyclops edax* (Crustacea: Copepoda). Biol. Bull. 170: 296-304.

PART THREE
Aging in *Drosophila*

Introduction

For many of the deepest problems in biology, *Drosophila*, and especially *D. melanogaster*, has been in the advance-guard of experimental progress. From the chromosome theory of inheritance to the study of mutagenesis to the population genetics of molecular variation to the molecular genetics of development, *Drosophila* has been at the leading edge of research. This should not be surprising. It was one of the first organisms ever adopted for research primarily because of its ease of handling and experimentation. As such, it is probably the oldest of all successful 'model organisms' in biological research.

Indeed, in aging research, *Drosophila* was taken up as a favored experimental system from the outset. Among all others, it was Raymond Pearl's series of papers, 'Experimental Studies on the Duration of Life', that established the study of aging in *Drosophila* in the 1920's. But the course of research on aging in *Drosophila* was not to follow the path to success blazed by so many other areas of research with the organism. Before 1980, the study of aging in *Drosophila* remained essentially murky. The key breakthrough was the creation of *Drosophila* with genetically postponed aging by selection in the laboratory. In the 1980's, several labs set about the selective creation and maintenance of stocks with postponed aging. It is such stocks that are the primary interest of the articles by Partridge and Barton, Graves and Mueller, Service and Fales, Arking *et al.*, Tyler *et al.*, and Fleming *et al.*

But there is another strategy by which interesting *Drosophila* can be created: transformation. Studies of transformed *Drosophila* indicate that it is possible to postpone aging in flies by transformation, at least for some loci, under some conditions. Stearns and Kaiser evaluate whether or not EF-1α can be used to postpone aging in *D. melanogaster*. The trickiness of this question reveals some of the subtlety that faces the use of transformation strategies in aging research.

M. R. Rose and C. E. Finch (eds.), Genetics and Evolution of Aging, 109–118, 1994.
© 1994 Kluwer Academic Publishers. Printed in the Netherlands.

Evolution of aging: testing the theory using *Drosophila*

L. Partridge & N. H. Barton
ICAPB, Division of Biological Sciences, Zoology Building, University of Edinburgh, West Mains Rd, Edinburgh EH9 3JT, UK

Received and accepted 22 June 1993

Key words: aging, senescence, lifespan, survival, *Drosophila*, evolution, fertility

Abstract

Evolutionary explanations of aging (or senescence) fall into two classes. First, organisms might have evolved the optimal life history, in which survival and fertility late in life are sacrificed for the sake of early reproduction or high pre-adult survival. Second, the life history might be depressed below this optimal compromise by the influx of deleterious mutations; since selection against late-acting mutations is weaker, deleterious mutations will impose a greater load on late life. We discuss ways in which these theories might be investigated and distinguished, with reference to experimental work with *Drosophila*. While genetic correlations between life history traits determine the immediate response to selection, they are hard to measure, and may not reflect the fundamental constraints on life history. Long term selection experiments are more likely to be informative. The third approach of using experimental manipulations suffers from some of the same problems as measures of genetic correlations; however, these two approaches may be fruitful when used together. The experimental results so far suggest that aging in *Drosophila* has evolved in part as a consequence of selection for an optimal life history, and in part as a result of accumulation of predominantly late-acting deleterious mutations. Quantification of these effects presents a major challenge for the future.

Introduction

The evolutionarily important features of aging (or senescence) are the declines in probability of survival and in fertility later in life (Charlesworth, 1980; Rose, 1991). *Drosophila* has determinate growth and little somatic repair in the adult, and the effects of aging are particularly obvious; after a period of high survival and fertility in early life, both traits show a decline (e.g. Partridge, 1988). The enormous value of *Drosophila* as a model organism for the study of aging is amply attested (Rose & Charlesworth, 1980; Service *et al.*, 1985; Service, 1987). The aim of this paper is to consider a basic issue: how can we test evolutionary theories of aging? We shall argue that there is as yet little quantitative assessment of the importance of the different evolutionary mechanisms, that quantitative testing of the theories is not simple and that both quantitative genetic studies and non-genetic

manipulations may be valuable for investigating the constraints on which the optimality theory of aging is based.

We first briefly outline evolutionary theories of aging, and then consider what kind of data would be required to evaluate their importance, in relation to the information so far available for *Drosophila*.

Evolutionary theories of aging

Because aging reduces the genetic contribution of individuals to future generations, it is not in itself favoured by natural selection. However, the natural selection which maintains survival and fertility becomes weaker through the life history because, even without aging, organisms are at risk of death and impaired fertility from disease, predation and accidents. As a cohort of individuals goes through adult life, its numbers and fertility will therefore decline, even in the absence of aging. If the genes affecting survival and fertility are to some extent age-specific in their effects, then those that are expressed later in the lifespan[1] will be subject to weaker selection because, by the time they are expressed, more of the original carriers will already have died or become infertile for other reasons (Medawar, 1946, 1952; Williams, 1957; Hamilton, 1966). Aging could then evolve, by two distinct mechanisms. First, it could appear later in life as a side effect of selection for an optimal life history. Second, an increased load of deleterious mutations on the late part of the life history could also produce senescence. We summarise below the more extensive discussion which we gave in Partridge and Barton (1993).

Optimality theory was originally developed to understand the evolution of adaptive traits, and it has been extensively used to understand the evolution of life histories – for example, traits such as age at first breeding, number of breeding attempts and level of reproductive effort at each (Stearns, 1992). The optimisation criterion has in general been maximisation of r, the intrinsic rate of increase, subject to the ecological and intrinsic constraints under which the organism operates. There must be a set of possible life histories, which is bounded by a surface that represents a compromise or 'trade-off' between different components of fitness. (It is important to realise that this trade-off surface represents the outer limits of what is possible: the population is not constrained to lie *on* the surface, and may be moved inside it by either genetic or experimental perturbations.) Selection to maximise fitness (r) will push the population to some point on that surface, which represents the optimal life history (Fig. 1). Such optimality theories have been shown to be consistent with at least some genetic models (Charnov, 1989; Charlesworth, 1980, 1990).

Selection for an optimal life history could produce senescence as the deleterious side-effect late in life of processes that are favourable early on. An antagonism between early and late performance could occur if, for instance, extra growth during the pre-adult period made low juvenile mortality im-

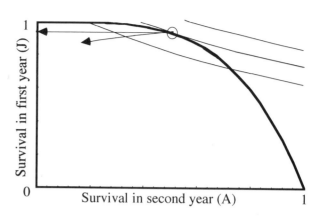

Fig. 1. A model of an organism with two age classes (after Fig. 2 of Partridge & Barton, 1993): this is the simplest scheme which can show senescence, but captures the essentials of more complicated life histories. Each female produces one daughter at each of two ages (m=1); the chance of survival from birth to age 1 is J, and from age 1 to age 2 is A. The heavy curve shows the maximum possible survival probabilities, and represents the tradeoff between adult and juvenile survival $(J + A^4 \leq 1)$. The light curves show contours of fitness, spaced at intervals $\Delta r = 0.1$; fitness is given by the discrete version of the Euler-Lotka equation, $1 = mJ e^{-r} + mJA e^{-2r}$. The optimal life history is where the furthest contour from the origin just touches the trade-off curve (open circle, J = 0.935, A = 0.505). The arrows show the reduction of survival below the optimal value caused by deleterious mutations at a rate U = 0.1. The horizontal arrow shows the effect of mutations that only reduce survival in the second year (A); when the net rate of mutation exceeds a threshold (U > 0.067), these accumulate indefinitely, and the life history becomes semelparous. The diagonal arrow shows the effect of mutations with a slight pleiotropic effect, such that survival in the first year is reduced by 1/16 as much as survival in the second year $((\delta J/J)=(\delta A/A)/16)$. Such mutations reduce survival to (0.866, 0.155) when U = 0.1.

possible, but yielded a longer-lived adult (Partridge & Fowler, 1992; Roper, Pignatelli & Partridge, in press). Similarly, extra reproduction can impair survival or subsequent fertility, by consuming resources, causing somatic damage, jeopardising repair mechanisms or exposing the organism to environmental injury, the well-known 'cost of reproduction' (Williams, 1966; Partridge & Harvey, 1985; Reznick, 1985; Bell & Koufopanou, 1986; Kirkwood & Rose, 1991). Such trade-offs will give the outer surface of the space of possible life-histories a negative slope. Since early survival and fertility necessarily have a greater effect on fitness than performance late in life, selection will produce an optimal life history which is weighted towards

---

[1] For a mutation to affect survival or fertility at a specific age, the gene does not have to be expressed at that age. For instance, a mutation affecting the development of heart muscle might only affect function at a later age.

early performance at the expense of late performance, and so involves aging.

The optimality theory of aging is sometimes referred to as being based on 'antagonistic pleiotropy'. This can be misleading, since optimisation does not *require* the existence of alleles with antagonistic effects on different parts of the life history. While such pleiotropy is to be expected, one could imagine that the optimum could be reached by the fixation of purely age-specific alleles; such a path would appear as a series of horizontal and vertical steps in Fig. 1. The life history that evolves depends on the set of possible genotypes, rather than on the pleiotropic effects of the alleles that take the population to the optimum. There is also some confusion between the optimality theory, which explains the mean location of the life history, and the idea that 'antagonistic pleiotropy' causes overdominance, and hence maintains variation in the life history (Rose, 1982, 1985; Rose, Service & Hutchinson, 1987).

Mutation pressure could also lead to the evolution of aging (Edney & Gill, 1968; Charlesworth, 1980, Partridge & Barton, 1993). Because the intensity of selection on mutants with effects later in life declines with their age of expression, alleles with a given deleterious effect will reach a higher frequency in mutation-selection balance the later the age at which they reduce fitness (e.g. arrows in Fig. 1). Strict age-specificity of mutational effects is perhaps unlikely; a varying age of onset of deleterious effects seems more plausible. The basic point none the less remains that mutations with predominantly late effects will be less strongly opposed by natural selection.

There is abundant evidence that populations are subject to an input of deleterious mutations, and that organisms cannot combine indefinitely high survival and fertility for both ecological and physiological reasons. It therefore follows that both processes must play some part in the evolution of aging. The task is to gain some insight into their relative contributions in different kinds of organisms, and into the physiological mechanisms involved. This requires that we are able to define when the rate of aging differs between individuals or populations.

How to measure the rate of aging

Aging is recognised as a decline in survival probability or fertility later in life. However, quantification of the rate of aging, or of changes in the rate of aging, is not straightforward. First, survival and fertility are not independent of one another, because of the cost of reproduction. Measures of aging that involve only one of the traits are therefore suspect. In addition, aging is not the only process that causes survival and fertility to change with age. For instance, growth or learning during the adult period may cause the values of both traits to increase for a time, and there is enormous variation between taxa in the shapes of their survival and fertility schedules (Finch, 1990).

The crucial measure of a life history is the product of survival and fertility, $l(x)m(x)$. This is the number of offspring which a newborn individual can expect to produce at age x. Since this is a continuous function of age, two life histories may differ in an infinite number of ways. If the difference in $l(x)m(x)$ is positive up to some age, and then becomes negative, then one can say that one life history involves more aging than the other. However, if two life histories cross over at several points, one cannot say that one involves more or less aging than the other.

A further problem is that if one life history always has lower $l(x)m(x)$ than another, when measured in the same environment, then it will lead to a lower rate of population growth. In nature, however, density would fall over time, leading either to extinction, or (more likely) to a compensating increase in fitness. Such density-dependence will ensure equilibrium at a lower density, and with higher $l(x)m(x)$. Whether the resulting life history involves more or less aging (or a more complicated pattern) depends on the way density affects reproduction and survival of different ages. Thus, there are serious practical problems in interpreting comparative and experimental observations, particularly when these are made at only a few ages, and without regard to density.

There are other complications in interpreting survival data from experiments on aging. One is that an alteration in mean lifespan is not necessarily attributable to an alteration in the rate of senescence. The reason is that the pattern of short-term risk may instead have been altered. For instance, increased reproduction in nature can be associated with increased impact of hazards such as predation (Calow, 1979), and hence reduced lifespan. Risks can also be physiological, if for instance reproduction can cause lethal heart attacks, but no long-term damage. Provided the hazard had been survived, the organism would not differ from one that had not taken the risk: all individuals would have the same expectations, regardless of their age and reproductive history. A useful analogy for this idea comes from the test tubes originally used by Medawar (1952) to illustrate why the intensity of selection changes with age. Once a cohort of test tubes has entered the laboratory, its numbers decline from breakage. This breakage is equivalent to death as a consequence of short-term risk; provided a tube survives an experiment, there are no lasting effects. A change in the rate of aging would then not be involved, because the intrinsic resistance to breakage of the tube would not be changed. If, instead, the tube developed hairline cracks, it would survive to another experiment, but its resistance to breakage would be compromised. Interestingly, it appears that reproduction may sometimes increase death rate by an analogous alteration in physiological risk (Partridge & Andrews, 1985; Partridge et al., 1986). It is therefore necessary to eliminate altered risk as a possible cause of altered schedules of survival before assuming the involvement of an altered rate of aging.

In some experiments and comparative studies on aging, the shapes of survival curves have been used to estimate parameters and hence assess the rate of aging. For instance, statistics such as the mortality rate doubling time have been calculated (Finch, 1990). Although these statistics are more informative than mean lifespans, there could be problems with their use. One is that they ignore the fertility schedule. In addition, an assumption behind analysis of the shapes of survival curves is that they reveal something about age-related changes within individuals. However, an individual can die only once. Survival curves are constructed using survival times from different individuals. If the population examined is heterogeneous with respect to an uncontrolled variable (such as, for instance, body size) that affects lifespan, then the shape of the survival curve could reflect the pattern of individual differences in that variable, as well as age-related changes within individuals (Vaupel &

Yashin, 1983; Rees & Long, in press). Such variation in quality between individuals could explain recent claims of slowing of the rate of aging at older ages (Carey et al., 1992; Curtsinger et al., 1992). The picture could become especially blurred in reversal experiments where, for instance, the survival rates of individuals with and without a history of reproductive activity are compared (e.g. Partridge & Andrews, 1985; Partridge et al., 1986). The survivors of reproductive activity could differ from the previously non-reproductive survivors in the mean and distribution of the uncontrolled variable, irrespective of any effect or lack of effect of reproductive activity on the rate of aging. Fertility is potentially a less ambiguous character in that, through repeated measurements, each individual can be used as its own quality-control.

Bearing in mind these issues of measurement, the aim is to discover the extent to which aging emerges as part of an optimal life history, and the extent to which the effects of mutation pressure reduces the value of life history traits below their optimal values.

Testing evolutionary theories of aging using *Drosophila*

Life history traits are of particular interest in part because their close relationship to fitness means that the optimality criterion can be clearly defined. However, as for any other trait, to test the life history for optimality, and to deduce the role of optimisation in the evolution of aging, it is necessary to know the constraints under which optimisation occurs (Parker & Maynard Smith, 1990). For life history traits, this requires measurements or theoretical deductions of the shapes of the trade-off curves which relate the maximum possible values of survival or fertility at different ages. These trade-off curves have not been measured for any organism, and there is some disagreement about the appropriate methods to use.

When attempting to measure trade-off curves, the aim is to discover the evolutionary options open to an organism. It has been argued (e.g. Lande, 1982; Reznick, 1985, 1992a, b; Charlesworth, 1980) that only quantitative genetic techniques are appropriate for this purpose. Other authors (e.g. Bell & Koufopanou, 1986; Partridge & Harvey

1985, 1988; Partridge & Sibly, 1991; Partridge, 1992; Lessells, 1991; Stearns, 1989, 1992) have argued that measurements of patterns of phenotypic plasticity can also be informative. The status of these arguments with respect to the study of aging has also been discussed (Rose, Service & Hutchinson, 1987).

Quantitative genetic techniques have been used to study aging in *Drosophila*, and both breeding designs and artificial selection have been used (e.g. Rose & Charlesworth, 1980; Rose, 1984; Luckinbill *et al.*, 1984). Both the optimality and the mutational theories of aging are consistent with the absence of standing genetic variance for the rate of aging. The optimal life history could in theory be realised by fixation of a single optimal genotype, while deleterious mutations could (at least in principle) become fixed by drift. On the other hand, both theories are also consistent with the presence of genetic variance, which could be attributable to mutation-selection balance or to various forms of balancing selection, including heterozygote advantage: the latter may be produced by 'antagonistic pleiotropy' between life history traits (Rose, 1982, 1985; Rose, Service & Hutchinson, 1987). A polygenic basis for aging has been found in *D. melanogaster* (Luckinbill *et al.*, 1988a; Hutchinson & Rose, 1991; Hutchinson, Shaw & Rose, 1991). However, the mechanisms maintaining this variation are not known, and are in general very difficult to discover for any quantitative trait (Barton & Turelli, 1989). The level of standing genetic variance for aging and the number of genes involved therefore do not at present allow assessment of the roles of optimality and mutation pressure in causing aging.

A fundamental difficulty with any attempt to measure trade-off curves using standing genetic variance is that there need be no direct or obvious connection between the pattern of genetic covariances and either the trade-off surface (which reflects the set of possible phenotypes) or the effects of spontaneous mutations. First, suppose that the population is distributed across the trade-off surface, and is close to the optimal life history. Then there will be little additive genetic variance in fitness, and if fitness were divided into just two components, these would be negatively correlated. However, when many fitness components are involved, the correlations between them will neces-

sarily be small, and may fluctuate in sign (Pease & Bull, 1988; Charlesworth, 1990). Given the difficulties of estimating genetic covariance matrices (Shaw, 1987), and of identifying all components of fitness (cf. Lande & Arnold, 1983), this approach seems problematic.

Another difficulty in interpreting genetic correlations is that the population may not lie on the trade-off surface, in which case the pattern of variation will not reflect the constraints on the phenotype. When variation in a set of quantitative traits is maintained by a balance between mutation and selection, the genetic covariances depend in a complex way on the covariances of effects of new mutations on the traits and on fitness, and on the pattern of selection (Lande, 1980; Turelli, 1985; Wagner, 1989; Charlesworth, 1990). Houle (1991) gives an elegant illustration of the problem. He supposes that loci may either affect the acquisition of resources or their allocation between different components of fitness. Mutation at 'acquisition loci' will produce positive correlations, while mutation at 'allocation loci' will produce negative correlations. While Houle (1991) finds that there must be a much higher mutation rate to the former class than to the latter if the net correlation is to be positive, he argues that such a disparity is likely. Supporting evidence comes from the mutation accumulation experiment of Houle *et al.* (1992), which showed that while the net mutation rate to alleles affecting homozygous fitness is comparable to that for homozygous viability, the rate of decline in mean fitness is considerably greater than for mean viability. This suggests that most mutations that reduce viability also reduce the values of other fitness components when homozygous.

If variation were maintained by balancing selecting, with constant fitnesses, then there would be no additive genetic variance in fitness at equilibrium. However, as noted above, the pattern of genetic covariances among fitness components would be unlikely to show a clear negative relation. Moreover, many genotypes (for example, homozygotes) might lie within the trade-off surface, so that the pattern of genetic variation would not necessarily reflect constraints. More complex mechanisms for maintaining fitness variation, based for example on spatial or temporal heterogeneity, would give additive variation, and would further blur the relation between genetic variation and constraints on fitness components.

These problems are likely to be acute in short-term breeding designs, which rely mainly on the properties of standing genetic variance. In addition, these tests are not particularly sensitive; very large sample sizes are needed to generate reasonable confidence limits on genetic correlations (Shaw, 1987) and biases can be introduced by deaths before late fitness is measured.

Long-term artificial selection may circumvent some of the difficulties with short-term studies. Genetic variation within a base population can be magnified as variation between selected lines, giving greater statistical power. More important, selection over tens of generations can take traits well beyond their original values, and in directions which could not be predicted from study of the base population. To put this another way, the genetic covariance matrix which determines the rate and direction of selection response is itself malleable (Turelli, 1988) because a limited number of genes with non-additive interactions may be involved, because of linkage disequilibria (Bulmer, 1980), and because new mutations can contribute to the response (Hill, 1982). Hence, the long term response may be best able to reveal the fundamental constraints on the character.

A contribution from new mutations is important not just because it allows sampling of a wider array of genetic variants than those present in the base population. There is evidence that mutations of major effect on quantitative traits frequently have substantial negative pleiotropic effects on fitness (Caballero, Toro & Lopez Fanjul, 1991; Mackay, Lyman & Jackson, 1992). However, during artificial selection on the quantitative trait, these negative effects on fitness can be removed by selection for modifiers (McKenzie & Clarke, 1988; Modi & Adams, 1991), so that these alleles can contribute to the long term response despite their unpromising initial effect on fitness. Responses and correlated responses to artificial selection may therefore be more likely to reflect the fundamental limits on the characters, rather than just the initial covariances in the base population.

Artificial selection has been extensively used to study the evolution of aging in *Drosophila*. The typical method has been selection on age at reproduction, in which only the eggs collected from flies of a certain age-band contribute to the next genera-

tion in the selection line. Nomenclatures vary, but in general lines propagated from 'young' adults have been compared with lines where the breeders are 'old'. This experimental design has produced abundant evidence for trade-offs between life history traits. Restriction of breeding to older adults would be expected to reduce aging; there is selection for higher longevity and higher fertility at old ages in these 'old' lines. If a response to selection occurs, and if aging in the original base population were attributable entirely to the presence of more or less age-specific deleterious mutations, then no immediate drop in survival or fertility would be expected to occur earlier in life. If trade-offs were important, then an immediate drop (a correlated response) would be predicted. This type of experiment in *Drosophila* has in general produced evidence for trade-offs, with lifespan and fertility late in life increasing in lines propagated from old adults, and either pre-adult survival or fertility of young adults showing a correlated decline (Rose, 1984; Luckinbill *et al.*, 1984; Partridge & Fowler, 1992).

Although these patterns of response suggest that trade-offs have been important in the evolution of aging in *Drosophila*, the use of artificial selection to study the issue has not been free of problems. It is possible inadvertently to apply direct selection to the characters whose correlated responses are of interest. For instance, in many selection experiments on age at breeding, there may have been unintended greater selection pressure for rapid development and reproductive maturation in the 'young' line flies with which the 'old' lines were compared. Because 'young' lines have often been kept with a two-week generation time, and after a development period of 10 or so days females take some days to reach peak egg-laying rates, those females maturing most rapidly will leave the most progeny. Evolution in the 'young' lines is therefore likely to be towards rapid development and early maturation, while in the 'old' lines, with their much longer generation time, this will not occur. This difference between the selection regimes could result in artefactual differences in early fertility between 'young' and 'old' lines, either because of an increase in early fertility in the former or because of a decrease in the latter. Instead of the difference appearing as a correlated response to selection for longevity or late fertility, it could therefore be a consequence of direct selection for early maturation in the 'young' lines (Partridge & Fowler, 1992; Roper *et al.*, 1993.)

So far, selection by age of breeding has in general been conducted by using flies from a particular band of ages as parents. It is perhaps not surprising that this type of experiment has revealed such consistent evidence for the importance of trade-offs. The reason is that this type of design often involves relaxation of selection on some life history traits. For instance, in 'old' lines, the selection maintaining fertility in early adult life is relaxed because the eggs produced during that time are discarded, so that their number is a selectively neutral trait. However, under the conditions in which the life history evolved, if an ecological change provided conditions permissive for extended survival and fertility, the flies would be selected by a later average age of breeding, but with selection for fertility in early life maintained. Under these conditions the response to selection might be achieved by a different set of alleles with different patterns of pleiotropy from those seen in selection experiments. It is interesting that one set of artificially selected lines in which selection for fertility in early life was maintained in the 'old' (K) lines showed evidence for effects of mutation accumulation in the 'young' (r) lines, but not for trade-offs (Mueller, 1987). However, differences in population density between the 'young' and 'old' lines in this study may have complicated the issue.

As has been frequently pointed out in work on the evolution of aging, all estimates of quantitative genetic variables by any method are sensitive to the environment in which they are measured; misleading gene-environment interactions can occur in an environment other than that in which the life history evolved (Service & Rose, 1985; Clark, 1987). This objection may apply to some extent to all published studies. Unless the base stock was kept for many years in the same conditions as those in which estimates of genetic parameters were made, or in which artificial selection was conducted and the lines assayed, problems could have arisen.

A potential alternative to quantitative genetic approaches for studying trade-offs in the evolution of aging is to use environmental manipulations to study the effects on one or more life history traits of manipulation of another. The use of artificial selection is in practice confined to a few tractable organisms, and is anyway not free of difficulties, for the reasons just given. Some of these difficulties apply also to experimental manipulations. For instance, the equivalent of gene-environment interaction can occur if the response to an environmental manipulation depends upon the environment in which it is conducted (e.g.. Partridge & Fowler, 1991; Chippindale *et al.*, in press). There is also a more basic difficulty of interpretation. Suppose, for example, that flies are reared on poor food, and so reduce their rate of egg laying, and perhaps change other life history traits. One could regard this as revealing a mechanical constraint on the life history, revealed by varying nutrition. Alternatively, one could regard the changes as an evolved response to bad conditions – that is, as reflecting the adapted plasticity of the phenotype (Reznick, 1985; Stearns, 1989; Gomulkiewicz & Kirkpatrick, 1992). If the organism had evolved to produce the maximum number of eggs possible, given its immediate environment, then the distinction would matter little. However, the optimal response depends not just on current circumstances, but on the conditional expectation of the future: the response to laboratory manipulations could not then be interpreted without knowing the sequences of environments which had been experienced in the evolutionary past.

Environmental manipulations are quicker than selection experiments, they can often allow independent manipulation of different variables (e.g.. Chapman, 1992) and they allow reversal experiments to be done, so that the timing of effects on survival and fertility can be measured. It is therefore worth enquiring whether artificial selection and experimental manipulations give at least approximately the same results.

The empirical data from *Drosophila* are so far reasonably encouraging, despite arguments to the contrary (Reznick, 1992a, b; Partridge, 1992). For instance, reduction of egg-production by X-irradiation, by removal of attractive oviposition sites and by reduced nutrition have all been shown to result in an extension of female lifespan (Maynard Smith, 1958; Lamb, 1964; Partridge, Green & Fowler, 1987; Chippindale *et al.*, 1993), while several breeding experiments have shown a negative genetic correlation between early fertility of females and their lifespan, and artificial selection by late age of reproduction has been shown to extend lifespan and to produce a correlated drop in early egg-production (Rose & Charlesworth, 1981; Rose, 1984; Luckinbill *et al.*, 1984).

Similarly, an experimental increase in the rate of re-mating by females reduced their lifespan (Fowler & Partridge, 1989; Chapman, 1992), and a reduction in female lifespan occurred as a correlated response to selection for increased remating rate (Trevitt, 1989). It is at present not clear if this cost of mating plays a part in the response to artificial selection by age at breeding. The survival of females selected by age at reproduction has been compared when they are kept either virgin or mated (e.g.. Luckinbill *et al.*, 1988b; Service, 1989; Partridge & Fowler, 1992). The aim was discover if the cost of mating seen in experimental manipulations played a part in producing the greater lifespan of 'old' line females. The reasoning was, if mating reduces lifespan, then a reduced mating rate would be expected as a correlated response to selection for increased lifespan, and enforced virginity would then be expected to remove at least some of the survival advantage for 'old' line females. All three studies found that virgins lived longer than mated females. The survival differences between flies selected by 'young' or 'old' age at reproduction persisted in virgins indicating that, if remating rate did differ between 'young' and 'old' females, it was not the only thing that did. Other candidate variables include metabolic rate, rate of yolk protein synthesis, egg-laying rate and the extent of allocation to somatic growth during development. Only in the experiment of Luckinbill *et al.* (1988b) did virginity reduce the survival advantage of 'old' line females, measured as a smaller difference in mean lifespan between selection regimes. It has been argued (Service, 1989; Reznick, 1992a, b), that this result indicates that the cost of mating may be evolutionarily irrelevant. However, remating rates were not measured in any of these experiments, and the interpretation ignores differences in fertility between selection lines. The elevated levels of egg-laying in 'old' line females for most of their lives would be expected to elevate remating rates in 'old' line females (Newport & Gromko, 1984; Trevitt, Fowler & Partridge, 1988; Harshman, Hoffmann & Prout, 1988). Furthermore, the response of survival rates to virginity in these experiments could be affected by a number of other differences between the selection lines. For instance, 'old' line females might be expected to show a lower response to

virginity if they were in better somatic condition than 'young' line females as a result of greater investment during development.

More data comparing the results of quantitative genetic approaches and experimental manipulations would be extremely valuable. The data so far suggest that both experimental manipulations and quantitative genetic measurements have a role to play, not least because they provide a check on each other's results. They could be made in few other organisms other than *Drosophila*. A combination of the two, as in the above experiments on the effects of virginity in selected lines, seems to be particularly powerful. The optimal life history for *Drosophila* has not been defined and, until it is, the extent to which aging is part of it will remain unknown. Progress with measuring the shapes of trade-off curves is required.

For understanding the evolution of aging, an additional requirement is to assess how much the observed life history is depressed below the (as yet unmeasured) optimal one by mutation pressure. However, the total rate of production of deleterious mutations and their pattern of age-specificity have not been measured, although progress in this direction is being made using mutation-accumulation experiments in *Drosophila* (Houle *et al.*, 1992). Mutation accumulation experiments will tend to underestimate mutation rates because, even where new mutations are held heterozygous against balancer chromosomes, it is impossible to eliminate natural selection altogether – dominant lethals can never be accumulated. However, in *Drosophila* it is possible to propagate mutation-accumulation lines using single males in each generation, and the resulting small effective population size means that all but dominant lethals are effectively selectively neutral (Houle *et al.*, 1992). Information on the pattern of age-specificity of mutational effects on survival probability and fertility for the mutations accumulated in these experiments would be extremely valuable.

Conclusions

The experimental results so far suggest that aging in *Drosophila* has evolved in part as a consequence of selection for an optimal life history, and in part as a result of accumulation of predominantly late-acting deleterious mutations. This conclusion is not surprising. Quantification of these effects presents a major challenge for the future. The most promising way of addressing the mutation accumulation theory may be direct analysis of the number and effects of spontaneous mutations. The trade-offs which are essential to the optimality theories may best be investigated by combining genetic and experimental manipulations.

References

Barton, N. H., 1990. Pleiotropic models of quantitative variation. Genetics 124: 773-782.

Barton, N. H. & M. Turelli, 1989. Evolutionary quantitative genetics: how little do we know? Ann. Rev. Genet. 23: 337-370.

Bell, G. & V. Koufopanou, 1986. The cost of reproduction. Oxf. Surv. Evol. Biol. 3: 83-131.

Bulmer, M. G., 1980. The Mathematical Theory of Quantitative Genetics. Oxford University Press, Oxford.

Caballero, A., M. Toro & C. Lopez-Fanjul, 1991. The response to artificial selection from new mutations in Drosophila melanogaster, Genetics 128: 89-102.

Calow, P., 1979. The cost of reproduction: a physiological approach. Biol. Rev. 54: 23-40.

Carey, J. R., P. Leido, D. Orozco & J. W. Vaupel, 1992. Slowing of mortality rates at older ages in large medfly cohorts. Science 258: 457-461.

Chapman, T., 1992. A cost of mating with males that do not transfer sperm in female Drosophila melanogaster. J. Insect Physiol. 38: 223-227.

Charlesworth, B., 1980. Evolution in Age-Structured Populations. Cambridge University Press, Cambridge.

Charlesworth, B., 1990. Optimization models, quantitative genetics and mutation. Evolution 44: 520-538.

Charnov, E. L., 1989. Phenotypic evolution under Fisher's fundamental theorem of natural selection. Heredity 62: 113-116.

Chippindale, A. K., A. M. Leroi, S. B. Kim & M. R. Rose, 1993. Phenotypic plasticity of life history mimics response to selection in Drosophila melanogaster: trade-offs between survival and reproduction. J. Evol. Biol. 6: 171-193.

Clark, A. G., 1987. Senescence and the genetic-correlation hang-up. Amer. Natur. 129: 932-940.

Curtsinger, J. W., H. H. Fukui, D. R. Townsend & J. W. Vaupel, 1992. Demography of genotypes: failure of the limited life-span paradigm in Drosophila melanogaster. Science 258: 461-463.

Edney, E. B. & R. W. Gill, 1968. Evolution of senescence and specific longevity. Nature 220: 281-282.

Finch, C. E., 1990. Longevity, senescence and the genome. University of Chicago Press, Chicago.

Fowler, K. & L. Partridge, 1989. A cost of mating in female fruitflies. Nature 338: 760-761.

Gomulkiewicz, R. & M. Kirkpatrick, 1992. Quantitative genetics and the evolution of reaction norms. Evolution 46: 390-411.

Hamilton, W. D., 1966. The moulding of senescence by natural selection. J. Theor. Biol. 12: 12-45.

Harshman, L. G., A. A. Hoffmann & T. Prout, 1988. Environmental effects on remating in Drosophila melanogaster. Evolution 42: 312-321.

Hill, W. G., 1982. Rates of change in quantitative traits from fixation of new mutations. Proc. Natl. Acad. Sci. (USA) 79: 142-145.

Houle, D., 1991. Genetic covariance of fitness correlates: what genetic correlations are made of, and why it matters. Evolution 45: 630-648.

Houle, D., D. K. Hoffmaster, S. Assimacopoulos & B. Charlesworth, 1992. The genomic mutation rate for fitness in Drosophila. Nature 359: 58-60.

Hutchinson, E. W., & M. R. Rose, 1991. Quantitative genetics of postponed aging in Drosophila melanogaster. I. Analysis of outbred populations. Genetics 127: 719-727.

Hutchinson, E. W., A. J. Shaw & M. R. Rose, 1991. Quantitative genetics of postponed aging in Drosophila melanogaster. II. Analysis of selected lines. Genetics 127: 729-737.

Kirkwood, T. B. L. & M. R. Rose, 1991. Evolution of senescence: late survival sacrificed for reproduction. Phil. Trans. Roy. Soc. Lond. B. 332: 15-24.

Lamb. M. J., 1964. The effects of radiation on the longevity of female Drosophila subobscura. J. Insect Physiol. 10: 487-497.

Lande, R., 1980. The genetic covariance between characters maintained by pleiotropic mutations. Genetics 94: 203-215.

Lande, R., 1982. A quantitative genetic theory of life history evolution. Ecology 63: 609-615.

Lande, R. & S. J. Arnold, 1983. The measurement of selection on correlated characters. Evolution 37: 1210-1226.

Lessells, C. M., 1991. The evolution of life histories, pp. 32-68. Behavioural Ecology: An Evolutionary Approach, edited by J. R. Krebs and N. B. Davies. Blackwell Scientific Publishing.

Luckinbill, L. S., R. Arking, M. J. Clare, W. C. Cirocco & S. A., Buck, 1984. Selection for delayed senescence in Drosophila melanogaster. Evolution 38: 996-1003.

Luckinbill, L. S., J. L. Graves, A. H. Reed & S. Koetsawang, 1988a. Localizing genes that defer senescence in Drosophila melanogaster. Heredity 60: 367-374.

Luckinbill, L. S., J. L. Graves, A. Tomkin & O. Sowirka, 1988b. A qualitative analysis of some life-history correlates of longevity in Drosophila melanogaster. Evol. Ecol. 2: 85-94.

Maynard Smith, J., 1958. The effects of temperature and of egg-laying on the longevity of Drosophila subobscura. J. Exp. Biol. 35: 832-842.

Mackay, T. F. C., R. F. Lyman & M. S. Jackson, 1992. Effects of P-element insertions on quantitative traits in Drosophila melanogaster. Genetics 130: 315.

McKenzie, J. A. & G. M. Clarke, 1988. Diazinon resistance, fluctuating asymmetry and fitness in the Australian sheep blowfly, Lucilia cuprina. Genetics 120: 213-220.

Medawar, P. B., 1946. Old age and natural death. Modern Quarterly 1: 30-56.

Medawar, P. B., 1952. An unsolved problem of biology. H. K. Lewis, London.

Modi, R. I. & J. Adams, 1991. Coevolution in bacterial-plasmid populations. Evolution 45: 656-667.

Mueller, L. D., 1987. Evolution of accelerated senescence in laboratory populations of *Drosophila*. Proc. Natl. Acad. Sci. USA 84: 1974-1977.

Newport, M. E. A. & M. H. Gromko, 1984. The effect of experimental design on female receptivity to remating and its impact on reproductive success in *Drosophila melanogaster*. Evolution 38: 1261-1272.

Parker, G. A. & J. Maynard Smith. Optimality theory in evolutionary biology. Nature 348: 17-33.

Partridge, L., 1987. Is accelerated senescence a cost of reproduction? Funct. Ecol. 1: 317-320.

Partridge, L., 1988. Lifetime reproductive success in *Drosophila*. Reproductive success, edited by T. R. Clutton-Brock, University of Chicago Press, Chicago.

Partridge, L., 1992. Measuring reproductive costs. Trends in Ecology and Evolution 7: 99.

Partridge, L. & R. Andrews, 1985. The effect of reproductive activity on the lifespan of male *Drosophila melanogaster* is not caused by an acceleration of ageing. J. Insect. Physiol. 31: 393-395.

Partridge, L. & N. H. Barton, 1993. Optimality, mutation and the evolution of ageing. Nature 362: 305-311.

Partridge, L. & K. Fowler, 1991. Non-mating costs of exposure to males in female *Drosophila melanogaster*. J. Insect Physiol. 36: 419-425.

Partridge, L. & K. Fowler, 1992. Direct and correlated responses to selection on age at reproduction in *Drosophila melanogaster*. Evolution 46: 76-91.

Partridge, L., K. Fowler, S. Trevitt & W. Sharp, 1986. An examination of the effects of males on the survival and egg-production rates of female *Drosophila melanogaster*. J. Insect. Physiol. 32: 925-929.

Partridge, L., A. Green & K. Fowler, 1987. Effects of egg-production and of exposure to males on female survival in *Drosophila melanogaster*. J. Insect Physiol. 33: 745-749.

Partridge, L. & P. H. Harvey, 1985. Costs of reproduction. Nature 316: 20-21.

Partridge, L. & P. H. Harvey, 1988. The ecological context of life history evolution. Science 241: 1449-1454.

Partridge, L. & R. Sibly, 1991. Constraints in the evolution of life histories. Phil. Trans. R. Soc. Lond. *B*332: 3-13.

Pease, C. M. & J. J. Bull, 1988, A critique of methods for measuring life-history trade-offs. J. Evol. Biol. 1: 293-303.

Rees, M. & M. J. Long, Germination biology and the ecology of annual plants. Amer. Nat. 139: 484-508.

Reznick, D., 1985. Costs of reproduction: an evaluation of the empirical evidence. Oikos 44: 257-267.

Reznick, D., 1992a. Measuring the costs of reproduction. Trends in Ecology and Evolution 7: 42-45.

Reznick, D., 1992b. Measuring reproductive costs: response to Partridge. Trends in Ecology and Evolution 7: 134.

Roper, C., P. Pignatelli & L. Partridge, 1993. Evolutionary effects of selection on age at reproduction in larval and adult *Drosophila melanogaster*. Evolution 47: 445-455.

Rose, M. R., 1982. Antagonistic pleiotropy, dominance and genetic variation. Heredity 48: 63-78.

Rose, M. R., 1984. Laboratory evolution of postponed senescence in *Drosophila melanogaster*. Evolution 38: 1004-1010.

Rose, M. R., 1985. Life history evolution with antagonistic pleiotropy and overlapping generations. Theor. Pop. Biol. 28: 342-358.

Rose, M. R., 1991. Evolutionary Biology of Aging. Oxford University Press, Oxford.

Rose, M. R. & B. Charlesworth, 1980. A test of evolutionary theories of senescence. Nature 287: 141-142.

Rose, M. R. & B. Charlesworth, 1981. Genetics of life history in *Drosophila melanogaster*. Sib analysis of adult females. Genetics 97: 173-186.

Rose, M. R., P. Service & E. W. Hutchinson, 1987. Three approaches to trade-offs in life history evolution, pp. 91-105 in Genetic Constraints on Adaptive Evolution, edited by V. Loeschcke, Springer-Verlag, Berlin.

Service, P. M., 1987. Physiological mechanisms of increased stress resistance in *Drosophila melanogaster* selected for postponed senescence. Physiol. Zool. 60: 321-326.

Service, P. M., 1989. The effect of mating status on lifespan, egg laying, and starvation resistance in *Drosophila melanogaster* in relation to selection on longevity. J. Insect Physiol. 35: 447-452.

Service, P. M., E. W. Hutchinson, M. D. MacKinley & M. R. Rose, 1985. Resistance to environmental stress in *Drosophila melanogaster* selected for postponed senescence. Physiol. Zool. 58: 380-389.

Service, P. M. and M. R. Rose. Genetic covariation among life-history components: the effect of novel environments. Evolution 39: 943-945.

Shaw, R. G., 1987. Maximum-likelihood approaches applied to quantitative genetics of natural populations. Evolution 41: 82-826.

Stearns, S. C., 1989. Trade-offs in life history evolution. Funct. Ecol. 3: 259-268.

Stearns, S. C., 1992. The Evolution of Life Histories. Oxford University Press, Oxford.

Trevitt, S., 1989. The costs and benefits of repeated mating in the female fruitly *Drosophila melanogaster* Meigen. Unpublished PhD Thesis, University of Edinburgh.

Trevitt, S., K. Fowler & L. Partridge, 1988. An effect of egg-deposition on the subsequent fertility and remating frequency of female *Drosophila melanogaster*. J. Insect Physiol. 34: 821-828.

Turelli, M., 1985. Effects of pleiotropy on predictions concerning mutation-selection balance for polygenic traits. Genetics 111: 165-195.

Turelli, M., 1988. Phenotypic evolution, constant covariances, and the maintenance of additive variance. Evolution 42: 1342-1348.

Vaupel, W. V. & A. I. Yashin, 1983. The deviant dynamics of death in heterogeneous populations. RR-83-1, International Institute for Applied Systems Analysis, Laxenburg, Austria.

Wagner, G., 1989. Multivariate mutation-selection balance with constrained pleiotropic effects. Genetics 122: 223-234.

Williams, G. C., 1957. Pleiotropy, natural selection, and the evolution of senescence. Evolution 11: 398-411.

Williams, G. C., 1966. Natural selection, the cost of reproduction, and a refinement of Lack's principle. Amer. Nat. 100: 687-690.

M.R. Rose and C.E. Finch (eds.), Genetics and Evolution of Aging, 119–129, 1994.
© 1994 *Kluwer Academic Publishers. Printed in the Netherlands.*

Population density effects on longevity

Joseph L. Graves Jr. & Laurence D. Mueller
Department of Ecology and Evolutionary Biology, University of California, Irvine, CA 92717, USA

Received and accepted 22 June 1993

Abstract

Population density, or the number of adults in an environment relative to the limiting resources, may have important long and short term consequences for the longevity of organisms. In this paper we summarize the way in which crowding may have an immediate impact on longevity, either through the phenomenon known as dietary restriction or through alterations in the quality of the environment brought on by the presence of large numbers of individuals. We also consider the possible long term consequences of population density on longevity by the process of natural selection. There has been much theoretical speculation about the possible impact of population density on the evolution of longevity but little experimental evidence has been gathered to test these ideas. We discuss some of the theory and empirical evidence that exists and show that population density is an important factor in determining both the immediate chances of survival and the course of natural selection.

Theory of population density and senescence

The study of the consequences of population density on the longevity of organisms was in fact first explored by scientists interested in problems in evolutionary ecology. In this field there had for some time been interest in understanding those ecological conditions which would favor two alternative life history patterns (Cole, 1954). Semelparity is one of these patterns which is characterized by a burst of reproduction shortly after sexual maturity, followed by rapid senescence and death. Semelparous life histories are observed in annual plants, salmon and black widows, in addition to many other diverse groups of organisms. Iteroparity, which is the pattern displayed by humans, fruit flies and many other organisms, is characterized by repeated episodes of reproduction after sexual maturity and thus a prolonged adult life stage.

One of the early attempts to understand the ecological pressures which may be important to determining which life history pattern might be most advantageous was made by MacArthur and Wilson (1967). In this work, MacArthur and Wilson suggest that natural selection will act in qualitatively different ways for populations kept at very high densities as opposed to those kept at very low densities. Much of the intuition and theory presented by MacArthur and Wilson was later expanded by several people including Pianka (1972). Predictions from these verbal theories were that under low density conditions rapid reproduction and thus early maturity and semelparity would be favored, while at high population density repeated reproduction and thus iteroparity should be advantageous.

Thus, these ecological theories suggest that low population density would generally accelerate senescence while high density would favor delayed reproduction and increased longevity. Many of the logical underpinnings of these verbal theories have been found to be faulty (see Mueller, 1991, for a review). Following the work of MacArthur and Wilson, Pianka, and others, formal theories which specifically took into account population age-structure and density were examined (Charlesworth, 1980). These theories (Charlesworth, 1980) showed that if density simply resulted in a constant increase in the rate of mortality, there would be no change in the form of selection relative to populations living at low density. However, if density-dependent natural selection acted only on pre-adult survival or fecundity, then it is possible that selec-

120

tion would favor increased longevity and delayed reproduction.

Evidence of density effects on longevity

In this section we focus on the experimental literature dealing with *Drosophila melanogaster*, which spans more than 60 years. While the effects we discuss are fairly well known, they have important implications for recent reports of decreasing rates of mortality with age (Carey *et al.*, 1992).

The earliest study with *Drosophila* (Pearl, Miner & Parker, 1927) used a protocol which has been repeated in more recent studies. A fixed number of adults is used to initiate the experiment, and the survival of this cohort of individuals is followed over time. An important aspect of this experimental protocol is that population density decreases over time since dead individuals are never replaced. As a result, the patterns of mortality from these types of experiments show a response which reflects both the aging process (which presumably increases rates of mortality) and declining density (which will presumably decrease rates of mortality).

The 200 control population in Figure 4 is an experiment like the one just described. From this experiment one can see that rates of mortality increase to a peak at age 11 days and then decline. There is a second increase in rates of mortality very late in life. These data late in life are not terribly accurate and result from the last few surviving adults. Obviously, when each of these last few adults die the estimated rate of mortality will be high (reaching 100% in the time interval at which the last adult dies). We have seen similar phenomena in our laboratory. Figures 1-2 show the mortality patterns for two different populations of *Drosophila melanogaster* (called B and O; see Rose, 1984b) which have experienced different selection regimes. More importantly, each population was initiated at two adult densities, 32 and 200.

Figures 1a,b show the mean and standard error of age-specific mortality rate for the five replicate B and O populations at densities of 32 and 200 flies per vial respectively. In figure 1a, at density 32, the initial mean rate of mortality is low for both the B and O populations; this rate begins to diverge at age 20 days when the B population begins to increase sharply. The corresponding sharp increase in the Os

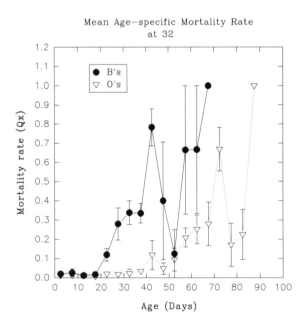
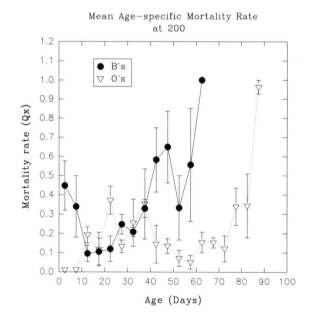

Fig. 1. (a) The fraction of the population dying during each 5 day period are shown for 5 replicated B (early reproduced) and five replicated O (late reproduced) populations kept at an initial density of 32 in a 8 dram vial. Sex ratio was 1:1 and dead individuals were not replaced during the experiment. (b) Similar data as in (a) except the density was initially 200 adults.

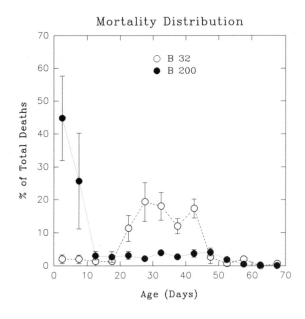
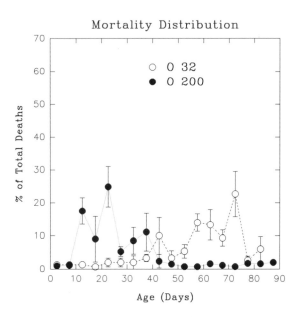

Fig. 2. (a) The same data as in figure 1 but plotted as a fraction of the total deaths for the B populations and (b) for the O populations.

does not occur until after day 40. At later ages the variance in mortality rates between lines becomes so large that accurate assessment of mortality rate is not possible; this is due to the fact that the few individuals left are not always dying in the same intervals for each line. Figure 1b shows the mortality rates for these same populations assayed at 200 flies per vial. In the B population the initial age-specific mortality rates are quite high. These fall after day 10 and then begin to increase afterward until all flies have died ($\approx$ day 60). The Os, on the other hand, show a similar low mortality rate as observed at density 32, a rise and fall between days 10 and 60, and a final increase until all flies have died ($\approx$ day 80).

Figures 2a and 2b are the percent of total deaths that occur in a time interval v. density treatment for the B and O lines respectively. They show that for the Bs at high density (200) that most deaths occur early on ($\leq$ day 10), and then constant percentages are observed until all flies have died, while at low density the distribution of deaths peaks at intermediate ages (days 20-50). The pattern for the Os is qualitatively similar, although the magnitude of death percentages per interval is similar in both the high and low density treatments, and the means of the distributions are considerably to the right of

those observed in the Bs. These figures seem to indicate that two processes may be at work to produce the observed mortality distributions in these stocks, both age-specific mortality factors (related to senescence) and density-dependent mortality factors (related to crowding resistance). The selection regime that produced the Os has been shown to have increased a number of stress resistance and physiological performance capacities of these stocks (Service *et al.*, 1985; Service 1987; Graves & Rose, 1990; Graves *et al.*, 1992). It is possible that these mechanisms of general stress resistance have also produced resistance to crowding effects in the Os relative to their B controls.

A third population (called K; see Mueller & Ayala, 1981a) kept at high adult densities (Fig. 3) again shows this same pattern. These data show that the patterns of mortality may also vary between males and females but larger samples would be required to establish this definitively. The mortality of females from 5-10 and 10-15 days was slightly higher than that of males, although not significantly so (t = 1.43, 1.76, with p = 0.288, 0.219 respectively).

Recently, Carey *et al.* (1992) have conducted experiments, using a protocol similar to the one used in the *Drosophila* experiments reviewed here,

122

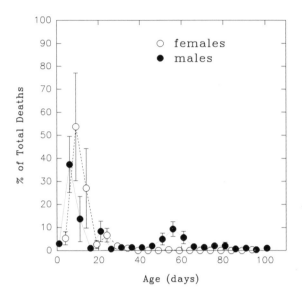

Fig. 3. The fraction of the population that died per 5 day period is shown for the K (high density) populations. Each experiment was initiated with a total of 200 adults, 100 of each sex.

with medflies (*Ceratitis capitata*). Thousands of medflies were put into population cages and the mortality of this cohort was followed over time; dead flies were not replaced, so density declined during the course of this experiment. Mortality reached a peak in middle ages and then fell at later ages. Carey *et al.* took this as evidence that these populations are not in fact senescencing, i.e. rates of mortality actually decreased at later ages rather than increased. However, these populations of medflies were also experiencing declining population density, which would be expected to cause rates of mortality to rise and then decline in a manner similar to what has been observed for the numerous *Drosophila* populations. In this light the Carey *et al.* results are more relevant to density-dependent mortality than they are to the process of senescence (Nusbaum *et al.*, 1993).

Over evolutionary time the manner in which adult populations are affected by density may affect the pattern of selection for either reproduction early or later in life. For example, one would expect that if densities always stayed high, then natural selection would favor earlier reproduction (Charlesworth, 1980). However, if populations faced an early density crisis, and survivors might expect lower overall densities and a renewal of resources, then selection for delayed reproduction might re-

sult. The rationale for the second scenario is two-fold. First, if all individuals at high density are reproducing early, than larval competition would be fierce. Any individual who could withhold reproduction and lay eggs after the population has crashed may actually reproduce more offspring over their life time. Second, if energy for survival of density stress is produced by shutting down reproduction, then the previous scenario is even more likely. Populations with lower reproductive effort retain greater energy reserves in their tissues (Service, 1987; Graves *et al.*, 1992). The evidence now suggests that stocks with different selection histories, when exposed to density stress, utilize energetic reserves that would otherwise be available for reproduction.

Mechanisms by which density affects longevity

In this section we consider mechanisms which have immediate effects on longevity as opposed to long term evolutionary effects.

Food limitation

One of the most obvious mechanisms by which population density may affect Darwinian fitness and hence longevity is by food limitation. The treatment of food as a limiting factor goes back to the earliest thinking about population density effects in ecological theory. Food as a limiting factor was the cornerstone of Malthus' thinking in the now infamous 'An Essay on the Principle of Population', in which Malthus made the erroneous claim that starvation in Ireland and England was due to the fact that food supplies had been exhausted (Malthus, 1798). Despite Malthus' sociological error, the fundamental logic of food limitation as an agent of selection is sound, and was crucial in the formulation of Darwin's principle of natural selection (Darwin, 1859). Modern theory has reformulated the idea of food limitation as a mechanism of density dependent population growth and in relation to life history trade-offs (e.g. Andrewartha & Birch, 1954; Tanner, 1966; Schoener, 1973; Mueller & Ayala, 1981a; Mueller, 1988; Mueller, Gonzalez-Candelas & Sweet, 1991).

Numerous examples exist in which both invertebrate and vertebrate animal populations have been shown to have their fertility lowered or mortality increased by food scarcity (*Drosophila*: Chiang & Hodson, 1954; insects in general; Klomp, 1964; snails: Eisenburg, 1966; Daphnia: Slobodkin, 1954; Frank, Boll & Kelly, 1957; Coleopteran beetles: Davis, 1945; sea urchins: Levitan, 1989; tits and goldcrests: Gibb, 1960; red grouse: Jenkins, Watson & Miller, 1970; birds and mammals in general: Lack, 1954). The competition for food generally has greater fitness effects on juveniles (fish: Beverton, 1962; voles: Hoffman, 1958; deer: Leopold, Sowls & Spencer, 1956; Mitchell, 1973; *Drosophila*: Prout & McChesney, 1985; moths: Gordon & Stewart, 1988; humans: Kulin *et al.*, 1984).

Dietary restriction

Most of the ecological and life history evolution literature has focused on population density effects in which food limitation causes both a depression in fecundity and an increase in mortality. However, gerontological research on laboratory rats and mice has shown that milder levels of dietary limitation (caloric, not nutrient) may actually extend longevity. This phenomenon has been called dietary restriction (DR).

Dietary restriction (DR) has a somewhat broad phylogenetic occurrence and is particularly powerful in its extension of mammalian longevity. The observation that dietary [caloric] restriction extends mammalian longevity was originally reported in laboratory rats (McKay *et al.*, 1935) and has also been demonstrated in a variety of organisms: flies, water fleas, fish, and mice (Ingram *et al.*, 1990). This has prompted a number of authors to speculate on the evolutionary origin of this response for mammals (Holliday, 1989; Harrison & Archer, 1988; Phelan & Austad, 1989).

The debate between Harrison and Archer (1988) and Phelan and Austad (1989) centered on how dietary restriction proved adaptive. Harrison and Archer (1988) argued for natural selection increasing the reproductive life span of mice that experienced periodic episodes of food shortage or drought. They proposed that dietary restriction benefits might be greater in a species with shorter reproductive life spans (therefore *Mus musculus* should benefit more by dietary restriction than *Peromyscus leocopus*, a species which already has a much greater reproductive life span). Phelan and Austad (1989) suggested that natural selection does not work in the way described above. They propose that life extension by dietary restriction is an incidental consequence of its effect on the timing and amount of reproduction. Thus their prediction was that dietary restriction should have the greatest impact on species with early and copious reproduction as compared to with late sexual maturity and relatively small amounts of energetic investment in reproduction. Comparative tests of the predictions of these hypotheses have yet to be conducted.

Experimental evidence exists to support the assertion that dietary restriction operates by fostering analogous trade-offs between growth, survival, and reproduction in a wide variety of organisms: mosquitoes fed on sucrose (Pena & Lavoipierre, 1960a,b;) spider (Austad, 1989); waterstriders (Kaitala, 1991); carabid beetles (Ernsting & Isaaks, 1991); rotifers (Robetson & Salt, 1981); and rats (Holehan & Merry, 1985a,b). Chippindale *et al.* (1993) have shown that dietary restriction (an environmental manipulation) can have the same phenotypic effect as selection for delayed reproduction (a genetic manipulation) in *Drosophila melanogaster*.

There are several striking results of the Chippindale *et al.* study. The study employed early-reproduced, control (B) lines, late-reproduced, postponed senescent lines (O), desiccation selected stocks (D) derived from Os and their controls (C). All stocks showed significant increases in longevity in the D.R. treatments except the (C) females. Longevity and fecundity in the (B) and (O) stocks that had shown the D.R. responses were negatively correlated. The fecundity of the (B) and (O) stocks recovered when their nutritional regimes were switched. Starvation resistance (positively correlated with longevity) and fecundity were also negatively correlated, as in the case of selection for delayed reproduction treatments (Rose, Graves & Hutchinson, 1990; Rose *et al.*, 1992).

This study is one of the most powerful demonstrations yet of phenotype responses to environmental variation paralleling a genetic response to selection. Similar results were found in Partridge (1987). The failure of Lebourg and Medioni (1991) and David (1971) to find D.R. responses in *Drosophila* is likely due to methodological differences, as described in Chippindale *et al.* These studies used food levels that suppressed both fecundity and longevity, thus causing a positive correlation between longevity and fecndity that would not have existed

124

at higher food levels. It has been demonstrated both theoretically and experimentally that environmental conditions may exist in which phenotypic correlations do not reflect genetic correlations (Service & Rose, 1985).

Thus, the D.R. results suggest that at some adult population densities, in a wide variety of species, longevity may actually be increased in contradiction to ecological theory. The antagonistic pleiotropy hypothesis predicts that this increase in longevity will occur at the expense of some component of early reproductive fitness. Models of density-dependent population growth have not yet taken into account this possibility.

Mechanisms of stress resistance and density in Drosophila melanogaster
High population density may also affect the ability of individuals to withstand stress of various sorts. These effects may then make individuals more susceptible to background causes of mortality, thus reducing longevity. The earliest controlled studies of the impact of crowding on longevity was made by Pearl and his colleagues (Pearl & Parker, 1922; Pearl, Miner & Parker, 1927) on *Drosophila melanogaster*. In one study *D. melanogaster* adults were kept at two different adult densities: 200 and 35. After 16 days some of the replicate tretments at density 200 were reconstituted to total densities of 200 (since many of the initial 200 adults had died). Samples of adult flies which had been placed at density 35 were also reconstituted to a density of 200. From Figure 3 it is apparent that the chances of survival are smaller for adults which had lived their first 16 days at a density of 200 compared to the adults raised at density 35, even when both experienced the same high density late in life. Of course from the control populations it is also clear that high density alone dramatically increases the chances of mortality, especially at young ages when these density differences were most profound.

Table 2 shows the effect of high population density on selected physiological performance characters in *D. melanogaster* stocks created by age-specific selection (B v. O, from Graves *et al.*, in prep). These data were derived from B and O stocks maintained at high density (200 per 8 dram vial) or low density (32 per 8 dram vial) for 24 h. The high density treatments for desiccation resistance, flight

Table 1. The effects of density on two physiological characters, desiccation resistance and starvation resistance, at two densities (low: 32 adults per vial, high: 200 adults per vial) and in two populations (B and O).

	B Low	B High	O Low	O High
Desiccation (h)	8.28	7.19	13.19	7.22
Standard error	0.78	0.38	1.22	0.33
Starvation (h)	40.37	31.05	58.69	52.75
Standard error	2.07	2.02	2.09	1.53

duration, and starvation resistance all showed a decline in performance regardless of stock. The O stock showed some density resistance for flight duration at 64; the mean flight duration was slightly higher than 32. The difference in starvation resistance for 24 h was not significant for the O stock.

However, the loss of physiological performance in the postponed senescent, long-lived O lines seemed to be more pronounced for desiccation resistance and flight duration. Graves *et al.* (1992) and Service (1987) demonstrate that the underlying energetic sources for these physiological characters are different. Glycogen has been shown to be the chief reserve accounting for differential performance in flight and desiccation, while lipids are central in starvation resistance. Wigglesworth (1949) has shown for *Drosophila* that during the course of starvation glycogen reserves get mobilized first; Graves *et al.* (in prep.) have confirmed this for these stocks. If density stress at some level involves mobilization of energetic reserves, then it is likely that glycogen reserves should be exhausted before lipids, and those physiological suites dependent on glycogen should be more density-sensitive than lipid suites. These data support that possibility.

Other mechanisms
The effects of density may act in specific ways that

Table 2. The effects of density on flight duration, at three densities and in two populations (B and O).

Density	32	64	200
B (min)	45.5	39.1	0
Standard error	7.94	3.55	0
O (min)	105	116	1.50
Standard error	8.95	14.0	0.42

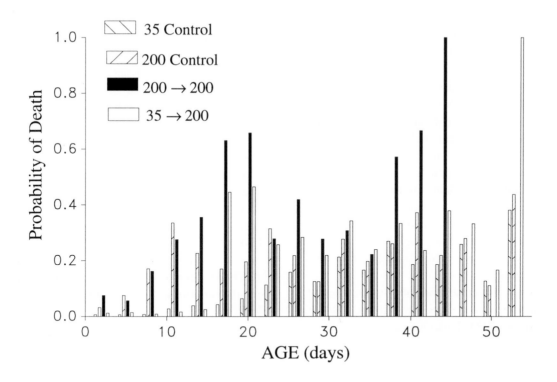

Fig. 4. The chances of death for adult *D. melanogaster* as a function of age and density (Pejarl, Miner & Parker, 1927). Two control populations were started at initial densities of 35 and 200 and survival of these cohorts followed over time. After 16 days adults from the 200 and 35 treatments were sampled and placed at a total density of 200. This dramatically increases the chances of dying but the population raised at 200 for the first sixteen days is more severely affected than the populations raised at a density of 35 during this same period.

are peculiar to certain groups of organisms. For instance, in crowded populations adults of the sea urchin, *Diadema antillarum*, may cannibalize each other (Levitan, 1989). Increased density does not always result in reductions in longevity. For instance, juvenile survivorship of the coral reef fish, *Dascyllus aruanus*, increased with increasing adult density. For this species adults may serve to provide juveniles with predator warning and thus reduce one source of mortality.

Density may also act to modify the environment and affect sources of mortality. In laboratory cultures of *D. melanogaster* there is a dramatic decline in survival rates of adults kept in crowded cultures (Mueller & Ayala, 1981b) and this decline is greater for females than for males. During the course of Mueller and Ayala's experiments (which lasted one week), eggs laid during the first day of the experiment hatched and grew. When the population density is high the large number of larvae

gives the food a very soft, sticky consistency. Adults easily get stuck in this material and drown. Females appear to be more adversely affected in these environments since they attempt to oviposit on the food surface while males seem content to live on the sides of the culture away from the food surface. Davis (1945) suggests that declines in longevity with increasing density of the coleopteran *Trogoderma versicolor* may be mediated by changes in the environmental substrate. Communicable disease is also expected to be spread more rapidly in crowded cultures and thus contribute to the decline in longevity (Hassell & May, 1989).

Population density and the evolution of longevity

Juvenile crowding
There are now two population genetic theories of aging with support from different experimental

systems. The antagonistic pleiotropy theory (Williams, 1957; Rose, 1991) suggests that genes with beneficial effects on early components of fitness, like fertility and survival, but deleterious effects late in life may nevertheless be favored by natural selection due to the greater weight natural selection puts on early fitness components. The mutation accumulation hypothesis (Medawar) suggests that deleterious mutations occur which may act to reduce survival or fertility late in life, but are only weakly disfavored by natural selection, and hence may accumulate to significant frequencies in most populations, giving rise to senescent phenotypes. A consequence of both of these theories is the prediction that natural populations may harbor genetic variation for senescence which can itself be the focus of selection.

There have now been several successful attempts to select for delayed senescence (Rose, 1984a; Luckinbill *et al.*, 1984; Partridge & Fowler, 1991) in the fruit fly *Drosophila melanogaster*. All of these experiments employed a similar protocol of progressively culturing populations from flies which had lived to a late age and then raising the larvae under a moderate to high density (50 or more larvae per vial). Several attempts at selecting for increased longevity have, however, been unsuccessful (Lints & Hoste, 1974; Lints *et al.*, 1979; Flanagan, 1980). These experiments differed from the previous ones in that the larvae were raised at a low density (10 larvae per vial). Is it possible that the density at which larvae are raised could have an effect on the ability to select for adult longevity?

The previous question can be partitioned into two related questions, one of which is easier to answer than the other. The first is, does the larval density affect adult longevity at the phenotypic level? This question was addressed in early work on *Drosophila* by Miller and Thomas (1958). They showed that larval crowding, which generally reduced the size of the resulting adults, had the effect of increasing adult longevity. An important distinction, then, is when does larval crowding ensue, and can we consider a larval density of 10/vial to be meaningfully different from a larval density of 50/vial? Miller and Thomas showed that larval densities of 5-20 larvae/vial showed no appreciable change in final adult size but at densities of 40, 60, 80 and 100 larvae/vial adult size showed a continuous decline. Thus, the larval densities used by Rose, Luckinbill

and Partridge can all be considered to fall in a range of densities in which the adult size is less than its maximum due to larval crowding.

A second, more difficult, question is, does the expression of additive genetic variation for longevity vary depending on the larval environment? The only practical way to test that proposition is to directly measure the additive genetic variation in the two environments or to select for the trait of interest in both environments and observe the change in phenotypic means of the populations due to this selection. In fact the latter experiment has been done by Luckinbill and Clare (1985) with the finding that selection for longevity was not successful when the larval rearing density was 10/vial but was successful at densities of 70, 120 or higher per vial. This effect is known as a genotype by environment interaction and implies that the outcome of selection for longevity will depend critically on the environmental conditions under which the selection is carried out. In this case at least one critical environmental factor is larval density. At this time there is insufficient information about the genetic and physiological mechanisms affecting longevity to offer a detailed explanation for this particular genotype by environment interaction.

Interactions between age-specific and density-dependent natural selection

The theories of density-dependent and age-specific natural selection have developed almost independently of each other (see Charlesworth, 1980, for exceptions). Nevertheless, it is still of some interest to determine if these different types of natural selection affect common sets of genes or physiological processes. Preliminary information is available from a comparison of *Drosophila* populations independently selected by age-specific and density-dependent selection (Mueller *et al.*, 1993). Mueller *et al.* (1993) examined traits which had responded to age-specific selection in populations that had been subject to density-dependent selection and vice-versa. In general there was little evidence that age-specific selection affected the traits that responded to density-dependent selection or vice versa.

Recently, Mueller (unpublished) has found that populations that have evolved at high larval densities are more tolerant to environmental urea than are controls. This is not too surprising since crowded larval cultures can have very high levels of urea. Perhaps more surprising was the observation that populations selected for increased longevity also have an increased ability to resist high levels of environmental urea (Leroi, unpublished), since these populations and their controls are raised at the same low density. These results suggest the possibility that these two types of selection may affect common sets of genes or physiological mechanisms. Further study of the mechanisms of urea tolerance may help determine these connections.

Summary

Increasing population density generally decreases longevity. These effects may be mediated through a variety of mechanisms such as reduction in food resources, alterations of the physical environment, and disease. Longevity is a trait that is molded by natural selection. By changing the manner in which natural selection acts, longevity may be increased in populations. The ability to successfully select for increased longevity in *Drosophila* is affected by the density regime used in the experiment. Current results with *Drosophila* also suggest that common physiological pathways may be involved in the process of adapting to high density and to prolonged lifespan.

Acknowledgements

We thank two anonymous referees for helpful comments on the manuscript. This work has been supported by US-PHS grant AG09970 to both authors.

References

Andrewartha, H. G. & L. C. Birch, 1954. Distribution and Abundance of Animals. University of Chicago Press, Chicago.

Austad, S. N., 1989. Life extension by dietary restriction in the bowl and doily spider, *Frontinella pyramitela*. Exp. Gerontol. 24: 83-92.

Beverton, R. J. H., 1962. Long-term dynamics of certain North Sea fish populations, pp. 242-259 in The Exploitation of Natural Animal Populations, edited by E. D. LeCren and M. W. Holgate. Wiley, N.Y.

Carey, J. R., P. Liedo, D. Orozco & J. W. Vaupel, 1992. Slowing of mortality rates at older ages in large medfly cohorts. Science 258: 457-461.

Charlesworth, B., 1980. Evolution in age-structured populations. Cambridge University Press.

Chiang, H. C. & A. C. Hodson, 1954. An analytical study of population growth in *Drosophila melanogaster*. Ecol. Monog. 20: 173-206.

Chippindale, A. K., A. M. LeRoi, S. B. Kim & M. R. Rose, 1993. Phenotypic plasticity and selection in *Drosophila* life-history evolution. I. Nutrition and the cost of reproduction. J. Evol. Biol. 6: 171-193.

Cole, L. C., 1954. The population consequences of life history phenomena. Q. Rev. Biol. 29: 103-137.

Darwin, C., 1859. On the Origin of Species. Reprinted by Harvard University Press.

David, J., J. Van Herrewege & P. Fouillet, 1971. Quantitative underfeeding of *Drosophila*: effects on adult longevity and fecundity. Exp. Gerontol. 6: 249-257.

Davis, M. B., 1945. The effect of population density on longevity in *Trogoderma versicolor* creutz (= *T. inclusa* Lec.). Ecology 26: 353-362.

Edney, E. B. & R. W. Gill, 1968. Evolution of senescence and specific longevity. Nature 220: 281-282.

Eisenburg, J. F., 1981. The Mammalian Radiations. The University of Chicago Press, Chicago.

Eisenburg, R., 1966. The regulation of density in a natural population of the pond snail *Lymnaea elodes*. Ecology 47: 889.

Ernsting, G. & J. A. Issaks, 1991. Accelerated aging: a cost of reproduction in the carabid beetle *Notiophilus biguttatus* F. Funct. Ecol. 5: 299-303.

Flanagan, J. R., 1980. Detecting early-life components in the determination of the age of death. Mech. Aging Develop. 13: 41-62.

Frank, P. W., C. D. Bell & R. W. Kelly, 1957. Vital statistics of laboratory cultures of *Daphnia pulex* DeGree as related to density. Physiol. Zool. 4: 287-305.

Gibb, J. A., 1960. Populations of tits and goldcrests and their food supply in pine plantations. Ibis 102: 163-208.

Gordon, R. M. & R. K. Stewart, 1988. Demographic characteristics of the stored product moth *Cadra cautella*. J. Anim. Ecol. 57: 627-644.

Graves, J. L. & M. R. Rose, 1990. Flight duration in *Drosophila melanogaster* selected for postponed senescence. Chapter 5. in Genetic Effects on Aging II., Telford Press, Caldwell, NJ.

Graves, J. L., E. C. Toolson, C. Jeong, L. N. Vu & M. R. Rose, 1992. Desiccation, flight, glycogen, and postponed senescence in *Drosophila melanogaster*. Physiol. Zool. 65: 268-286.

Hamilton, W. D., 1966. The moulding of senescence by natural selection. J. Theor. Biol. 12: 12-45.

Harrison, D. E. & J. R. Archer, 1988. Natural selection for extended longevity from food restriction. Growth Dev. Aging 52: 207-211.

Hassell, M. P. & R. M. May, 1989. The population biology of host-parasite and host-parasitoid associations. Chapter 22, in Perspectives in Ecological Theory, edited by J. Roughgarden, R. M. May and S. A. Levin. Princeton University Press, Princeton, N.J.

Hoffman, R. S., 1958. The role of reproduction and mortality in population fluctuations of voles (Microtus). Ecol. Monog.

128

28: 79-109.

Holehan, A. M. & B. J. Merry, 1985a. Modification of the oestrous cycle hormonal profile by dietary restriction. Mech Aging and Dev. 32: 63-76.

Holehan, A. M. & B. J. Merry, 1985b. The control of puberty in the dietary restricted female rat. Mech Aging and Dev. 32: 179-191.

Holehan, A. M. & B. J. Merry, 1986. The experimental manipulation of aging by diet. Biol. Rev. Camb. Phil. Soc. 61: 329-368.

Holiday, R., 1989. Food, reproduction and longevity: is the extended lifespan of calorie-restricted animals an evolutionary adaptation? Bio Essays 10: 125-127.

Ingram, D. K., R. Weindruch, E. W. Spangler, J. R. Freeman & R. L. Walford, 1987. Dietary restriction benefits learning and motor performance of aged mice. J. Gerontol. 42: 78-81.

Luckinbill, L. S. & M. J. Clare, 1985. Selection for life span in *Drosophila melanogaster*. Heredity 55: 9-18.

MacArthur, R. H. & E. O. Wilson, 1967. The theory of island biogeography. Princeton University Press.

Malthus, T., 1798. An Essay on the Principle of Population. Reprinted by MacMillan, New York.

McKay, C. M., L. A. Maynard, G. Sperling & L. L. Barnes, 1939. Retarded growth, life span, ultimate body size and age changes in the albino rat after feeding diets restricted in calories. J. Nutr. 18: 1-13.

Medawar, P. B., 1952. An unsolved problem in biology. H. K. Lewis, London.

Miller, R. S. & J. L. Thomas, 1958. The effects of larval crowding and body size on the longevity of adult *Drosophila melanogaster*. Ecology 39: 118-125.

Mitchell, B., 1973. The reproductive performance of wild Scottish red deer, *Cervus elaphus*. J. Repro. Fert. 19: 271-285.

Morris, J. G., 1991. Nutrition, in Environmental and Metabolic Animal Physiology. Wiley-Liss, New York.

Mueller, L. D., 1987. Evolution of accelerated senescence in laboratory populations of *Drosophila*. Proc. Natl. Acad. Sci. USA. 84: 1974-1977.

Mueller, L. D., 1988. Density-dependent population growth and natural selection in food limited environments: the *Drosophila* model. Am. Nat. 132: 786-809.

Mueller, L. D., 1991. Ecological determinants of life-history evolution. Phil. Trans. R. Soc. Lond. B 332: 25-30.

Mueller, L. D. & F. J. Ayala, 1981a. Trade-off between r-selection and K-selection in *Drosophila* populations. Proc. Natl. Acad. Sci. U.S.A. 78: 1303-1305.

Mueller, L. D. & F. J. Ayala, 1981b. Fitness and density dependent population growth in *Drosophila melanogaster*. Genetics 97: 667-677.

Mueller, L. D., J. L. Graves & M. R. Rose, 1993. Interactions between density-dependent and age-specific selection in *Drosophila melanogaster*. Functional Ecology (in press).

Mueller, L. D., F. Gonzalez-Candelas & V. F. Sweet, 1991. Components of density-dependent population dynamics: models and tests with *Drosophila*. Amer. Nat. 137: 547-475.

Nusbaum, T. J., J. L. Graves, L. D. Mueller & M. R. Rose, 1993. Letters to Science. Science 260: 1567.

Partridge, L., 1987. Is acclerated senescence a cost of reproduction? Funct. Ecol. 1: 317-320.

Partridge, L. & Fowler, 1992. Direct and correlated responses to

selection on age at reproduction in *Drosophila melanogaster*. Evolution 46: 76-92.

Pearl, R. & S. L. Parker, 1922. Experimental studies on the duration of life. IV. Data on the influence of density of population on duration of life in *Drosophila*. Amer. Nat. 56: 312-321.

Pearl, R., J. R. Miner & S. L. Parker, 1927. Experimental studies on the duration of life. XI. Density of population and life duration in *Drosophila*. Amer. Nat. 61: 289-318.

Pena de Grimaldo, E. & M. M. J. Lavoipierre, 1960a. Efecto de la fertilazacion sobre la ovopostura de los mosquitos *Aedes aegypti* variedad queenlandensis, con algunas observaciones sobre anomalias y viabilidad de los hevos retindos por los mosquitos esteriles fecundizados a diferentes intervalos despues de la comida de sangre. Rev. Iberica Parasitol. 20: 163-176.

Pena de Grimaldo, E. & M. M. J. Lavoipierre, 1960b. Longevidad de los mosquitos *Aedes aegypti* variedad queenlandensis fecundados y no fecundados, alimentados con sangre o privados de ella; y dejados en qyuno; o preveidos de aqua; de solucion de azucar; o de aqua y solucion de azucar. Rev. Iberica Parasitol. 20: 39-52.

Phelan, J. P. & S. N. Austad, 1989. Natural selection, Dietary Restriction, and Extended longevity. Growth, Dev. and Aging.

Pianka, E., 1972. r- and K-selection or b and d selection? Am. Nat. 106: 581-588.

Prout, T. & F. McChesney, 1985. Competition among immatures affects their adult fertility: population dynamics. Am. Nat. 126: 521-558.

Robertson, J. R. & G. W. Salt, 1981. Responses in growth, mortality, and reproduction to variable food levels by the rotifer, *Aspolanchia girodi*. Ecology 62: 1585-1596.

Rose, M. R., 1984a. Laboratory evolution of postponed senescence in *Drosophila melanogaster*. Evolution 38: 1004-1010.

Rose, M. R., 1984b. Evolutionary route to Methusalah. New Scientist 103: 15.

Rose, M. R., 1985. Life history with antagonistic pleiotropy and overlapping generations. Theor. Popul. Biol. 28: 342-358.

Rose, M. R., 1991. Evolutionary Biology of Aging. Oxford University Press, New York.

Rose, M. R., J. L. Graves & E. W. Hutchinson, 1990. The use of selection to probe patterns of pleiotropy in fitness characters. Chapter 2. In: Insect Life Cycles: Genetics, Evolution, and Coordination. Springer-Verlag. Berlin.

Rose, M. R., L. N. Vu, S. Park & J. L. Graves, 1992. Selection on stress resistance incrases longevity in *Drosophila melanogaster*. Exop. Gerontol. 27: 241-250.

Schoener, T., 1973. Population growth regulated by intraspecific competition for energy or time. Theor. Popul. Biol. 4: 56-84.

Service, P. M., 1987. Physiological stress mechanisms of increased stress resistance in *Drosophila melanogaster*. selected for postponed senescence. Physiol. Zool. 60: 321-326.

Service, P. M., 1989. The effect of mating status on life span, egg laying, and starvation resistance in *Drosophila melanogaster*. in relation to selection for longevity. J. Insest Phys. 35: 447-452.

Service, P. M., E. W. Hutchinson, M. D. MacKinley & M. R. Rose, 1985. Resistance to environmental stress in *Droso-

phila melanogaster. selected for postponed senescence. Physiol. Zool. 58: 380-389.

Service, P. M. & M. R. Rose, 1985. Genetic covariation among life-history components: The effect of novel environments. Evolution 39: 943-945.

Slob, A. K., S. J. M. Vreeburg & J. J. van der Werff ten Bosch, 1979. Body growth, puberty and under nutrition in the male guinea pig. Br. J. Nutr. 41: 231-237.

Slobodkin, L. B., 1954. Population dynamics in *Daphnia obscura* Kurz. Ecol. Monog. 24: 69-88.

Tanner, J. T., 1966. Effects of population density on growth rates of animal populations. Ecology 45: 733-745.

Wigglesworth, V. B., 1949. The utilization of reserve substances in *Drosophila* during flight. J. Exp. Biol. 26: 150-163.

Williams, G. C., 1957. Pleiotropy, natural selection, and the evolution of senescence. Evolution 11: 398-411.

M. R. Rose and C.E. Finch (eds.), Genetics and Evolution of Aging, 130–144, 1994.
© 1994 *Kluwer Academic Publishers. Printed in the Netherlands.*

Evolution of delayed reproductive senescence in male fruit flies: sperm competition

Philip M. Service[1] & Amanda J. Fales[2]
[1] *Department of Biological Sciences, Northern Arizona University, Box 5640, Flagstaff, AZ 86011*
[2] *T. H. Morgan School of Biological Sciences, University of Kentucky, Lexington, KY 40506, USA*

Received 22 June 1993   Accepted 22 June 1993

Key words: Drosophila, evolution, reproduction, senescence, sperm competition

Abstract

Populations of *Drosophila melanogaster* that had been subjected to long-term selection favoring either delayed or rapid senescence were compared with respect to age-specific components of male reproductive success involving sperm competition. These components of reproductive success were divided into those related to sperm 'defense' (protection of sperm from other males), and into those related to sperm 'offense' (ability to mate with previously mated females and to displace the sperm of other males). Males were tested at four ages ranging from 1-2 d to 5-6 wk after eclosion. Several aspects of sperm defense capability showed clear evidence of senescent decline. Furthermore, males from populations selected for delayed senescence were superior to males from control (rapid senescence) populations with regard to components of sperm defense. The superiority of males from populations with delayed senescence either increased as a function of male age, or was present at all ages tested. These results indicate that the rate of reproductive senescence in male *D. melanogaster* can be altered in predictable directions by artificial selection. There were no differences between selection regimes with regard to sperm offense, and most components of sperm offense did not show clear evidence of senescence. The improved late-age reproductive success of males from populations selected for delayed senescence did not appear to entail any cost or trade-off at early ages with respect to the reproductive traits examined in these experiments.

Introduction

Evolutionary changes in patterns of senescence in the laboratory

In terms of evolutionary and population genetic theory, senescence describes the decline of age-specific reproductive success and survivorship that occurs with advancing age. Williams (1957) proposed that senescence may be due to pleiotropic genes with antagonistic age-specific effects, that is, increased fitness at young ages in return for rapid decline in age-specific reproductive success and short life span. Alleles with beneficial early-age effects but deleterious late-age effects may have a net selective advantage because the force of selec-

tion in age-structured populations declines at later ages (Hamilton, 1966; Charlesworth, 1980). Alternatively, senescence may be due to mutation accumulation – the accumulation of alleles with deleterious late-age effects but without corresponding pleiotropic advantages at earlier ages (Rose & Charlesworth, 1980). Because the force of selection declines with advancing age, as already noted, such alleles may rise to relatively high frequency by mutation-selection balance (Charlesworth, 1980).

Regardless of the underlying mechanism of gene action, it should be possible to manipulate patterns of senescence through artificial selection by changing the force of selection acting on reproductive success at different ages. Selection regimes that favor reproductive success at older ages increase mean female and male longevity and increase the fecundity of older females in laboratory populations of *Drosophila melanogaster* (Rose & Charlesworth, 1981; Luckinbill *et al.*, 1984; Rose, 1984; Partridge & Fowler, 1992). That is, senescence is delayed in populations that are selected for increased late-life fitness relative to populations that are maintained under selection regimes that favor early-life fitness or that relax selection on late-life fitness.

In *Drosophila*, evolutionary studies of reproductive senescence have been concerned mostly with age-specific female reproductive success, which can be estimated fairly well by fecundity. Determination of reproductive success in males is more problematic because it is generally more difficult to determine the actual number of progeny fathered by males under conditions resembling those in laboratory cultures. Male reproductive success can be decomposed into components such as competitive and non-competitive mating ability, and into components related to sperm competition. Selection for delayed senescence increases the reproductive success of older males when measured either as mating success with virgin females under competitive and non-competitive conditions, or as the ability to recover from exhaustive mating bouts (Service, 1993). Selection for delayed senescence also increases the reproductive success of older males when measured in a way that combines the effects of mating success, fertility, and sperm competition (Roper, Pignatelli & Partridge, 1993). Both of these sets of experiments clearly demonstrate that senescence of at least some components of reproductive success in male *Drosophila* can be delayed by appropriate selection. Service (1993) also found that males from populations selected for delayed senescence showed a decrease in competitive mating ability at very young ages, but no trade-off was observed in non-competitive mating success or in ability to recover from an exhaustive mating bout. Roper, Pignatelli & Partridge (1993) also did not observe a reduction in their inclusive measure of reproductive success in young males from lines selected for delayed senescence.

In this paper, we present the results of further analyses of age-specific male reproductive success in populations of *D. melanogaster* that have been selected for divergent rates of senescence. The present studies concern the correlated effects of such selection on components of male reproductive success that involve sperm competition.

Female remating and sperm competition in
D. melanogaster

Female *D. melanogaster* store sperm. However, a female may mate with several males during her lifetime. If a female remates before utilizing all stored sperm, then sperm competition (Parker, 1970) will occur if the 'first' male suffers a loss in the number of progeny that he sires and/or if the 'second' male loses progeny to the first male. Studies of sperm competition in *Drosophila* have concentrated on losses by first males. Such losses have been well documented and have generally been referred to as sperm 'displacement' (e.g., Lefevre & Jonsson, 1962; Boorman & Parker, 1976; Prout & Bundgaard, 1977; Gromko & Pyle, 1978), or sperm 'predominance' (Gromko, Gilbert & Richmond, 1984).

Despite the fact that the loss of first-male progeny due to female remating is well established, the importance of sperm competition in *Drosophila* has been a subject of considerable argument. This debate has centered around the question of whether females tend to remate only after most stored sperm have been utilized – 'sperm dependence' of remating – (Manning, 1962, 1967; Gromko, Newport & Kortier, 1984; Newport & Gromko, 1984), or whether remating occurs relatively rapidly and before first-male sperm have been substantially depleted (Lefevre & Jonsson, 1962; Prout & Bundgaard, 1977; Harshman, Hoffman & Prout, 1988). Under experimental conditions of continuous confinement, as is the case in laboratory culture, remating tends to occur relatively rapidly (e.g., Lefevre & Jonsson, 1962). If experimental conditions are designed to mimic the intermittent remating opportunities that are presumed to be more 'natural', remating tends to be sperm-dependent (e.g., Newport & Gromko, 1984).

Sperm-competitive ability can be partitioned into 'defensive' and 'offensive' components. The defensive ability of a first male might be crudely quantified by the likelihood that a female will remate before using all of his stored sperm. A more refined measure might assess the degree of sperm dependence of remating. Females mated to first males with greater defensive ability would not remate until fewer sperm remained in storage. Given that remating occurs, the sperm-defensive capability of a first male might further be quantified by the ability of his sperm to resist displacement or to reduce the fertilization success of second-male sperm. Similarly, the offensive capability of a second male could be quantified by his ability to induce a female to remate before she has utilized all stored sperm. And, given that the female remates, offensive capability might further be quantified by the ability of the second-male to displace the sperm of the first male.

In this paper, we examine the effects of selection for delayed senescence on components of age-specific male reproductive success related to sperm defense by first males and offense by second males. We predict that populations selected for delayed senescence should be superior in both defense and offense at older ages when compared to control populations that have been selected for rapid senescence. At younger ages, males from populations selected for delayed senescence may be inferior to those from control populations, if age-specific trade-offs within reproductive fitness components have occurred.

Materials and methods

Populations

The populations used for these experiments have been described in detail elsewhere (Rose, 1984; Service *et al.*, 1985; Service, 1993). Their histories and culture conditions will be summarized here. Eight fly populations were studied. Four of these, denoted B_2–B_5, were maintained in discrete generations of 2-wks length. The remaining four populations, O_2–O_5, were maintained in discrete generations of 10-wks length. All eight populations were derived from the same ancestral population, denoted IV. The present experiments were conducted during 1988-89 at the University of Kentucky. At the time of these experiments, there had been approximately 200 generations of the B populations and 30 ten-week generations of the O populations.

Before establishment of the B and O populations, the IV population from which they were derived had been maintained at large size (> 1,000 individuals) in the laboratory for about 130 two-week generations. The IV and B populations were under

strong selection for early-life fitness, and selection was absent for fitness components that would have been expressed at ages greater than five or six days of adult age. All populations were maintained at 25 °C. At that temperature, the minimum development time from oviposition to adult eclosion is about eight days. The O populations, conversely, were under strong selection for late-life fitness: that is, selection for increased number of viable zygotes produced 70 d after the start of a generation. Thus, selection pressures in the IV and B populations were designed to maximize the rate of senescence while those in the O populations were designed to delay senescence, relative to the IV and B populations. These selection procedures resulted in relative postponement of senescence in the O populations as measured by changes in longevity and age-specific fecundity patterns (Rose, 1984) and in male mating ability (Service, 1993). Mean adult life span of O-population males is almost 50% longer than that of B-population males (Service, 1989).

The B populations are the 'controls' or baseline to which the O populations are to be compared. The B populations were maintained in exactly the same way as the ancestral IV population. If the IV population was at, or close to, an evolutionary equilibrium prior to establishment of the B and O populations, then little further selective change should have occurred in the B populations. Using the 'base' IV population as a control would not remove the potential problem of a shifting baseline with which to compare the O populations: continued evolution due to consistently applied selection for early-life fitness and relaxed selection for late-life fitness is as likely to have occurred in the IV as in the B populations. Replication of the B and O populations makes it possible to distinguish the effects of selection (applied consistently to the populations within each selection regime) from the effects of random genetic drift.

In order to evaluate sperm utilization by doubly-mated females, it was also necessary to use females and males carrying an easily observed phenotypic marker. We used the common recessive eye-color mutant, *sepia* (*se*). *Sepia* and Oregon-R (wild-type) flies were obtained from the Mid-America Drosophila Stock Center, Department of Biological Sciences, Bowling Green State University, Bowling Green, OH, USA. These two stocks were crossed, and virgin *se* homozygotes were extracted from the

F4 generation in order to establish the *sepia* stock used for these experiments. This *sepia* stock was maintained in the same manner as the B and IV populations for approximately four months prior to beginning the present experiments.

Procedures for rearing and maintaining experimental flies

Experimental males were produced by two generations of controlled-density rearing. For each population, six vials (95 × 25 mm), containing 60 eggs each, were established with eggs collected from the stock cultures. The resulting adults were collected as virgins. From these adults, 40 male/female pairs per population were allowed to mate and lay eggs. From each pair, ten eggs were collected, and the eggs from six pairs were transferred to a single rearing tube (i.e., a total density of 60 eggs per tube). There were also six rearing tubes per population in this generation. Males emerging from these rearing tubes were the experimental males. During earlier experiments (Service, unpublished), it became apparent that the egg-adult development time of the O-population flies was approximately 12 h longer than that of B-population flies. Therefore, egg collection for the second generation of controlled-density rearing was offset so that B-population eggs were collected approximately 9 h after O-population eggs. This helped to synchronize eclosion of flies from both selection regimes. Experimental males from each tube (also including the females from that rearing tube) were transferred to a fresh food tube three times per week for the duration of the experiment. *Sepia* females and males were obtained by rearing larvae at moderate, but not precisely controlled, densities. In general, CO_2 anesthesia was used for transfer of flies and for virgin collections. However, CO_2 was not used on flies during the 24 h preceding mating or remating tests. All experimental flies, as well as the stocks they were derived from, were maintained on a 24:0 light:dark regime.

Sperm defense (B and O-population males used as first males)

B and O-population males (experimental males) were tested at four ages: 24-36 h (0 wk), and 1, 3, and 6 wk after eclosion. Males were put into individual vials with standard food medium 24 h prior to mating tests. Mating tests began when a virgin *sepia* female was aspirated into a vial containing an experimental male. Zero-wk-old males were virgins. In order to further ensure that 0-wk-old males of both selection regimes were the same age after eclosion, for that age we used only males that eclosed during the same 12-h interval. For all ages, mating tests were conducted at approximately the same time of day, and tests were set up to avoid any systematic differences between populations and selection regimes in the time at which they began. For each population at each age, 12-16 males were used as first males. Vials were observed continuously to ensure that copulation occurred, and the time to initiation and duration of copulation were recorded for every pair. (In a few instances copulations did not readily occur, and it was necessary to replace either the experimental male with another or to replace the *sepia* female.) After completion of copulation, experimental males were discarded. The ages of the *sepia* virgin females at the time of the first mating varied from 2-4 to 6-7 d after eclosion, depending on the age of the experimental males.

On the 6th day after mating, each *sepia* female was given an opportunity to remate by aspirating a single *sepia* male (the 'second' male) into the same vial in which the first mating had occurred, and in which the female had been laying eggs. These second males were 3-5 to 7-10 days old, again depending on the age of the first males. Second males were not virgins, but were separated from females 24 h before being used in remating tests. Remating tests were watched continuously for 4.5 h. If remating occurred, the time to begin and the duration of copulation were recorded. Females that remated were transferred to fresh food vials immediately, and then four more times at intervals of 2 d, and were finally discarded 14 d after remating. For the three youngest ages of experimental males, all adult progeny produced by females from eggs laid after remating were scored for eye-color phenotype in order to determine paternity. Because a large number of the females that were first mated to 6-wk-old experimental males remated, we scored progeny only from a sample of that group of remated females. Random samples of six remated females were chosen from each population except O_2, for which only five females remated.

The 6-d interval between the first and second mating tests was chosen based on the results of Gromko & Pyle (1978). We wished to obtain approximately 50% remating frequency overall (in order to increase our chances of detecting possible treatment effects), and to have a reasonably short duration for the remating period so that it would be feasible to observe copulations. Thus the 6-d interval was used primarily for reasons of experimental convenience. We do not mean to imply that this procedure accurately reflects the conditions in the stock cultures in which females are immediately and continuously exposed to remating opportunities.

Sperm offense (B and O-population males used as second males)

Virgin *sepia* females were individually mated to *sepia* males (first males), one female/male pair per vial. These flies were 3-10 d old. The males were removed from the vials after approximately 8 h and discarded. The females were retained in the mating vials for four more days, and successful insemination by first males was determined by subsequent appearance of larvae. Four days after mating with first males, each inseminated *sepia* female was given an opportunity to remate with either a B or O-population male (second male). Remating tests took place in the same vials as were used for first matings, and second males were aspirated into the vials. Remating tests lasted 3 h and vials were observed continuously. If a female remated, the time to begin copulation and the duration of copulation were recorded. After mating, remated females were immediately transferred to fresh food vials. These females were then transferred to fresh vials five more times at 2-d intervals, and were discarded 16 d after remating. All adult progeny produced by females from eggs laid subsequent to remating were scored for eye-color phenotype and counted.

B and O males were tested at four ages: 24-36 h (0 wk), and 1, 3, and 5 wk after eclosion. For each population at each age, 12 males were used. The choice of a 4-d interval between first and second matings and a 3-h length for the remating tests was again based on practical considerations. We hoped that, overall, about 50% of the females would remate.

Statistical analysis

The experimental design was a three-factor analysis of variance with two 'main' effects (selection regime and age) and one 'nested' effect (population within selection regime). Selection regime and age were fixed effects and population was a random effect. In order to determine appropriate ratios of mean squares for F tests, expressions for the expected mean squares of all model terms were derived by the procedures of Dunn & Clark (1974), and these are shown in Table 1. The SAS GLM procedure (SAS Institute Inc., 1990) was used for all ANOVAs, and Type III sums of squares were used for all F tests. In general, analyses were unbalanced either because different numbers of males were tested per population/age combination or because a variable fraction of females remated, that fraction depending upon population and age of experimental males. Unequal sample sizes in hierarchical ANOVAs may mean that the customary mean-square ratios (Table 1) result in inexact tests (Sokal & Rohlf, 1981). 'Corrected' F ratios can be obtained using Satterthwaite's approximation

Table 1. Expected mean squares and F ratios for analyses of variance. $i = 2$ is the number of selection regimes; $j = 4$ is the number of populations per selection regime; $k = 4$ is the number of ages at which males were tested; and n is the number of observations per population per age. n varies according to the dependent variable under consideration.

Effect		df	EMS	F ratio
a	Selection regime	i-1	$\sigma_e^2 + cn\sigma_b^2 + bcn\sigma_a^2$	MS_a/MS_b
b	Population (Sel. reg.)	$i(j$-1$)$	$\sigma_e^2 + cn\sigma_b^2$	MS_b/MS_e
c	Age	k-1	$\sigma_e^2 + n\sigma_{bc}^2 + abn\sigma_c^2$	MS_c/MS_{bc}
ac	Sel. reg. × Age	$(i$-1$)(k$-1$)$	$\sigma_e^2 + n\sigma_{bc}^2 + bn\sigma_{ac}^2$	MS_{ac}/MS_{bc}
bc	Pop. × Age	$i(j$-1$)(k$-1$)$	$\sigma_e^2 + n\sigma_{bc}^2$	MS_{bc}/MS_e
e	Residual	$ijk(n$-1$)$	σ_e^2	

(Sokal & Rohlf, 1981), which is available in the GLM procedure. We computed 'uncorrected' F values using the mean-square ratios shown in Table 1 and also 'corrected' values using Satterthwaite's approximation. In all cases the F-values of model effects differed only slightly between the two methods, and there were no changes in general significance levels. We report only the F values obtained from the simple mean-square ratios (Table 1).

Homogeneity of variances was examined by Bartlett's X^2. The Wilk-Shapiro W statistic (SAS UNIVARIATE procedure, SAS Institute Inc., 1988) was used to test residuals for conformity to the normal distribution. Analyses in which the variances were heterogeneous ($P[X^2] < 0.05$) and/or residuals were not normal ($P[W] < 0.05$) are noted. Suitable transformations were sought in cases where the data were non-normal and/or heteroscedastic. The use of transformations is noted in the Results.

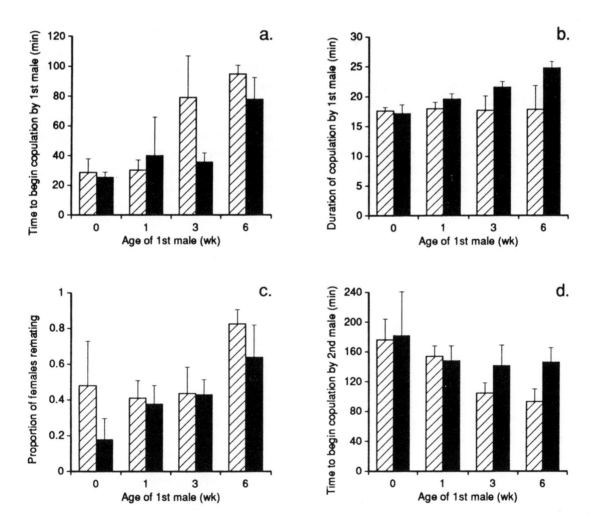

Fig. 1. Copulation times and proportion of females remating for sperm defense experiment (B and O males used as first males). Data are selection-regime means: bars with diagonal lines – B selection regime; solid bars – O selection regime. Error bars are 2 standard errors of the mean of the population averages for each combination of selection regime and age. Note (1) that the variable plotted on the vertical axis is different for each graph and the vertical scales differ; and (2) that all plotted data are untransformed although some variables were transformed for the statistical analyses summarized in Table 2.

136

Results

Sperm defense (B and O-population males used as first males)

Time to begin copulation by first and second males, duration of copulation by first and second males, and proportion of females remating were analyzed (Fig. 1a-d; Table 2). The time for first males to begin copulation increased significantly with age (Fig. 1a). There was a significant interaction between selection regime and age, with the time to mate increasing more rapidly for B population males. The duration of copulation by first males also increased significantly with age (Fig. 1b). Additionally, there was a significant effect of selection regime – O-population males copulated for longer times – and the increase in mating duration with age was greater for O than for B males. The proportion of females that remated increased as a function of the age of first males (Fig. 1c). Overall, females that were first-mated to O males were less likely to remate than were females that were first-mated to B males. Time to begin copulation by second males decreased with the age of first males, and did so more rapidly when B males were used as first males

than when O males were used as first males (Fig. 1d). There was no effect of selection regime, population, or age of first male on the duration of copulation by second (*sepia*) males (Table 2).

Progeny production by females after remating is shown in Figure 2 and a summary of the analyses is presented in Table 3. The total number of progeny produced by females after remating increased with the age of first males (Fig. 2a). There was no effect of selection regime. The number of first-male progeny produced after remating was affected by the age of first males, although there was no clear age-related trend (Fig. 2b). Overall, O males sired more progeny after females remated than did B males. The proportion of the post-remating progeny that was sired by first males was affected by the age of the first males although, again, there was no clear age-related trend (Fig. 2c). There was also a significant effect of selection regime, with O males siring a higher proportion of post-remating progeny. For first males that were 3-wk old, the number (Fig. 2b) and proportion (Fig. 2c) of first-male progeny sired after remating appear to be anomalously low. We have no explanation other than unknown laboratory artifact for this result. Because the data for the 3-wk-old first males may have had a disproportion-

Table 2. Analyses of time to copulate, duration of copulation, and proportion of females remating for sperm defense experiment (B and O males used as first males). Table entries are *F* values, calculated according to the *MS* ratios in Table 1. For each combination of population and age, the proportion of females remating is a single datum. For that variable, therefore, the analysis is truncated because there is no residual representing replicate observations within population × age. The analysis reduces to a 2-way fixed-effect factorial ANOVA. *N* is the total number of observations for an analysis, *W* is the Wilk-Shapiro statistic, and X^2 is the Bartlett's test statistic. The *df* for the Bartlett's tests are 31, except 7 in the case of proportion of females remating.

Effect	Time to mate [a] (1a)†	Duration of mating [b] (1b)†	Proportion of females remating (1c)†	Time to remate (1d)†	Duration of remating (not shown)
Selection regime	4.69	9.60*	6.68*	5.00	0.09
Population (Sel. reg.)	1.51	2.68*	–	0.46	1.07
Age of first male	55.71***	6.32**	12.46***	4.13*	1.39
Sel. reg. × Age	7.71**	6.46**	1.85	3.51*	0.32
Pop. × Age	0.98	1.08	–	0.67	1.05
N	423	424	32	202	201
R^2	0.361	0.216	0.674	0.218	0.146
Wilk-Shapiro *W*	0.978*	0.988	0.978	0.969*	0.996
Bartlett's X^2	53.05**	74.59**	6.41	17.33	35.70

† data shown in this figure: * $P < 0.05$; ** $P < 0.01$; *** $P < 0.001$
(a) data transformed to natural logarithm; (b) data transformed to square root

137

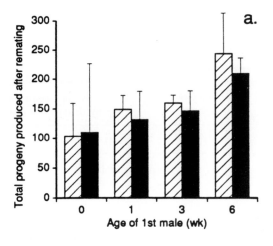

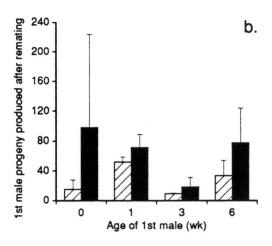

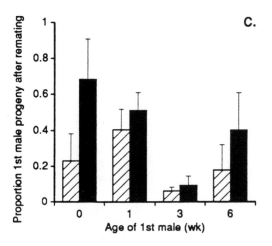

Fig. 2. Progeny production after remating by females in the sperm defense experiment (B and O males used as first males). See Figure 1 legend for additional information. Statistical analyses for these data are summarized in Table 3.

ate effect on these two analyses, we repeated the analyses without that data. For number of first-male progeny, the selection-regime effect was still significant ($F_{[1, 6]}$ = 26.04, $P < 0.01$), but there was no longer an effect of first-male age ($F_{[2, 12]}$ = 0.88, NS). Similarly, for the proportion of post-remating progeny that was sired by first males, the selection-regime effect remained significant ($F_{[1, 6]}$ = 12.80, $P < 0.05$), but the effect of age was not significant ($F_{[2, 12]}$ = 1.45, NS). Therefore, there appears to be no effect of first-male age on these two characters. No other model effects in either analysis were substantially affected by the exclusion of the data for 3-wk-old first males.

Sperm offense (B and O-population males used as second males)

The results of the sperm offense experiment are presented in Tables 4 and 5, and Figure 3. For none of the variables examined was there any difference between selection regimes, nor was there an interaction between selection regime and age. Only for proportion of females remating with second males was there clear evidence of a decline in the mating ability of second males with advancing age (Fig. 3a). There was also an effect of age in the analysis of the proportion of post-remating progeny that were sired by second males (Fig. 3d). However, there is no obvious age-related trend. We investigated several data transformations for this latter analysis, including the angular and several power transformations (y^2, y^3, etc.). We present the analysis for y^{10} (Table 5) because that transformation appreciably reduced the degree of non-normality and eliminated significant heterogeneity of variances. However, all transformations produced qualitatively similar results. For the untransformed data, the effect of age was not significant ($F_{[3, 17]}$ = 2.81, $P = 0.07$).

Table 3. Analyses of progeny production and paternity after remating for sperm defense experiment (B and O males used as first males). Table entries are F values, calculated according to the *MS* ratios in Table 1. N is the total number of observations for an analysis, W is the Wilk-Shapiro statistic, and X^2 is the Barlett's test statistic. The *df* for the Bartlett's tests are 31.

	Dependent variable		
	Total number of progeny after remating	Total number of 1st-male progeny after remating[a]	Proportion of post-remating progeny sired by first male[b]
Effect	(2a)†	(2b)†	(2c)†
Selection regime	0.02	35.07***	15.31**
Population (Sel. reg.)	2.01	0.47	0.98
Age of first male	10.19***	7.32**	10.18***
Sel. reg. × Age	0.97	2.39	2.28
Pop. × Age	0.93	0.74	0.80
N	168	168	157
R^2	0.322	0.277	0.335
Wilk-Shapiro W	0.976	0.963**	0.977
Bartlett's X^2	28.59	26.66	48.66*

† data shown in this figure: * $P < 0.05$; ** $P < 0.01$; *** $P < 0.001$
(a) data transformed to natural logarithm; (b) data transformed to square root

Table 4. Analyses of proportion of females remating, and time to begin and duration of remating for sperm offense experiment (B and O males used as second males). Explanation of table entries and *df* for the Bartlett's tests are as for Table 2

	Dependent variable		
	Proportion of females remating	Time to remate	Duration of remating
Effect	(3a)†	(3b)†	(not shown)
Selection regime	0.01	0.71	1.04
Population (Sel. reg.)	–	1.41	0.49
Age of second male	8.42***	2.73	0.86
Sel. reg. × Age	0.37	0.48	0.15
Pop. × Age	–	1.35	0.91
N	32	133	133
R^2	0.523	0.302	0.167
Wilk-Shapiro W	0.935	0.970	0.982
Bartlett's X^2	3.03	19.47	27.67

† data shown in this figure: *** $P < 0.001$

Table 5. Analyses of progeny production and paternity after remating for sperm offense experiment (B and O males used as second males). Explanation of table entries and the *df* for the Bartlett's tests are as for Table 3.

	Dependent variable		
	Total number of progeny after remating [a]	Total number of 2nd-male progeny after remating [a]	Proportion of post-remating progeny sired by second male[b]
Effect	(not shown)	(3c)†	(3d)†
Selection regime	2.56	1.01	0.07
Population (Sel. reg.)	0.98	1.95	1.48
Age of second male	1.16	2.78	3.45*
Sel. reg. × Age	0.33	0.12	1.60
Pop. × Age	0.84	0.85	1.09
N	133	133	131
R^2	0.216	0.273	0.323
Wilk-Shapiro W	0.977	0.970	0.963*
Bartlett's X^2	16.38	16.85	38.68

† data shown in this figure: * $P < 0.05$; (a) data transformed to square root; (b) data transformed to 10th power

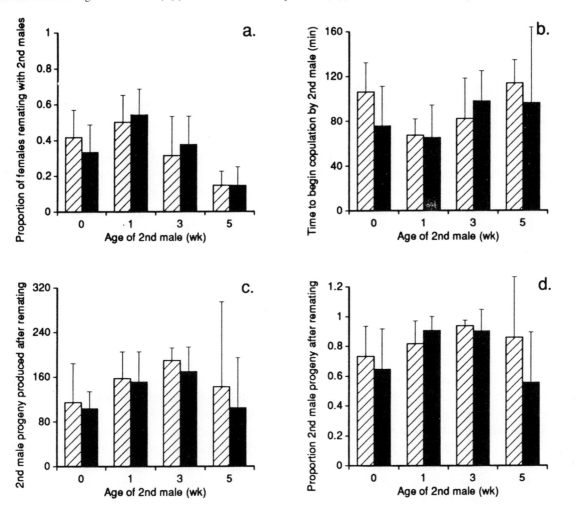

Fig. 3. Components of male reproductive success for the sperm offense experiment (B and O males used as second males). See Figure 1 legend for additional information. Statistical analyses for these data are summarized in Tables 4 and 5.

Discussion

Evolution of senescence

Several of the components of male reproductive success that we examined showed clear evidence of senescent decline. These included time to begin copulation by first males (Fig. 1a), the proportion of females remating when experimental (B and O) males were used as first males (Fig. 1c), the time for second males to begin copulation when experimental males were used as first males (Fig. 1d), and the ability of experimental males, when used as second males, to induce females to remate (Fig. 3a). Two other traits showed significant trends with age, but their relationship to senescent decline is less clear. First, duration of copulation by first males increased with first male age, and the effect was primarily seen in the O males (Fig. 1b). Because O-population males performed better (i.e., in the direction of superior reproductive success) than B males in all other traits for which there was an effect of selection regime, the increase in copulation time by O males may function to retard their decline with age in other components of reproductive success (see below). Second, in the sperm defense experiment, the total number of progeny produced by females after remating increased with first-male age (Fig. 2a). Perhaps second males transferred more sperm and/or accessory fluid to females that mated with older first males.

For two traits, there were significant interactions between selection regime and age, and O males were superior to B males at older ages: (1) when experimental males were used as first males, the rate of increase with age in time to begin copulation was slower for O males than for B males (Fig. 1a); and (2) when experimental males were used as first males, the time to begin copulation by second males decreased more slowly when females were first mated to O males than when they were first mated to B males (Fig. 1d). For these two traits, therefore, the rate of senescent decline was slower in populations maintained through reproduction of older individuals. For three traits, the O males appeared to be generally superior to the B males (i.e., had higher reproductive success) over the range of ages tested, and in none of these cases was there an interaction between selection regime and age: (1) the proportion of females that remated with second males (Fig. 1c); (2) the number of first-male progeny that were sired after females mated with second males (Fig. 2b); and (3) the proportion of the progeny sired by females after they had remated (Fig. 2c). For these traits, selection for fitter old adults also resulted in fitter young adults. Selection for delayed senescence produced superior old males, but did not alter the rate of senescent decline *per se*. Taken as a whole, the results for the sperm-defense experiment (Figs. 1a, c, d; 2b, c) show that senescence of at least some aspects of reproductive success in male *D. melanogaster* can be altered by appropriate selection procedures. These results are consistent with two other studies of *D. melanogaster* males (Roper, Pignatelli & Partridge, 1993; Service, 1993).

It is somewhat surprising that there was little effect of male age when B and O males were used as second males (sperm offense experiment). Superficially, this suggests that the characters examined do not show senescent decline. However, this may simply be a consequence of the particular procedures used to assay characters in this experiment. For example, if we had tested males at older ages, or if we had used a different time interval for remating tests, we might have seen a stronger age effect.

The traits examined in these experiments do not show trade-offs between early and late ages. For example, the superior sperm defense capability at older ages that has evolved in O males relative to B males has not been accompanied by a diminution of sperm defense capability at younger ages. Superficially, these results would appear to support the mutation accumulation rather than the antagonistic pleiotropy hypothesis for the evolution of senescence. However, it is doubtful that the present experiments shed much light on this issue. There is no necessity for pleiotropic effects to occur between early and late-age expression of the same life history trait (Rose, 1985), and it is possible that trade-offs did occur between the traits examined in these experiments and other, unexamined traits, e.g. early-age competitive mating ability (Service, 1993). Furthermore, even if there are underlying trade-offs among a multivariate set of life history characters, some pairs of traits will exhibit positive genetic correlations at equilibrium (Charlesworth, 1990). It is difficult, if not impossible, to predict which pairs of life history traits will show negative correlations without detailed knowledge of the underlying genetic and metabolic architecture (Houle, 1991). Additional selection experiments (e.g., Service, Hutchinson & Rose, 1988) might help to distinguish between the roles of antagonistic pleiotropy and mutation accumulation in the evolution of male reproductive senescence in these populations.

Evolution of sperm-competitive ability

Significant selection-regime effects and interactions between selection regime and age provide evidence that several aspects of sperm-competitive ability have evolved in these populations. O males were superior to B males in sperm defense. Females were less likely to remate when O males were used as first males (Fig. 1c). For remating females, time to remate was greater if O males were used as first males than if B males were first males (Fig. 1d). If females remated, the number and proportion of first male progeny that were produced after remating was greater if the first male was from the O selection regime (Fig. 2b, c).

Paternity was determined by examination of adult progeny. Therefore, it is possible that the superiority of O males in terms of number and proportion of progeny could have been due to greater egg-adult viability or to greater larval competitive ability (against *se se*) of O/*sepia* hybrids relative to B/*sepia* hybrids. We tested the possibility of differences between selection regimes in egg-adult viability using 'pure' (non-hybrid) flies by transferring eggs at controlled densities of 60 per rearing tube for two generations, as described above. We counted all adults that subsequently emerged. There was no effect of selection regime ($F = 0.005$, NS) or of population within selection regime ($F = 1.343$, NS), although the possibility still remains that O/*sepia* hybrids have greater viability than B/*sepia* hybrids. We did not test for differences between selection regimes in competitive ability of hybrids against homozygous *sepia* flies. However, there was no difference between selection regimes in the total number of progeny produced by females after remating (Table 3). Therefore, if the hybrids differed in their ability to compete against *sepia* larvae under the conditions of this experiment, the differences would not have been due to any systematic variation in larval density between selection regimes. Using similar selection procedures, Roper, Pignatelli & Partridge (1993) found that populations maintained by reproduction of old individuals had inferior larval competitive ability relative to populations that were maintained by reproduction of young adults, when both selection regimes were competed against a marker stock. This result is opposite to that which could account for the present results.

Given the differences between selection regimes in the sperm defense experiment and in previous experiments (Service, 1993), the absence of any difference between selection regimes in sperm offense capability was surprising. At a minimum, we expected that older O males would have been superior to older B males in ability to mate with previously mated females (Fig. 3a, b). In another study, males did show a response to selection for increased remating speed at young ages in both sexes (Pyle & Gromko, 1981). Prout & Bundgaard (1977) observed inter-strain differences in the sperm displacement ability of second males. Although they measured displacement ability differently than we did, their results suggest that this trait (Fig. 3c, d) could also have evolved in our populations. Again, it is possible that our results are an artifact of the procedures used to assay sperm offense. Alternatively, it is possible that selection for improved offensive capability at older ages has been weak in the O selection regime, has been opposed by selection on correlated traits, or that there has been insufficient genetic variance for components of sperm offense (Gromko & Pyle, 1978; Gromko & Newport, 1988).

Additional experiments will be required to elucidate the mechanisms underlying the greater sperm defense capability of O males relative to B males. For example, if female remating is sperm-dependent (Manning, 1962; Gromko, Newport & Kortier, 1984; Newport & Gromko, 1984), there are several alternative but not mutually exclusive explanations for the lower remating frequency of females that were first mated to O males (Fig. 1c): (1) females mated to O males used sperm less rapidly than females mated to B males; (2) O males transferred more sperm, on average, than did B males; (3) the threshold number of sperm below which remating took place was higher for females first mated to O males than for females mated to B males. Alternatively, Hihara (1981) has suggested that female remating depends not on the amount of sperm but on the amount of male accessory gland fluid in female storage organs. Under this hypothesis, the lower remating frequency of females that were first mated to O males could have resulted if O males transferred more accessory gland fluid to females or if the fluid was used more sparingly by females.

The greater number (Fig. 2b) and proportion (Fig. 2c) of first-male progeny produced subsequent to remating by females that were first mated to O males suggests the possibility that O-male sperm were less subject to displacement than were B-male sperm. This argument assumes that the sperm-dependence thresholds for remating by females were the same regardless of the selection regime of the first male. Alternatively, O-male sperm may have remained viable in the female reproductive tract longer than did B-male sperm, or there were interactions between the sperm of experimental males and *sepia* second males that gave a relative advantage to O-male sperm over B-male sperm.

Adult lifespan in the B selection regime stock cultures is no more than 6 d and probably closer to 3-4 d, on average. Given that females may not be receptive to males until approximately 1 d after eclosion, the length of time during which females are vulnerable to remating is probably only about 2-3 d, on average. It is possible that many females mate only once, and that males direct most of their attention to virgin females that are newly maturing throughout the brief period when adults are present in the cultures. Under these conditions, selection for sperm-defensive ability (effective for periods of up to six days post-mating, as assayed in these experiments) may be weak. However, it is possible that short-term sperm-defensive ability is under strong selection. Short-term mechanisms include cuticular compounds and components of the male accessory gland fluid that are transferred to females during courtship and mating and that reduce the attractiveness of females to other males (Scott & Richmond, 1985; Scott, Richmond & Carlson, 1988), or that reduce female receptivity to other males (Manning, 1967; Burnet *et al.*, 1973; Hihara, 1981; Chen *et al.*, 1988).

Adult lifespan in the O selection regime is at least 8 wk. Under these conditions, there are no virgin females after the first few days of each generation and females are exposed to remating oppor-

tunities for the long term. Evolutionary pressures should be strong on males to defend their own sperm and to increase their offensive capability against the sperm of other males. These pressures should be particularly intense during the last few weeks of each generation. Selective improvement of defensive capability at old ages has, however, also resulted in improvements at young ages (Figs. 1c; 2b, c). It is not clear why improved sperm-offensive ability does not seem to have evolved in the O selection regime relative to the B controls. As mentioned above, this might reflect lack of genetic variance. The difference between the evolutionary responses of sperm defense and offense indicates that these traits are genetically and physiologically distinct.

Conclusion

We predicted that selection for delayed senescence should result in a relative enhancement of male reproductive success at later ages in comparison to populations selected for faster senescence. This prediction was borne out by the greater sperm defense capability of males of the O selection regime relative to males of the B selection regime, particularly at older ages. These results are consistent with other studies of male and female *D. melanogaster* that have shown that rates of senescence can be modified in predictable ways by selection (Rose & Charlesworth, 1981; Luckinbill *et al.*, 1984; Rose, 1984; Luckinbill & Clare, 1985; Mueller, 1987; Partridge & Fowler, 1992; Roper, Pignatelli & Partridge, 1993; Service, 1993). We observed no correlated responses to selection that would indicate a trade-off between increased late-life fitness and decreased early-life fitness in males. This result is similar to those for most other components of male age-specific reproductive success (Roper, Pignatelli & Partridge, 1993; Service, 1993). However, trade-offs may have occurred in other, unexamined fitness components. Our results indicate that at least some aspects of sperm-competitive ability in *D. melanogaster* are subject to senescent decline, and can be modified evolutionarily.

Acknowledgements

M. Turelli, T. Prout, L. Harshman, and J. Rawls provided advice and help in facilitating this research. T. Prout suggested the use of the terms 'sperm defense' and 'sperm offense'. We thank K. Nishikawa, A. Sih, J. Krupa, and T. Miller for comments on various drafts of this paper. We also thank the Department of Genetics, University of California, Davis, and the School of Biological Sciences, University of Kentucky. This research was supported by U.S. Public Health Service fellowship F32 AG05394 to P.M.S.

References

Boorman, E. & G.A. Parker, 1976. Sperm (ejaculate) competition in Drosophila melanogaster, and the reproductive value of females to males in relation to female age and mating status. Ecol. Entomol. 1: 145-155.

Burnet, B., K. Connolly, M. Kearney & R. Cook, 1973. Effects of male paragonial gland secretion on sexual receptivity and courtship behaviour of female Drosophila melanogaster. J. Insect Physiol. 19: 2421-2431.

Charlesworth, B., 1980. Evolution in age-structured populations, Cambridge University Press, Cambridge, U.K.

Charlesworth, B., 1990. Optimization models, quantitative genetics, and mutation. Evolution 44: 520-538.

Chen, P.S., E. Stumm-Zollinger, T. Aigaki, J. Balmer, M. Bienz & P. Böhlen, 1988. A male accessory gland peptide that regulates reproductive behavior of female D. melanogaster. Cell 54: 291-298.

Dunn, O.J. & V.A. Clark, 1974. Applied Statistics: Analysis of Variance and Regression, John Wiley & Sons, New York, NY.

Gromko, M.H., D.G. Gilbert & R.C. Richmond, 1984. Sperm transfer and use in the multiple mating system of Drosophila, pp 371-426 in Sperm Competition and the Evolution of Animal Mating Systems, edited by R.L. Smith, Academic Press, London.

Gromko, M.H. & M.E.A. Newport, 1988. Genetic basis for remating in Drosophila melanogaster. II. Response to selection based on the behavior of one sex. Behav. Genet. 18: 621-632.

Gromko, M.H., M.E.A. Newport & M.G. Kortier, 1984. Sperm dependence of female receptivity to remating in Drosophila melanogaster. Evolution 38: 1273-1282.

Gromko, M.H. & D.W. Pyle, 1978. Sperm competition, male fitness, and repeated mating by female Drosophila melanogaster. Evolution 32: 588-593.

Hamilton, W.D., 1966. The moulding of senescence by natural selection. J. Theor. Biol. 12: 12-45.

Harshman, L.G., A.A. Hoffman & T. Prout, 1988. Environmental effects on remating in Drosophila melanogaster. Evolution 42: 312-321.

Hihara, F., 1981. Effects of the male accessory gland secretion on oviposition and remating in females of Drosophila melanogaster. Zool. Mag. 90: 307-316.

Houle, D., 1991. Genetic covariance of fitness correlates: what genetic correlations are made of and why it matters. Evolution 45: 630-648.

Lefevre, G., Jr. & U.B. Jonsson, 1962. Sperm transfer, storage, displacement, and utilization in Drosophila melanogaster. Genetics 47: 1719-1736.

Luckinbill, L.S., R. Arking, M.J. Clare, W.C. Cirocco & S.A. Buck, 1984. Selection for delayed senescence in Drosophila melanogaster. Evolution 38: 996-1003.

Luckinbill, L.S. & M.J. Clare, 1985. Selection for life span in Drosophila melanogaster. Heredity 55: 9-18.

Manning, A., 1962. A sperm factor affecting the receptivity of Drosophila melanogaster females. Nature 194: 252-253.

Manning, A., 1967. The control of sexual receptivity in female Drosophila. Anim. Behav. 15: 239-250.

Mueller, L.D., 1987. Evolution of accelerated senescence in laboratory populations of Drosophila. Proc. Natl. Acad. Sci. U.S.A. 84: 1974-1977.

Newport, M.E.A. & M.H. Gromko, 1984. The effect of experimental design on female receptivity to remating and its impact on reproductive success in Drosophila melanogaster. Evolution 38: 1261-1272.

Parker, G.A., 1970. Sperm competition and its evolutionary consequences in the insects. Biological Reviews 45: 525-567.

Partridge, L. & K. Fowler, 1992. Direct and correlated responses to selection on age at reproduction in Drosophila melanogaster. Evolution 46: 76-91.

Prout, T. & J. Bundgaard, 1977. The population genetics of sperm displacement. Genetics 85: 95-124.

Pyle, D.W. & M.H. Gromko, 1981. Genetic basis for repeated mating in Drosophila melanogaster. Am. Nat. 117: 133-146.

Roper, C., P. Pignatelli & L. Partridge, 1993. Evolutionary effects of selection on age at reproduction in larval and adult Drosophila melanogaster. Evolution 47: 445-455.

Rose, M.R., 1984. Laboratory evolution of postponed senescence in Drosophila melanogaster. Evolution 38: 1004-1010.

Rose, M.R., 1985. Life history evolution with antagonistic pleiotropy and overlapping generations. Theor. Pop. Biol. 28: 342-358.

Rose, M. & B. Charlesworth, 1980. A test of evolutionary theories of senescence. Nature 287: 141-142.

Rose, M.R. & B. Charlesworth, 1981. Genetics of life history in Drosophila melanogaster. II. Exploratory selection experiments. Genetics 97: 187-196.

SAS Institute Inc., 1988. SAS Procedures Guide, Release 6.03 Edition, SAS Institute, Inc., Cary, NC.

SAS Institute Inc., 1990. SAS/STAT Users Guide, Version 6, 4th ed. SAS Institute Inc., Cary, NC.

Scott, D. & R.C. Richmond, 1985. An effect of mate fertility on the attractiveness and oviposition rates of mated Drosophila melanogaster females. Anim. Behav. 33: 817-824.

Scott, D., R.C. Richmond & D.A. Carlson, 1988. Pheromones exchanged during mating: a mechanism for mate assessment in Drosophila. Anim. Behav. 36: 1164-1173.

Service, P.M., 1989. The effect of mating status on lifespan, egg laying, and starvation resistance in Drosophila melanogaster in relation to selection on longevity. J. Insect Physiol. 35: 447-452.

Service, P.M., 1993. Laboratory evolution of longevity and reproductive fitness components in male fruit flies: mating ability. Evolution 47: 387-399.

Service, P.M., E.W. Hutchinson, M.D. MacKinley & M.R. Rose,1985. Resistance to environmental stress in Drosophila melanogaster selected for postponed senescence. Physiol. Zool. 58: 380-389.

Service, P.M., E.W. Hutchinson & M.R. Rose, 1988. Multiple genetic mechanisms for the evolution of senescence in Drosophila melanogaster. Evolution 42: 708-716.

Sokal, R.R. & F.J. Rohlf, 1981. Biometry, 2nd. W.H. Freeman, San Francisco.

Williams, G.C., 1957. Pleiotropy, natural selection, and the evolution of senescence. Evolution 11: 398-411.

M.R. Rose and C.E. Finch (eds.), Genetics and Evolution of Aging, 145–160, 1994.
© 1994 *Kluwer Academic Publishers. Printed in the Netherlands.*

Genetic and environmental factors regulating the expression of an extended longevity phenotype in a long lived strain of *Drosophila*

Robert Arking[1,2], Steven P. Dudas[1] & George T. Baker, III[3,4]
[1] *Department of Biological Sciences and*
[2] *Institute of Gerontology, Wayne State University, Detroit, MI 48202, USA*
[3] *Gerontology Research Center, National Institute on Aging, Baltimore, MD, USA*
[4] *Shock Aging Research Foundation, Silver Spring, MD 20905, USA*

Received and accepted 22 June 1993

Key words: aging, extended longevity, genetics of aging, *Drosophila*, gene-environment interactions, gene regulation, antioxidants

Introduction

More than a century ago, August Weismann (1891) asked what factors accounted for the significant differences in life span observed between different species and among different members of the same species. Although Raymond Pearl (1928) and others began developing the answer to his question, the task was soon dropped in favor of more tractable problems in genetics, evolution, embryology, biochemistry and the like. Evolutionary and population genetic investigations allow us to view the life span as an integral part of the life history of an organism and are valuable in that they allow the identification of probable physiological 'trade-offs' (Kirkwood & Rose, 1991). Although such evolutionary analyses permit the identification of the fundamental cause of aging, they do not allow one to identify the specific developmental and physiological mechanisms which are actually responsible for the expression of specific lifespans (Clark, 1990; Mayr, 1961; Rose, 1991). Since what appears to be the same phenotype – that of extended longevity – may actually be brought about by different mechanisms, then it becomes important to characterize each set of proximate mechanisms in some detail.

There are many plausible and testable theories of aging (Hayflick, 1985). Yet theory alone has proven to be an insufficient guide to the analysis of the physiological and genetic mechanisms regulating aging and longevity. For example, Nicholas Hoffman (1989 and in preparation) has demonstrated by means of standard population genetic analyses that no one of the current evolutionary theories that he tested can adequately explain all of the physiological and behavioral life history differences which he observed between our normal (R) and long lived (L) strains. He measured a suite of life history characters including fecundity and fertility of same and different aged parents, survivorship of virgins, and negative geotactic responses. Each of the factors studied appear to be under the control of different sets of genes with different temporal patterns of action in both sexes of both strains. For example, he found that the fecundity data supported the antagonistic pleiotropy hypothesis but that the fertility data was consistent with both the mutation accumulation and the age of onset hypotheses. No one theory appeared capable of explaining all of his data. The reports of Partridge and Fowler (1992) and of Service, Hutchinson and Rose (1988) on independently derived long lived strains of *Drosophila* are in agreement with these findings. One implication of these reports is that a better understanding of the aging processes will most likely be achieved through the adoption of an empirical and mechanistic approach towards the analysis of longevity.

Event dependency of the ELP

The long lived L strains were derived from t

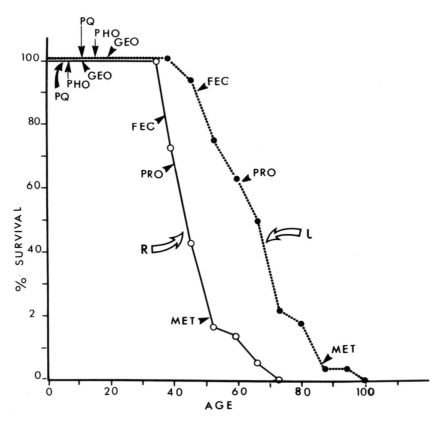

Fig. 1. The age at which the performance of the animals in each strain was 50% of peak values as defined for each of the six tests was calculated and plotted as a function of survival. See text for discussion. PQ = resistance to exogenous paraquat; PHO = phototactic behavior; GEO = geotactic behavior; FEC = female fecundity; PRO = in vivo incorporation of labeled amino acids; and MET = metabolic rate. (Modified after Figure 6 of Arking & Wells, 1990).

normal lived R strain (itself synthesized from wild caught Michigan flies) by selection for delayed female reproduction over a period of 22 generations (Luckinbill *et al.*, 1984; Luckinbill & Clare, 1985). A complete description of the various long and short lived strains obtained during the selection experiment has been presented elsewhere (Arking, 1987a). In this article, we define senescence to refer to that portion of the adult life span characterized by a progressive loss of functional capability as defined by the biomarker analysis described below. Longevity refers to the mean and/or maximum life span of defined populations, while aging refers to ʾanges occurring in the biological processes un- ʾving the time dependent increase of mortality. ʾ latter two terms can be used more or less ʾngably.

ʾve demonstrated that the R and L strains ʾe same sequence of age-related physio- ʾs (Arking & Wells, 1990). We used ʾstimate physiological ages of the R ʾnals throughout their adult life. A

comparison of the chronological and physiological ages in the two strains clearly shows a shift in the temporal pattern of expression of these biomarkers, but does not show any change in the sequence with which they appeared (Fig. 2). Since the strains differ only in the chronological age of expression of the six biomarkers examined, then the two strains comply with the Casarett rules (see Arking & Dudas, 1989, for discussion) and provide us with a valid tool with which to conduct a comparative genetic analysis of aging. Biomarker analysis also allows us to conclude that the extended longevity phenotype (ELP) is a result of a delayed onset of senescence in the L strain relative to the R strain, rather than the result of some change in either the sequence or temporal manifestation of these events which operationally define the senescent process. In these two strains, then, the processes regulating the onset of senescence are separable from those processes responsible for each of the senescent steps themselves. Since the onset of senescence is delayed in the L strain, the molecular genetic events regulating this onset must take place early in life, prior to the age of onset of the first observed biomarker event in the R strain. Given the set of biomarkers analyzed in these experiments, this means that these molecular genetic events must occur sometime prior to the loss of the initial senescent event (i.e. loss of resistance to exogenous paraquat) by the R strain, which occurs not later than seven days of adult life (Arking *et al.*, 1991). Data presented below confirm and extend the concept that the regulation of aging in this strain is an event dependent, developmental process.

Genetic analysis

The goal of any genetic analysis is to characterize the genetic system responsible for the observed phenotype. Although our genetic analysis has just begun, we have identified some of the roles of the three major chromosomes, suggested the nature of the genetic circuitry regulating the expression of the ELP in our strains, and provided an initial description of the additive and non-additive components. These experiments have been previously described (Wells *et al.*, 1987; Buck *et al.*, 1993a,b).

Chromosome substitution experiment

One goal was to understand the role which each chromosome played in the genetic regulation of longevity in our strains. To accomplish this task, we measured the lifespan of a number of isochromosomal lines which together contained all possible heterozygous and homozygous combinations of L and R chromosomes. Quantitative measurements of the effect on longevity of each chromosome type (R vs L), as a function of dosage and as a function of other chromosome combinations in the genotype, provided insight into the specific role each chromosome played in regulating the expression of the ELP.

Understanding the response of each of these isochromosomal lines is critical to any interpetation of the genetic mechanisms involved; thus a brief description of the chromosome notation used is presented. In this notational system, the three sequential positions in the notation represent the first, second, and third major chromosomes while the numerical value (0, 1, or 2) present at each position represents the number of chromosomes of that set which are derived from the L parental strain. For example, the *012* line would have the following chromosome composition: *R/R*; *L/R*; *L/L* females and *R/Y*; *L/R*; *L/L* males. The parental strains are noted as *000* (R) and *222* (L).

All 27 possible combinations of the three major L and R chromosomes (see Fig. 2 for listing) combinations were constructed and 3,875 males and females of defined chromosome composition were raised under high density (HD, > 50 eggs/vial) conditions and their lifespans and fecundities measured (Buck *et al.*, 1993a). Analysis of the data with a three way fixed effect factorial analysis of variance (ANOVA) with compensation for unequal sample sizes showed that, for both sexes, all three of the major chromosomes had an effect on the expression of the ELP but that the first (c1) and third (c3) chromosomes exhibited the major effects. At first glance, this seemed like a typical polygenic, multi-chromosomal trait in which a quantitative trait depends on the additive effects of a large number of genes, each of which has but a small individual effect. Such a polygenic interpretation was put forth by Luckinbill *et al.* (1988), following their initial report that only one gene was responsible for the expression of the ELP (Luckinbill *et al.*, 1987). However, a more detailed statistical analysis supports a different interpretation (Buck *et al.*, 1993a).

Both parametric (Dunnett's) and non-parametric (Kolmogorov-Smirnov) tests were used to determine whether a particular sex and genotype had a mean life span statistically the same as, or different from, that of each of the two control lines (i.e. the *000* (R) and the *222* (L) strains). This procedure allowed the determination of the statistical similarity of each line relative to both controls. Both types of analyses yielded the same interpretation. Most, but not all, of the lines are statistically similar to one of the control lines and statistically dissimilar from the other control line. For ease of comparison, these results are shown in Figure 2 where they have been arranged according to whether the survival data for each particular chromosome had a mean life span which was statistically greater than, less than, identical to, or intermediate between that of the *000* (R) and the *222* (L) controls. The data for males showed the same classification into the same statistically discrete lifespan groups.

A comparison of the *200*, *020*, and *002* ger

148

types shows that only the homozygous L type c3 (c3^L) is capable, by itself, of allowing the expression of the ELP. In addition, a comparison of the life spans of the *002*, *001*, and *000* genotypes suggests that the c3^L genes involved are recessive. However, this is not the complete story, for a comparison of the *002*, *012*, and *022* genotypes shows that the presence of even one L type c2 (c2^L) represses the life span enhancing effect of homozygosity for the c3^L. But not all isogenic lines with a c2^L display a shortened life span. For example, the *212* line has both a c2^L and is homozygous for the L type c1 (c1^L). In this case, the life shortening effect of the c2^L is negated by the c1^L. Thus the net conclusion of these data is that the expression of the longevity enhancing recessive genes on c3^L are repressed by c2^L, unless the c2^L genes are themselves repressed by c1^L. Thus the c3^L genes will be ex-

pressed in either the mutual presence of c1^L and c2^L, or in the absence of c2^L (Buck *et al.*, 1993a). These recessive genes on c3^L presumably represent the additive genetic variance component of the ELP, while the non-additive genetic components of the ELP probably involve the epistatic effects illustrated by the interactions of c1^L and c2^L.

Isozyme analysis

During the course of a survey of various metabolically important isozymes (Dudas *et al.*, in preparation), we obtained independent evidence illustrating the existence of epistasis in these strains. The homodimeric enzyme alcohol dehydrogenase (ADH) maps to 50.1 on c2. The enzyme plays a key role in the detoxification of ethanol and other alcohols (David *et al.*, 1976) and has other important

EFFECTS OF CHROMOSOMAL SUBSTITUTION ON FEMALE LONGEVITY

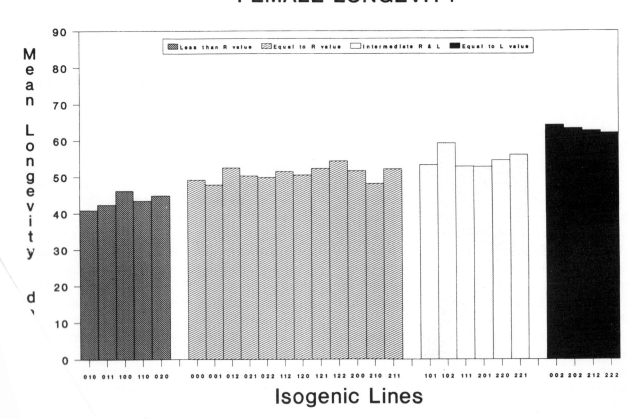

ome composition and the mean life span are shown for females of each of the 27 isogenic lines analyzed by Buck for details. (From Buck *et al.*, 1993a, Heredity 71: 11-22, with permission).

GENOTYPE   **BANDING PATTERN**

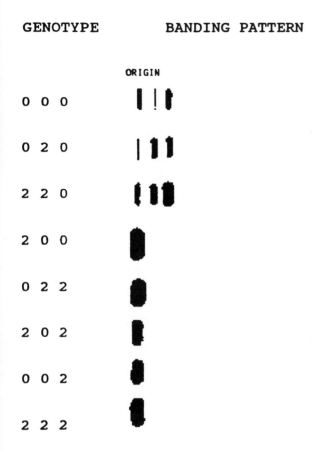

ORIGIN

0 0 0

0 2 0

2 2 0

2 0 0

0 2 2

2 0 2

0 0 2

2 2 2

Fig. 3. A computer scan of the ADH specific bands produced by the indicated genotypes after non-denaturing gel electrophoresis and specific visualization of the ADH-carbonyl-NAD bands. The ADH gene is on c2. The number of bands produced by any genotype (with the exception of *200*) is dependent on the identity of the c3, not the c2. Any genotype with a R type c3 yields three bands; any genotype with an L type c3 yields one band. (Conditions were 1% agarose, 0.3M Tris/HCl, pH 8.6, 17 V/cm; from Dudas *et al.*, in preparation).

metabolic roles (Heinstra & Geer, 1991; Oudman *et al.*, 1991). The several isochromosomal lines differ in the number of distinct bands visible after non-denaturing agarose gel electrophoresis and visualization of ADH specific bands by the ADH-NAD-formazan procedure (Everse *et al.*, 1971). The data summarized in Figure 3 show that the number of bands formed by the ADH locus on c2 is apparently under the control of gene(s) on c3. If the c3 is derived from the L strain (c3^L), then only one electrophoretic band is seen regardless of the source of the c2. If the c3 is derived from the R strain (c3^R), then three electrophoretic bands are seen in all lines except the *200*. These observations provide specific evidence for the epistatic effect of c3 on c2. More

importantly, they provide direct evidence, independent of the chromosome substitution studies discussed above, of the role which epistatic effects play in altering patterns of gene product expression in these strains. Such effects play a critical role in the expression of the ELP, as is implicit in the life span data from which their existence was initially deduced (see Fig. 2; also Buck *et al.*, 1993a), and as is demonstrated by the gene expression data discussed below. Such interacting gene-enzyme systems are not uncommon in *Drosophila*. There exists, for example, a known trans-acting gene on c3 (R^{26-43}) which exerts a post-translational control on the *in vivo* stability of ADH (King & McDonald, 1987). That effect does not appear to be same as the one we describe here, since the R^{26-43} gene affected the level of Adh protein but was not reported to affect the banding patterns. Coincidentally, this gene maps to the same chromosomal region to which we have localized the c3 longevity assurance genes (Buck *et al.*, 1993a). As an example of the reciprocal epistatic interaction, Graf and Ayala (1986) reported the case of a gene on c2 which regulated the amount of a c3 gene product, Cu-Zn superoxide dismutase (cSOD). Finally, the existence of multiple trans-acting effects with important metabolic consequences is implicit in the data of Clark (1990). These non-additive genetic effects probably provide an important source for modulating the expression of the ELP.

Environmental analysis

The goal of these experiments was to provide a quantitative and qualitative assessment of the types of environmental factors involved in regulating the expression of the ELP.

Some environmental factors have no effect on expression of the ELP

The life span of poikiothermic animals can be significantly affected by changes in the ambient temperature (David, 1989). We examined the effects of ambient temperature on adult life span in order to determine if manipulation of this factor would affect the regulation of longevity in our strains (Arking *et al.*, 1988). The results were instructive. First, animals raised at 18 °C have a mean and maximu

150

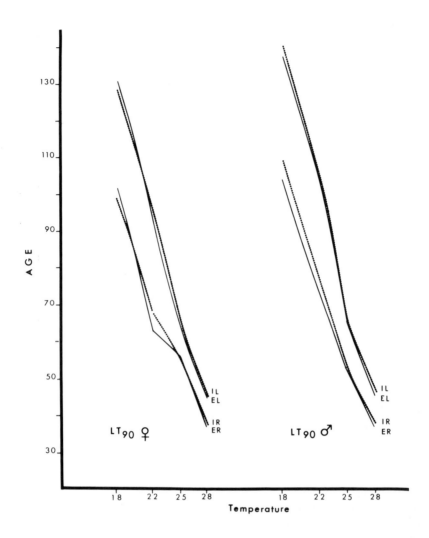

Fig. 4. A graphical representation of the effects of ambient temperature on adult longevity. The I group of both the L and R strains were allowed to develop at 25 °C and then spent their adult life at the indicated temperature. The E group of both the L and R strains spent their entire developmental and adult life span at the one indicated temperature. LT_{90} = the chronological age by which 90% of the animals in the original cohort have died. (From Arking *et al.*, 1988. Experimental Gerontology *23*: 59-76, with permission).

life span which is two to three times as great as it is for genetically identical animals raised at 28 °C. Thus the adult life span is changed by ambient ˆemperature in the manner that one would expect on ˆe basis of past work. However, it is also quite ˆrent that, while these temperature manipula- affect the absolute value of the life span ˆach strain, they have no effect on the rela- ˆpans between strains (Fig. 4). Tempera- ˆnts affect the duration of all stages of ˆy in both strains in a non-specific and ˆner. The length of life of the same ˆus test temperatures is expanded ˆve to the 25 °C control. Temper-

ature alteration is an immediate effect that may not have long term consequences, as is suggested by the fact that preimaginal development from egg to adult can be doubled in time (from 12.5 to 25 days) by incubation at different temperatures, but this need not result in a corresponding change in the adult longevity when the adults are subsequently raised at the same temperature (see Table 1 of Arking *et al.*, 1988). Temperature manipulation does not appear to bring about any alteration in the relative timing of the events that constitute the senescent process in either strain. On the other hand, the genetic differences between the strains are known to involve a specific process (the delayed onset of

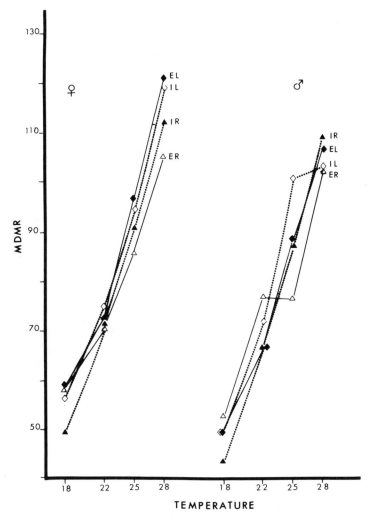

Fig. 5. The variation in the mean daily metabolic rate (MDMR) as a function of sex, strain, developmental conditions and adult ambient temperature. Abbreviations as in Figure 4. (From Arking *et al.*, 1988. Experimental Gerontology *23*: 59-76, with permission).

senescence in the L strain) and appear to be expressed at all temperatures tested. From these findings, we conclude that the effects of temperature and genetic manipulations are independent of one another, and exert their effect on life span via different processes.

The effect of temperature is evidently manifested through a generalized alteration in the metabolic rate. There is an inverse relationship between ambient temperature and mean daily metabolic rate (MDMR) such that animals raised at 18 ˚C have a significantly lower MDMR than do animals of the same strain raised at 28 ˚C (Fig. 5). However, there is no statistically significant difference in the MDMR between the R and L strains raised at the same temperature, as is illustrated by the data of Figure 6. Temperature modulates the life span of

Drosophila by affecting metabolic rate, and consequently it affects the absolute length of time between the important physiological events in both strains. It does not affect only the genetic mechanisms regulating the relative time of onset of these events in the L_A strain, and thus it does not regulate specifically the expression of the ELP. The genes operative in the L strain modulate longevity by affecting, in a specific manner, some process other than metabolic rate.

Some environmental factors have very specific effects on expression of the ELP effect of larval density

Larval density is known to affect life span in wild strains of *D. melanogaster* (Miller & Thomas,

152

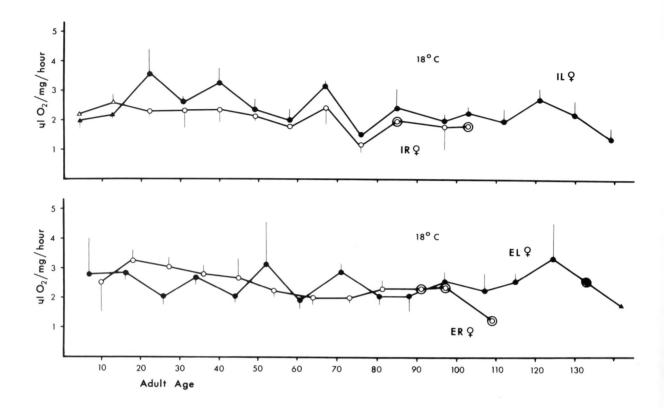

Fig. 6. The top panel shows the measured metabolic rates at 18 °C throughout the lifetimes of the L and R females raised under I conditions. The bottom panel shows the measured metabolic rates throughout the lifetimes of the L and R females raised under E conditions. The vertical bars are the standard deviations. Symbols without error bars indicate when population numbers had fallen so low that replicate samples could not be tested. Other abbreviations as in Figure 4. (From Arking *et al.*, 1988, Experimental Gerontology *23*: 59-76, with permission).

1955; Lints & Lints, 1971; Economos & Lints, 1984; Zwaan, Bijlsma & Hoekstra, 1991). Our selected strains also exhibit density dependent life spans (Clare & Luckinbill, 1985; Wells *et al.*, 1987). This density dependency has been analyzed in some detail (Buck *et al.*, 1993b). As shown in Figure 7, genotypes previously shown to express the ELP when raised under HD (high density, < 50 eggs/vial) conditions no longer do so when raised under LD (low density, 10 eggs/vial) conditions. Under LD conditions, the long lived genotypes (*222, 002*) are not statistically different from the normal genotypes (*020, 200*). There is an effect of larval density on most genotypes (Fig. 7), the LD causing a mean decrease ranging up to 7.4 days in the mean life span of normal lived genotypes (see Table 4 of Buck *et al.*, 1993b). It has been previously shown that LD conditions decreased the life

span in adults of several different wild type strains by about the same length of time (Miller & Thomas, 1955; Lints & Lints, 1971; Zwaan, Bijlsma & Hoekstra, 1991). Thus there appears to be a generalized response of *Drosophila* longevity to larval density which alters adult longevity by about 5-7 days. However, the genotypes which express the ELP show a mean decrease in their LD life span of 16.9 days (Buck *et al.*, 1993b), a response which is statistically different from that of all other genotypes, and is comparable to the 14 day difference between the 222-HD and 222-LD life spans done on sister strains to ours (Clare & Luckinbill, 1985). In contrast to temperature, larval density affects different genotypes in different ways. Larval density is apparently a specific environmental modulator of the ELP.

EFFECT OF LARVAL DENSITY ON FEMALE MEAN LIFESPAN IN VARIOUS ISOGENIC LINES

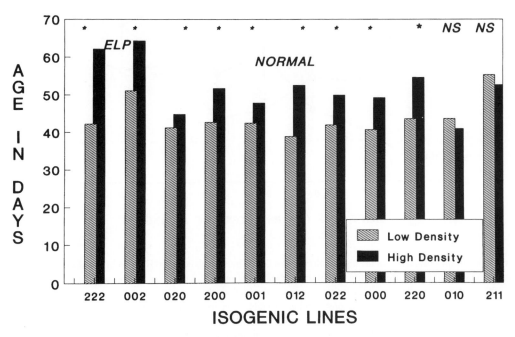

Fig. 7. The mean life span of females from some representative isogenic lines assayed under HD and LD larval density conditions is shown. Note that genotypes which express the ELP when raised under HD conditions not only fail to express it when raised under LD conditions, but their mean life span falls to a level comparable to that of the non-longlived genotypes. See text for discussion. (Abbreviations: * = difference between the HD and LD mean lifespan is significant (P < .05); NS = difference between the HD and LD mean lifespan is not significant (P >.05); ELP = longlived strains/isogenic lines chosen from the long-lived group of Fig. 2; Normal = strains/isogenic lines chosen from the three non-longlived groups of Fig. 2).

Timing of larval density effect. Density shift experiments were done in a manner analogous to that of temperature shift experiments in order to determine the timing of the critical period for larval density. Developing *222* animals were shifted from HD to LD conditions (the H-L shift) or from LD to HD conditions (the L-H shift) at defined times during the second and third larval instars and pupal stages. These density shifted larvae were then assayed at five days of adult life for paraquat resistance to determine the presence or absence of the ELP. We have previously shown that paraquat reliably identifies populations which express the ELP from the normal lived populations which do not express the ELP (Arking *et al.*, 1991). As shown in Figure 8, there was a significant increase in the L-H shift at 60 h, which signals the beginning of the critical period. The significant plateau in the H-L shift at 120 h signals the end of the critical period. Taken together, these two sets of data delineate a

critical period in larval development which begins no later than 60 h after oviposition and which ends no later than 120 h after oviposition, and wherein the animals are sensitive to the effects of density. Developing larvae must be exposed to HD conditions during this time frame if the ELP is to be expressed. These data confirm our earlier conclusion, based on the biomarker analysis, that the expression of the ELP as measured by paraquat resistance is a developmental phenomenon.

Potential mechanism(s) of the density effect. Larval density affects adult body weight (Buck *et al.*, 1993b). In both sexes and strains, the LD body weight is higher than the HD body weight. These differences in body weight are associated with a significant alteration in adult life span. Such data present us with several possible physiological mechanisms by which the effects of larval density might be translated into an increased adult longev-

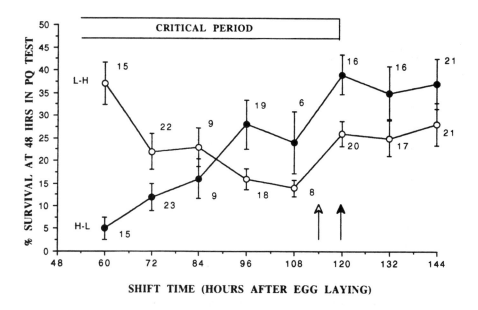

Fig. 8. Determination of the critical period during which time the larvae must be exposed to HD conditions if they are to exhibit the ELP. Timed L strain larvae were started at one density condition, shifted to the other at the indicated time, and assayed with the paraquat test (Arking *et al.*, 1991) at 5 days of adult life to see if they expressed the ELP. Each point represents the mean ± SEM of three independent replicate experiments. The numbers indicate the total number of vials, each containing ten animals, used for each indicated shift. The open arrow indicates the estimated time when the larvae stop feeding, the solid arrow indicates the observed time of pupation. Open circles = LD to HD shifts. Solid circles = HD to LD shifts. The reasons underlying the increase in the L-H curve at 120 hrs are not clear. (From Buck *et al.*, 1993b. Heredity 71: 23-32, with permission).

ity. First, the larval HD conditions may give rise to the presence in the food of higher titers of chemical substances (excretory, hormonal, etc.) not present in LD conditions and which act as inducers of gene action in the developing larvae. A secondary assumption would be that these induced changes in gene expression would then have profound effects in later stages of the life cycle. The existence of such chemicals has been demonstrated in several organisms, including *Drosophila* (Boetella *et al.*, 1985; Crowl & Covich, 1990). We know that the chemical composition of the LD food appears to be different in its reducing ability from that of the HD food. We also know that there exist early and obvious changes in specific mRNA levels between L-HD and L-LD larvae (see below). Second, the density effect may simply be a matter of dietary restriction. Such a regime has long been known to be effective in vertebrates (Weindruch *et al.*, 1986) and has been shown by Austad (1989) to be effective in spiders as well. Finally, hormesis may be involved. Experiments designed to distinguish between these several possibilities in our strains are currently underway.

Genes and gene products critically associated with the expression of the ELP

The identification of the genes involved in the regulation of life span has proven to be an elusive biological problem. Our experimental system allows us to manipulate critical genetic and environmental variables. Any gene product(s) that is critically involved in the expression of the ELP must track with the phenotype as it is expressed or not expressed as a result of these manipulations. We have arranged our various genotypes and density conditions into a 'decision matrix' (Fig. 9) which allows us to determine just which patterns of gene activity track with and are operationally inseparable from the expression of the ELP. We have constructed, for each genotype-density combination listed in Figure 9, a series of developmental Northerns and dot blots containing the mRNAs expressed from second instar larvae through 13 day old adults, inclusive. This system allows us to identify critically important expressed gene products on the basis of empirical data, and not just on the basis of theoretical assumptions.

Our data (Dudas, 1993 and in preparation) suggest the following conclusions:

1. The longevity differences between the strains cannot be attributed to allelic differences at known structural loci.

2. Most of the quantitative and temporal differences in gene expression observed at the mRNA level between the various genotype-density combinations of Figure 9 are not correlated with the expression of the ELP. This observation empirically demonstrates the necessity of the decision matrix. The use of only the R and L strains, even using replicate sister strains of each, if examined only under one set of chromosomal/density conditions, would not have allowed us to sort out strain effects from density effects from longevity effects, and thus would not have let us objectively decide which changes in gene expression were important to the expression of the ELP and which were not.

3. Significant density induced elevations in ADH mRNA levels are readily apparent by the early 3rd instar larvae, a result which is consistent with the critical period data of Figure 8. The density induced increase in Adh activity is found in all genotypes examined to date. It therefore represents a specific example of a ubiquitous density induced alteration in gene expression which is not, however, correlated with the expression of the ELP.

4. The 5 day old long lived adult shows significant quantitative alterations in a number of mRNAs whose protein products have known antioxidant (catalase, superoxide dismutase, xanthine dehydrogenase) and metabolic (elongation factor 1a-F1 and F2) activities. The enzyme activities for superoxide dismutase, catalase, and glutathione-S-transferase (another enzyme with antioxidant activity) also show significant increases in activity at these same stages. These changes are not just strain or density dependent but are correlated with the ELP. Our previous data have shown that this 5-7 day period is the age when the L strain animal demonstrates a maximum level of resistance to exogenous paraquat (Arking et al., 1991), shows the maximum expression of phototactic and geotactic behaviors (Neslund & Arking, 1990; Arking & Wells, 1990), and is also the minimum age at which the delay in the onset of senescence can be detected (Arking & Wells, 1990). The 5-7 day old R strain adult, on the other hand, does not show these changes in gene expression, has effectively lost its resistance to exogenous paraquat, and shows a significantly decreased expression of its phototactic and geotactic behaviors (Neslund Wells & Arking, 1989; Arking & Wells, 1990). Again, independent experiments have led to compatible conclusions and demonstrate that the changes in gene activity lead to a functional phenotypic change in the response of the animal to its environment.

5. The 002-HD animals show a somewhat different alteration in their patterns of gene expression relative to their 022-HD and 220-HD controls. In these animals, the ELP appears to result from a moderate increase in some antioxidant gene activities in the adult, coupled with moderate increases in elongation factor-1α RNAs during the late pupal and early adult stages. A comparison of the 222-HD and 002-HD expression patterns suggests that epistatic chromosome interactions can significantly alter gene regulatory processes, and hence alter the exact nature of the

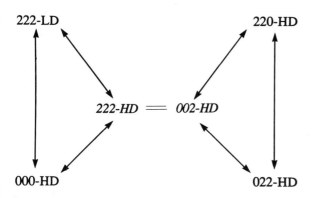

Fig. 9. The decision matrix. Only the two italized genotypes (L-HD and 002-HD) express the ELP. Gene products critically important to the expression of the ELP should be the same within these two strains but quite different between these two strains and any of the other genotype-density combinations shown. For example, the 222-HD/222-LD comparison reveals the effects of larval density on gene expression patterns. In a similar manner, the 222-HD/000-HD comparison reveals the effects of non-specific strain difference; while the 222-HD vs (222-LD and 000-HD) comparison reveals the difference in gene expression specifically associated only with the expression of the ELP. The three isochromosomal line comparisons were designed to reveal the effects of each of the above chromosomes on the gene expression patterns of c3.

physiological mechanisms involved in the expression of the ELP.

Thus, relative to the controls, it appears as if there is a specific and coordinated regulation of the temporal and quantitative expression of a number of structural genes which is only observed in adults exposed to HD and able to express the ELP. The observed differences between the 222-HD and 002-HD strains with respect to their antioxidant activities was in fact accurately predicted several years ago by the paraquat test (Arking *et al.*, 1991). Identifying the known genes critically involved in this process will allow us to empirically identify the physiological mechanisms involved in the expression of the ELP in these strains, and perhaps allow us to devise segmental interventions. However, the more important genetic task confronting us is the identification and characterization of the genetic regulatory elements which modulate the expression of these (and other) critically important structural genes in such a way as to bring about the expression of the ELP. The model presented in Figure 9 is a first step towards the attainment of this important goal.

A model of the genetic circuitry involved in the expression of the ELP

The data presented above have made it apparent that aging is a genetically determined developmental process which can be significantly modulated by manipulation of specific environmental parameters. The epistatic repressive effects of $c1^L$ and $c2^L$ were discussed earlier, as was the presence on the $c3^L$ of longevity enhancing recessive genes. The following additional points may be deduced from the data and are incorporated into the model:

1. The difference between the R and L strains doesn't reside in their structural genes (i.e. their longevity assurance genes, or LAGs) but rather should be sought in the regulatory mechanisms that modulate the LAG expression so as to yield an ELP. Our data suggests, for example, that the two strains have the same Cu-Zn superoxide dismutase gene but that they express it differently.

2. The response of the genotypes to HD conditions must be transduced through some sort of density response elements (DRE). Since almost all genotypes respond in a positive manner to density, then the DRE must be present in all genotypes.

$\rightarrow$

Fig. 10. A diagrammatic representation of the genetic model summarized in the text and described in detail in Buck *et al.*, (1993a,b). The model is illustrated using the *000, 222, 220, 012* and *002* females under HD and/or LD conditions, since all relevant components and interactions may be seen in these genotypes. The three pairs of lines represent, from left to right, chromosomes c1, c2 and c3. The *LAG* refers to the various structural genes involved in the aging process and modulated by the DRE (density response elements, common to all genotypes) and the LRE (longevity response elements, specific only to those genotypes with a $c3^L$). For the sake of simplicity, the diagram only shows the *LAG* localized on the right arm of c3; neither this nor the c3 location of DRE should be interpreted literally. *Sup-3* (*suppressor of c3*) is located on $c2^L$ and represses the longevity enhancing effect of the $c3^L$ LRE and/or LAG, probably the latter. *Sup-2* (*suppressor of c2*) is located on $c1^L$ and represses the inhibitory effect of *sup-2*. The pluses or minuses indicate the effect of the signal in question upon the indicated target. Truncated arrows indicate the inhibition of that particular signal; complete arrowheads indicate an effective signal. In Panel A, the *000* female is raised under LD conditions. The DRE are not activated and the LAGs function at a level that yields a short life span (i.e. significantly shorter than the R-HD value). This represents a baseline level. In Panel B, the *000* female is raised under HD conditions. The DRE is activated by HD and activates the LAGs to the extent necessary to produce an R type life span. In Panel C, the *222* female is raised under HD conditions. The *sup-3* gene, which would normally inactivate the *DRE* gene, is itself inactivated by the homozygous *sup-2* genes on $c1^L$; the *DRE* gene is now capable of being activated by the HD signal. The DRE activates the LRE present only on the $c3^L$, which in turn somehow enhances the LAGs' expression, thus leading to the expression of the ELP. In Panel D, the *222* female is raised under LD conditions. There is no effective HD signal. Neither the DRE nor the LRE are activated and the activities of the *LAG* are maintained at a baseline level, resulting in a short life span. Panel E suggests that the absence of the HD signal allows the *220* female to also express its LAGs at a baseline level, and phenotypically express a short life span. Panel F depicts the same genotype raised under HD conditions. The presence of the HD signal activates the DRE which subsequently induces the LAGs to function at an R type level and pattern. This eventually results in an R type life span. In Panel G, the *sup-3* gene is no longer inactivated since this animal has no $c1^L$, and hence no *sup-2* genes. The effects of the LRE and/or the LAG are inactivated despite the presence of an active DRE. This leads to a normal life span. In panel H, absence of both the $c1^L$ and $c2^L$ results in the removal of the epistatic chromosomal interactions which account for much of the non-additive genetic variance associated with longevity. In their absence, and in the presence of the HD signal, the LRE and LAG interact so as to yield elevated levels and patterns of gene activity. This eventually results in an extended life span. No attempt has been made in this model to account for the stage-specific difference in LAG expression between 222-HD and 002-HD (see text).

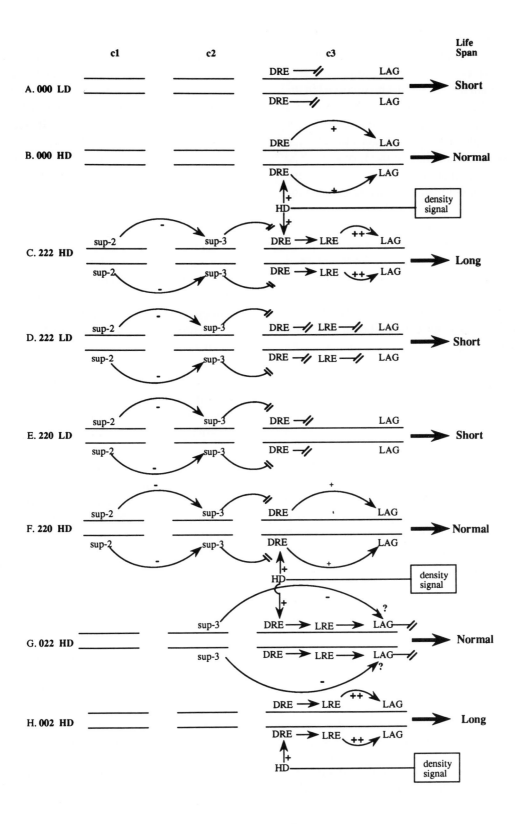

Presumably, the density dependent increase in Adh activity in both R and L strains is one example of the effect of the DRE.

3. When exposed to HD, only the L and the 002 genotypes express the ELP. Therefore some element other than the DRE must be involved in the response of these two genotypes. We term these hypothesized genetic elements the longevity response elements (LRE). Our data indicate that they are probably located on the L type c3. The data also suggest that these elements act in some presently unknown manner so as to enhance the effect of the DRE on the LAGs.

4. The same set of LAGs (i.e. the antioxidant defense system genes) are overexpressed in the two strains that do express the ELP, but their pattern of overexpression differs between the two strains.

A diagrammatic representation of the situation empirically described by our data is shown in Figure 10. The development of such a model facilitates the formulation of testable hypotheses regarding the effects of mutations on the genes suggested as being involved in the regulation of life span. For example, our data suggest the existence of a gene(s) on c2^L (termed here as *suppressor of c3, sup-3*) which can act as a dominant repressor of the genes on c3^L necessary for the expression of the ELP. The model predicts that mutational inactivation of the *sup-3* gene should result in the derepression of the c3^L genes and the subsequent expression of the ELP in the *022* genotype, a genotype in which the ELP is not normally expressed (see figure legend for details). As another example of the utility of the model, the existence of the DRE could be verified by obtaining a density-insensitive mutant, while the existence of the LRE could be verified by obtaining a mutant on c3^L which was density sensitive but unable to express the ELP. Such mutational experiments are now underway.

Comparative analysis: a multiplicity of phenotypes

Successful selection for three independently derived long lived strains of *Drosophila* has been reported by several laboratories (e.g. Rose & Charlesworth, 1980, 1981; Luckinbill *et al.*, 1984; Rose, 1984; Arking, 1987a; Partridge & Fowler, 1992). A comparison of the data available for each of these strains suggests that each strain appears to be characterized by its own particular constellation of physiological mechanisms. The long lived 'O' strains selected by Rose (1984) are reported to have an increased resistance to starvation and dessication (Service *et al.*, 1985; Rose *et al.*, 1992), a decreased metabolic rate and an increased lipid content (Service, 1987), a higher wet weight as well as an increased glycogen content (Graves *et al.*, 1992), and a decreased early fertility, all relative to their B strain controls. Partridge and Fowler (1992) found their long lived 'old' strains had an equivalent early but an increased late fertility and fecundity, a longer developmental time, and a greater body weight, all relative to their 'young' controls. Our L$_A$ strain has been shown to have a decreased early but greater late fertility (Arking, 1987b), an equivalent metabolic rate (Arking *et al.*, 1988), a low and similar resistance to starvation or dessication, an increased glycogen content, a decreased lipid content, and a faster development time, all relative to its R$_A$ control strain (Force & Arking, in preparation). In a somewhat different approach, Hoffman and Parsons (1989, 1992) selected lines for dessication resistance and found that these lines also had an increased adult longevity, an increased resistance to starvation and other stresses, an equivalent development time, and an equivalent lipid content, all relative to their controls. It seems reasonable to conclude that these several strains may have utilized different combinations of physiological mechanisms to alter their physiology in such a manner so as to extend their longevities. The phenotypes are similar; the physiological mechanisms involved may be different. If this should prove to be the case, then it will be of interest to determine whether precisely the same selection pressures would always lead to an extended longevity strain with the same sets of altered physiological mechanisms. If identical phenotypes with different physiologies are obtained, then the detailed predictions built into various evolutionary scenarios may represent an insufficiently broad range of potential outcomes for the testing of the proximal mechanisms postulated by the particular scenario, and may lead to the acceptance of erroneous conclusions. Should there be only a low correlation between the genes responsible for the proximal mechanisms of extended longevity and the genes responsible for various diagnostic characters, then one might expect a shifting suite of secondary characters to be associated with longevity (see Rose, 1991). Perhaps such a situation underlies the results obtained by Hoffman (1989) in his analysis of our strains. Should there be a multiplicity of extended longevity phenotypes, then this might imply that the ELP may be regulated at several different levels. Determining the points of similarity and difference between the several phenotypes will require detailed investigation. Theoretically based conclusions based on only one set of sister strains may be difficult to extrapolate to other strains.

Summary and conclusions

We have demonstrated that the expression of the ELP in our strains is the outcome of a genetically determined, environmentally modulated, event dependent, developmental process. Given the appropriate genetic and environmental conditions, we observe an early acting temporal progression of alterations in specific gene activity patterns which appear to give rise to functional phenotypic changes. The observed patterns are consistent with the interpretations drawn from our chromosome substitution and biomarker experiments. The interaction of specific environmental and genetic factors is sufficient to explain the observed plasticity of longevity in our L strain. Independently derived long lived strains may have altered different combinations of physiological mechanisms so as to give rise to a statistically equivalent ELP. Theoretically based conclusions obtained from only one set of sister strains may be difficult to extrapolate to other strains. Future work will involve the experimental verification of the genetic-environmental circuitry discussed here, using novel molecular genetic techniques to define, characterize, and isolate the genes involved in the expression of the ELP.

Acknowledgements

It is a pleasure to acknowledge the efforts of Steven Buck, Allen Force, Susan LaGrou, Christine Neslund, and Michael Nicholson during the course of these experiments. The efforts of Angela Berrios, Sharon Skorupski, and Robert Wells are also much appreciated. The work was supported by the Glenn Foundation, the Nathan Shock Foundation, the WSU Institute of Gerontology, and by various grants from WSU.

References

Arking, R., 1987a. Successful selection for increased longevity in *Drosophila*: Analysis of the survival data and presentation of a hypothesis on the genetic regulation of longevity. Exp. Gerontol. 22: 199-220.

Arking, R., 1987b. Genetic and environmental determinants of longevity in *Drosophila*, pp. 1-22 in Evolution of Longevity in Animals: A Comparative Approach, edited by A. D. Woodhead and K. H. Thompson. Plenum Press, New York.

Arking, R., S. Buck, R. A. Wells & R. Pretzlaff, 1988. Metabolic rates in genetically based long lived strains of *Drosophila*. Exp. Gerontol. 23: 59-76.

Arking, R. & S. P. Dudas, 1989. A review of genetic investigations into aging processes in *Drosophila*. J. Amer. Geriatrics Soc. 37: 757-773.

Arking, R. & R. A. Wells, 1990. Genetic alteration of normal aging processes is responsible for extended longevity in *Drosophila*. Develop. Genetics 11: 141-148.

Arking, R., S. Buck, A. Berrios, S. Dwyer & G. T. Baker, III., 1991. Elevated paraquat resistance can be used as a bioassay for longevity in a genetically based long-lived strain of *Drosophila*. Develop. Genetics 12:

Austad, S. N., 1989. Life extension by dietary restriction in the bowl and doily spider, *Frontilnela pyramitela*. Exp. Gerontol. 24: 83-92.

Boetella, L. M., A. Moya, M. C. Gonzalez & J. L. Mensua, 1985. Larval stop, delayed development and survival in overcrowded cultures of *Drosophila melanogaster*: effect of urea and uric acid. J. Insect Physiol. 31: 179-185.

Buck, S., R. A. Wells, S. P. Dudas, G. T. Baker, III, & R. Arking, 1993a. Chromosomal localization and regulation of the longevity determinant genes in a selected strain of *Drosophila melanogaster*. Heredity 71: 11-22.

Buck, S., M., Nicholson, S. P. Dudas, G. T. Baker, III & R. Arking, 1993b. Larval regulation of adult longevity in a genetically selected long lived strain of *Drosophila melanogaster*. Heredity 71: 23-32.

Clare, M. & L. Luckinbill, 1985. The effects of gene-environment interaction on the expression of longevity. Heredity 55: 19-29.

Clark, A. G., 1990. Genetic components of variation in energy storage in *Drosophila melanogaster*. Evolution 44: 637-650.

Crowl, T. A. & A. P. Covich, 1990. Predator-induced life-history shifts in a freshwater snail. Science 247: 949-951.

David, J. R., C. Bocquet, M. Arens & P. Foullet, 1976. Biological role of alcohol dehydrogenase in the tolerance of *Drosophila melanogaster* to aliphatic alcohols. Utilization of an ADH-null mutant. Biochem. Genet. 14: 989-977.

Dudas, S. P., 1993. Molecular genetic investigation of the extended longeveity phenotype of a long lived strain of *Drosophila melanogaster*. A dissertation submitted to the Graduate School of Wayne State University, May 1993.

Economos, A. C. & F. A. Lints, 1984. Growth rate and life span in *Drosophila*. III. Effect of body size and developmental temperature on the biphasic relationship between growth rate and life span. Mech. Ageing and Develop. 27: 153-160.

Everse, J., E., Zoll, L. Kahon & N. Kaplan, 1971. Addition products of diphosphopyridine nucleotides with substrates of pyridine nucleotide-linked dehydrogenases. Bioorg. Chem. 1: 207.

Graf, J.-D. & F. J. Ayala, 1986. Genetic variation for superoxide dismutase level in *Drosophila melanogaster*. Biochemical Genetics 24: 153-168.

Graves, J. L., E. C. Toolson, C. Jeong, L. N. Vu & M. R. Rose, 1992. Dessication, flight, glycogen, and postponed senescence in *Drosophila melanogaster*. Physiol. Zool. 65: 268-286.

Hayflick, L., 1985. Theories of biological aging. Exp. Gerontol. 20: 145-159.

Heinstra, P. W. H. & B. W. Geer, 1991. Metabolic control analysis and enzyme variation: nutritional manipulation of the flux from ethanol to lipids in *Drosophila*. Mol. Biol. Evol. 8: 703-708.

Hoffmann, R. N., 1989. The effects of delayed mating on components of fitness, life span, and the geotactic behavior of *Drosophila melanogaster* selected for a postponed senescence. A dissertation submitted to the Graduate Faculty in Biology of the City University of New York.

Hoffmann, A. & P. Parsons, 1989. Selection for increased desiccation resistance in *Drosophila melanogaster*: additive genetic control and correlated responses for other stresses. Genetics 122: 837-845.

Hoffmann, A. & P. Parsons, 1993. Selection for adult dessication resistance in *Drosophila melanogaster*: Fitness components, larval resistance and stress correlations. Biol. J. Linn. Soc. 48: 43-54.

King, J. J. & J. F. McDonald, 1987. Post-translational control of alcohol dehydrogenase levels in *Drosophila melanogaster*. Genetics 115: 693-699.

Kirkwood, T. B. L. & M. R. Rose, 1991. Evolution of senescence: late survival sacrificed for reproduction. Phil. Trans. R. Soc. Lond. (B) 332: 15-24.

Lints, F. A. & C. V. Lints, 1971. Influence of preimaginal environment on fecundity and ageing in *Drosophila melanogaster* hybrids. II. Developmental speed and life span. Exp. Gerontol. 6: 427-445.

Lints, F. A. & H. M. Soliman (ed.), 1988. *Drosophila* as a model organism for ageing studies. Blackie and Son Ltd, Glasgow.

Lunckinbill, L. S., R. Arking, M. J. Clare, W. C. Cirocco & S. A. Buck, 1984. Selection for delayed senescence in *Drosophila melanogaster*. Evolution 38: 996-1004.

Luckinbill, L. S. & M. J. Clare, 1985. Selection for life span in *Drosophila melanogaster*. Heredity 55: 9-18.

Luckinbill, L. S., M. J. Clare, W. L. Krell, W. C. Cirocco, & P. Richards, 1987. Estimating the number of genetic elements that defer senescence in *Drosophila*. Evolutionary Ecology 1: 37-46.

Luckinbill, L. S., J. L. Graves, A. H. Reed & S. Koetsawang, 1988. Localizing genes that defer senescence in *Drosophila melanogaster*. Heredity 60: 367-374.

Mayr, E., 1961. Cause and effect in biology. Science 134: 1501-1506.

Miller, R. S. & J. L. Thomas, 1958. The effects of larval crowding and body size on the longevity of adult *Drosophila melanogaster*. Ecology 39: 118-125.

Neslund, C. M., R. A. Wells & R. Arking, 1989. Behavior genetic analysis of a long lived strain of *Drosophila melanogaster*. The Gerontologist 28: 226A (abstr.).

Oudman, L., W. Van Delden, A. Kamping & R. Bijlsma, 1991. Polymorphism at the Adh and αGpdh loci in *Drosophila melanogaster*: effects of rearing temperature on developmental rate, body weight, and some biochemical parameters. Heredity 67: 103-115.

Partridge, L. & K. Fowler, 1992. Direct and correlated responses to selection on age at reproduction in *Drosophila melanogaster*. Evolution 46: 76-91.

Pearl, R., 1928. The Rate of Living. University of London Press, London.

Rose, M. R., 1991. Evolutionary Biology of Aging. Oxford University Press, New York.

Rose, M. R., 1984. Laboratory evolution of postponed senescence in *Drosophila melanogaster*. Evolution 38: 1004-1010.

Rose, M. R. & B. Charlesworth. A test of evolutionary theories of senescence. Nature 287: 141-142.

Rose, M. R. & B. Charlesworth, 1981. Genetics of life history in *Drosophila melanogaster*. II. Exploratory selection experiments. Genetics 97: 187-196.

Rose, M. R., L. N. Vu, S. U. Park & J. L. Graves, Jr., 1992. Selection on stress resistance increases longevity in *Drosophila melanogaster*. Exp. Gerontol. 27: 241-250.

Service, P. M., E. W. Hutchinson, M. D. Mackinley & M. R. Rose, 1985. Resistance to environmental stress in Drosophila melanogaster selected for postponed senescence. Physiol. Zool. 58: 380-389.

Service, P. M. 1987. Physiological mechanisms of increased stress resistance in *Drosophila melanogaster* selected for postponed senescence. Physiol. Zool. 60: 321-326.

Service, P. M., E. W. Hutchinson & M. R. Rose. Multiple genetic mechanisms for the evolution of senescence in *Drosophila melanogaster*. Evolution 42: 708-716.

Weindruch, R., R. L. Walford, S. Fligiel & D. Guthrie, 1986. The retardation of aging in mice by dietary restriction: Longevity, cancer, immunity and lifetime energy intake. Journal of Nutrition. 116: 641-654.

Weismann, A., 1891. Essays upon Heredity and Kindred Biological Problems, 2nd ed., vol. 1. Clarendon Press, Oxford.

Wells, R. A., S. Buck, R. Ali, O. Marzouq & R. Arking, 1987. Localization of the longevity genes in *D. melanogaster*. The Gerontologist 27: 149A. (abstr.).

Zwaan, B., R. Bijlsma & R. Hoekstra, 1991. On the developmental theory of ageing. I. Starvation resistance and longevity in *Drosophila melanogaster* in relation to pre-adult breeding conditions. Heredity 66: 29-39.

M.R. Rose and C.E. Finch (eds.), Genetics and Evolution of Aging, 161–167, 1994.
© 1994 Kluwer Academic Publishers. Printed in the Netherlands.

The effect of superoxide dismutase alleles on aging in *Drosophila*

Robert H. Tyler, †, Hardip Brar, Meena Singh, Amparo Latorre[1], Joseph L. Graves[2], Laurence D. Mueller, Michael R. Rose & Francisco J. Ayala
Department of Ecology and Evolutionary Biology, University of California, Irvine, CA 92717, USA
[1] *Present address: Departamento de Genetica, Facultad de Biologia, Universidad de Valencia, Spain*
[2] *Author for correspondence*

Received and accepted 22 June 1993

Key words: superoxide dismutase, aging, *Drosophila*, evolutionary genetics, senescence

Abstract

The effects of superoxide dismutase on aging were tested using two differt experimental approaches. In the first, replicated populations with postponed aging were compared with their controls for frequencies of electrophoretic alleles at the SOD locus. Populations with postponed aging had consistently greater frequencies of the allele coding for more active SOD protein. This allele was not part of a segregating inversion polymorphism. The second experimental approach was the extraction of *SOD* alleles from different natural populations followed by the construction of different *SOD* genotypes on hybrid genetic backgrounds. This procedure did not uncover any statistical effect of *SOD* genotype on longevity or fecundity. There were large effects on longevity and fecundity due to the family from which a particular *SOD* genotype was derived. To detect the effects of *SOD* genotypes on longevity with high probability would require a ten-fold increase in the number of families used.

Introduction

Studies of the temporal correlates of the aging process do not necessarily indicate the causal mechanisms controlling it, while treatments that shorten life span may kill as a result of novel pathologies (Maynard Smith, 1966; Rose, 1991). For this reason, other experimental strategies are required to unravel the mechanisms that normally control aging. Postponed aging arising from selection for survival to, and reproduction at, later ages requires genetic mitigation of normal aging mechanisms, and therefore constitutes a model system of choice for the analysis of the genetics of aging (Hutchinson & Rose, 1987).

Choice of a good model system for the genetics of aging has not, however, led to immediate breakthroughs in causal understanding. Some electrophoretic studies of the genetic basis of selectively postponed aging in *Drosophila melanogaster* have already been performed (Luckinbill et al., 1989). Some significant correlations between postponed aging and the loci affecting energetic metabolism were found by Luckinbill et al., though that work suffered from a lack of replication. Only one or two selected populations were compared to their control populations. In addition, the founding population sizes were fairly small. Fortunately, these problems are easily remedied using the larger set of stocks created by Rose (e.g. 1984), in which there are five selected lines and five controls, the entire set of stocks coming from a large wild-caught sample of *D. melanogaster* from the endemic Ives population of South Amherst, Massachusetts.

A more profound problem with electrophoretic comparisons of postponed-aging stocks with their controls is that of linkage disequilibrium. When postponed-aging stocks are created by selection from a common founding population, the particular linkage disequilibria of that population will be in common among all derivatives undergoing selection. Allelic state correlations between physically proximal loci will be maintained for some part of the selection process, causing nearby loci to undergo parallel changes of allele frequency in all selected stocks. Therefore, stocks with selectively postponed aging are not necessarily, by themselves, reliable systems for inferring the causal involvement of particualar loci in aging.

One potential solution to this difficulty is to use stocks with postponed aging to identify initial candidate genes, genes that might be involved in postponed aging, or that might change in allelic composition because of linkage disequilibrium alone. Then genotypes at this locus can be arbitrarily assembled from different natural populations, and tested for their effects upon aging (cf. Serradilla & Ayala, 1983). In this way, the problem of the original linkage relations in the founding population before selection can be overcome. In the present paper, we apply this approach to one locus of great interest to gerontologists, that coding for Cu, Zn superoxide dismutase. This free radical scavenging molecule is of interest because of its role in catalyzing the conversion of superoxide radicals to hydrogen peroxide, the latter then undergoing conversion to water due to the action of catalase (McCord & Fridovich, 1969). Since superoxide radicals are highly damaging for macromolecules such as protein, lipid, and nucleic acid, the cumulative effects of such free radicals have long been proposed as a major factor in aging (e.g. Harman, 1956). For this reason, we were particularly interested in applying the combination of stocks with postponed aging and sampled alleles from wild populations to test for the effects of superoxide dismutase alleles upon aging. In natural populations of *D. melanogaster*, Cu, Zn superoxide dismutase has F ('fast') and S ('slow') allelomorphs, with S being rarer but of greater *in vitro* activity (Lee, Misra & Ayala, 1981). On the free radical theory of aging, this allelomorph should have elevated frequencies in populations that are longer-lived. In addition, the S allele should give rise to increased longevity, on average, in flies of SS or FS genotype, relative to FF genotypes. The present article tests these predictions. We find that some of these predictions are borne out, but others are not.

Materials and methods

Postponed aging stocks

The postponed aging populations used in the present study were derived from a long-standing wild population that had been sampled in 1975 (Rose, 1984). All fly culture proceeded at 25 °C with abundant food, little crowding, and moderate humidity. In Feb. 1980, one generation of the base population was used to found ten outbred populations, five of which (called B_1-B_5) were kept under the same conditions as the base population, with two-week discrete generations. The other five were cultured using females of increasingly greater ages as the generations progressed, such that by late 1981 all reproducing females had to attain 70 days of age (Rose, 1984). Such populations are designated O_1-O_5, with no correspondence between B and O subscripts. Since 1981, cohorts sampled from the O populations have had significantly greater mean and maximum longevities, relative to B populations (Rose, 1984; Hutchinson & Rose, 1991).

In addition to this longevity difference, B and O populations have a number of other differences which indicate that O flies live longer because of postponed aging. These differences include increases in: tethered flight duration (Graves & Rose, 1990), spontaneous locomotion at later ages (Service, 1987), 24-h egg-laying at later ages (Rose, 1984), and resistance to stress (Service *et al.*, 1985). These findings strongly suggest that the longer-lived O populations indeed possess postponed aging, through both general and late-age enhancements in physiological functions. Derivatives of the O populations have also been subjected to relaxed selection (R populations), in which cultivation returned to 2-week generations (Service, Hutchinson & Rose, 1988). The phenotypes of these R populations have largely returned to those of the control B populations (Graves *et al.*, 1992; and unpub. data), including mean longevity. If superoxide dismutase is a causal factor postponing aging, then the electrophoretic profile of the R populations should be identical to that of the B populations.

Extraction of SOD genotypes from natural populations

D. melanogaster females were collected from Culver City, Los Angeles County, in southern California and El Rio Vineyard, San Joaquin Country, in northern California. From each population, 20 isofemale lines were made homozygous for each of both the F and S superoxide dismutase alleles. This was done by inbreeding and electrophoretic screening over multiple generations.

These inbred lines were then crossed with each other in pairs to create lines of known SOD genotype but hybrid background. This procedure mitigates problems of inbreeding depression. Such crosses were performed by crossing in each case 20 males from one line with 20 females from a different line. Four different genotypes were constructed: *FF, SS, FS,* and *SF,* where the latter pair of genotypes differ according to the maternal genotype.

Assay procedures
All electrophoretic procedures used the methods described in Ayala *et al.* (1972). Polytene chromosomes were extracted from larvae of B and O populations and stained. Surveys were made of 29-55 different chromosomes for the presence of inversion heterozygotes on each arm. All assays of life-history phenotypes used control-density larval rearing, with 25-50 larvae per vial. Adult females were kept together with two males, in vials of banana-molasses food for the Culver City samples and corn meal-flour food for the El Rio samples. Fecundity

assays were made periodically, using a charcoal-colored high-agar medium without yeast. Vials were examined each day for survival of the adults.

In these experiments there were block effects which confounded two sources of experimental variation. The first of these is the population of origin and the second is laboratory handling. The assays for each population were conducted in different laboratories. Thus, different personnel, incubators and type of food were used for collecting data from each population. Consequently, all statistical analyses treated results from El Rio and Culver City as two blocks.

Results

Allele frequencies in postponed-aging stocks
As shown in Table 1, the S allele is significantly more frequent in O populations, averaging about 25%, with no S alleles detected in the B population samples. Indeed, this average frequency of S in the O populations is greater than has been observed before for this allele in natural populations of *Drosophila* (Smit-McBride, Moya & Ayala, 1988; Singh, Hickey & David, 1982 and unpub. data). While these samples suggest that the B populations entirely lack the S allele, further sampling of these populations has produced a few instances where flies from B populations carry the S allele heterozygously. The R populations have the same profile as the B sample shown here, the S allele being absent. All of these findings are in accord with the free-radical theory of aging, in that the more active allele is significantly more frequent in populations with postponed aging.

But the conclusion that SOD is causally related to aging could be erroneous if the S allele is tightly linked to an allele which itself postpones aging. The total number of generations separating B from O populations is over 300, the two population-types have been replicated five-fold, and the census population sizes have been in the thousands. These experimental design features facilitate attainment of linkage equilibrium at unselected loci. Nonetheless, if the S allele were associated with an inversion polymorphism, recombination could be too weak to prevent linkage disequilibrium between the S allele and a selected allele at another locus within the same inversion. Table 2 shows the results from

Table 1. Frequency of the S('slow') allele in B and O populations for Cu,Zn superoxide dismutase (SOD). Only two electromorphs were detected. Electrophoresis was performed according to Ayala *et al.* (1972); n is the number of individuals sampled.

Line	n	FF	FS	SS	SFrequency
B$_1$	50	50	0	0	0.00
B$_2$	48	50	0	0	0.00
B$_3$	50	50	0	0	0.00
B$_4$	24	24	0	0	0.00
B$_5$	24	24	0	0	0.00
Average allele frequency					0.00

Line	n	FF	FS	SS	SFrequency
O1	48	26	17	5	0.28
O2	48	39	8	1	0.10
O3	48	22	15	11	0.39
O4	49	20	26	3	0.33
O5	48	36	9	3	0.16
Average allele frequency standard error					0.25 ± 0.05

Table 2. Number of polytene chromosome arms sampled and inversions observed for B and O populations.

| Line | Chromosome arm | | | | | |
	X	2R	2L	3R	3L	Inversions
B_1	38	48	48	48	48	None
B_2	30	41	41	41	41	None
B_3	30	41	41	41	41	None
B_4	36	50	50	50	50	None
B_5	36	48	48	48	48	None
O_1	35	49	49	49	49	None
O_2	28	44	44	44	44	None
O_3	29	40	40	40	40	None
O_4	36	52	52	52	52	One on 2R
O_5	40	55	55	55	55	None

a study of the cytology of 2205 polytene chromosomes. Only one inversion variant was detected (In(2R)NS), on the right arm of chromosome 2 in the O_4 population. Moreover, the superoxide dismutase locus is located on chromosome 3, in any case. Therefore, the only remaining countervailing hypothesis would be linkage disequilibrium with a favored allele that is physically very close to SOD on chromosome 3. Any such allele is unlikely to

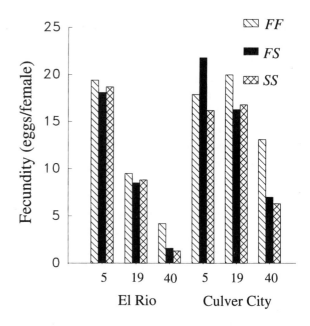

Fig. 2. Female fecundity for each SOD genotype at three different ages 5, 19 and 40 days in the El Rio and Culver City populations.

have the same linkage relationship with SOD in the unrelated California populations of *D. melanogaster* discussed below.

Effects of SOD genotype on aging phenotypes
The results of analyses of variance for longevity and age-specific fecundity are shown in Tables 3

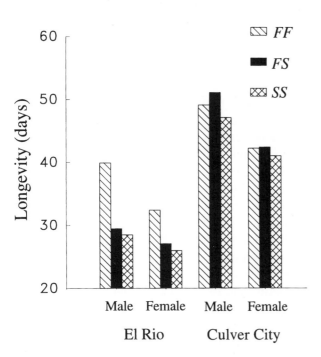

Fig. 1. The mean longevity for each SOD genotype as a function of sex and population.

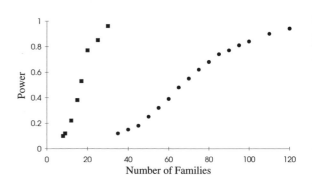

Fig. 3. The power of detecting significant genotype effects on longevity (circles) and fecundity (squares) as a function of the number of families sampled within each genotype. These calculations utilized the longevity data from the El Rio population to estimate the differences in longevity between genotypes and the variance between families and the fecundity data at day 40 from Culver City.

Table 3. Results of the analysis of variance (ANOVA) on longevity. Population was treated as a block effect.

Source of variation	Sum of squares	Degrees of freedom	Mean square	F	Significance
Population	139,030	1	139,030		
Genotype	1,697	2	849	0.255	NS
Sex	25,418	1	25,418	80	<0.001
Family (Genotype)	56,475	17	3,322	10.5	<0.001
Sex by Genotype	219	2	109	0.03	NS
Error	907,114	2,867	316		

NS = not significant

and 4 respectively. Throughout these analyses, it is apparent that there is little significant effect of genotype upon aging patterns (Figs. 1-2). To the extent to which there is a directional effect, the *F* allele is associated with increased longevity, although not significantly so.

In these experiments within each genotypic class were nested several families (5-10). In every case (Tables 3-4) there was a statistically significant effect of families on either longevity or fecundity. Genotypic effects were assessed by placing the family mean square in the denominator of the F ratio. Consequently, it appears that the number of families sampled will have a great impact on the statistical power of these tests. We should add that ANOVAs on the separate experimental results from El Rio or Culver City do not reveal significant

effects of genotype due to the large between-family variation.

In Figure 3 the statistical power of the longevity assay on the El Rio population and the day 40 fecundity assay on the Culver City population are shown. These particular examples were chosen since they show the greatest differences between genotypes. For instance, the power of the ANOVAs on fecundity at day 5 and 19 is still below 20%, even when 100 families are sampled within each genotype.

These results suggest that the genetic background that is unique to each family is causing variation that obscures any differences that the SOD locus might produce. Except for the possibility of late fecundity, samples of about 100 families per genotype would be required to insure a high

Table 4. Results of the analysis of variance (ANOVA) on fecundity. The ANOVAs were carried out on fecundity at day 5, 19 and 40 separately. Population was treated as a block effect.

Source of variation	Sum of squares	Degrees of freedom	Mean square	F	Significance
Day 5 Fecundity					
Population	347	1	347		
Genotype	12,325	2	6,163	2.76	NS
Family (Genotype)	37,910	17	2230	10.67	<0.001
Error	357,348	1,709	209		
Day 19 Fecundity					
Population	26,413	1	26,413		
Genotype	749	2	374	0.428	NS
Family (Genotype)	14,884	17	876	5.76	<0.001
Error	228,493	1,503	152		
Day 40 Fecundity					
Population	6,360	1	6,360		
Genotype	1,536	2	768	3.40	0.06
Family (Genotype)	3,842	17	226	2.89	<0.001
Error	53,281	681	78		

NS = not significant

probability of detecting differences between genotypes of the magnitude observed in this study.

Discussion

The present study finds that more active superoxide dismutase alleles are associated with postponed aging in laboratory stocks. We have failed to demonstrate that such alleles directly increase life span or later fecundity, but this failure arose from a lack of experimental power. That is to say, this negative result is not a pertinent refutation of the free radical mechanism for aging. The lack of inversion polymorphisms involving *SOD* indicates that alleles at the locus are not in linkage disequilibrium with distal loci. However, linkage disequilibrium with physically proximal loci could nevertheless be the factor responsible for the association between superoxide dismutase and postponed aging. Luckinbill *et al.* (1989) found no difference in *SOD* allele frequencies between their long- and short-lived populations of *D. melanogaster*. Luckinbill's study employed two replicates of each treatment, and thus has less statistical power then the present study. Moreover, the frequency of the *S* allele is low in natural populations ($\approx 1\%$), so the *S* allele may have been absent in the original founding sample, or lost by chance during the early generations of selection.

Of great relevance to the present study is the finding of Reveillaud, Niedzwiecki and Fleming (1991), who transformed *D. melanogaster* stocks with highly active bovine SOD DNA to which a strong actin promoter had been linked. In addition to greatly increasing SOD activity levels in transformed flies, adult life span was significantly increased among replicate transformants. Although this increase was of no greater magnitude than 10%, it constitutes a powerful demonstration that more active SOD can give rise to increased lifespan.

Comparing the results with (i) allele frequencies of postponed aging stocks, (ii) alleles extracted from natural populations, and (iii) transformation experiments with extremely active SOD genes (Reveillaud, Niedzwiecki & Fleming, 1991), some conclusions suggest themselves. These disparate results can be reconciled if it is supposed that the *FS SOD* electromorphs have differential effects on aging that are smaller in magnitude than the differential effects between bovine and dipteran *SOD*. If bovine *SOD* transformation can increase life span by at most 10%, relative to dipteran *SOD*, then the effect of the normal dipteran allelic difference might be about 2-4%. The problem for experiments in which alleles are extracted from natural populations is that the hybrid lines constructed from the isofemale inbreds have substantial differences in life span arising from different accidents of fixation in their ancestral inbred lines. Such differences among lines would thus swamp the effect of longevity differences between genotypes at the *SOD* locus, because the latter differences are so small.

From this line of argument, two final inferences can be drawn. Firstly, this paper provides additional evidence supporting the involvement of *SOD* in postponing, and thus controlling, aging in *Drosophila*, particularly in light of the bovine DNA transformation results. Secondly, the extraction of *SOD* alleles from natural populations has been unable to demonstrate a quantitative effect of this locus on aging phenotypes due to variation in genetic backgrounds between families. Transformation experiments might be preferable in any such tests.

Acknowledgements

Lisa M. Chiu and Sarah S. Kwak provided considerable technical help in the performance of these experiments. This research was supported in part by grants PHS AG06346 and AG09970 from the National Institute of Aging (USA) to MRR and the Department of Energy (USA) to FJA, and in part by the estate of Robert H. Tyler. We are grateful to the Tyler family and attornies for their support, cooperation, and encouragement.

References

Ayala, F. J., J. R. Powell, M. L. Tracey, C. A. Mourao & S. Perez-Salas, 1972. Enzyme variability in the *Drosophila willistoni* group. IV. Genic variation in natural populations of *Drosophila willistoni*. Genetics 70: 113-139.

Graves, J. L. & M. R. Rose, 1990. Flight duration in *Drosophila melanogaster* selected for postponed senescence, pp. 59-65 in Genetic effects on aging. II, edited by D. E. Harrison. Telford Press, Caldwell, N.J.

Graves, J. L., E. C. Toolson, C. Jeong, L. N. Vu & M. R. Rose 1992. Desiccation, flight, glycogen, and postponed senescence in *Drosophila melanogaster*. Phys. Zool. 65: 268-286.

Harman, D., 1956. Aging – a theory based on free radical and radiation chemistry. J. Gerontol. 11: 298-300.

Hutchinson, E. W. & M. R. Rose, 1987. Genetics of aging in insects. Rev. Biol. Res. Aging 3: 63-70.

Hutchinson, E. W. & M. R. Rose, 1991. Quantitative genetics of postponed aging in *Drosophila melanogaster*. I. Analysis of outbred populations. Genetics 127: 719-727.

Lee, Y. M., H. P. Misra & F. J. Ayala, 1981. Superoxide dismutase in *Drosophila melanogaster*. Biochemical and structural characterization of allozyme variants. Proc. Natl. Acad. Sci. USA 78: 7052-7055.

Luckinbill, L. S., T. A. Grudzien, S. Rhine & G. Weisman, 1989. The genetic basis of adaptation to selection for longevity in *Drosophila melanogaster*. Evol. Ecology 3: 31-39.

Maynard Smith, J., 1966. Theories of aging, pp. 1-35 in Topics in the Biology of Aging, edited by P. L. Krohn. Interscience, NY.

McCord, J. M. & I. Fridovich, 1969. Superoxide dismutase: An enzymatic function for erythrocuprin. J. Biol. Chem. 224: 6049-6055.

Reveillaud, I., A. Niedzwiecki & J. E. Fleming, 1991. Expression of bovine superoxide dismutase in *Drosophila melanogaster* augments resistance to oxidative stress. Mol. Cell. Biol. 11: 632-40.

Rose, M. R., 1984. Laboratory evolution of postponed senescence in *Drosophila melanogaster*. Evolution 38: 1004-1010.

Rose, M. R., 1991. Evolutionary Biology of Aging. Oxford University Press, NY.

Serradilla, J. M. & F. J. Ayala, 1983. Ecological and evolutionary divergence in five species of *Drosophila*. Z. zool. Syst. Evolut.-forsch. 21: 194-200.

Service, P. M., 1987. Physiological mechanisms of increased stress resistance in *Drosophila melanogaster* selected for postponed senescence. Physiol. Zool. 60: 321-326.

Service, P. M., E. W. Hutchinson, M. D. MacKinley & M. R. Rose, 1985. Resistance to environmental stress in *Drosophila melanogaster* selected for postponed senescence. Physiol. Zool. 58: 380-389.

Service, P. M., E. W. Hutchinson & M. R. Rose, 1988. Multiple genetic mechanisms for the evolution of senescence in *Drosophila melanogaster*. Evolution 42: 708-716.

Singh, R. M., D. A. Hickey & J. David, 1982. Genetic differentiation between geographically distant populations of *Drosophila melanogaster*. Genetics 101: 235-256.

Smit-McBride, Z., A. Moya & F. J. Ayala, 1988. Linkage disequilibrium in natural and experimental populations of *Drosophila melanogaster*. Genetics 120: 1043-1051.

M.R. Rose and C.E. Finch (eds.), Genetics and Evolution of Aging, 168–182, 1994.
© 1994 *Kluwer Academic Publishers. Printed in the Netherlands.*

Use of recombinant inbred strains to map genes of aging

Linda K. Dixon
Center for Developmental and Health Genetics, Pennsylvania State University, University Park, PA 16804, USA

Received and accepted 22 June 1993

Key words: recombinant inbreds, quantitative trait loci, aging

Abstract

Recombinant inbred strains have been used in a number of organisms for segregation and linkage analysis of quantitative traits. One major advantage of the recombinant inbred (RI) methodology is that the genetic identity of individuals within a strain permits replicate measures of the same recombinant genotype. Such replicability is important for traits such as aging in *Drosophila*, where phenotypic expression is highly influenced by different environmental conditions. RI strain methodology has an added advantage for DNA marker-based linkage analysis of traits measured over the lifespan of the organism. The DNA can be extracted from individuals of the same genotype as those measured in a longevity study. In this paper an argument is presented for the use of a set of recombinant inbred strains to map the quantitative trait loci involved in the aging process in *Drosophila*. A unique use of a set of stable, transposable molecular markers to trace the quantitative trait loci involved is suggested.

Introduction

A major challenge in understanding genetic influences during aging is to identify and locate the relevant genes on the linkage map. There are a number of benefits to be gleaned from this information. Knowledge of gene location will allow for the possibility of identifying and molecularly cloning a particular gene of interest. Further studies will allow for the assessment of the role of each specific gene as well as its interaction with other genes. The gene products of structural genes will be determined, and regulatory genes which affect expression of specific loci will be discovered. Evolutionary relationships of genes will be uncovered. Ultimately it may be possible to alter specific genes which have direct effects on the process of aging.

Aging is a complex phenomenon influenced by many genes as well as a plethora of environmental factors. Rare single-gene disorders which have a severe effect on aging in humans and other organisms have given us little insight about normal aging. Genetic diseases which are rare and lead to severe pathologies are unlikely to be responsible for the decline in fitness components characteristic of the normal aging process. Therefore, while the identification and analysis of specific age-related genes is a goal, the techniques of classical Mendelian discrete-group analysis are not generally useful in the study of aging.

Until recently, molecular genetic techniques have been employed primarily in pursuit of single-gene effects, but classical major-gene effects are difficult to localize in most organisms for the complex process of aging. It seems more reasonable to posit that such complex processes, including those related to aging, are influenced by many genes, each with small effects across a population, as well as by environmental factors. Quantitative genetic models that recognize both multiple-gene and environmental influences provide a powerful theoretical perspective on genetic and environmental origins of variability for such complex phenotypes.

Techniques for identifying genes responsible for small but identifiable portions of the variance have been limited. However, there is a recognition of the importance of studying the effects of such polygenic gene action in a wide array of continuously distributed phenotypes.

The challenge now is to use the new genetics to climb from the DNA level at the bottom, up to the analysis of variation, whose genetics and corresponding gene products have hardly been defined. ... The still greater challenge is to deal with diseases and other variations that are not simple Mendelian traits but are multifactorial, a term that basically implies that there are inherited components, but these are not well defined and may be due to a mixture of the effects of several genes and the environment. Heart disease, mental disease, cancer, and complex normal variation, including behavior as well as general physical attributes, all fall in this category. The molecular challenge, which has as its goal the characterization of the whole human genome, is to use this knowledge to unravel the genetic contributions to complex human traits (Bodmer, 1986, p. 1).

Traditional quantitative approaches have been used to show that the process of aging has polygenic underpinnings. Approaches such as selection studies, family analyses, and strain difference studies have given us insight into the nature of the variance involved in the determination of lifespan as well as interrelationships among various correlated aspects of the aging process. Quantitative genetic methods are indirect in the sense that they assess components of variation in a population – that is, they do not assess DNA variation directly. The actual number, location, or products of the genes remain unknown. The merger of quantitative genetics and molecular genetics can capture the strengths of each; the synergism created by this merger has great potential for advancing our understanding of complex phenotypes such as age-related traits.

Molecular mapping of the quantitative trait loci (QTL) involved in the determination of a phenotype is dependent upon identifying a multitude of informative genetic markers which are highly polymorphic and distributed throughout the genome. In addition, the choice of population used for mapping is important for the efficiency and accessibility of the mapping information. In this paper it will be argued that 1) techniques are available to identify informative genetic markers to study aging in *Drosophila melanogaster* and that 2) recombinant inbred (RI) strains of flies generated from the mating of highly inbred, polymorphic progenitor lines provide certain advantages over other mapping populations to complete the task.

Analysis of quantitative trait loci

Quantitative genetics
The quantitative genetic foundation was built in the early part of this century as a solution to the problem of reconciling Mendelian genetics with continuous distributions. When Mendel's work with dichotomous characteristics in the pea plant was rediscovered at the turn of the century, it provoked controversy between Mendelians and biometricians. The biometricians felt that the laws of heredity described by Mendel could not apply to complex characteristics because, unlike the discontinuous pea plant characteristics studied by Mendel, complex phenomena nearly always involve a continuous distribution. The resolution to the controversy came when it was understood that a normal distribution is observed when a number of genes of small effects influence a trait, even though each gene operates exactly as Mendel hypothesized. In 1918, when Ronald Fisher published this theory, quantitative genetics was born.

The focus of quantitative genetics is on individual differences in a population, and it suggests methods such as comparisons among family members, including parent-offspring, full sibs, and half sibs, to partition variance in a population into genetic and environmental components of variance. In experimental populations directed selection and inbred strain comparisons can also be used to add further understanding to the causal effects of the variance components (Plomin, DeFries & McClearn, 1990).

Although a few major-gene effects on aging have been reported, most authors agree that aging is a multifactorial system. In *Drosophila*, for example, there is an abundance of evidence for the presence of quantitative genetic variation (Rose, 1991); inbred lines of *Drosophila* differ widely for longevity and reproductive aging (Gowen & Johnson, 1946; Clarke & Maynard Smith, 1955; Rose, 1984a); artificial selection studies for postponed aging in

Drosophila have been successful in more than one laboratory (Wattiaux, 1968; Rose, 1984b; Luckinbill *et al.*, 1984); quantitative measurements indicate substantial additive genetic variance for most age-related characters (Hutchinson & Rose, 1991; Hutchinson, Shaw & Rose, 1991). Although it will be useful experimentally if major-gene influences are found for aging, a polygenic perspective will be more generally applicable in approaching genetic contributions to complex phenotypes. This is especially so in terms of an attempt to identify genetic factors that each account for only a small amount of variance in the population.

Molecular genetics
Quantitative genetic techniques do not assess genetic variation directly. This has been a source of great strength because the quantitative methods address the 'bottom line' of genetic influences on the phenotype. That is, these methods assess the total impact of genetic variability of any kind, regardless of its molecular source. Until recently, it has not been possible to assess genetic variation for quantitative traits directly. However, advances in molecular genetics have made it possible to identify gene markers that can be screened for the contribution of genes in that chromosomal location, independently and jointly, to the variance of complex phenotypes. This can occur even for traits where dozens of genes each contribute small portions of variance in the population and where environmental influences are substantial. The significance of this approach is that it permits the identification of associations between specific molecular markers and the quantitative character. This is an approach that will open new vistas for research on genetics and aging.

Genetic markers. An ideal set of molecular markers useful for the localization of quantitative trait loci would have a number of characteristics (Dietrich *et al.*, 1992). First, the markers must be highly abundant and evenly distributed, so that a large portion of the genome (ideally, the entire genome) can be followed simultaneously in a cross. Second, the markers must be highly polymorphic, so that one could follow the markers in any cross between laboratory strains or specific individuals. Thirdly, the markers should be easily typed, so that scoring would be rapid relative to generation time. Finally, for purposes of replication and confirmation, the

markers should be easily disseminated, so that any laboratory would have ready access to them. Over the years molecular markers have been used which have some of these characteristics, and recently markers have been identified which fit all the criteria.

The first genetic marker used to associate a gene with a specific chromosome was the *Drosophila white* eye color mutant reported by Morgan in 1910. Soon afterward linkage associations of six genes on the X chromosome were reported (Sturtevant, 1913). The notion of visible gene markers was expanded in the 1940s and 1950s with the use of biochemical techniques. The biochemical markers, like the rare visible mutants, were generally present at mutation rate frequencies in natural populations. They led to drastic phenotypic effects in the organism and were of limited use in linkage studies. The discovery of isozymes led to a new class of genetic markers. Isozymes are variant forms of an enzyme which are present in natural populations at higher than mutation rate frequencies. They were exploited extensively in the 1960s and 1970s by population and developmental geneticists because the level of polymorphism was so high relative to single gene mutations. Yet isozyme polymorphisms also had limited use as gene markers for determining linkage relationships. They were not highly abundant or evenly distributed in the genome; they were time-consuming to type, and only a few could be followed in a given cross.

A new era in gene markers for mapping began in 1980 with the discovery of restriction fragment length polymorphisms (RFLPs) (Botstein *et al.*, 1980). RFLP refers to inherited differences in DNA which result in differences in the lengths of the fragments produced by cleavage with restriction enzymes. When the DNA at any given genomic site differs among individuals, this leads to stretches of DNA being cut differentially by a given restriction endonuclease. The resultant fragments made by the restriction endonuclease digestion are of varying sizes among the different individuals and are easily detectable on an electrophoretic gel. RFLPs are genetic markers which are abundant in a population and often evenly distributed throughout the genome. However, they have a number of disadvantages when used in gene mapping. There is often a low rate of polymorphism among specific individuals or strains of interest; it is time-consuming and tedious to type individuals for RFLPs; before carrying out the procedure it is necessary to seek out and request specific DNA probes from widely scattered sources. Despite these limitations, RFLP genetic maps were made for a variety of organisms, including humans (see White *et al.*, 1990; Stephens *et al.*, 1990), mouse (see Copeland & Jenkins, 1991), and many plant species (see Vallejos, Sakiyama & Chase, 1992; Tanksley *et al.*, 1989).

Recent advances in molecular genetics have made it possible to detect a new class of DNA markers found to be highly polymorphic and randomly distributed throughout the genome. These markers, called simple sequence repeats (SSRs) or microsatellites, consist of simple, tandemly repeated mono-to tetranucleotide sequence motifs flanked by unique sequences. The unique sequences serve as primers for polymerase chain reaction (PCR) amplification. SSRs occur frequently in most eukaryotic genomes (Hamada, Petrino & Takunaga, 1982; Tautz, 1989; Stallings *et al.*, 1991). The polymorphisms are due to naturally occurring length variations of the repeat motifs in different individuals or different inbred lines of experimental organisms. Such simple sequence length polymorphisms (SSLPs) were reported first in human genomes (Weber & May, 1989; Weber, 1990); and later in mouse (Love *et al.*, 1990; Cornall *et al.*, 1991; Todd *et al.*, 1991) and rat (Hilbert *et al.*, 1991; Jacob *et al.*, 1991).

The SSR markers have all the attributes of ideal genetic markers: they are abundant, highly polymorphic, easily typed, and easily disseminated. Genetic linkage maps consisting of simple sequence repeats have been reported for a number of organisms, including mouse (Dietrich *et al.*, 1992) and rat (Serikawa *et al.*, 1992) and will likely become a method of choice for gene mapping of quantitative trait loci.

As an example of the use of this technique, in the Love *et al.* (1990) study, 88% of the sequences showed size variations between different inbred strains of mice as well as the wild strain, *Mus spretus*; 62% of the sequences had three or more alleles. Using BXD RI strains of mice, these investigators were able to verify published chromosomal locations for some genes and to map other loci. The beauty of this technique is that it is efficient and easy to use, particularly since access to a probe requires only the sequences of the primers flanking the repeats. Moreover, the microsatellites may be analyzed without radiolabeling during the polymerase chain reaction (PCR), and they are identified directly by electrophoresis.

In addition to all the above-mentioned genetic markers, there is a special class of polymorphic DNA markers in *Drosophila* and *C. elegans* which can be used in lieu of the simple sequence repeats. These are the stable transposable elements found in abundance in *Drosophila* and *C. elegans* populations. The special features of these elements will be discussed below in the section on *Drosophila* as a model system to study the quantitative loci affecting aging.

Genetic markers and QTL. Until very recently geneticists interested in quantitatively varying characteristics have not had available the genetic markers densely spaced throughout the genome necessary to carry out the appropriate crosses. Some early studies did detect linkage to putative quantitative trait loci. As early as 1923, Sax reported an association between seed size, a quantitatively varying characteristic, and seed coat pigmentation, a single gene characteristic. However, the basic approach of identifying associations between genetic markers and quantitatively varying traits was not developed until much later (Thoday, 1961). Since then, a number of such associations have been found [as reviewed by Thompson & Thoday, 1979], but the mechanisms for the action of quantitative trait loci remain elusive.

Early QTL/gene marker associations were made using visible single gene mutants as markers. More recent examples are based on electrophoretic variants. In a study of the tomato, associations between 12 polymorphic enzyme loci and four quantitative traits were explored (Tanksley, Mendina-Filho & Rick, 1982). At least five of the genetic markers were associated with each of the quantitative traits. Evidence for epistasis was found by analyzing interactions between genetic markers.

With the advent of RFLPs (and later SSLPs), denser genetic marker maps became available. These DNA polymorphisms have been used as markers in studies of QTL in a variety of plants including maize (Edwards, Stuber & Wendel, 1987) and tomato (Paterson *et al.*, 1988; Paterson *et al.*, 1991), but only recently in mammals, e.g. studies in the rat on hypertension (Jacob *et al.*, 1991; Hilbert *et al.*, 1991). A study of maize was quite successful in identifying genetic markers linked to quantitative traits (Edwards, Stuber & Wendel, 1987). Analyses of associations between 20 markers and 72 quantitative traits found significant associations for each of the quantitative traits. The average trait was significantly influenced by almost two-thirds of the genetic markers. In concert, the genetic markers predicted between 8% and 40% of the phenotypic variation for each of a subset of 25 traits evaluated in this manner. This proportion was greater for some traits than for others. Single marker loci accounted for between 3% and 16% of the phenotypic variation. Furthermore, heterozygosity (as measured by the markers) was significantly associated with variation in many traits. In contrast to the study by Tanksley, Mendina-Filho and Rick (1982), there was no epistasis.

Two studies (Jacob *et al.*, 1991; Hilbert *et al.*, 1991) illustrate the use of SSLPs in rat to study hypertension, a quantitatively inherited trait. Jacob *et al.* (1991) reported a major locus effect on blood pressure identified by QTL analysis. In this study they also report a possible second locus 'weakly' linked to a gene on a different chromosome. The authors indicate this is only a partial genetic dissection of the genes involved.

Quantitative genetics and molecular genetics
The goal of an RI/QTL program of research on aging is to identify genetic markers associated with age-related processes, employing a battery of diverse age-related phenotypes in order to increase the likelihood of finding such associations. The usefulness of a molecular approach for the study of associations with age-related phenotypes has been demonstrated for postponed aging in *Drosophila* (Rose *et al.*, 1991) and for longevity in mice (Gelman *et al.*, 1988). A broad sampling of loci from *Drosophila* stocks exhibiting postponed aging and control stocks using 2-D protein gel electrophoresis revealed a number of proteins (as yet unidentified) which differ between the selected and control stocks (Rose *et al.*, 1991). In an attempt to identify the genes involved, Rose and colleagues are carrying out a one-dimensional protein electrophoretic survey of the postponed-aging stocks. The rationale is to identify proteins which were polymorphic in the control stocks, but have become fixed in all the five selected stocks. This would indicate that a given locus must either be important in the selection response or linked to a locus important in the response. In a structural gene marker survey of female mice from 20 different BXD recombinant inbred strains, Gelman *et al.* (1988) found a large number of genetic markers correlated with survival. However, since in mice there are only a small number of RI strains available per set, QTLs of limited effect size cannot be detected with reliability. The usefulness of the RI approach for age-related traits will be increased in an organism such as *Drosophila*, where a large number of RI strains can be generated per set.

Recombinant inbred strains

Recombinant Inbred (RI) strains are inbred lines derived from sib mating the progeny of individual members of the same genetically segregating F_2 generation until homozygosity is achieved. To ensure homozygosity of approximately 97%, with sib-mating at least 20 to 22 generations of inbreeding is carried out within each line. (In plants or self-fertilizing animals, consecutive selfing is possible, and homozygosity is achieved in fewer generations.) All the strains derived independently from the same two progenitor inbred lines are called an RI set. They are called *recombinant* strains because parts of chromosomes from the progenitor strains have recombined in the F_2 generation. With sib mating, there will continue to be recombination in later generations, albeit at diminishing levels, until homozygosity is achieved.

For each locus at which the alleles of the progenitor inbred strains differ, approximately half of the RI strains should be homozygous for the allele of one progenitor strain and the other half should be homozygous for the allele of the other progenitor. This is because the alleles from one or the other progenitor strain become fixed at each locus during inbreeding. Thus, if a single gene is responsible for a trait that differs between two parental strains, approximately half of the RI strains should be like one parent and half like the other. There should be no intermediate phenotypes if just one locus is involved. If a major gene is detected, linkage can be determined by matching the allelic pattern for the locus across the RI strains (called the strain distribution pattern, SDP) to the SDPs for previously mapped markers. If the SDP for the new locus is the same as the SDP for a mapped marker, the new locus must be closely linked to the marker because genes at linked loci become fixed or lost together during the creation of the RI strains. A single recombination between marker SDPs indicates the presence of different loci. The number of RI strains that show recombination between the two loci indexes the distance between the loci.

Use of RI strains

RI strain methodology was initiated by Bailey (1971) and Taylor (1976a) for the mapping of single gene traits in mice. Bailey (1981) and Taylor (1989) have discussed the theoretical aspects underlying the use of RI strains. Although this mapping methodology has its most widespread acceptance and use in mouse genetics, plant geneticists also have developed a number of RI sets for use in mapping genes in wheat, peas, and maize (see Burr & Burr, 1991). RI strain methodology may be further expanded for the molecular mapping of quantitative traits governed by QTL.

Advantages of RI strain analysis
In the search for genes involved in aging, RI strains have some considerable advantages compared to conventional crosses, such as F_2 or backcross populations. First, once homozygosity has been achieved, same-sex individuals within a line are of the same genotype. Thus, it is possible to repeat measures on the same genotype, albeit the observations will be made on different individuals. This means the same genotype for which lifespan data are available will also be available for a study of correlated measures of interest. In addition, genetic marker typing for that line can be done separately from the specific individual being measured on the other parameters. Accurate measurements of phenotypic variability are possible, since all individuals within a line are genotypically identical. This replicability is especially important for traits such as aging, where phenotypic expression is highly influenced by different environmental conditions.

Second, the RI lines may be propagated indefinitely. It follows that the same population can be used for mapping by different researchers. Once the genotypes of an RI series have been determined,

these data are available for other investigators. The data are cumulative; new information may be added to the existing database so that the map will be continually refined.

Third, the RI set is an array of genotypes that have undergone multiple rounds of meiosis before homozygosity is reached. As a result of this breeding regime, linked genes have a greater probability of recombination in RI lines than in F_2 or backcross populations. This was shown by Haldane and Waddington (1931) in their study of inbreeding and linkage. Compared to a backcross, the density of crossovers is quadrupled in a recombinant inbred strain produced by sib mating. This property of RI lines is important in QTL mapping. More accurate map resolution of quantitative trait loci for a population of a given size is possible with an RI set than for backcross progeny (Lander & Botstein, 1989).

Disadvantages of the use of RI strains
The major disadvantages of RI lines are practical rather than theoretical. First, it takes considerable time and expense to produce the lines. For example, using *Drosophila* with a two-week generation time, it takes 44 weeks to develop a set of homozygous RI lines beginning with the progenitor inbreds. This assumes that one has on hand two inbred lines which differ in an array of gene markers to use as the progenitors.

Second, due to inbreeding depression, the viability of the inbred strains will be reduced and maintenance will be difficult. Since Drosophilae show extensive inbreeding depression, up to 60% in some populations (Dobzhansky, Spassky & Tidwell, 1963; Mukai & Yamaguchi, 1974), one must begin the generation of the lines with many more than are needed for the final analysis. To counter this problem, it is possible to make replicates of the mating for each line, realizing that only one mating pair will be needed for the propagation of the line in each generation. By definition, there will be selection for viability (and likely also for fecundity) in each generation.

Third, once the lines are established, care must be taken to ensure their continued availability. This means replicates of the lines must be placed in separate locations, and backup stocks from the previous generation must be available.

Fourth, only those loci at which the progenitor strains differ can be analyzed. Thus, it is important that the two progenitor lines have detectable differences in as many chromosomal sites as possible. For any new RI set being constructed the progenitor lines should be chosen to maximize the molecular differences.

Finally, other practical considerations include the low but measurable risk of heterozygosity at any locus in individuals of an inbred strain, the possibility of an accumulation of mutations when stocks are maintained for a considerable time, and, particularly for *Drosophila*, the possibility of the existence of inversion heterozygosity in the F_1 which would prevent crossing over. However, all these risks are predictable and can be tested experimentally.

RI strains and QTL

The value of RI strains for genetic analysis of quantitative traits would be minimal if the approach were limited to use in detecting single-gene effects, because quantitative traits rarely show huge single-gene effects. Indeed, the interesting aspect of quantitative inheritance is to explain how loci with small effects act in consort to influence a complex phenotype.

Although RI series were developed primarily to identify and map single genes (Bailey, 1971), it was recognized early on that RI strains 'should be especially useful for the analysis of complex characters' (Taylor, 1976a, p. 118). Of the scores of studies employing RI strains, however, few have investigated QTL when the RI strain distribution pattern for a trait is continuously distributed rather than displaying the bimodal pattern expected of a major gene. Two exceptions include a study of regulation of the expression of endogenous mammary tumor proviruses (Traina-Dorge et al., 1985) and a study of longevity (Gelman et al., 1988). The results of the study by Traina-Dorge et al. are instructive conceptually in relation to QTL analyses: Even though the presence of the provirus is necessary for the transcription of virus-specific RNA, nonviral QTL were found that influence expression of viral RNA. In the Gelman et al. (1988) study a number of chromosomal regions were significantly correlated with survival.

However, a correlational approach can be exploited using RI strain means to investigate QTL associations between genetic markers and quantitative traits (Plomin *et al.*, 1991). RI means for a quantitative trait can be correlated with a marker scored as 0 and 1, representing alleles from the two progenitor strains. Significant correlations indicate associations between markers and quantitative traits, and they also index the strength of the associations. The Pearson product-moment correlation is equivalent to a t-test that compares the means of a quantitative measure for strains that differ for a marker. A weak association could be the result of tight linkage to a QTL with small effect or loose linkage to a QTL with large effect; the existence of other markers near the index marker can provide additional information with which to interpret a significant correlation. Because 5% of such correlations are expected to be significant by chance when $p < .05$, the number of significant associations needs to be interpreted in relation to the number expected on the basis of chance alone. Neumann (1992) and Belknap (1992) have proposed different methods for protection against false positives. They suggest a Bonferroni correction for multiple comparisons. This use of a more stringent α level focuses attention on the most likely QTL sites.

Other statistical procedures have been proposed to identify QTL. For example, in order to circumvent the typical problem that RI strains do not fall into two discrete groups, Briles *et al.* (1986) simply compared the phenotypic means of strains with each parental allele using a nonparametric two-sample rank test. Lander and Botstein (1989) have proposed a technique that they call QTL interval mapping, which incorporates information about distance between markers in a maximum-likelihood LOD score analysis, although the approach is available only for F_2 and backcross populations. However, an argument can be made for the simplicity and directness of the simple correlation approach which conveys information about the strength of associations as well as their significance. When significant associations are found, they can be presented in relation to the *Drosophila* linkage map (Merriam *et al.*, 1991) in order to determine the extent to which significant associations are clustered on a particular chromosome.

There are some limitations to RI QTL analysis. Démant and Hart (1986) argue that RI strains are not useful for quantitative analysis when a large number of genes is involved in the expression of a trait, since the same phenotype may be caused by more than one genotype. Moreover, even a relatively small number of genes leads to a large number of possible genotypes (2^n, where n = the number of genes involved in the trait). Thus, the probability of a particular genotype not being present at all in a given set of RIs may be rather large, if the number of lines within the set is small. (For example, if the number of involved genes is 3, with only 16 RI sets, the probability of a particular genotype not being present is 0.12.) They argue that this is a major limitation when there is a limited number of lines available in an RI set. Démant and Hart cite examples of hypothetical situations where spurious results are obtained because certain genes act in a negative fashion, cancelling the effect of genes with an equal but positive effect. (It should be noted that such a situation likewise could not be detected as additive genetic variation by more traditional quantitative genetic analysis.) Moreover, they point out that the influence of nonadditive (epistatic) gene interaction can cause distortion of the correlation between genotype and phenotype. Indeed, RI strains will be most useful in identifying effects of gene action which are additive (in the formal sense). Clearly, some genes may be missed by this method, while those contributing to the nonadditive genetic variance may not be detected.

Three critical differences distinguish the RI QTL approach from conventional single-gene use of RI strains. First, QTL associations can be detected when a phenotype shows a continuous rather than bimodal distribution across RI strains. Thus, there is no need to attempt to categorize the distribution into groups that are more or less like the progenitor strains, as is done in conventional RI analysis. Such continuous distributions are likely to be obtained for age-related processes. Second, the RI QTL association approach does not require that the progenitor strains differ for the quantitative measure, as illustrated in analyses of published data for alcohol-related measures in mice (McClearn *et al.*, 1991). Third, although a small number of strains is adequate for segregation and mapping of single gene effects, QTL association analyses require large numbers of strain comparisons in order to increase the power to detect QTL of small effect. In this respect *Drosophila* is an ideal organism, since it is relatively inexpensive and technically easy to maintain large numbers of strains.

A very important feature of the RI QTL association approach is the cumulative nature of RI work. Because RI strains are inbred, individuals are comparable across time:

> The lifetime of the genotype is not limited by the lifetime of the individual. Many observations can be made on one genotype even though the observations must be made on different individuals. This property allows the accumulation of linkage data obtained by different investigators to form an ever expanding linkage analysis based on the same set of genetically assorted genotypes. Once the genotypes of an RI series have been typed for a trait, those data are available for any other investigator to compare his typing data; they do not have to be retyped for the earlier traits. (Bailey, 1981, p. 226).

Once an RI series is characterized for a particular quantitative trait, all markers previously typed for the RI strains become available in the search for QTL associations. Furthermore, any quantitative characterization of RI strains can be re-analyzed in the future for associations with newly discovered markers. Finally, dependent measures for which RI strains have been characterized can be compared to measures in all previous studies using the same RI series in multivariate RI QTL analysis of genetic correlations among traits assessed in different studies.

Statistical considerations

A result of cited research with important implications for mapping studies is that the distribution of R^2 values for the regression of the traits on the markers may be skewed, with 'far greater frequencies of loci accounting for very small than for large R^2 values'. Nonetheless, about 5% of the loci each predicted greater than 5% of the variance of at least one of the phenotypes (Edwards, Stuber & Wendel, 1987, p. 118). This finding implies that considerable statistical power is needed to detect associations between gene markers and complex phenotypes because most associations account for only a small percentage of variance. Such power can be attained with a large number of RI strains within a set.

The goal of the use of RI QTL analyses in a study of aging is to identify single and multiple associations between genetic markers and age-related related responses using an appropriate RI set of strains. It is possible to assess these associations using a simple t-test that compares the means of each measure for strains that differ for the marker. The Pearson product-moment correlation is equivalent to this t-test but has the advantage of providing an immediate index of the strength of the association. Power considerations are well understood (Cohen, 1988). For example, with 100 RI strains in the set plus the two progenitor inbred lines, a significant association between a gene marker and an age-related measure can be detected when the correlation exceeds .2, a correlation that accounts for 4% of the variance of the phenotype (p < .05, two-tailed).

Moreover, the power estimates are conservative because a hierarchical analysis of variance can be employed that is considerably more powerful than the usual analysis of mean differences. In the usual t-test or correlation, the RI strain mean is the unit of analysis and the test of significance compares the mean difference between strains that show the probe and strains that do not, using as an error term the variance between strains. The hierarchical approach (Meyers, 1980) is more powerful because it also incorporates variance within strains. Most importantly, larger sample size within RI lines increases the power to detect mean effects of genetic markers in this analysis. In part, this increase in power is due to the well-known phenomenon of the reduction of the standard error of the mean. However, less well known is the power that accrues in a hierarchical analysis of variance by increasing the N within groups. It has been shown, for example, that this hierarchical approach is much more powerful than attempting to increase power by increasing the p value and then replicating the result (Games, 1970). The point for the proposed analyses is that the use of a hierarchical approach that considers variance within strains capitalizes on the use of large sample sizes within strains to increase the power to detect marker-behavior associations that account for small amounts of variance.

Drosophila, aging and QTL

Drosophila is a model system for the use of RI strain methodology to map quantitative trait loci involved in the aging process. The 44-week generation time necessary to create an RI set is less than for many other experimental organisms. A large number of strains can be generated for an RI set. This means considerable power for the statistical tests to be carried out. As previously noted, an RI set of 100 strains will allow for the detection of genes which account for a very small proportion of the total variance.

The RI set serves as a permanent equivalent of a sample of homozygous genotypes generated from recombinant events in a certain population. Thus, repeatability studies on that population sample are possible. This is especially important in studies of the complex genetics of lifespan. Studies in both Drosophila (Rose, 1991) and mouse (Dear et al., 1992) show that genes which influence lifespan need to be studied in many genetic backgrounds and varying environmental conditions.

In the choice of progenitor inbred lines to begin the RI set, it is important to consider a number of factors. It is not necessary to have strains which differ in longevity (McClearn et al., 1991). If unrelated strains do not differ in longevity, it still is likely that the genetic basis of the strain phenotypes is different. For example, it is likely that the phenotypic similarity is based on the balance of input from very different QTLs within each strain. Thus, when the progenitors are mated, and recombinant genotypes developed, the recombinant genotypes will contain different proportions of the positive and negative alleles fixed within the parentals. Indeed, the finding of a greater range of phenotypes in RIs derived from inbreds which do not differ in that phenotype is a commonplace finding and considered to be prima facie evidence of a polygenic system.

It is crucial, however, that the progenitor lines differ highly in a large number of molecular markers randomly dispersed throughout the genome. This is necessary to follow inheritance throughout the genome by tracing the markers between parents and offspring. Thus, the degree of polymorphism limits the utility of the RIs. Indeed, a major problem with RFLPs as markers is the lack of polymorphism among strains of interest.

A goal for detecting QTL is to have a high resolution map with closely spaced markers. This increases the likelihood that a QTL is sufficiently close to a marker to make detection feasible. The cumulative and integrative nature of RI work is a distinct advantage in attaining this goal. As available, new marker data on the strains can be added to the data base, and any new phenotypic measurements on the strains may be related to the accumulated data on markers. Thus, an RI set of strains becomes more valuable with increasing use.

Inbreeding in Drosophila

Any proposal to use inbred lines for a study of aging in *Drosophila* must address the issue of inbreeding depression. The problem occurs because *Drosophila* is by nature outbred. When inbred in the laboratory, flies typically exhibit a reduction in lifespan due to rare deleterious alleles which become homozygous at a particular locus. Conversely, the usual outcome of mating between two highly inbred strains is an F_1 population with an extended lifespan. This is the familiar phenomenon known as hybrid vigor, or heterosis. During the 20 or more rounds of inbreeding required to form the recombinant inbreds, most of the loci will become homozygous once more. It might be argued that the only genes identified using an inbred strain approach would be those responsible for inbreeding depression.

While it may be agreed that in *Drosophila* inbreeding depression is preeminent, there are strategies one might consider to deal with this problem. One possibility is to choose as progenitors two strains which, when crossed, do not exhibit hybrid vigor. This would, of course, necessitate the measurement of the lifespans of a number of F_1 hybrids. The argument is as follows: hybrid vigor occurs when the effects of recessive alleles fixed in one strain are attenuated in the hybrid by dominant alleles from the other strain. Hence, in the absence of heterosis, it can be argued that the two strains share many of the same recessive alleles and, when crossed, would not differ from the parentals. The argument would not imply that the alleles causing inbreeding depression are absent, but rather because they are 'in common', the effect of other alleles influencing the trait can now be detected. One could then argue that the loci thereafter identified will be those that affect normal aging.

It would be useful if previous research had demonstrated F_1 hybrids which do not exhibit hybrid vigor. A pertinent example from the recent litera-

ture in *Drosophila* is that of Curtsinger *et al.* (1992). The mean longevities of four inbred lines and their F1 hybrids were reported. Although hybrid vigor was the most common finding, this was not the case for the cross between strains 2 and 4. In the results of replicate F_1s from this cross, the lack of hybrid vigor for longevity held not only when the average lifespan across three replicates was considered, but also when lifespan within each replication was separately considered. This finding lends confidence to the feasibility of identifying inbred strains which do not exhibit hybrid vigor as progenitors for the generation of an RI set. In another example in which inbred strains and their F1 hybrids were used to study drug susceptibility in mice, the usual finding was hybrid vigor for decreased susceptibility (Taylor, 1976b). However, Taylor also reported that one F_1 hybrid between two highly inbred strains was as susceptible to isoniazid-induced seizures as the parental strains.

Isochromosome lines

There is another method available in *Drosophila* for creating lines of flies with homozygous chromosomes (or portions of chromosomes). By this method, previously heterozygous chromosomes (or portions of them) are made homozygous by the use of balanced-lethal marker stocks. There are a number of protocols available outlining suggested crosses to produce homozygosity (see Grigliatti, 1986). Such lines can be designed so that the genetic background is either homozygous or heterozygous. In the former case viability problems similar to those of inbred lines are encountered. In either case a selection of background genotype is necessary.

The creation of isochromosome lines has certain advantages over sib-mated breeding schemes. There are fewer generations needed to attain homozygosity, and the resultant lines can be more defined genetically, since they are specifically created. For example, one set of lines could be homozygous only for the left arm of chromosome 3. However, there is labor-intensive scoring needed to create the lines. Moreover, the resultant stocks are often weak and have viability problems similar to sib-mated inbreds.

DNA markers in Drosophila

Although PCR-generated simple sequence length polymorphisms are ideal genetic markers, as yet there is no SSLP reference map for *Drosophila*. The question arises, do such microsatellite polymorphisms exist in *Drosophila*? Certainly simple sequence repeats exist in *Drosophila* (Tautz & Renz, 1984); sites are distributed over most euchromatic regions in both autosomes and the X chromosome, although there is a significantly higher density of sites along the X chromosome (Pardue *et al.*, 1987; Lowenhapt, Rich & Pardue, 1989). Although polymorphisms for these microsatellites have not yet been reported, given their extensive distribution across species, it is likely that the sites may be polymorphic in *Drosophila* as in other eukaryotes.

Although presently there is no SSR map of the *Drosophila* genome, there do exist other genetic markers which have excellent properties for use in gene mapping. These are the stable, randomly distributed, transposable elements which are highly polymorphic in *Drosophila* and *C. elegans* populations. In *Drosophila*, stable transposable elements can be physically located on the polytene salivary gland chromosomes within the resolution of a single band. It must be noted that such resolution is not high level, since the average size of a *Drosophila* chromosomal band is approximately 200 Kb.

In *C. elegans*, Williams *et al.* (1992) have reported a mapping scheme based on the use of transposable element polymorphisms. After the positioning of individual, randomly cloned elements within the genomic physical map, they then developed PCR assays for a subset of the elements selected to give markers well-distributed throughout the genome. Thus by using this method, once an initial genomic localization of the elements is achieved, even greater site resolution can be attained.

The *Drosophila* genome contains about 3000 copies of 30-50 families of transposable elements (Biémont, 1992). These families include both stable and unstable (such as P) elements. Only the stable elements would be useful in a mapping study. Although the insertion sites for these elements apparently are dispersed at random along the chromosome, some transposable elements do insert in a site-specific fashion. Examples of stable elements which insert into the chromosomes without obvious site specificity are *copia* and *mdg-1* (Inouye, Yuki & Saigo, 1984) and likely *roo* (Montgomery, Charlesworth & Langley, 1987).

In particular, *copia* and *copia*-like elements (such as *mdg*-1 and *roo*) are useful for mapping purposes. The *roo* (B104) element (Finnegan & Fawcett, 1986) is present in high copy numbers per genome (Montgomery, Charlesworth & Langley, 1987) is polymorphic in many populations and is highly stable (Charlesworth & Lapid, 1989). A study by Mackay *et al.* (1992) illustrates the stability of this element. From a highly inbred line, 25 sublines were begun and maintained over 100 generations by ten randomly chosen mating pairs each generation. The Southern digest of samples from generation 98 of the 25 sublines was probed with *roo*. There were at least 35 *roo* copies per genome, representing 35 separate loci. All were identical across the lines. *Copia* and *mdg-1*, a *copia*-like element, are also stable even after severe environmental stresses. No changes of these elements occurred even after heat shock, dichlorvos, hydrogen peroxide or ecdysterone treatments (Arnault *et al.*, 1991).

For *Drosophila*, a small number of markers can sweep a fairly large portion of the genome. Assuming that N genetic markers are randomly distributed in a genome of D cM, then the proportion of the genome lying within d cM of a marker is approximately

$$1 - e^{-2Nd/D} \text{ (Jacob } et\ al., 1991).$$

The genetic length of the *Drosophila* genome is approximately 280 map units (X = 66 cM; 2nd chromosome = 108 cM; 3rd chromosome = 106 cM) (Lewin, 1990, p. 50). With only 100 markers, 97% of the genome will lie within 5 cM; 88% will lie within 3 cM and 51% will lie within 1 cM.

The generality of the RI approach is also of import. Once the molecular map has been defined, the RI set will be available for studies of any quantitative trait of interest.

Age-related phenotypes

In any study of aging, there is one valid phenotypic measure: survival or age at death. However, there are related parameters which are of interest, even though they may be only effects of aging rather than attributes which directly relate to the mechanisms of the aging process. In *Drosophila* there have been attempts to identify physiological and behavioral attributes which are measures of aging. Although many of these measures are not straightforward with respect to the aging process, there is some reliability in specific settings. This suggests the feasibility of mapping the QTL defining such traits and analyzing their relationship to lifespan.

A study of the quantitative trait loci involved in aging could include a broad coverage of the domain of age-related phenotypes. The choice of measures to be employed involves selection of an appropriate battery. Since extensive work has been carried out on a number of *Drosophila* populations, including wild-type strains as well as postponed-aging selected lines to determine appropriate physiological and behavioral measures of aging, a test battery for the RI series might include traits consistently shown to be related to aging in *Drosophila*. These include survival, fecundity (Rose & Charlesworth, 1981), locomotor activity (Ganetzky & Flanagan, 1978), starvation resistance, desiccation resistance, ethanol-vapor tolerance (Service, 1987; Service *et al.*, 1985), flight duration (Graves, Luckinbill & Nichols, 1988; Luckinbill *et al.*, 1988), and lipid content (Service, 1987). All or a selection of these measures constitute an appropriate test battery.

The set of RI strains will be very useful in dealing with the perennial problem that occurs when studying genetic correlations in relatively small numbers of inbred lines. In that situation, genetic correlation that can be demonstrated in random breeding populations can be obscured by the fixation of unlinked alleles which also affect the trait. For example, an association of considerable interest in the study of longevity in *Drosophila* is that between fecundity and lifespan. In lines selected for postponed aging, early fecundity is much depressed while later fecundity is elevated compared to controls (Luckinbill *et al.*, 1984; Rose, 1991). However, inbred lines often give phenotypic correlations which are not indicative of the genetic correlations which exist between two traits (Rose, 1984a; Clark, 1987). Thus, the initial progenitor lines may not be useful indicators of the 'true' genetic correlations which exist for fecundity and longevity. However, even if the relationship between two traits does not exist in the inbred progenitors due to a spurious phenotype correlation, the RI

analysis, which is essentially a chromosomal sampling procedure, could give some insight into the QTL relationships between two traits such as longevity and fecundity. With a large number of RI lines it should be possible to detect the genetic correlations which persist across lines by identifying those traits which have QTL in common. Traits that are unlinked will presumably be randomized in a set as large as the one proposed. Thus, RI QTL analysis may be useful in defined genes which lead to the genetic correlations between fecundity and longevity which have been observed.

The nature of quantitative trait loci

Mackay, Lyman and Jackson (1992) review various proposals to account for the action of the loci which affect variation in quantitative traits. They point out that QTL have been considered to act in a variety of ways: as alleles having major effects on related traits, as genes which have only minor effects on a quantitative character, as loci which are modifiers of major gene expression, and as loci which have a spectrum of allelic effects, ranging from major effects to small 'isoallele' effects of major loci. QTL maps and the identification of the loci involved will enable us to begin to distinguish among these alternatives.

The localization of genes into linkage groups is a basic thrust of geneticists. The ability to study the effects of any gene is advanced greatly when the gene can be identified. To know the DNA structure, the location relative to other genes in the chromosome, and to be able to manipulate that gene in a variety of different genetic and environmental backgrounds is to gain an understanding that cannot be achieved by a more general description of its effects. When the quantitative trait loci for a variety of traits are precisely mapped, we will be closer to an understanding of the molecular basis of quantitative variation.

Acknowledgements

I wish to thank my colleagues, Gerald E. McClearn and Robert Plomin, for their useful comments and suggestions on the manuscript. Also, I thank David Blizard for many stimulating discussions on RI strain methodology, and Betty Asendorf for help with manuscript preparation.

References

Arnault, C., A. Heizmann, C. Loevenbruck & C. Biémont, 1991. Environmental stresses and mobilization of transposable elements in inbred lines of Drosophila melanogaster. Mut. Res. 248: 51-60.

Bailey, D. W., 1981. Recombinant inbred strains and bilineal congenic strains, pp. 223-239 in The Mouse in Biomedical Research, Vol. 1, edited by H. L. Foster, J. D. Small & J. G. Fox. Academic Press, New York.

Bailey, D. W., 1971. Recombinant-inbred strains: An aid to finding identity, linkage, and function of histocompatibility and other genes. Transplantation 11: 325-327.

Belknap, J. K., 1992. Empirical estimates of Bonferroni corrections for use in chromosome mapping studies with the BXD recombinant inbred mouse strains. Beh. Gen. 22: 677-684.

Biémont, C., 1992. Population genetics of transposable elements: a Drosophila point of view. Genetica 86: 67-84.

Bodmer, W. F., 1986. Human genetics: The molecular challenge. Cold Spring Harbor Symposia on Quantitative Biology 51: 1-13.

Botstein, D., R. L. White, M. Skolnick & R. W. Davis, 1980. Construction of a genetic map in man using restriction fragment length polymorphisms. Am. J. Hum. Genet. 32: 314-331.

Briles, D. E., W. H. Benjamin, W. J. Huster & B. Posey, 1986. Genetic approaches to the study of disease resistance: With special emphasis on the use of recombinant inbred mice. Current Topics in Microbiology and Immunology 124: 21-35.

Burr, B. & F. A. Burr, 1991. Recombinant inbreds for molecular mapping in maize: theoretical and practical considerations. TIG, Vol. 7, No. 2, 55-60.

Charlesworth, B. & A. Lapid, 1989. A study of ten families of transposable elements on X chromosomes from a population of Drosophila melanogaster. Genet. Res. 54: 113-125.

Clark, A., 1987. Senescence and the genetic correlation hang-up. Am. Nat. 129: 932-940.

Clarke, J. M. & J. Maynard Smith, 1955. The genetics and cytology of Drosophila subobscura XI. Hybrid vigor and longevity. J. Genetics 53: 172-180.

Cohen, J., 1988. Statistical Power Analysis for the Behavioral Sciences, 3rd ed. Academic Press, New York.

Copeland, N. G. & N. A. Jenkins, 1991. Development and applications of a molecular genetic linkage map of the mouse genome. Trends Genet. 7: 113-118.

Cornall, R. J., T. J. Aitman, C. M. Hearne & J. A. Todd, 1991. The generation of a library of PCR-analyzed microsatellite variants for genetic mapping of the mouse genome. Genomics 10: 874-881.

Curtsinger, J. W., H. H. Fukui, D. R. Townsend & J. W. Vaupel, 1992. Demography of Genotypes: Failure of the Limited Life-Span Paradigm in Drosophila melanogaster. Science 258: 461-463.

Dear, K. B. G., M. Salazar, A. L. M. Watson, R. S. Gelman, R. Bronson & E. J. Yunis, 1992. Traits that influence longevity in mice: a second look. Genetics 132: 229-239.

Démant, P. & A. A. M. Hart, 1986. Recombinant Congenic Strains - A New Tool for Analyzing Genetic Traits Determined by More Than One Gene. Immunogenetics 24: 416-422.

Dietrich, W., H. Katz, S. E. Lincoln, H-S. Shin, J. Friedman, N. C. Dracopoli & E. S. Lander, 1992. A genetic map of the mouse suitable for typing intraspecific crosses. Genetics 131: 423-447.

Dobzhansky, Th., B. Spassky & T. Tidwell, 1963. Genetics of natural populations. XXXIV. Adaptive norm, genetic load, and genetic elite in Drosophila pseudoobscura. Genetics 48: 1467-1485.

Edwards, M. D., C. W. Stuber & J. F. Wendel, 1987. Molecular-marker-facilitated investigations of quantitative-trait loci in maize. I. Numbers, genomic distribution and types of gene action. Genetics 116: 113-125.

Finnegan, D. J. & D. H. Fawcett, 1986. Transposable elements in Drosophila melanogaster, pp. 1-62 in Oxford Surveys on Eukaryotic Genes, Vol. 3, edited by N. Maclean. Oxford University Press, New York.

Games, P. A., 1970. Contemporary perspectives on the measurement of human abilities in Research in Psychology: Readings for the Introductory Course, edited by B. L. Kintz & J. L. Brunning. Scott-Foresman, New York.

Ganetzky, B. & J. R. Flanagan, 1978. On the relationship between senescence and age-related changes in two wild-type strains of Drosophila melanogaster. Exp. Gerontol. 13: 189-196.

Gelman, R., A. Watson, R. Bronson & E. Yunis, 1988. Murine chromosomal regions correlated with longevity. Genetics 118: 693-704.

Gowen, J. W. & L. E. Johnson, 1946. On the mechanism of heterosis. I. Metabolic capacity of different races of Drosophila melanogaster for egg production. Amer. Nat. 80: 149-179.

Graves, J. L., L. S. Luckinbill & A. Nichols, 1988. Flight duration and wing beat frequency in long- and short-lived Drosophila melanogaster. Insect Physiol. 34: 1021-1026.

Grigliatti, T., 1986. Mutagenesis, pp. 39-58 in Drosophila: a Practical Approach, edited by D. B. Roberts, IRL Press, Oxford-Washington DC.

Haldane, J. B. S. & C. H. Waddington, 1931. Inbreeding and linkage. Genetics 16: 357-374.

Hamada, H., M. G. Petrino & T. Takunaga, 1982. A novel repeated element with z-DNA-forming potential is widely found in evolutionary diverse eukaryotic genomes. Proc. Natl. Acad. Sci USA 79: 6465-6469.

Hilbert, P., K. Lindpaintner, J. S. Beckmann, T. Serikawa, F. Soubrier, C. Dubay, P. Cartwright, B. de Bouyon, C. Julier, S. Takahasi, M. Vincent, D. Ganten, M. Georges & G. M. Lathrop, 1991. Chromosomal mapping of two genetic loci associated with blood-pressure regulation in hereditary hypertensive rats. Nature 353: 521-529.

Hutchinson, E. W. & M. R. Rose, 1991. Quantitative genetics of postponed aging in Drosophila melanogaster. I. Analysis of outbred populations. Genetics 127: 719-727.

Hutchinson, E. W., A. J. Shaw & M. R. Rose, 1991. Quantitative genetics of postponed aging in Drosophila melanogaster. II. Analysis of selected lines. Genetics 127: 729-737.

Inouye, S., S. Yuki & K. Saigo, 1984. Sequence-specific insertion of the Drosophila transposable genetic element 17.6. Nature 310: 332-333.

Jacob, H. J., K. Lindpaintner, S. E. Lincoln, K. Kusumi, R. K.

Bunker, Y-P. Mao, D. Ganten, V. J. Dzau & E. S. Lander, 1991. Genetic mapping of a gene causing hypertension in the stroke-prone spontaneously hypertensive rat. Cell 67: 213-224.

Lander, E. S. & D. Botstein, 1989. Mapping Mendelian factors underlying quantitative traits using RFLP linkage maps. Genetics 121: 185-199.

Lewin, B., 1990. Genes IV. Oxford Press, New York.

Love, J. M., A. M. Knight, M. A. McAleer & J. A. Todd, 1990. Towards construction of a high resolution map of the mouse genome using PCR-analysed microsatellites. Nucleic Acids Res. 18: 4123-4130.

Lowenhapt, K., A. Rich & M. L. Pardue, 1989. Nonrandom distribution of long mono- and dinucleotide repeats in Drosophila chromosomes: Correlations with dosage compensation, heterochromatin, and recombination. Mol. Cell. Biol. 9: 1173-1182.

Luckinbill, L. S., R. Arking, M. J. Clare, W. C. Cirocco & S. A. Buck, 1984. Selection for delayed senescence in Drosophila melanogaster. Evolution 38: 996-1003.

Luckinbill, L. S., J. L. Graves, A. Tomkiw & O. Sowirka, 1988. A qualitative analysis of some life-history correlates of longevity in Drosophila melanogaster. Evol. Ecol. 2: 85-94.

Mackay, T. F. C., R. F. Lyman & M. S. Jackson, 1992. Effects of P element insertions on quantitative traits in Drosophila melanogaster. Genetics 130: 315-332.

Mackay, T. F. C., R. F. Lyman, M. S. Jackson, C. Terzian & W. G. Hill, 1992. Polygenic mutation in Drosophila melanogaster: estimates from divergence among inbred strains. Evolution 46: 300-316.

McClearn, G. E., R. Plomin, G. Gora-Maslak & J. Crabbe, 1991. The gene chase in behavioral science. Psychological Science 2: 222-229.

Merriam, J., M. Ashburner, D. L. Hartl & F. C. Kafatos, 1991. Toward cloning and mapping the genome of Drosophila. Science 254: 221-225.

Meyers, J. L., 1980. Fundamentals of Experimental Design, 3rd ed. Allyn Bacon, New York.

Montgomery, E., B. Charlesworth & C. H. Langley, 1987. A test for the role of natural selection in the stabilization of transposable element copy number in a population of Drosophila melanogaster. Genet. Res. 49: 31-41.

Mukai, T. & O. Yamaguchi, 1974. The genetic structure of natural populations of Drosophila melanogaster. XI. Genetic variability of local populations. Genetics 76: 339-366.

Neumann, P. E., 1992. Inference in linkage analysis of multifactorial traits using recombinant inbred strains of mice. Beh. Gen. 22: 665-676.

Pardue, M. L., K. Lowenhaupt, A. Rich & A. Nordheim, 1987. dC-dA$_n$ dG-dT$_n$ sequences have evolutionary conserved chromosomal locations in Drosophila with implications for roles in chromosome structure and function. EMBO J. 6: 1781-1789.

Paterson, A. H., S. Damon, J. D. Hewitt, D. Zamir, H. D. Rabinowitch, S. E. Lincoln, E. S. Lander & S. D. Tanksley, 1991. Mendelian factors underlying quantitative traits in tomato: comparison across species, generations, and environments. Genetics 127: 181-197.

Paterson, A. H., E. S. Lander, J. D. Hewitt, S. Peterson, S. E. Lincoln & S. D. Tanksley, 1988. Resolution of quantitative traits into Mendelian factors by using a complete linkage map of restriction fragment length polymorphisms. Nature 355: 721-726.

Plomin, R., G. E. McClearn, G. Gora-Maslak & J. M. Neiderhiser, 1991. Use of recombinant inbred strains to detect quantitative trait loci associated with behavior. Behavior Genetics 21: 99-116.

Plomin, R., J. C. DeFries & G. E. McClearn, 1990. Behavioral Genetics: A Primer. W. H. Freeman, New York.

Rose, M. R., 1984a. Genetic covariation in Drosophila life history: Untangling the data. Am. Nat. 123: 565-569.

Rose, M. R., 1984b. Laboratory evolution of postponed senescence in Drosophila melanogaster. Evolution 38: 1004-1010.

Rose, M. R., 1991. Evolutionary Biology of Aging. Oxford University Press, New York.

Rose, M. R. & B. Charlesworth, 1981. Genetics of life history in Drosophila melanogaster. I. Sib analysis of adult females. Genetics 97: 173-186.

Rose, M. R., F. J. Ayala, G. S. Spicer, R. H. Tyler & J. E. Fleming, 1991. Genetics of postponed aging in Drosophila melanogaster: AGE 14: 136 (abstract).

Sax, K., 1923. The association of size differences with seed-coat pattern and pigmentation in Phaseolus vulgaris. Genetics 8: 552-560.

Serikawa, T., T. Kuramoto, P. Hilbert, M. Mori, J. Yamada, C. J. Dubay, K. Lindpainter, D. Ganten, J-L. Guenet, G. M. Lathrop & J. S. Beckmann, 1992. Rat gene mapping using PCR-analyzed microsatellites. Genetics 131: 701-721.

Service, P. M., 1987. Physiological mechanisms of increased stress resistance in Drosophila melanogaster selected for postponed senescence. Physiol. Zool. 60: 321-326.

Service, P. M., E. W. Hutchinson, M. D. MacKinley & M. R. Rose, 1985. Resistance to environmental stress in Drosophila melanogaster selected for postponed senescence. Physiol. Zool. 58: 380-389.

Stallings, R. L., A. F. Ford, D. Nelson, D. C. Torney, C. E. Hildebrand & R. K. Moyzis, 1991. Evolution and distribution of (GT)$_n$ repetitive sequences in mammalian genomes. Genomics 10: 807-815.

Stephens, J. C., M. L. Cavanaugh, M. I. Gradie, M. L. Mador & K. K. Kidd, 1990. Mapping the human genome: current status. Science 250: 237-244.

Sturtevant, A. H., 1913. The linear arrangement of six sex-linked factors in Drosophila, as shown by their mode of association. J. Exp. Zool. 14: 43-59.

Tanksley, S. D., H. Medina-Filho & C. M. Rick, 1982. Use of naturally-occurring enzyme variation to detect and map genes controlling quantitative traits in an interspecific backcross of tomato. Heredity 49: 11-25.

Tanksley, S. D., N. D. Young, A. H. Paterson & M. W. Bonierbale, 1989. RFLP mapping in plant breeding: new tools for an old science. Bio Technology 7: 257-264.

Tautz, D., 1989. Hypervariability of simple sequences as a general source for polymorphic DNA markers. Nucleic Acids Res. 17: 6463-6471.

Tautz, D. & M. Renz, 1984. Simple sequences are ubiquitous repetitive components of eukaryotic genomes. Nuc. Acids Res. 12: 4127-4138.

Taylor, B. A., 1976a. Development of recombinant inbred lines

182

of mice. Behavior Genetics 6: 118 (abstract).

Taylor, B. A., 1976b. Genetic analysis of susceptibility to isoniazid-induced seizures in mice. Genetics 83: 373-377.

Taylor, B. A., 1989. Recombinant inbred strains, pp. 773-789 in Genetic Variants and Strains of the Laboratory Mouse, 2nd ed., edited by M. F. Lyon and A. G. Searle. Oxford University Press, New York.

Thoday, J. M., 1961. Location of polygenes. Nature 191: 368-370.

Thompson, J. N., Jr. & J. M. Thoday, editors, 1979. Quantitative Genetic Variation. Academic Press, New York.

Todd, J. A., T. J. Aitman, R. J. Cornall, S. Ghosh, J. R. S. Hall, C. M. Hearne, A. M. Knight, J. M. Love, M. A. McAleer, J. Prins, N. Rodrigues, M. Lathrop, A. Pressey, N. H. Delarato, L. B. Peterson & L. S. Wicker, 1991. Genetic analysis of autoimmune type 1 diabetes mellitus in mice. Nature 351: 542-547.

Traina-Dorge, V. L., J. K. Carr, J. E. Bailey-Wilson, R. C. Elston, B. A. Taylor & J. C. Cohen, 1985. Cellular genes in the mouse regulate in trans the expression of endogenous mouse mammary tumor viruses. Genetics 11: 597-615.

Vallejos, C. E., N. S. Sakiyama & C. D. Chase, 1992. A molecular marker-based linkage map of *Phaseolus vulgaris L.* Genetics 131: 733-740.

Wattiaux, J. M., 1968. Cumulative parental effects in *Drosophila subobscura*. Evolution 22: 406-421.

Weber, J. L., 1990. Human DNA polymorphisms based on length variations in simple-sequence tandem repeats. Genome Analysis 1: 159-181.

Weber, J. L. & P. E. May, 1989. Abundant class of human DNA polymorphisms which can be typed using the polymerase chain reaction. American Journal of Human Genetics 44: 388-396.

White, R., J-M. Lalouel, M. Lathrop, M. Leppert, Y. Nakamura & P. O'Connell, 1990. Human genetic linkage maps. pp. 5.134-5.157 in Genetic Maps: Locus Maps of Complex Genomes. 5th ed. Book 5: Human Maps, edited by J. O'Brien. Cold Spring Harbor Laboratory Press, Cold Spring Harbor, N.Y.

Williams, B. D., B. Schrank, C. Huynh, R. Shownkeen & R. H. Waterston, 1992. A genetic mapping system in *Caenorhabditis elegans* based on polymorphic sequence-tagged sites. Genetics 131: 609-624.

M.R. Rose and C.E. Finch (eds.), Genetics and Evolution of Aging, 183–198, 1994.
© 1994 *Kluwer Academic Publishers. Printed in the Netherlands.*

The effects of enhanced expression of elongation factor EF-1α on lifespan in *Drosophila melanogaster*

IV. A summary of three experiments

Stephen C. Stearns & Marcel Kaiser
Zoology Institute, Rheinsprung 9, CH-4051 Basle, Switzerland

Received and accepted 22 June 1993

Key words: lifespan, elongation factor, *Drosophila*, life history evolution, genetic manipulation, tradeoffs, genetic correlations

Abstract

This paper summarizes three experiments on the genetic manipulation of fitness components involved in the evolution of lifespan through the introduction of an additional copy of the gene for elongation factor EF-1α into the genome of *Drosophila melanogaster*. The first experiment checked a prior claim that enhanced expression of elongation factor increased the lifespan of virgin male fruitflies. It used inbred stocks; three treatment and three control lines were available. The second experiment put one treatment and one control insert into different positions on the third chromosome, then measured the influence of six genetic backgrounds on treatment effects in healthier flies. The third experiment put six treatment and six control inserts into the genetic background whose lifespan was most sensitive to the effects of treatment in the second experiment, then measured the influence of insert positions on treatment effects in healthy flies.

The treatment never increased the lifespan of virgin males. It increased the lifespan of mated females in inbred flies reared to eclosion at 25°, reduced it in the positions experiment, and made no difference to lifespan in the backgrounds experiment. When it increased lifespan, it reduced fecundity. In inbred flies and in the positions experiment, the treatment reduced dry weight at eclosion of females. Marginal effects of gene substitutions on tradeoffs were measured directly. The results suggest that enhanced expression of elongation factor makes local changes within the bounds of tradeoffs that are given by a pre-existing physiological structure whose basic nature is not changed by the treatment.

Introduction

For decades, the fruitfly *Drosophila melanogaster* has been a model system for research on lifespan in which evolutionary and molecular paradigms have shaped the questions posed and the experiments done to answer them (Rose, 1991). The evolutionary view is that aging results from genes that have positive or neutral effects early in life and negative effects late in life (Williams, 1957). Such genes accumulate because selection is stronger on traits expressed earlier in life (Hamilton, 1966; Charlesworth & Williamson, 1975). A substantial tradition of selection experiments has led to the conclusion that lifespan in *Drosophila* trades off with compo-

nents of fitness expressed early in life, but which early fitness component is involved has depended on the experiment. Both early fecundity (Rose & Charlesworth, 1980, 1981; Rose, 1984; Luckinbill *et al.*, 1984; Hillesheim & Stearns, 1992) and weight at eclosion (Hillesheim & Stearns, 1992) have been found to be negatively correlated with lifespan. Partridge and Fowler (1992) found that flies selected both for longer life and for increased fecundity late in life eclosed later and were slightly heavier at eclosion than flies selected for increased fecundity early in life.

Molecular manipulations of lifespan in *Drosophila* were first done by Shepherd *et al.* (1989). They inserted an additional copy of the gene for elongation factor EF-1α into the genome of inbred flies on a P-element plasmid. They claimed that enhanced expression of elongation factor significantly increased the lifespan of virgin males; the difference in lifespan between the longest lived treatment line and the longest lived control line was a larger percentage of average lifespan at higher temperature – 33% at 29.5° and 18% at 25° ($p < 0.01$). This appeared to initiate a new approach to research on lifespan and life history evolution that would provide a substantial increase in the precision with which one can dissect the molecular and physiological causes of variation and covariation in fitness components. However, there were two reasons to be skeptical of the claim.

First, the original experiment was not well controlled and not replicated at all. The insert was flanked by the promoter and termination sequences for heat shock protein so that the additional copy of elongation factor would be expressed at high temperature. This was done to provide an internal control for the effects of different insert positions. The 'control' lines constructed with the same P-element plasmid differed from the treatments in four respects, not one: lack of elongation factor, insert position, length of insert, and slightly different genetic background (the lines were derived from an inbred stock but were not homozygous). The only unconfounded comparison allowed by such a manipulation is the two-way treatment × temperature interaction effect in a design that accounts for three-way line × treatment × temperature effects through adequate replication. The original experiment compared just one treatment line with one control line, a design without replication that does not allow one to separate the effects of treatment, position, length of insert, and genetic background. The design used by Shepherd et al. has one treatment (29.5°) and one control (the same flies measured at 25°), for the control line does not count. Building in the internal control provided by temperature-sensitive expression of the gene at the same insert position was a good idea, but when line × treatment × temperature interaction effects are significant, it can only be exploited if several treatment inserts are compared with several control inserts.

Second, implicit in such genetic manipulations is the assumption that single genes can have large effects on lifespan. Otherwise there would be no point in doing the manipulation. While evolutionary theory does not explicitly rule out such genes, lifespan, like other fitness components, behaves in selection experiments as though it were determined by many genes each of which has small phenotypic effects. If enhanced expression of elongation factor were sufficient, by itself, to produce a significant effect on lifespan, then the gene for elongation factor would have larger phenotypic effects than the average gene thought to affect a quantitative trait.

We do not know the function of the 'average gene thought to affect a quantitative trait', but the function of elongation factor is known. Elongation factor is present in all organisms and as a gene family in eukaryotes. It docks the incoming aminoacyl-tRNA correctly on the ribosome and catalyses the transition in which one amino acid is added to the peptide (Webster & Webster, 1982; Darnell, Lodish & Baltimore, 1990). Thus elongation factor is necessary for protein synthesis. In *Drosophila melanogaster* the gene exists in two copies: F2, which is translated mostly in pupae, and F1, which is thought to be a housekeeping gene needed in all growing cells. There is only one copy of F1 present in normal cells (Hovemann et al., 1988). The synthesis of EF-1α protein sharply decreases with age in *Drosophila* and precedes the decrease in total protein synthesis by a few days (Webster & Webster, 1983). So far as is known, EF-1α exists as a single, fixed, homozygous allele. Such a fixed housekeeping gene would not contribute to quantitative variation.

Thus the manipulation of a gene like elongation factor should alter the physiological processes associated with lifespan and might not produce the sort of genetic variation present in selection experiments. Therefore we thought it would be worth repeating and extending the work of Shepherd et al. to make sure they were observing a real effect and, if they were, to try to understand better the conditions under which such effects can be elicited. The details are presented in three papers; this article summarizes them and makes points best drawn from comparisons of them.

We posed the following questions: (1) Does enhanced expression of elongation factor really increase lifespan in the lines used by Shepherd et al. (1989) when their experiments are repeated with adequate replication of lines (Stearns, Kaiser, & Hillesheim, 1993)? Does it also affect other fitness components, especially fecundity, age at eclosion, and weight at eclosion? If it does, are the effects larger than would be expected for a gene contributing to a quantitative trait? (2) Do the effects of elongation factor on lifespan – and other fitness components – depend upon the genetic background into which the plasmid is inserted and upon the position of the insert (Kaiser & Stearns, 1993)? (3) What impact does enhanced expression of elongation factor have on tradeoffs between lifespan and other fitness components (Stearns & Kaiser, 1993)?

Methods

We first describe the methods common to all three experiments, then those only used in certain experiments. Treatment lines consisted of *D. melanogaster* transformed with a P-element containing the ry^+ marker plus the *F1* copy of the gene for elongation factor *EF-1α* flanked by initiation and termination sequences for the heat shock protein. Control lines had been similarly manipulated but lacked the gene for elongation factor; their insert was shorter and at a different position. All measurements were made at 25° and at 29.5° to allow us to compare traits with and without the enhanced expression of elongation factor at the higher temperature. Flies were reared from first instar larvae gathered within eight hours of hatching and then kept at standard density. Ages were calculated from the mid-point of egg-laying, which lasted four hours. When experiments had to be run in sequence rather than parallel, treatment and control lines were run together, so that differences among sequences were absorbed into the effects of replicates (individuals or vials) within lines. To measure the longevity of virgins, ten vials per line and sex were established with ten two-day-old flies and given one drop of fresh yeast. For mated females, ten vials per line were established with ten two-day-old virgin females and 15 two-day-old males. Three times a week the flies were transferred to new vials and the number of dead flies was recorded until the last fly died. In vials with mated females, males were replaced if there were fewer males than females in the vial.

To measure developmental time and dry weight at eclosion, ten vials per line each received 12 larvae. Three days later the number of newly pupated larvae was recorded every four hours. When all larvae had pupated, the vials were placed in an eclosion fractionator (Stearns *et al.*, 1987) that collected freshly eclosed flies at six hour intervals. After their sex was determined, they were dried at 50° for three h and weighed to 0.01 mg.

To measure lifetime egg production, 30 vials per line were established with one female and two males per vial. The laying surface was replaced daily by a new one with a drop of fresh yeast. Dead males were removed and replaced by a young virgin male from the same line. The number of eggs laid by each female in 24 h was counted daily until the last female died. Fecundity early in life was defined as the number of eggs laid per day from the 4th to the 14th day after eclosion. This was 13-23 days after birth at 25° and 11-21 days after birth at 29°. Fecundity late in life was defined as the number of eggs laid from the 15th to the 5th day before the death of the last female. This was 32-42 days after birth at 25° and 23-33 days after birth at 29.5°.

Experiment I – Inbred flies
Three treatment and three control lines that Shepherd *et al.* had produced were adapted to our laboratory conditions for four generations. Because one control was lost at high temperature because of high preadult mortality, comparisons of lifespans and fecundities were made with three treatments and two controls. Tests for other traits used all six lines. The lines used by Shepherd *et al.* (1989) are labelled E3 (experimental) and C3 (control) in these experiments. In this experiment flies used to measure fecundity and longevity were held at 25° until eclosion to make our results comparable to those of Shepherd *et al.* and because larval mortality was high at 29.5° in these inbred flies. Longevity and fecundity were measured with 40 flies per line, age and weight at eclosion with 24 flies per line.

Experiment II – Effects of backgrounds
The treatment and control lines were obtained from a jumpstart cross in which one arranges for P elements already inserted in the genome to jump to new insert positions (Cooley, Kelley & Sprading, 1988; Robertson *et al.*, 1988; Bellen *et al.*, 1989). The use of balancers ensured that the only surviving flies carried the plasmid at a random position on the third chromosome (Fig. 1). After one fly with the desired marker combinations was crossed again with the balancer stock, a brother-sister cross be-

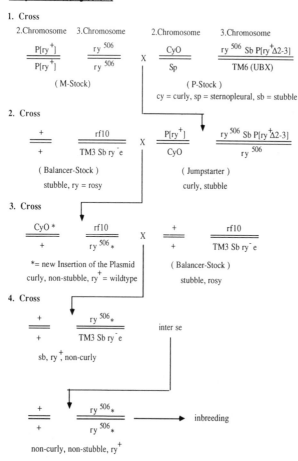

Fig. 1. The jumpstart crossing scheme. In the flies tested, half
the genome came from the jumpstart cross. Its first chromosome
was the same in all flies to the degree that inbreeding had made
it homozygous; its second chromosome was the same in all flies;
its third chromosome was the same in all flies except that the
treatment flies had the EF-insert and the control flies had the
control insert. In the other half of the genome, the chromosomes
were the same within genetic backgrounds to the degree that
brother-sister mating since 1987 had made them homozygous.
They differed among genetic backgrounds.

tween flies with the correct markers yielded flies
homozygous for the plasmid construct. The test
lines were obtained by outcrossing the one inbred
treatment and one inbred control line with six different inbred lab stocks to yield six pairs of outbred
treatment and control lines with the genetic backgrounds to be tested. Genetic backgrounds were
provided by isofemale lines held in the laboratory
since 1987 and maintained by full-sib mating for
about 60 generations. In the experiments on backgrounds and positions larvae and pupae were reared
at the same temperature as the adults.

Experiment III – Effects of insert positions

The test flies were obtained by outcrossing the six
inbred treatment and six inbred control lines from
the jumpstart cross with one inbred lab stock to
yield six pairs of heterozygous treatment and control lines with the positions to be tested. Thus all
flies tested had the same genetic background, Line
5, whose lifespan had been most sensitive to the
effects of enhanced expression of elongation factor
in the previous experiment. All inserts were on the
third chromosome, but the control inserts were not
in the same positions on the third chromosome as
the treatment inserts.

Comparison of experimental designs

Table 1 compares the four experiments that have
been done on the effects of enhanced expression of
elongation factor on fitness traits in *Drosophila*.
The major differences among the experiments are
these: (a) Replication – one control and one treatment line in Shepherd's experiment, three controls
(one was lost for longevity and fecundity at 29.5°)
and three treatments in our experiment on Shepherd's lines, six controls and six treatments in our
experiments on backgrounds and positions. (b)
Dosage of the additional gene for elongation factor
– two additional copies in Shepherd's experiment
and our experiment on his lines, one additional
copy in the backgrounds and positions experiments. (c) Flies tested – Shepherd tested virgin
males only, we tested virgin males and females and
mated females in all three experiments. (d) Genetic
backgrounds – different to an unknown but probably small degree in every line in Shepherd's experiment and our experiment on his lines, six pairs of
backgrounds matching treatment and control lines
in the backgrounds experiment, one background in
the positions experiment. (e) Insert positions – different in every line in Shepherd's experiment and
our experiment on his lines with some on Chromosome II and some on Chromosome III; one treatment position and one control position in the backgrounds experiment, both on Chromosome III; six
treatment positions and six control positions in the
positions experiment, all on Chromosome III. (f)
Genetic state of the flies – inbred in Shepherd's our
experiment and our experiment on his lines, outbred in the backgrounds and positions experiments.
(g) Temperature at which the larvae and pupae were
reared – 25° in Shepherd's experiment and our ex-

Table 1. Design of experiments on the effects of elongation factor on fitness components in *Drosophila melanogaster*.

Design element		Paper		
	Shepherd	Expt. I Repeat	Expt. II Backgrounds	Expt. III Positions
Control lines	1	3 (2 at 29.5 °)	6	6
Treatment lines	1	3	6	6
Flies tested:	Virgin males	Virgin males, virgin females, and mated females		
Backgrounds	2	6	6	1
Insert positions	2	6	1	6
extra *EF – 1*α	2 copies	2 copies	1 copy	1 copy
Larvae and pupae	25 °	25 °	25 & 29.5 °	25 & 29.5 °
Sample size per line:				
Longevity	300	40	100[1]	100[1]
Fecundity	–	40	30	30
Eclosion	–	24	120[2]	120[2]
Sample size total:				
Longevity	300	1160	7200	7200
Fecundity	–	411	720	720
Eclosion	–		2384[3]	2268[3]

[1] For each class: virgin males, virgin females, and mated females.
[2] Number starting.
[3] Number of larvae surviving to eclosion out of 2400.

periment on his lines, 25° and 29.5° in the backgrounds and positions experiments. (h) Total sample sizes – small in Shepherd's experiment and our experiment on his lines, large in the backgrounds and positions experiments.

Statistical analysis

In all experiments, comparisons of treatment and control lines had to take into account variation among replicates and differences in average lifetime between 25 °C and 29.5 °C. We used two methods to analyse that variation: factorial ANOVAs on time to pupation, time to eclosion, dry weight at eclosion, fecundities, and lifespans; and Cox regressions (Cox, 1972) on time to pupation, time to eclosion, and lifespans. Cox regressions measure differences in rates. When the two methods differed on level of significance, we have reported the less significant level. The statistical model had three main factors – treatment, temperature, and background (or position) – and all interactions. Thus where i: 1..2 indexes treatment α, j: 1..2 indexes temperature β, k: 1..6 denotes background (or position) γ, and l: 1..n denotes either vials (n = 10 per line per treatment, temperature,

and background) or individuals (n ≈ 700 for fecundity, ≈ 2400 for lifespan), the model was:

$$y = \mu + \alpha_i + \beta_j + \gamma_k + (\alpha\beta)_{ij} + (\alpha\gamma)_{ik} +$$

$$(\beta\gamma)_{jk} + (\alpha\beta\gamma)_{ijk} + \epsilon_{ijkl}$$

In all three experiments the only test for treatment effects that was not confounded by the effects of position or length of insert is the comparison of treatment with control for *differences* in the expressions of the traits at 25° and 29.5°. This is tested by the treatment × temperature interaction effect (Fig. 2). Because that effect should measure a percentage rather than an absolute difference, and to normalize distributions, all data were log-transformed for the ANOVAs.

The ANOVAs for lifespan and fecundity were done using individual MS as the denominator for the F-test. In measuring lifespan, flies were reared in groups of ten individuals per vial, but as deaths occurred, flies were moved each day to hold densities at ten per vial insofar as possible. Vial MS was inappropriate for testing effects on lifespan because flies had been moved among vials. For fecundity,

188

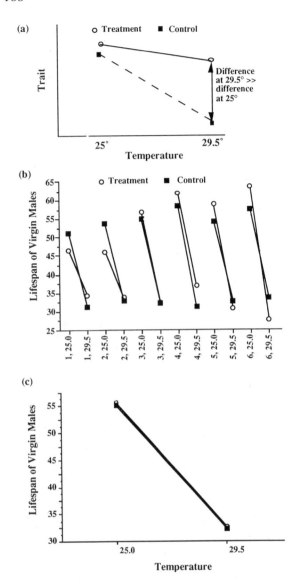

Fig. 2. The importance of line × treatment × temperature interaction effects in these experiments. (a) Increased expression of
elongation factor at higher temperature might lengthen life, but
all flies have shorter lives at higher temperature. One could test
for the effects of treatment by comparing the difference between
treatment and control at 29.5° with that at 25°; (b) That simple
test is inappropriate, for different lines have different reactions
to the change in temperature. These produce three-way interaction effects, here depicted for position × treatment × temperature effects on the lifespan of virgin males; (c) The line effects
cancel each other out in this two-way treatment × temperature
interaction.

with one female per vial, individual and vial MS
were equivalent. For time to pupation, time as pupae, time to eclosion, and weight at eclosion, flies
were grouped 12 to a vial with ten vials per line.
The ANOVAs were done with vials nested within

treatment, temperature, and line, and vial MS was
used as the denominator in the F test.

In the experiment on backgrounds, lifespans
were tested for flies that lived more than eight
days, for the distribution of lifespans was bimodal
with a small group of flies that died early in life
(5-6% of the total).

Line × treatment × temperature interaction effects are the key to interpreting these experiments.
While increased expression of elongation factor at
higher temperature might lengthen life, all flies
have shorter lives at higher temperature. The simplest situation, apparently the implicit assumption
of Shepherd et al. (1989), is that there is a small
difference in the lifespan of treatments and controls
at 25°, a larger difference at 29.5°, and one can test
for the effects of treatment by comparing the difference between treatment and control at 29.5° with
that at 25° (Fig. 2a). However, that simple test is
inappropriate for two reasons. First, in some lines
treatments and controls have different lifespans at
25° (there appears to be some 'leakage' in the expression of elongation factor at 25°). Second, different lines have different reactions to the change in
temperature. These produce three-way line × treatment × temperature effects. Fig. 2b depicts the
position × treatment × temperature effects on the
lifespan of virgin males in the positions experiment
(p = 0.0001 in the ANOVA on log-transformed
data). If one worked only with one pair of positions,
one could reach any of the possible conclusions.
For example – Line 1: treatment has shorter life at
25°, longer life at 29.5°; conclusion: increased expression of elongation factor lengthens lifespan.
Line 6: Treatment has longer life at 25°, shorter life
at 29.5°; conclusion: increased expression of elongation factor shortens lifespan. Line 3: not much
difference at either temperature; conclusion: treatment has no effect. In this particular case, the line
effects cancel each other out in the mean treatment
× temperature interaction (Fig. 2c); the correct
conclusion: the treatment effect is not significant
(p = 0.1894). For this reason, we present the details
on interactions for the most significant effects.

Results

Treatment effects

The treatment × temperature interaction effect cap-

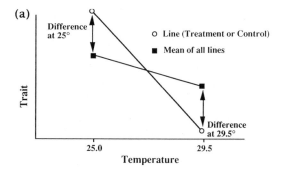

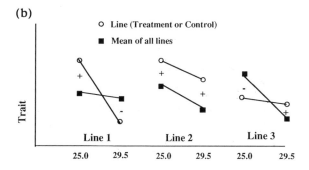

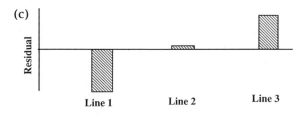

Fig. 3. The calculation of the residuals. (a) First the deviation of each line from the overall mean was calculated at each temperature; (b) These deviations express line × temperature interaction effects. Line 1 was larger than the mean at 25° and smaller at 29.5°; the effect of increased temperature was a larger reduction in the value of the trait in Line 1 than was experienced on average. Line 2 was consistently larger than the mean at both temperatures, just slightly more so at 29.5°. Line 3 was smaller than the mean at 25° and larger at 29.5°; (c) The residuals are calculated by subtracting the difference at 25° from the difference at 29.5°. If a residual is negative, increased temperature reduced the trait. If a residual is positive, increased temperature increased the trait, relative to the overall mean. These residuals isolate the temperature effect on each line. The average of the residuals for the treatment lines is the overall response of treatment to increased temperature; the average of the residuals for the control lines is the overall response of the control lines to increased temperature. If the difference between the average treatment and average control effect is significant, that will be expressed in the treatment × temperature interaction effect in the ANOVA.

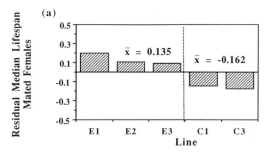

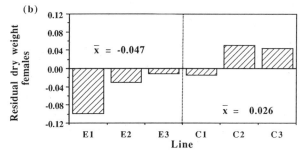

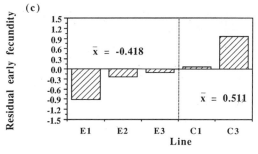

Fig. 4. Plots of residuals for significant treatment × temperature effects in Expt. I. Treatment lines are on the left and are labelled E1, E2 ... Control lines are on the right and are labelled C1, C2 ... The significance of the treatment × temperature interaction effect is reported: (a) Median lifespan of mated females (p < 0.0001 by ANOVA, p < 0.0001 by Cox regression); (b) dry weight at eclosion of females (p < 0.0004 by ANOVA); (c) early fecundity (p < 0.0001 by ANOVA).

tures the significance of an increase or decrease in the value of a trait at higher temperature and corresponds to the purported enhanced expression of elongation factor at higher temperature. For the major significant treatment × temperature effects, plots of residuals are given depicting both the magnitude of the effects in treatment and control and the variation among lines. Fig. 3 describes how these residuals were calculated.

Table 2 reports the magnitude and direction of significant treatment × temperature interaction effects in all three experiments. In Expt. I on inbred flies, the treatment affected females but not males. It increased the lifespan of mated females by 30.0%

Table 2. Magnitude and direction of the treatment × temperature interaction effects in the three experiments. * Not significant when tested by 3-way MS but significant when tested by vial or individual MS.

Trait	Expt. I Biocentre	Expt. II Backgrounds	Expt. III Positions
Time to pupation	NS	NS	NS
Time as pupae			
Males	NS	NS	NS
Time to eclosion			
Males	NS	(– 1.1%)*	NS
Females	NS	NS	NS
Dry weight at eclosion			
Females	– 7.3%	NS	(– 8.0%)*
Fecundity			
Lifetime	– 77%	NS	NS
Early	– 93%	NS	NS
Lifespan			
Virgin males [1]	NS	(+ 6.8%)*	NS
Mated females	+ 30.0%	NS	NS

[1] The relatively longer lifespan at higher temperature resulted from a decrease in the absolute lifespan of the treatment flies at lower temperature with no difference at higher temperature.

(Fig. 4a), decreased their dry weight at eclosion by 7.3% (Fig. 4b), reduced their lifetime fecundity by 77%, and reduced their early fecundity by 93% (Fig. 4c).

In the backgrounds experiment on heterozygous flies, the treatment reduced time to eclosion of males by 1.1% but had no clear effects on lifespan or fecundity. In the positions experiment, the treatment reduced the dry weight at eclosion of females by 8% (Fig. 5).

The relative magnitude of the effects of treatment

The relative magnitude of the different effects on pupation rate, eclosion rate, and mortality rate can be estimated from the coefficients in the Cox regressions. Those coefficients weight the impact of each effect on instantaneous rates. Table 4 compares the largest treatment × temperature interaction effects in all three experiments on those rates, using temperature, background, and position effects as the standard of comparison where appropriate. Effects of enhanced expression of EF-1α, as detected in the treatment × temperature interaction, were largest on lifespan, where they ranged from 16% to 136% of the temperature effect. Effects of enhanced expression of EF-1α on time to pupation ranged from 2.1% to 2.5% of the temperature effect. The effect of enhanced expression of EF-1α on time to eclosion was 3.5% of the temperature effect in Expt. II. Effects of treatment were always smaller than effects of background (7% to 18%) or position (13% to 16%).

Summary of treatment effects on individual traits

In no case did the treatment increase the lifespan of virgin males. The relative increase in lifespan at higher temperature in the backgrounds experiment in fact resulted from an absolute decrease in treatment lifespan at lower temperature and no difference at higher temperature. When the treatment increased the lifespan of mated females in Expt. I, it reduced weight at eclosion and fecundity. When it reduced weight at eclosion in the positions experiment, it also reduced lifespan.

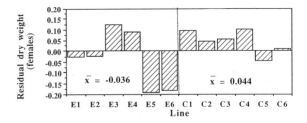

Fig. 5. Plots of residuals for significant treatment × temperature effects in Expt. III (effects of insert position). Treatment lines are on the left and are labelled E1, E2 ... Control lines are on the right and are labelled C1, C2 ... The significance of the treatment × temperature interaction effect is reported: dry weight at eclosion of females (p = 0.0001 by ANOVA).

Table 3. Relative magnitude of the treatment × temperature interaction effect inferred from the absolute value of the coefficients in the Cox regression.

Expt	Trait	Effect expressed as		
		% of T° effect	% of Back effect	% of Position effect
I	Lifespan of mated females	136%		
II	Time of pupation	2.5%	6.7%	
II	Time to eclosion (females)	3.5%	18%	
III	Lifespan of mated females	16%		13%
III	Time to pupation	2.1%		15%

Table 4. Analysis of the effects of treatment on relationships between pairs of traits calculated from the line means. All the effects reported were significant at the p<0.01 level by the Tukey-Kramer test on the difference in slopes; SAS Type II regressions. The only cases reported are those in which the relationship was negative at both temperatures or in which the treatment changed a positive relation to a negative one (*).

Expt	Relation	Slopes		Diff
		Cont	Treat	
		25°	or 29.5°	
a)	Slopes of relations compared between treatment and control (lumping temperatures).			
II	Weight at eclosion vs. late fecundity	−3847.23	−4888.10	−1040.86
b)	Slopes of relations compared between treatment and control within temperature.			
II	Early fecundity vs. lifespan at 25 ° *	6.07	−3.14	−9.21
	Age at eclosion vs. late fecundity at 25 ° *	428.65	−59.92	−488.57
	Weight at eclosion vs. late fecundity at 29.5 °	−2786.77	−785.64	2001.13
III	Weight at eclosion vs. late fecundity at 25 °	−3870.41	−1810.62	2059.79
	Early fecundity vs. lifespan at 29.5 ° ° *	5.19	−17.43	−22.62
	Weight at eclosion vs. late fecundity at 29.5 °	−2378.19	−1239.46	1138.74

Effects of mating and temperature

In all three experiments, both mating and temperature shortened the lives of females, but the effects were not additive. Mated females did not suffer as great a reduction in lifespan at high temperature as did virgin females (reduction in median lifespans of females from 25° to 29.5° in Expt. I: virgins −15 days = 41% of lifespan at 25°, mated −1 day = 6%; in Expt. II: virgins −35 days = 56%, mated −11 days = 29%; Expt. III: virgins −13 days = 28%, mated −7 days = 16%).

Background effects

Expt. II revealed significant effects of genetic background (Kaiser & Stearns, 1993). Backgrounds varied significantly in time to pupation; time spent as pupae by males; time to eclosion (developmental time); dry body weights; lifetime, early, and late fecundity; mortality rates; and mean and median lifespans. To compare traits measured on different scales (e.g. milligrams and days), we converted the range among backgrounds into a percentage of the appropriate mean. This range was 3.0% of time to pupation, 3.3% of time spent as pupae by males, 4.0% of time spent as pupae by females, 2.1% of time to eclosion for males, 2.7% of time to eclosion for females, 12% of dry body weight at eclosion for males, 8.8% of dry body weight at eclosion for

females, 39% of total lifetime fecundity, 33% of early fecundity, 79% of late fecundity, 19% of virgin male lifespan, 13% of virgin female lifespan, and 32% of mated female lifespan. For all traits except female dry weight at eclosion, background effects were larger than treatment effects as judged by the ratio of the mean squares. The ratio of background to treatment mean square was 0.83-2.65 for traits measured early in life (time to pupation, time spent as pupae, time to eclosion, dry weight at eclosion), but for fecundity (2.40-29.57) and lifespan (4.52-372.30) it was much higher.

The background × treatment, background × temperature, and background × treatment × temperature interaction effects were significant in 21 of 39 cases.

Position effects

Expt. III revealed significant position effects (Kaiser & Stearns, 1993). Lines varied significantly for time to pupation, time spent as pupae by females (not males), time to eclosion (developmental time) for both sexes, dry body weights of females (not males), lifetime (but not early or late) fecundity, mortality rates and mean lifespans. The range among insert positions was 2.9% of time to pupation, 3.2% of time spent as pupae by males, 5.6% of time spent as pupae by females, 2.7% of time to

(a)

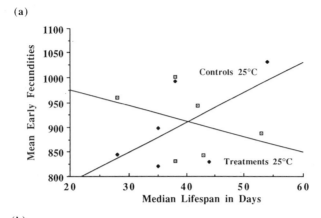

(b)

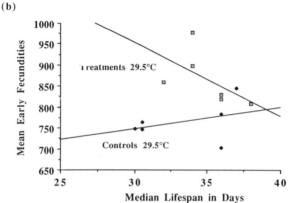

Fig. 6. Two of the four cases in which the treatment changed a relationship from positive to negative, creating a tradeoff involved in lifespan where none had existed in the control: (a) early fecundity vs. lifespan at 25° in Expt. II; (b) early fecundity vs. lifespan at 29.5° in Expt. III.

eclosion for males, 4.7% of time to eclosion for females, 4.0% of dry body weight at eclosion for males, 6.5% of dry body weight at eclosion for females, 22% of total lifetime fecundity, 16% of early fecundity, 52% of late fecundity, 14% of virgin male lifespan, 7.0% of virgin female lifespan, and 24% of mated female lifespan. For time spent as pupae by females (ratio = 4.21), lifespan of virgin males (ratio = 4.51), and lifespan of mated females (ratio = 1.39), position effects were larger than treatment effects as judged by the ratio of the mean squares. That ratio was 0.09-0.53 for the other traits.

The position × treatment, position × temperature, and position × treatment × temperature interaction effects were significant in 23 of 39 cases.

Tradeoffs

There are six tradeoffs between components of fit-

ness expressed early and late in life and measured in these experiments that bear on the evolution of lifespan: (1) early fecundity vs. lifespan, (2) age at eclosion vs. lifespan, (3) weight at eclosion vs. lifespan, (4) early fecundity vs. late fecundity, (5) age at eclosion vs. late fecundity, (6) weight at eclosion vs. late fecundity. Several criteria could be used to assess the significance of these six tradeoffs. They include:

First criterion. Did the treatment significantly increase one trait and significantly decrease the other as measured in the residuals that are tested by the treatment × temperature interaction effect? By this criterion, only a few tradeoffs were found. In Expt. I, the treatment significantly increased the lifespan of mated females, reduced weight at eclosion, and reduced lifetime and early fecundity. In Expt. III, it significantly reduced the lifespan and the dry weight at eclosion of mated females but did not change lifetime, early, or late fecundity. In Expt. I weight at eclosion and fecundity covaried positively (both were reduced). No tradeoffs of lifespan with fitness components were detected in Expt. II.

Second criterion. Did the effects of treatment significantly change the slope of the regression of one trait on the other? This question can be asked at both temperatures and in the two-dimensional plots of residuals. All combinations of lifetime fecundity, early fecundity, late fecundity, lifespan, age at eclosion, and weight at eclosion were examined in mated females. All combinations of age and weight at eclosion and lifespan were examined in virgin males and females. Relations were compared (a) between treatment and control (lumping temperatures), (b) between temperatures (lumping treatments), (c) between treatment and control within each temperature, (d) between treatment and control in the residuals expressing treatment × temperature interaction effects. Only the effects reported in Table 4 were significant at the p < 0.01 level by the Tukey-Kramer test on the difference in slopes. We chose 0.01 as the significance level to compensate for the effect of multiple unplanned comparisons.

In four cases the treatment made a positive relation negative, establishing a tradeoff in the treatment where none had existed in the control: age at eclosion vs. late fecundity (Expt. III, lumping temperatures), early fecundity vs. lifespan at 25° (Expt. II) (Fig. 6a), age at eclosion vs. late fecundity at 25° (Expt. II), and early fecundity vs. lifespan at 29.5° (Expt. III) (Fig. 6b).

Third criterion. Did the changes from control to treatment within line all lie within an envelope of variation describing a tradeoff in the residuals? This can only be properly applied in Expt. II, on backgrounds, where there was one treatment insert and one control insert. The analysis is best described by an example, the tradeoff between mean lifetime fecundity and median lifespan. The pairs of controls and treatments can be connected by arrows (Fig. 7a). Three of these arrows (from controls 4, 5, and 6 to the corresponding treatments) describe a tradeoff in one direction: the treatment increased residual lifespan and decreased residual fecundity. Three (from controls 1, 2, and 3 to the corresponding treatments) represent changes in the other direction: the treatment reduced residual lifespan and increased residual fecundity.

There are two methods of summarising these changes; both are depicted in Fig. 7b, where the

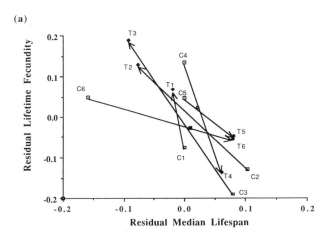

(a)

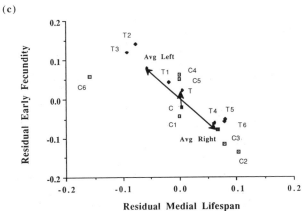

(c)

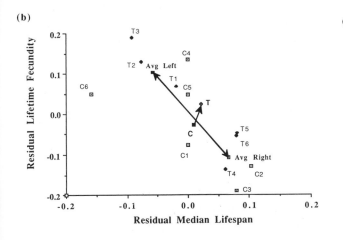

(b)

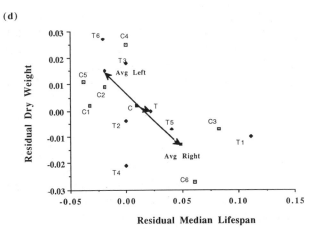

(d)

Fig. 7. Analysis of the details ot tradeoffs in the residuals for Expt. II (a) The changes ftom control to treatment are plotted for each line; (b) There are two methods of summarising the changes. The first – the change from mean control to mean treatment – is misleading, for it produces the short arrow pointing in a direction in which none of the changes occurred. The second – the change from the mean left end of an arrow to the mean right end – expresses the pattern in the data much better; (c) Similar efffects for the tradeoff early fecundity vs. lifespan in mated females; (d) Similar effects for the tradeoff dry weight at eclosion vs. lifespan in virgin males.

short arrow pointing up and to the right describes the change between the average control residual and the average treatment residual. It misses the consistency of the changes in all six backgrounds, which are all contained within an envelope running from the upper left to the lower right. The longer line running from the upper left to the lower right describes the change from the average left end to the average right end of an arrow in Fig. 7a, irrespective of treatments and controls. The left average residual lifespan is significantly smaller than the right one, and the upper average residual fecundity is significantly larger than the lower one.

The problem with the analysis of means becomes clear: there are two legitimate senses in which a difference between treatment and control can describe a change corresponding to a tradeoff – reduced lifespan and increased fecundity, or increased lifespan and reduced fecundity. When both types of change are present, they counteract each other, and adding all changes together wipes out significant differences and leads to the misleading conclusion that the treatment had no effect on the tradeoff. It would be more accurate to say: all effects of treatment were constrained to occur within the broader limits of a tradeoff, but the direction of the change caused by the treatment depended on the genetic background.

One can judge the significance of such a relation by two criteria: (a) t-tests on the differences between the left and right and upper and lower ends of the lines connecting treatment and controls, and (b) the significance of the treatment $\times$ line interaction for each trait. In the background experiment, two tradeoffs were significant for the first criterion but not for the second: lifespan of mated females (t-test, $p = 0.0074$; treat $\times$line, $p = 0.1357$) vs. total fecundity (t-test, $p = 0.003$; treat $\times$ line, $p = 0.0004$) (Fig. 7b) and lifespan of virgin males (t-test, $p = 0.0118$; treat $\times$ line, $p = 0.2013$) vs. dry weight at eclosion (t-test, $p = 0.006$; treat $\times$ line, $p = 0.0001$) (Fig. 7d). Only one tradeoff was significant by both criteria: lifespan of mated females vs. early fecundity (t-test, $p = 0.0024$; treat $\times$ line, $p = 0.0038$) (Fig. 7c). Thus in the backgrounds experiment, where no tradeoffs associated with lifespan were detected by other methods, there were changes occurring in such tradeoffs, but they were hidden in the summary statistics because the effects cancelled each other out.

Discussion

Does enhanced expression of elongation factor increase lifespan?

Enhanced expression of elongation factor increased the lifespan of mated females by 30% in Expt. I on inbred flies reared to eclosion at 25°. It had no effects on lifespan in Expt. II (backgrounds), where the flies were heterozygotes with half the genome derived from a wild population. It reduced the lifespan of mated females by 15% in Expt. III (positions), which was done in the genetic background whose lifespan was most sensitive to the effects of elongation factor in Expt. II. In no case did enhanced expression of elongation factor increase the lifespan of virgin males. Thus the genetic manipulation of EF-1α extended the lifespan of mated females in inbred flies reared to eclosion at 25° and had either no effect on (Expt. II) or reduced (Expt. III) the lifespan of mated females in outbred flies reared to eclosion at both temperatures. These results do not rule out an as yet unanalyzed interaction between the genetic state of the flies – inbred or outbred – and whether or not the flies were reared to eclosion at 25° or at both temperatures.

Does enhanced expression of EF-1α effect other fitness components?

The treatment reduced time to pupation in Expt. II and time to eclosion in both sexes in Expt. II and in males in Expt. III. It decreased weight at eclosion in females in Expts. I and III. In Expt. I it reduced early and lifetime fecundity. In Expt. I it increased the lifespan of mated females; in Expt. III it reduced that lifespan. Overall, the effects of elongation factor on age and weight at eclosion were more consistent than its effects on lifespan or fecundity. This genetic manipulation produced interesting effects on the components of fitness that are most important in life history evolution and the evolution of lifespan. Further applications of recombinant DNA methods to the analysis of the causes of lifespan and of tradeoffs among fitness components would certainly be justified.

Were treatment effects large?

The treatment $\times$ temperature interaction effects on fitness components were quite large in inbred flies – 30% differences were observed for lifespan, up to 93% for fecundity – and smaller but still impressive for single gene effects in outbred flies – about 1-2% for measures of developmental time, up to 8% for dry weight at eclosion, up to 15% for lifespan of mated females.

Comparisons to temperature, background, and position effects

There is another approach to the question about the magnitude of treatment effects: one can compare the changes attributed to treatment with those attributed to temperature, genetic background, and insert position (cf. Table 4). Effects of enhanced expression of EF-1α on lifespan were quite large in comparison to temperature effects – from 16% to 136% of the change in lifespan caused by the difference between 25° and 29.5°. Any genetic manipulation of a fitness component in a fruitfly that has as large an effect as temperature must be considered as having very significant impact. Effects of treatment were consistently smaller than effects of backgrounds and positions. This has two implications, one for experimental design and one for the spread of such an insert as a mutant allele in a natural population.

The implication for experimental design is that such experiments must contain adequate representation of genetic backgrounds and insert positions, for if the effect of treatment is usually smaller than the effect of background or position, its significance can only be detected and properly assessed with adequate replication. This comment is strengthened by the observation of pervasive interaction effects. The degree to which a trait responded to enhanced expression of EF-1α at higher temperature depended very strongly on the genetic background and the position of the insert. Such interaction effects make it impossible to conclude whether the treatment has any effect at all in an experiment done with only one insert position in only one genetic background (cf. Shepherd *et al.*, 1989).

The implication for the spread of such an insert as a mutant allele is that its interactions with genetic backgrounds, generated by mating and recombination, and its movement into new positions in the genome, generated by transposon jumping, will have effects on its fate that are as large as or larger than its direct effect on fitness components. This makes it virtually impossible to predict from experiments like these what would be the fate of a genetically manipulated fruitfly in a natural population, for in such a case much would depend on interaction effects with unknown genetic backgrounds and unknown insert positions. This has rather pessimistic implications for our ability to predict the consequences of releasing genetically manipulated organisms into natural populations.

Tradeoffs

Tradeoffs between fitness components expressed early and late in life are central to the evolutionary theory of lifespan. In this series of experiments, several were detected. Which traits traded off depended on the experiment and the criterion used to detect the tradeoff.

First criterion. The most straightforward criterion is significant, opposite effects of treatment on two traits. Where treatment effects are judged by the significance of the treatment $\times$ temperature interaction, that criterion detected a tradeoff between the lifespan of mated females and their total and early fecundity in Expt. I, no tradeoff associated with lifespan in Expt. II, and a positive association, not a tradeoff, between lifespan of females and weight at eclosion in Expt. III.

Second criterion. The treatment also had significant effects on the slopes of relationships between two traits. Those relationships can be calculated for treatment and control lumping temperatures together, for each temperature separately, and in the residuals that expression the treatment $\times$ temperature interaction effects. In all three experiments combined, there were 108 such relations that could be tested (Stearns & Kaiser, 1993). Enhanced expression of elongation factor had significant effects on relations that could be described as measured in seven cases (Table 4).

In three cases the treatment changed a positive relation in the control into a negative relation in the treatment. Those cases involved age at eclosion and early fecundity on the one hand and late fecundity and lifespan on the other, and all occurred in Expt. II or Expt. III. This suggests that in healthy flies increases in lifespan associated with enhanced expression of elongation factor entail costs that must be paid in age and weight at eclosion and early fecundity. However, the fact that the treatment also erased some tradeoffs that had existed in the control makes clear that the total impact of treatment on tradeoffs is complex. Whether the treatment would increase the fitness of flies in a natural population could only be assessed in an experiment that followed the frequency of the treatment insert in a population under natural conditions.

The correlations across lines in Expt. II are genetic correlations generated by the epistatic interactions between one treatment insert position, one control insert position, and six genetic backgrounds. Thus they are not additive genetic correlations. Nevertheless, we found it interesting that a manipulation that is equivalent to introducing a single mutant into a population (Expt. II) can change such genetic correlations from positive to negative and from negative to positive. We did not expect that epistatic genetic correlations would be so labile that fixing a single new mutation could change their sign.

Third criterion. Recall that no tradeoffs relevant to lifespan were detected in Expt. II in the treatment × temperature interaction effects (Table 2). When we analyzed the changes in each genetic background from control to treatment in the residuals, we discovered that they were sometimes constrained to lie within an envelope that appeared to be determined by processes on which the treatment itself had little effect. This was not true for all relations that we examined; in some of them the changes occurred in a confusing variety of directions. But there were three relations in the residuals in which most of the differences between control and treatment lines described changes in one direction or the other along a tradeoff (cf. Fig. 7a): total fecundity of mated females vs. lifespan, early fecundity of mated females vs. lifespan, and dry weight at eclosion of virgin males vs. lifespan. In these cases, changes were occurring in tradeoffs due to epistatic interactions between treatment inserts and genetic backgrounds, but those changes were hidden in the summary statistics because the effects cancelled each other out.

Similar effects may be hidden in the other experiments but cannot be detected because the logic of the experimental design does not allow one to connect a single treatment with a single control line in either Expt. I or Expt. III, where every treatment insert was at a different position from every control insert. We suspect that such effects may exist in

other organisms; this is, to our knowledge, the first experiment done that could have detected them.

The physiology and genetics of tradeoffs. Several research groups working on lifespan in *Drosophila* have concluded that interactions between genetics and physiology may be the next thing that sheds light on the evolution of lifespan (M. Rose, L. Partridge, pers. comm.). Two aspects of the results reviewed here suggest that genetic changes are carried out within the bounds of a physiological framework that determines the tradeoffs among fitness components. The origin and causes of that framework may have to be understood before we can understand the evolution of lifespan itself.

The first point can be seen by comparing tradeoffs in Expt. I with those in Expt. III. In Expt. I, increased lifespan of mated females was associated with reduced fecundity in the residuals. In Expt. III, reduced lifespan of mated females was associated with reduced dry weight at eclosion in the residuals. The trait with which lifespan was significantly associated changed with the type of fly tested. The second point can be seen in the analysis of changes from control to treatment in the residuals in Expt. II, which revealed hidden, compensatory changes in tradeoffs. Both suggest that the tradeoff existed independent of the treatment, which could make a change within an existing structure but could not modify the structure itself. We think that structure is physiological.

A difference between the experiments on inbred and outbred flies suggests that temperature physiology may also be involved in the generation of tradeoffs. In the experiment on inbred flies, the larvae and pupae used to measure fecundity and lifespan were all reared at 25° so that the measurements would be comparable with those made be Shepherd *et al.* (1989), who had reared all flies to eclosion at that temperature because pre-eclosion mortality was high in those inbred flies at 29.5°. In the backgrounds and positions experiments, flies were reared at both temperatures for all traits. In the first experiment, effects on fecundity and lifespan were much larger than in the latter two experiments, but effects on larval and pupal traits were not. In the latter two experiments, effects on time to pupation, time as pupae, and time to eclosion were significant, but there were no effects on fecundity, and effects on lifespan were less dramatic.

Table 5. The experiments in which tradeoffs associated with aging have been detected in *Drosophila melanogaster.*

Tradeoff detected	Experiment						
	Luckinbill[1]	Rose[2]	Partridge[3]	Hillesheim[4]	Expt. I	Expt. II	Expt. III
Early fecundity/ lifespan	yes	yes	no	yes	yes	yes [6,7]	yes [6]
Early fecundity/ late fecundity	yes	yes	no	yes	no	yes [7]	no
Size at eclosion/ lifespan	–[5]	–[5]	yes	yes	no	yes [7]	no
Developmental time/ lifespan	–[5]	–[5]	yes	–[5]	no	no	no
Larval competitive ability/lifespan	–[5]	–[5]	yes	–[5]	–[5]	–[5]	–[5]
Developmental time/ late fecundity	–[5]	–[5]	no	–[5]	no	no	yes [6]
Weight at eclosion/ late fecundity	–[5]	–[5]	no	–[5]	no	yes [8]	yes [8]

[1] Luckinbill *et al.* (1984). [2] Rose (1984). [3] Partridge & Fowler (1992). [4] Hillesheim & Stearns (1992). [5] Did not look. [6] Change of sign of tradeoff caused by treatment. [7] Tradeoff detected in the detailed analysis of residuals. [8] Tradeoff present, treatment reduced slope, which remained negative.

Comparison with other experiments on tradeoffs involved in lifespan. Table 5 compares the tradeoffs that have been found in four selection experiments on traits involved in lifespan and in these three experiments. On balance, these experiments support the view that early fecundity and lifespan do trade off, as suggested by Luckinbill *et al.* (1984), Rose (1984), and Hillesheim and Stearns (1992). However, they also support the idea that age and size at eclosion are involved in tradeoffs on lifespan, as suggested by Partridge and Fowler (1992). The general conclusion remains: early and late fitness components do trade off, but the ones involved depend on the experiment and how the tradeoffs are measured.

Strategies of molecular manipulation of life histories. Two problems with the interpretation of the results of these experiments suggested improvements in the design of molecular manipulations. First, the treatment insert was 2 kb longer than the control insert. This confounded the treatment effects and led to particular difficulties in interpreting the direct effects of treatment in the positions experiment. A better control would be created by *in situ* deactivation of the treatment insert, resulting in a control insert of precisely the same position and length and almost the same sequence as the treatment insert. This method is available in yeast but not yet in *Drosophila*.

Second, interaction effects are hard to interpret. They make the statistical analysis more complicated and harder to communicate. It would be better to design inserts that would be expressed at one lower temperature, within the normal physiological range of *D. melanogaster*; and vary the insert dosage for a sample of insert positions, e.g. one and four inserts, treatment and control, on each of three third chromosomes (three positions, same for treatment and control, achieved by *in situ* mutation) inserted into each of four genetic backgrounds. That would yield 48 lines to be tested, the same number that was tested here in both experiments combined, and it would give more complete information on position × background interaction effects.

Conclusion

Genetic manipulation of a gene of known function can significantly affect lifespan in *Drosophila*, and it can lead to surprisingly large changes in other fitness components, including age and weight at eclosion and fecundity. Such manipulations also provide promising new information on the nature of tradeoffs.

Acknowledgements

These experiments would not have been possible without the precise and reliable work of Anni Mislin, Barbara Sykes, Victor Mislin, and Hanni Zingerli in the laboratory. We are very grateful for their support. Our research was supported by the Swiss Nationalfonds (Nr. 31-28511.90).

References

Bellen, H. J., C. J. O'Kane, C. Wilson, U. Grossniklaus, R. K. Pearson & W. J. Gehring, 1989. P-element-mediated enhancer detection: a versatile method to study development in *Drosophila*. Genes and Development 3: 1288-1300.

Charlesworth, B. & J. A. Williamson, 1975. The probability of the survival of a mutant gene in an age-structured population and implications for the evolution of life-histories. Genet. Res. 26: 1-10.

Cooley, L., R. Kelley & A. Sprading, 1988. Insertional mutagenesis of the Drosophila genome with single P-elements. Science 239: 1121-1128.

Cox, D. R., 1972. Regression models and life-tables (with discussion). J. Roy. Stat. Soc. B 34: 187-220.

Darnell, J., H. Lodish & D. Baltimore, 1990. Molecular cell biology. 2nd Ed. W. H. Freeman, New York.

Hamilton, W. D., 1966. The moulding of senescence by natural selection. J. Theor. Biol. 12: 12-45.

Hillesheim, E. & S. C. Stearns, 1992. Correlated responses in life history traits to artificial selection for body weight in *Drosophila melanogaster*. Evolution 46: 745-52.

Hovemann, B., S. Richter, U. Walldorf & C. Ciepluch, 1988. Two genes encode related cytoplasmic elongation factors EF-1a in *Drosophila melanogaster* with continuous and stage specific expression. Nucleic Acids Res. 16: 3175-94.

Johnson, T. E., 1990. Increased life-span of *age-1* mutants in *Caenorhabditis elegans* and lower Gompertz rate of aging. Science 249: 908-912.

Kaiser, M. & S. C. Stearns, 1993. Effects on fitness components of enhanced expression of elongation factor EF-1α in *Drosophila melanogaster* II. Genetic backgrounds and insert positions. Under review.

Luckinbill, L., R. Arking, M. G. Clare, W. C. Cirocco & S. A. Buck, 1984. Selection for delayed senescence in *Drosophila melanogaster*. Evolution 38: 996-1003.

Partridge, L. & K. Fowler, 1992. Direct and correlated responses to selection for age at reproduction in *Drosophila melanogaster*. Evolution 46: 76-91.

Robertson, H. M., C. R. Preston, R. W. Phillis, D. M. Johnson-Schlitz, W. K. Benz & W. R. Engels, 1988. A stable source of P-element transposase in *Drosophila melanogaster*. Genetics 118: 61-470.

Rose, M., 1984. Laboratory evolution of postponed senescence in *Drosophila melanogaster*. Evolution 38: 1004-1010.

Rose, M. R., 1991. Evolutionary biology of aging. Oxford University Press, Oxford.

Rose, M. R. & B. Charlesworth, 1980. A test of evolutionary theories of senescence. Nature 287: 141-142.

Rose, M. R. & B. Charlesworth, 1981. Genetics of life history in *Drosophila melanogaster*. II. Exploratory selection experiments. Genetics 97: 187-196.

SAS Institute, 1985. SAS User's Guide: Statistics. SAS Institute, Cary, North Carolina, 957 p.

SAS Institute, 1991. SAS Technical report P-127. SAS/STAT Software: The PHREG procedure. Version 6. SAS Institute, Cary, North Carolina. 59 p.

Shepherd, J. C. W., U. Walldorf, P. Hug & W. J. Gehring, 1989. Fruit flies with additional expression of the elongation factor EF-1α life longer. Proc. Natl. Acad. Sci. USA 86: 7520-1.

Sokol, R. R. & F. J. Rohlf, 1981. Biometry. 2nd Ed. W. H. Freeman, San Francisco.

Stearns, S. C., T. Diggelmann, M. Gebhardt, H. Bachmann & R. Wechsler, 1987. A device for collecting flies of precisely determined post-hatching age. Drosophila Information Service 66: 167-169.

Stearns, S. C., 1992. The evolution of life histories. Oxford University Press, London. 248 p.

Stearns, S. C., M. Kaiser & E. Hillesheim, 1993. Effect on fitness components of enhanced expression of elongation factor EF-1α in *Drosophila melanogaster*. I. The contrasting approaches of molecular and population biologists. Am. Nat. (in press).

Stearns, S. C. & M. Kaiser, 1993. Effects on fitness components of enhanced expression of elongation factor EF-1α in *Drosophila melanogaster*. III. Analysis of tradeoffs between fitness components. Under review.

Webster, G. C. & S. L. Webster, 1982. Effects of age on the post-initiation stages of protein synthesis. Mech. Aging Dev. 18: 369-378.

Webster, G. C. & S. L. Webster, 1983. Decline in synthesis of elongation factor on (EF-1) precedes the decreased synthesis of total protein in aging *Drosophila melanogaster*. Mech. Aging Dev. 22: 121-128.

Williams, G. C., 1957. Pleiotropy, natural selection, and the evolution of senescence. Evolution 11: 398-411.

M.R. Rose and C.E. Finch (eds.), Genetics and Evolution of Aging, 199–214, 1994.
© 1994 Kluwer Academic Publishers. Printed in the Netherlands.

Two-dimensional protein electrophoretic analysis of postponed aging in *Drosophila*

James E. Fleming[1]*, Greg S. Spicer[1], Roger C. Garrison[2] & Michael R. Rose[2]
[1] *Linus Pauling Institute of Science and Medicine, 440 Page Mill Road, Palo Alto, CA 94306, USA*
[2] *Institute of Molecular Medical Sciences, 460 Page Mill Road, Palo Alto, CA 94306, USA*
[3] *Department of Ecology and Evolutionary Biology, University of California, Irvine, CA 92717, USA*
* *Address for correspondence: Department of Biology, Eastern Washington University, Cheney, WA 99004, USA*

Received and accepted 22 June 1993

Key words: aging, *Drosophila*, electrophoresis, evolution, longevity

Abstract

Five populations of *Drosophila melanogaster* that had been selected for postponed aging were compared with five control populations using two-dimensional protein gel electrophoresis. The goals of the study were to identify specific proteins associated with postponed aging and to survey the population genetics of the response to selection. A total of 321 proteins were resolvable per population; these proteins were scored according to their intensity. The resulting data were analyzed using resampling, combinatoric, and maximum parsimony methods. The analysis indicated that the populations with postponed aging were different from their controls with respect to specific proteins and with respect to the variation between populations. The populations selected for postponed aging were more heterogeneous between populations than were the control populations. Maximum parsimony trees separate the selected populations, as a group, from their controls, thereby exhibiting a homoplastic pattern.

Introduction

The identification of the genes involved in the control of aging has been one of the major goals of gerontology. One method for the pursuit of this goal is the analysis of the genetic basis of accelerated aging or shortened lifespan. This approach has led to the identification of metabolic deficiencies that can hasten death, such as those of some *Drosophila* mutants (Bozcuk, 1981). However, deterioration and death in such stocks may depend on the introduction of novel pathologies (Maynard Smith, 1966; Hutchinson & Rose, 1987) rather than an acceleration in the mechanisms of normal aging. While this certainly need not always occur, the problem of ambiguity as to whether or not it has occurred in any particular case undermines the reliability of inferences concerning the mechanistic basis of aging from genetic analyses of shortened lifespan.

An alternative approach is to create stocks with genetically postponed aging. This has been done by means of mutagenesis and selection on recombinant inbreds in *Caenorhabditis elegans*, a nematode (Johnson & Hutchinson, 1990). It has also been achieved by selection on outbred *Drosophila* populations (Rose, 1989). A particularly important feature of the *Drosophila* research has been the replication of these selection experiments between laboratories (Rose, 1984; Luckinbill *et al.*, 1984), in repeated experiments (Rose & Charlesworth, 1981; Rose, 1984), and within experiments (Rose, 1984; Luckinbil *et al.*, 1984; Hutchinson & Rose, 1991). The *Drosophila* postponed-aging stocks as a group thus will not be overly dependent on the specific genetic accidents arising from mutagenesis, linkage disequilibrium, and the like in individual lines.

There are two additional advantages to the use

of these stocks. The first is that they have been generated by the use of natural selection in the laboratory. Normal *Drosophila* cultures are strongly selected for high early fertility, with negligible selection for later fertility. The *Drosophila* stocks with postponed aging have been created by selection for later reproduction, thereby producing an increase in the force of natural selection at later ages (Rose, 1991, Ch. 1, 3). These stocks are thus directly interpretable in terms of the evolutionary theory of aging, and may be used to test this theory. This makes findings obtained from them of relevance both to mechanistic gerontology and to evolutionary biology.

The second advantage to working with these stocks is that they have been characterized in terms of a number of biological variables: life-history (e.g. Rose, 1984; Luckinbill *et al.*, 1984), morphology (Rose *et al.*, 1984; Luckinbill *et al.*, 1988), physiology (Service *et al.*, 1985; Service, 1987), and quantitative genetic transmission (Clare & Luckinbill, 1985; Hutchinson & Rose, 1991). One-dimensional protein electrophoresis has been used to test for the involvement of specific loci in postponed aging (Luckinbill *et al.*, 1989; Tyler *et al.*, this volume), and significant results have been obtained for a few loci.

One-dimensional electrophoretic surveys of proteins allow the detection of genetic associations between biochemically characterized proteins and postponed aging, but not other types of protein. To overcome this limitation, the present study used two-dimensional protein electrophoresis to survey a significant fraction of the polypeptides found in *Drosophila*.

Material and methods

Stocks

All populations employed derive ultimately from an endemic *Drosophila melanogaster* population from Massachusetts, studied by Ives (e.g. 1970) for more than 40 years. In 1975, a sample from this population was established in the laboratory and kept free of systematic inbreeding for five years. In 1980, ten populations were derived from this base population, five then selected for postponed aging (01-05) and five controls (B1-B5). The O stocks underwent selection for late fertility, while the B

stocks were selected for early fertility, as was the ancestral population. This type of selection is expected to generate increased O stock life spans and did (Rose, 1984). Recent data (unpub.) indicate a longevity increase in the O stocks of 60-80%, along with numerous other biological changes. A cytological point of note for the present study is that these populations largely lack inversion polymorphism, within or between populations (Tyler *et al.*, this volume). This makes it unlikely that allele frequencies change in parallel because of pervasive linkage disequilibrium ranging over many loci.

Two-dimensional gels and radiolabelling

Flies were labelled with [^{35}S]-methionine (specific activity 1151 Ci/mmole; 1 Ci = 37 GBq; New England Nuclear) for 18 hours at 25 °C as described in Fleming *et al.* (1988). This labelling procedure is limited in the number of proteins detected compared with silver staining, which could label one thousand or more proteins per gel. However, for this initial survey, it was decided to handle a smaller number of spots on the gels, to reduce the difficulty of scoring. The electrophoretic technique was essentially that of O'Farrell (1975), making use of the Anderson Iso-Dalt system (Anderson & Anderson, 1978). After the labelling period, 50 flies of each population were homogenized in 200 microliters of O'Farrell's lysis buffer A in a dounce homogenizer. The isoelectric focusing gel contained LKB ampholines (pH range 3.5-10, lot 99). Ten microliter samples (approximately 10^6 dpm) were applied under the electrolyte at the negative ends of the gels and then run at 500 V for 20 h. Each gel was loaded with the same amount of protein so as to minimize variation in spot migration due to protein loading, which is common for two-dimensional gel electrophoresis. The O populations took up slightly less label per milligram of protein than the B population, but relative spot intensity is adjusted by different exposure times of the film, as in Fleming *et al.* (1986). All first dimension samples were run simultaneously in the same apparatus under identical conditions. The voltage was then increased to 1000 V for an additional hour. Immediately after this, the gels were removed from the glass tubes and frozen in Anderson's equilibration buffer at -76 °C.

The second-dimension electrophoretic procedure was based on O'Farrell's original technique. All second-dimension gels were run simultaneously in the Anderson ten-gel Dalt tank using the Anderson Iso-Dalt system (Anderson & Anderson, 1978). These gels consisted of 10% acrylamide and 1% NaDodSO$_4$. After electrophoresis, gels were dried with a Hoeffer model SE1150 slab-gel dryer. The dried gels were exposed to a Kodak XAR-2 x-ray film. Three different exposures of these gels were carried out in order to score lower intensity spots. These exposure times were 3 days, 9 days, and 13 days. The resulting autoradiographs were developed with a Konica automatic film processor. The gel shown in Figure 1 is a nine-day exposure.

Gels were scored as previously described by Fleming *et al.* (1986). The autoradiograph from population B$_1$ was arbitrarily chosen as the master

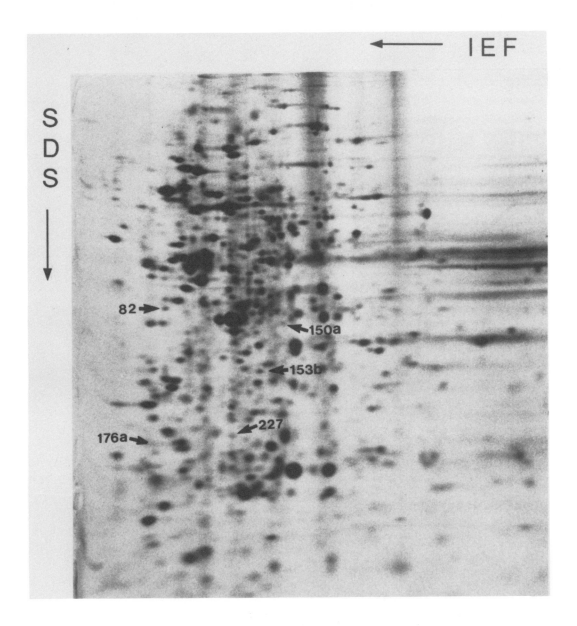

Fig. 1. Autoradiograph of a two-dimensional gel electrophoretic pattern of soluble [^{35}S]-methionine-labelled polypeptides from adult *Drosophila melanogaster*, strain B1. Several of the proteins that are significantly differentiated between the B and O stocks are indicated by their corresponding numbers from Table 1. IEF, isoelectric focusing dimension: SDS, sodium dodecyl sulfate, molecular weight dimension.

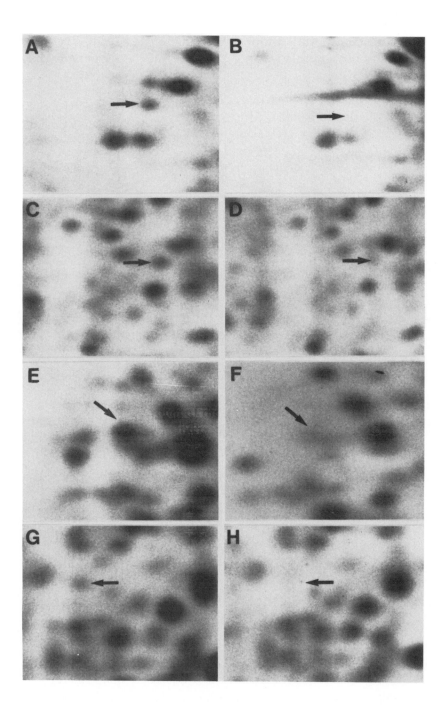

Fig. 2. Regions of the autoradiographs selected from the B and O stocks that show several of the proteins that are significantly differentiated. It is evident from these exposures that some of the differences may be quantitative. A, spot ('sp') 82, stock ('st') B1; B, sp 82, st 02; C sp 153b, st B1; D, sp 153b, st 05; E sp 176a, st B3; F, sp 176a, st 02; G, sp 227, st B1; H, sp 227, st 02.

gel. Each autoradiogram from the ten gels was then divided into nine equal sections and the protein spots were scored in each section for each gel beginning with the high molecular weight acidic proteins and ending with the low molecular weight basic proteins. Spots near the edge of the gel at the low molecular weight end were not scored. The protein scoring was carried out independently by three individuals in order to minimize variation due to any one individual's scoring.

Initially, all gels were compared for the presence or absence of spots relative to this gel. Relative quantitative differences were scored after the spots were numbered and the initial comparison of presence/absence status was made. For semi-quantitative comparisons, spots were given scores ranging from 0 to 5. A value of 3 represents the 'normal' abundance for a particular spot as determined from the master gel. Values of 4 and 5 were assigned to spots with apparently increased quantity, while values of 1 and 2 were assigned to spots with apparently decrease quantity, values 5 and 1 representing more extreme differences. A value of 0 means that a particular spot was not detected on that gel. The majority of spots were scored using the three-day exposure autoradiograph. Weaker spots were scored from the nine-day and thirteen-day exposures. This permitted accurate scoring for the weakest spots, where their presence was concerned. For the most abundant spots, the three-day exposure allowed the best scoring, because none of these spots were obscured by overexposure at that time, as they were in the nine-day or thirteen-day exposure autoradiograms. The nonlinearity of the X-ray film was not considered an important issue for the semi-quantitative comparisons, because only relative values were assigned to spots based on comparison with spot size on the master gel. These relative values were also assigned only when they were easily scored. Since the O populations incorporated less radiolabel than the B populations, it is possible that some of the proteins that received scores of zero in the O population might show up on longer exposures of the gel. However, as noted by Garrels (1979), quantitative differences of this magnitude can be considered significant qualitative differences in protein expression nonetheless. Illustrative spots are shown in Figure 2.

Results

A total of 321 abundant proteins were resolved and coded by number. The 109 that exhibited heterogeneity over populations were analyzed for patterns of differentiation among populations. Please note that, in population genetics generally, the term 'differentiation' refers to statistical separation of populations, not biological development or its effects.

Resampling analysis of quantitative scores
The problem facing any inference from these scores is the large amount of between-population differentiation within each of the B and O groups, because some differences between Bs and Os could thus be the result of chance alone. This problem can be alleviated by comparing the observed data with all possible resamplings of the data from a statistical universe consisting of the original gel scores rearranged in every possible combination of 'pseudo-Os' and 'pseudo-Bs', each a group of five populations. For example, one such arrangement would be B1, B3, B5, O1, and O2 in the first group of five populations, with B2, B4, O3, O4, and O5 in the second. There are 3,628,800 unique permutations of this kind. However, only membership in the two distinct groups of five matters, so the permutations reduce to 252 resampling classes, each consisting of 14,400 equivalent permutations.

The data for these 252 resamplings were analyzed for comparison with the particular pattern observed. Three different statistics were analyzed in this manner: (i) mean protein intensity score for each group of five populations, original or resampled; (ii) variance in protein intensity scores over the group of five populations, original or resampled; and (iii) the difference between the total protein intensity scores of the two groups of five populations, original or resampled. Each will be discussed in turn.

The resampling pattern for the mean protein intensity scores placed the B and O means at opposite extremes of the distribution, with the B mean (2.57) the highest case of all and the O mean (2.08) part of a doubleton tied for the lowest value. The original B populations and O populations have mean scores that seem unlikely to have been so extremely differentiated by chance alone, given that they are located at the two extremes of the resampling distribution. The B population is located as a singleton

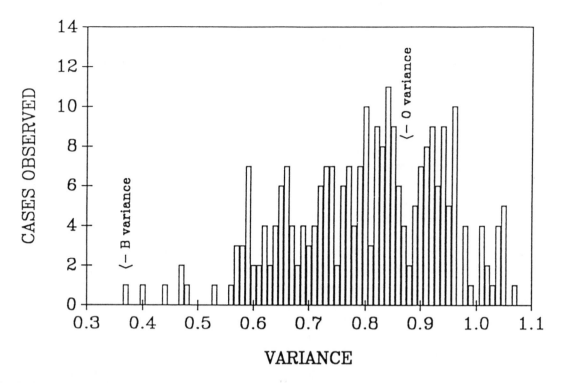

Fig. 3. The between-population variances of all 252 samplings of five populations from the original set of ten populations. The B and O sets of populations are included in this group of 252, and their between-population variances are indicated.

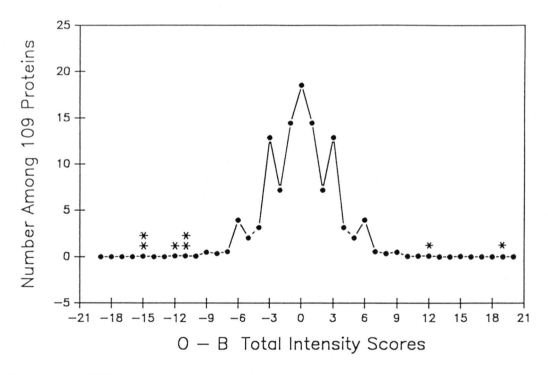

Fig. 4. The distribution of differences between total protein intensity scores of two groups of five populations each sampled from the original ten populations. All 252 possible resamplings were performed. These give the expectations for the number of extreme differences expected to arise by chance. The asterisks indicate seven proteins for which the observed difference between O and B total intensity scores is greater than 10. A difference this great or greater is expected to occur in less than one protein out of 109 polymorphic proteins in a random sample constructed from these same data. The occurrence of seven such proteins out of 109 polymorphic proteins is therefore not likely to be purely accidental.

at its extreme, while the O population is part of a doubleton. If there is just one population at either extreme, then the likelihood of those being the B and O populations is 2(1/252)(1/251), assuming uniformity of probability of occurrence over the distribution (this is a conservative assumption). If one of the extremes is a doubleton and the other a singleton, then the probability of B and O populations being located at the extremes is 2[1/252 + 1/251] (1/250), which is less than 0.0001. Therefore, the B populations have average protein intensity that is significantly greater than that of the O populations.

Figure 3 shows the distribution of average variances over the 109 proteins in the resampled data. The B populations have the lowest average variance, which has a probability of 1/252 of occurring by chance alone, assuming uniform likelihood over the distribution. However, the O populations are located well within the distribution of resampled variances. This seems to indicate that the B populations are particularly lacking between-populations, within-treatment heterogeneity, while the O populations are not.

Figure 4 shows the distribution of the total intensity scores over five pseudo-O populations minus the total pseudo-B protein score, which we will define as 'DT'. Extreme values of DT indicate proteins that may be associated with postponed aging. In the resampled data, the frequency of DT values greater than nine in magnitude is 0.0045, so nine constitutes a threshold for probabilities less than 0.01. Out of 109 proteins, 0.49 proteins would be expected to have this degree of differentiation, by chance alone. Seven proteins had DT values this extreme: 22a, 73a, 82, 150a, 153b, 227, and 230.

An alternative method of data analysis is to consider the resampling of these proteins individually, and then assign a probability to the observed protein differentiation pattern *per protein*. When this is done for the proteins that were found to be significant in the overall resampling analysis, the DT values of proteins 22a, 82, 153b, 227, and 230 have probabilities of less than 0.01 of occurring by chance, while the DT values of proteins 73a and 150a have probabilities of less than 0.05 of occurring by chance, in two-tailed tests. This qualitatively corroborates the results of the overall analysis, though the specific significance values are reduced in two cases.

One concern with these data is that these proteins may be differentiating as a result of overall differentiation in mean protein intensity scores. That is, there might be a global effect on protein intensities, rather than a protein-specific one. Of course, the global effect could alternatively be a side-effect of individual substitutions proceeding in the same direction. In any case, if the significance thresholds for the protein differentiation scores (DT) are moved by the 2.5 units required to allow for the difference in five individual B and O protein intensity differences, proteins 22a and 150a are no longer significantly differentiated at the 0.01 level. However, they are still significantly differentiated at the 0.05 level. In any case, the parsimony analysis given below also indicates that proteins 22a and 150a are statistically associated with the differentiation of B and O populations.

Combinatoric analysis

A different method of analysis can be used if the data are reduced to presence/absence scores of 1/0, respectively. Thus the present combinatoric analyses of protein scores ignore quantitative variation in spot size, given only that the protein is present. Note also that this analysis ignores the preceding resampling analysis, and therefore is not limited to considering the proteins already found to be differentiated.

There are exactly five proteins scored with five 1s and five 0s: 82, 111a, 200d, 225, and 227. Two out of five of these cases involve five Bs of state 1 and five 0s of state 0. The probability of either of these cases arising by chance, out of the 252 equally likely combinations of five 1s and five 0s, is 1/126, since there are only two possible combinations of five 1s and five 0s with this differentiation pattern. The probability of getting two or more cases like proteins 82 and 227 out of five is 0.00062. Therefore, proteins 82 and 227 are not likely to have differentiated by chance alone.

A similar calculation for proteins 73a, 150a, and 176a, which are only one spot away from a similar five-present/five-absent spot differentiation, indicates that the probability that they are differentiated by chance alone is 0.000416. But the parallel calculation for protein 200d, which is *two* spots away from five-present/five-absent differentiation, reveals that such a differentiation pattern has a probability of 0.669 of being observed by chance alone.

Table 1. Resampling analysis of the pattern of concordance between proteins in patterns of within-treatment, between-populations differentiation. The 'steps' indicate how many proteins along the ordered list of proteins, observed or shuffled, concordance was checked. Note that monomorphic proteins are included in the observed and resampled data, giving rise to most cases of concordance. In particular, this is responsible for the greater concordance among the B populations, since more of these populations exhibit an absence of change in protein score. The hypothesis test is for differences between observed and resampled mean number of concordants (*, $P < 0.05$; **, $P < 0.01$), so that *either* too much or too little concordance is detected.

Steps	Observed concordants	B Resampled Mean	Variance	Significance of difference	Steps	Observed concordants	B Resampled Mean	Variance	Significance of difference
1	238	227.5	6.95	**	293	14	20.59	5.10	**
2	231	226.88	6.667		294	14	19.76	4.56	**
3	228	226.11	6.78		295	13	19.16	4.74	**
4	227	225.28	7.45		296	13	18.47	4.66	*
5	224	224.66	7.54		297	14	17.66	4.40	
6	223	223.84	7.39		298	13	17.03	4.09	*
7	222	223.18	7.39		299	11	16.32	4.30	**
8	224	222.58	7.62		300	10	15.55	4.03	**
9	226	221.85	7.29						
10	222	221.20	7.91		1	176	158.58	15.89	**
11	226	220.32	8.29	*	2	167	158.20	15.41	*
12	223	219.66	8.21		3	172	157.67	15.78	**
13	227	219.18	9.20	*	4	170	157.16	16.52	**
14	225	218.33	8.03	*	5	165	156.64	17.69	*
15	223	217.48	8.57		6	167	156.12	16.57	**
					7	162	155.84	16.51	
145	123	125.41	6.25		8	169	155.07	16.61	**
146	123	124.73	5.93		9	161	154.53	17.18	
147	123	124.22	6.17		10	158	154.25	17.36	
148	124	123.27	5.33		11	161	153.67	15.64	*
149	121	122.59	5.42		12	159	153.13	17.38	
150	122	121.76	5.18		13	160	152.68	16.83	*
151	120	121.22	5.20		14	166	152.06	18.54	**
152	119	120.47	4.89		15	154	151.86	17.35	
153	120	119.87	4.47						
154	119	119.10	4.80		145	89	87.71	11.49	
155	120	118.35	4.26		146	88	86.91	12.24	
					147	85	86.57	10.47	
245	51	54.78	9.01		148	88	86.01	11.40	
246	48	53.86	9.37	*	149	85	85.53	9.81	
247	49	53.20	9.49		150	83	85.01	9.86	
248	46	52.44	9.16	*	151	83	84.35	10.00	
249	49	51.69	8.86		152	83	84.03	10.01	
250	47	51.08	8.42		153	87	83.55	9.45	
251	46	50.25	8.49		154	88	82.94	9.90	
252	45	49.40	8.19		155	79	82.79	9.66	
253	43	48.91	9.39	*					
254	43	48.12	9.33		245	31	38.20	11.58	*
255	42	47.46	8.95	*	246	30	37.59	11.81	*
					247	28	36.87	11.17	**
285	21	26.24	5.80	*	248	29	36.50	11.64	*
286	20	25.49	5.80	*	249	30	36.06	11.93	
287	20	24.75	5.50	*	250	25	35.36	11.55	**
288	18	24.19	5.40	**	251	26	35.14	11.15	**
289	17	23.48	5.80	**	252	27	34.40	11.74	*
290	16	22.61	5.04	**	253	25	34.23	10.97	**
291	16	21.90	5.68	*	254	25	33.51	11.87	*
292	16	21.33	5.32	*	255	22	32.99	10.73	**

Table 1. Continued.

Steps	Observed concordants	B Resampled Mean	Variance	Significance of difference
285	13	18.34	8.30	
286	10	17.70	6.73	**
287	11	17.32	7.60	*
288	9	16.89	7.29	**
289	8	16.40	7.48	**
290	9	15.75	6.57	**
291	7	15.25	6.32	**
292	6	14.93	6.88	**
293	5	14.27	6.27	**
294	6	13.80	6.10	**
295	4	13.20	5.77	**
296	4	12.93	5.50	**
297	2	12.34	5.60	**
298	3	11.94	4.90	**
299	2	11.38	4.79	**
300	1	10.43	5.09	**

Thus there are only five proteins that have significant differentiation *when scoring is by presence/ absence only*, and the data can be analyzed combinatorially: 73a, 82, 150a, 176a, and 227.

Valid inference of B/O differentiation depends on comparisons of multiple populations of each type, as opposed to comparisons of single populations of each type. Five wholly independent comparisons of this kind can be made with the presence/ absence scoring, each comparison using one of the five B_i/O_i pairs. When this is done, comparisons of single pairs over all 321 proteins yield an average of 21 proteins for which differentiation in single pairs is not sustained over the ten populations, 4.4 cases for which the single pair is an adequate guide to the general pattern, and 0.6 proteins where there is overall differentiation, but not in the pair examined. The reason these disparities arise is the differentiation of populations within the selected and control groups, that is, the within-treatment variance.

The resampling and combinatoric methods of analysis give two different groups of proteins that could be involved in postponed aging: 22a, 73a, 82, 150a, 153b, 227, 230 and 73a, 82, 150a, 176a, 227, respectively. Of the second group, only 176a is absent from the first. However, the DT score for 176a is nine, for which the resampling probability is only about 0.05. The first-group proteins missing

from the second group, 22a, 153b, and 230, consist of cases with only two or three populations lacking the protein, cases that have high probability of occurring by chance when the data are reduced to presence/absence scores.

Effects of proximity on differentiation
One of the natural concerns facing any analysis of data from two-dimensional protein electrophoresis is the degree to which the protein spots represent genetically interdependent phenotypes. That is, perhaps differentiation detected in one spot is genetically associated with that of another spot, so that both are only markers of one underlying genetic change.

We tested for this possibility using protein intensity scores from the full set of 321 protein spots. Again, the data were extensively resampled, but in this case resampling shuffled the data by rows, so that different proteins were placed in juxtaposition in the table of intensity scores. We then tested how often adjacent protein scores matched over populations of a given type, B or O, in both the original data and the resampled. One thousand resamplings, obtained by 321 shufflings each, were calculated; the total number of possible combinations was 321!, since order mattered, an astronomical number. The proteins were originally labelled in rough sequence, so that protein spots 1 and 2 are near each other, while 1 and 200 are far apart. The most obvious test is to check whether adjacent proteins from the original data are more likely to resemble each other than randomly shuffled data. In both B and O data, immediately adjacent protein scores resemble each other more than they do in any of the shuffled cases. (See Table 1) This seems to indicate that there may be some genetic associations between adjacent proteins in the original data.

But Figure 5 reveals that the actual pattern is more complex than that. The graph displays the pattern of correspondence between proteins a variable number of steps away from each other in the data table. Proteins that are within 100 steps of each other generally resemble each other more than expected by chance, while proteins 200 to 300 steps apart resemble each other less than expected by chance. Table 1 indicates the outcome of statistical testing from the resampling.

The obvious interpretation of these results is that the areal resemblance effect is a result of variation

208

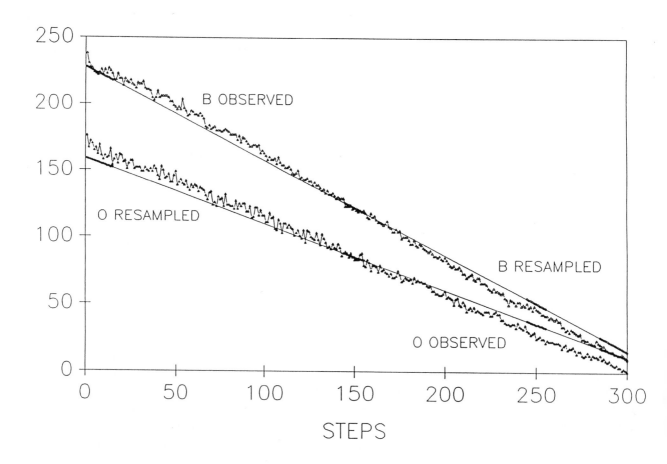

Fig. 5. The number of concordant patterns of between-population differentiation among B and O stocks. The thick solid line gives the observed number in the B stocks. The dashed line gives the observed number in the O stocks. The boxed and starred segments give the areas of resampling, as outlined in the text. The thin solid lines give the interpolated resampled pattern between regions actually resampled. Evidently, the observed patterns of concordance indicate that nearby proteins tend to resemble each other in between-population differentiation more than expected by chance, with distant proteins tending to resemble each other less. See Table 2 for numerical detail.

in autoradiographic intensities during autoradiogram preparation. There are several factors that could be at work: different gel thicknesses, variation in proximity of the photographic plate to the gel, and so on. In combination, they appear to make nearby protein intensities resemble each other. In terms of scientific inference, effects like these would constitute sources of noise that would obscure the detection of differences between populations, since they would vary within treatments, and thus obscure between-treatment differentiation. But such effects would not lead to spurious inferences of consistent differentiation.

Tree analysis

The data were analyzed using both phenetic and

parsimony approaches. The phenetic analyses were conducted using the NTSYS-pc programs of Rohlf (1990), and the parsimony analyses were performed using the PAUP program of Swofford (1989).

The presence/absence data were analyzed phenetically by calculating a genetic distance. It is difficult to homologize loci among populations using two-dimensional electrophoresis, so the generally employed genetic distance methods could not be employed. Instead, we used the association coefficient $F = 2n_{xy} / (n_x + n_y)$, where n_x and n_y are the number of protein spots scored for populations x and y, respectively, and n_{xy} is the total number of shared protein spots between populations x and y (Aquadro & Avise, 1981; cf. Dice, 1941; Sokal &

Table 2. Genetic similarity and distance values calculated from the 321 protein spots scored in the two-dimensional electrophoretic study. The values below the diagonal are the genetic similarity values (F), which are derived from the Dice association coefficient, and the values above the diagonal are the genetic distance values (–ln F). See text for more detail concerning the calculation of these genetic measures.

	B1	B2	B3	B4	B5	O1	O2	O3	O4	O5
B1	–	.0195	.0212	.0261	.0279	.0434	.0330	.0544	.0712	.0528
B2	.9807	–	.0081	.0197	.0214	.0369	.0231	.0443	.0573	.0426
B3	.9791	.9919	–	.0147	.0198	.0421	.0182	.0391	.0592	.0410
B4	.9742	.9805	.9854	–	.0148	.0508	.0266	.0479	.0646	.0428
B5	.9724	.9788	.9804	.9853	–	.0423	.0250	.0464	.0560	.0377
O1	.9575	.9638	.9588	.9505	.9585	–	.0304	.0451	.0546	.0541
O2	.9675	.9771	.9820	.9738	.9753	.9701	–	.0238	.0543	.0326
O3	.9470	.9567	.9616	.9532	.9546	.9559	.9764	–	.0702	.0404
O4	.9313	.9444	.9426	.9374	.9456	.9468	.9472	.9322	–	.0611
O5	.9486	.9583	.9599	.9581	.9630	.9474	.9680	.9604	.9408	–

Sneath, 1963; Sneath & Sokal, 1973). To convert this association coefficient into a genetic distance, we have used a logarithmic transformation to linearize the distance measure, to give –ln F. This genetic distance is analogous to Nei's (1972) genetic distance, and has the same interval. The present genetic distance measure was then clustered using UPGMA (Sokal & Sneath, 1963), with the 'Find' option to detect ties.

The F and –ln F values are given in Table 2. The mean genetic similarity among the ten populations is 0.962, with a range of 0.931 to 0.992. However, there is a substantial difference between the B and O groups in the amount of genetic differentiation. The B lines show a within-group mean genetic similarity of 0.981, but the O lines, with a mean of 0.954, show much greater within-group genetic distances. Figure 6 gives the phenogram resulting from the UPGMA clustering of values given in Table 2. The cophenetic correlation for this tree is 0.923, which is in the very good fit range (Rohlf, 1990).

One problem with the phenetic analysis just presented is that it requires a uniform rate of change over populations, an unlikely assumption given the known generation numbers of B and O populations, as well as the disparity in genetic distances within the B and O groups just presented. Accordingly, a parsimony analysis of these data was also performed. The analysis was performed using the branch-and-bound algorithm, which guarantees that the most parsimonious tree(s) will be found (Hendy & Penny, 1982). Both the presence/absence

data and the multistate data were analyzed, the latter with the character states ordered. To assess confidence limits for the branching patterns, a bootstrap analysis was also performed (Felsenstein, 1985; Sanderson, 1989). A total of 300 replications were performed using the branch-and-bound algorithm. The result is presented as a majority-rule consensus tree (Margush & McMorris, 1981), which shows the most frequently occurring branching orders.

The presence/absence parsimony analysis produced three equally most parsimonious trees, which together define the strict consensus tree given in Figure 7a. These three trees have length 105, with

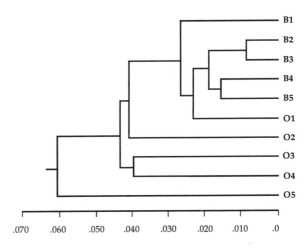

Fig. 6. Phenetic analysis. Phenogram produced when the binary data are analyzed by clustering the genetic distance values of Table 2 with UPGMA. The cophenetic correlation is 0.923.

210

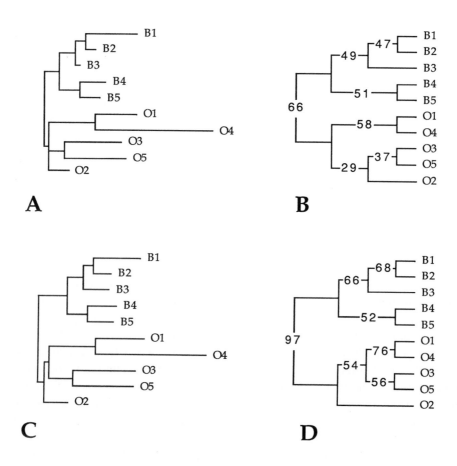

Fig. 7. Parsimony trees. A. Strict consensus tree based on the three most parismonious trees for the binary data set. Branches are drawn proportional to their length. B. Majority rule consensus tree from the bootstrap of the binary data set. The numbers on the nodes represent the frequency of occurrence. C. Most parsimonious tree from the ordered multistate data. Branches are drawn proportional to their length. D. Majority rule consensus tree from the bootstrap of the ordered multistate data. The numbers on the nodes represent the frequency of occurrence.

an overall consistency index of 0.667 and a consistency index excluding uninformative characters of 0.527. The bootstrap majority-rule consensus tree is topologically identical to one of the most parsimonious trees, and is shown in Figure 7b. Note that only one of the bifurcations, that between B and O lines, has a confidence level greater than 65%, though the 66% confidence level of that branching does not inspire particularly great certainty. In any case, the other branchings are of such low occurrence that little confidence can be placed in them. Table 3 gives the proteins that are important in determining the node separating the B group from the O group. Three different optimization methods

were used to calculate this node: minimum F-value, accelerated transformation, and delayed transformation (Swofford & Maddison, 1987). With three different calculation methods and three equally parsimonious trees, the number of trees in which a given protein can contribute to the resolution of B and O groups ranges between one and nine. The ordered multistate data produce a unique most parsimonious tree, given in Figure 7c. The tree has a length of 379, with an overall consistency index of 0.675 and a consistency index excluding uninformative characters of 0.595. The bootstrap majority-rule consensus tree is topologically identical to the most parsimonious tree, which is frequently not the

Table 3. Protein spots that define the separation of the B and O groups of flies as determined by the parsimony analysis of the binary data set. Spots are listed in order of their importance. See text for more details.

Protein spot number	Number of trees present	Optimiza-tion[1]	Consistency index
82	9	F, A, D	1.000
227	9	F, A, D	1.000
176A	9	F, A, D	.500
73A	7	F, A, D	.500
150A	3	D	.500
22A	3	A	.333
168	3	A	.333
200D	2	D	.333

[1] F = minimum f-value transformation optimization; A = Accelerated transformation optimization; D = Delayed transformation optimization.

case (Felsenstein, 1985; Sanderson, 1989), and is shown in Figure 7d. The split between the B and O populations arises 97% of the time. However, once again the levels of significance for the rest of the tree are much lower, only three of the six branches

Table 4. Protein spots that define the separation of the B and O groups of flies as determined by the parsimony analysis of the multistate data set. Spots are listed in order of their importance, as defined by their consistency index. See text for more details.

Protein spot number	Number of trees present	Optimiza-tion (see Table 3)	Consistency index
54	3	F, A, D	1.000
82	3	F, A, D	1.000
227	3	F, A, D	1.000
230	3	F, A, D	1.000
20c	1	D	0.750
45c	3	F, A, D	0.750
153b	3	F, A, D	0.667
150a	1	D	0.600
176a	3	F, A, D	0.600
73a	3	F, A, D	0.500
155a	3	F, A, D	0.500
181a	3	F, A, D	0.500
197b	3	F, A, D	0.500
209a	3	F, A, D	0.500
18	3	F, A, D	0.429
22a	3	F, A, D	0.429
198a	1	A	0.375
200d	1	D	0.375
168	1	A	0.333

exceeding the 65% confidence level. Table 4 gives the proteins involved in the B and O group split from the multistate tree analysis. Recall that there is only one most parsimonious tree per calculation method, so the three different calculation methods give a maximum of three trees on which a protein can be placed at the B-O node.

Discussion

Resolving proteins that postpone aging

One of the elementary questions that can be asked of the present study is whether or not it indicates those loci or proteins that are causally involved in postponing aging. The answer to this can only be somewhat equivocal, partly because of the different results obtained with different methods and partly because of fundamental population-genetic limitations.

The resampling study delimited a set of seven proteins that are statistically associated with the differentiation of stocks having postponed aging, based on the total score for B and O groups: 22a, 73a, 82, 150a, 153b, 227, and 230. Combinatoric analysis of presence/absence data found the following proteins: 73a, 82, 150a, 176a, and 227. Protein 176a would also have been regarded as significantly different in the resampling analysis at a significance level of about 0.05. Thus the combinatoric analysis essentially reveals a subset of the proteins detected in the resampling analysis of the multistate data. Tables 4 and 5 give the proteins found to differentiate B and O groups in the tree analyses. Ordered by position number, Table 3 gives proteins 22a, 73a, 82, 150a, 168, 176a, 200d, and 227. The only two new proteins in this list are 168 and 200d, though it is notable that these are among the lowest ranked proteins in Table 3, with respect to both incidence over maximum parsimony trees and consistency index. The DT scores are –9 and –7 for proteins 168 and 200d, respectively, the former being marginal for statistical significance in the resampling analysis. Table 4 gives a total of 19 proteins, the ones not already listed being 18, 20c, 45c, 54, 155a, 181a, 197b, 198a, and 209a. It is noteworthy that Table 4 includes all ten proteins identified by the other methods.

Thus there are four different lists of proteins associated with B and O stock differentiation. All four of these lists indicate the involvement of proteins 73a, 82, 150a, 176a, and 227. If one relaxes stringency to presence in three out of the four lists, then protein 22a should be added. If two out of four lists is regarded as sufficient, then proteins 153b, 168, 200d, and 230 are to be added to those involved in differentiation. For now it seems most prudent to confine further consideration to proteins delineated by at least three out of four methods of data analysis: 22a, 73a, 82, 150a, 176a and 227.

Statistical association between protein differences and B-O differentiation does not necessarily imply causal involvement. One point is that there is a non-trivial possibility that one or two statistically associated proteins are only associated by chance, this being inherent in any statistical testing procedure. In addition, other detected proteins might not be causally involved in the physiological changes postponing aging, instead being products of alleles in tight linkage disequilibrium with alleles at the loci causally involved in the response to selection. In this case, the alleles detected here would at least be a guide to the location of causally important alleles. Another possibility is that differentiated proteins could be undergoing changes as a result of the effects of genetic changes at other loci that then give rise to modification of the detected protein at some point after transcription of the protein's gene sequence. Such modification could be a pleiotropic effect of the original genetic change which is unrelated to the effect of that change upon aging.

However, we would argue that while there may be reasonable doubt concerning the causal involvement of any of the proteins adduced here, what has been accomplished is a considerable reduction in the number of candidate proteins worthy of further investigation. That is to say, further tests of six proteins is a considerably more feasible project than further tests of 321 proteins. This is not to say that only these six proteins are of interest. Undoubtedly our methods have failed to identify numerous other loci that associated with postponed aging in our longer-lived stocks.

Pattern of evolution

One of the interesting features of the data is that there is an abundance of variance between populations within selection regimes. Part of this variation probably arises from 'regional' autoradiogram intensity variation, as revealed by the correlation between protein intensities over regions (Fig. 5). However, the variation between B populations is strikingly less than that between O populations, as shown in Figure 3, suggesting that genuine biological differences play a role in determining this type of variation. Evidently, the O populations have differentiated from each other over the course of the experiment more than B populations have. Yet the O populations have undergone many fewer generations, about 50 O generations versus more than 200 B generations at the time the samples were obtained. Thus a simple drift hypothesis is not an obvious explanation for these results, unless the O populations have had sufficiently lower effective population sizes than the B populations, which is not known at present. An alternative hypothesis might be one of heterogeneous responses to selection in the different O populations. If this has in fact arisen, then some of the inconsistent differentiation of the O populations may be due to the heterogeneous responses of particular loci in individual populations to selection for postponed aging, responses that are not general over all O populations. However, the present data can do little to test the validity of this conjecture, which must await the analysis of specific proteins in greater detail than the present study.

Number of loci

One of the more interesting questions in research on the genetics of aging is the number of loci that control aging. Previous work in *Drosophila* has used the segregation indices for F_1, F_2, and back-cross progeny (e.g. Luckinbill *et al.*, 1987; Hutchinson & Rose, 1990). Such studies appear to have had little statistical power, and cannot distinguish the number of loci involved from infinitely many loci.

The present data also provide an extremely crude perspective on this question. Suppose that each of the six robustly differentiated proteins is so differentiated because of genetic differentiation at a distinct locus. This may be a linked locus or it may be a locus that modifies the transcription or post-synthetic alteration of the labelled protein. Thus, this locus does not have to be the locus coding for the primary sequence of the protein. Granting this assumption of genetic distinctness, then our findings suggest that 6 out of about 300 loci were involved in postponing aging, or 2%. If the genome size is

10,000 to 20,000 transcribed loci in *Drosophila*, and our loci are a representative sample, then the total number of loci that can postpone aging in these flies is about 200 to 400. The assumptions used to obtain these numbers are of course speculative, and we do not wish to defend them. However, these results at least suggest the possibility that on the order of 10^2 loci may be involved in the control of aging in *D. melanogaster*. This further suggests the necessity of methods that can identify loci in the face of considerable polygenic variation, where the genetics of aging are concerned.

Acknowledgements

We are grateful to W. M. Fitch, R. R. Hudson, R. E. Lenski, and L. D. Mueller for suggestions concerning the analysis of the data. We thank S. Dinan and M. Hartsuck for technical assistance. This research was supported in part by a US Public Health Service grant (AGO6346) to MRR and donations provided to the Linus Pauling Institute of Science and Medicine.

References

Anderson, N. G. & N. L. Anderson, 1978. Analytical techniques for cell fractions. XXI. Two-dimensional analysis of serum and tissue proteins: multiple isoelectric focusing. Anal. Biochem. 85: 331-340.

Aquadro, C. F. & J. C. Avise, 1981. Genetic divergence between rodent species assessed by using two-dimensional electrophoresis. Proc. Natl. Acad. Sci. 78: 3784-3788.

Bozcuk, A. N., 1981. Genetic longevity in *Drosophila*. V. The specific and hybridized effect of *rolled*, *sepia*, *ebony* and *eyeless* autosomal mutants. Exp. Gerontol. 11: 103-112.

Clare, M. J. & L. S. Luckinbill, 1985. The effects of gene-environment interaction on the expression of longevity. Heredity 55: 19-29.

Dice, L. R., 1941. Measures of the amount of ecologic association between species. Ecology 26: 297-302.

Felsenstein, J., 1985. Confidence limits on phylogenies: an approach using the bootstrap. Evolution 39: 783-791.

Fleming, J. E., P. S. Melnikoff & K. G. Bensch, 1984. Identification of mitochondrial proteins on two-dimensional electrophoresis gels of extracts of adult *Drosophila melanogaster*. Biochim. Biophys. Acta 802: 340-345.

Fleming, J. E., E. Quaottrocki, G. Latter, J. Miquel, R. Marcuson, E. Zuckerkandl & K. G. Bensch, 1986. Age-dependent changes in proteins of *Drosophila melanogaster*. Science 231: 1157-1159.

Fleming, J. E., J. K. Walton, R. Dubitsky & K. G. Bensch, 1988. Aging results in an unusual expression of *Drosophila* heat shock proteins. Proc. Nat. Acad. Sci. USA 85: 4099-4103.

Garrels, J., 1979. Two-dimensional gel electrophoresis and computer analysis of proteins synthesized by clonal cell lines. J. Biol. Chem. 254: 7961-7977.

Hendy, M. D. & D. Penny, 1982. Branch and bound algorithms to determine minimal evolutionary trees. Math. Biosci. 59: 277-290.

Hutchinson, E. W. & M. R. Rose, 1987. Genetics of aging in insects. Rev. Biol. Res. Aging 3: 62-70.

Hutchinson, E. W. & M. R. Rose, 1990. Quantitative genetic analysis of postponed aging in *Drosophila melanogaster*, p. 66-87 in Genetic effects on aging II, edited by D. A. Harrison. Telford, Caldwell, N.J.

Hutchinson, E. W. & M. R. Rose, 1991. Quantitative genetics of postponed aging in *Drosophila melanogaster*. I. Analysis of outbred populations. Genetics 127: 719-727.

Ives, P. T., 1970. Further studies of the South Amherst population of *Drosophila melanogaster*. Evolution 24: 507-518.

Johnson, T. E., 1987. Aging can be genetically dissected into conponent processes using long-lived lines of *Caenorhabditis elegans*. Proc. Natl. Acad. Sci. USA 84: 3777-3781.

Johnson, T. E. & E. W. Hutchinson, 1990. Aging in *Caenorhabditis elegans*: Update 1988. Rev. Biol. Res. Aging 4: 15-27.

Luckinbill, L. S., M. J. Clare, W. L. Krell, W. C. Cirocco & S. A. Buck, 1987. Estimating the number of genetic elements that defer senescence in *Drosophila melanogaster*. Evolution 38: 996-1003.

Luckinbill, L. S., T. A. Grudzien, S. Rhine & G. Weisman, 1989. The genetic basis of adaptation to selection for longevity in *Drosophila melanogaster*. Evol. Ecology 3: 31-39.

Luckinbill, L. S., J. L. Graves, A. H. Reed & S. Koetsawang, 1988a. Localizing the genes that defer senescence in *Drosophila*. Heredity 60: 367-374.

Luckinbill, L. S., J. L. Graves, A. Tomkiw & O. Sowirka, 1988b. A qualitative analysis of some life-history correlates of longevity in *Drosophila melanogaster*. Evol. Ecol. 2: 85-94.

Maynard Smith, J., 1966. Theories of aging, p. 1-35 in Topics in the biology of aging, edited by P. L. Krohn. Interscience, New York.

Margush, T. & F. R. McMorris, 1981. Consensus n-trees. Bull. Math. Biol. 43: 239-244.

Nei, M., 1972. Genetic distances between populations. Am. Nat. 106: 283-292.

O'Farrell, P. H., 1975. High resolution two-dimensional electrophoresis of proteins. J. Biol. Chem. 250: 4007-4021.

Ohnishi, S., M. Kawanishi & T. I. Watanabe, 1983. Biochemic phylogenies of *Drosophila*: protein differences detected by two-dimensional electrohoresis. Genetica 61: 55-63.

Parker, J., J. Flanagan, J. Murphy & J. Gallant, 1981. On the accuracy of protein synthesis in *Drosophila melanogaster*. Mech. Age. Dev. 16: 127-139.

Rohlf, F. J., 1990. NTSYS-pc. Numerical taxonomy and multivariate analysis system (ver. 1.60). Exeter Pub., Setauket, New York.

Rose, M. R., 1984. Laboratory evolution of postponed senescence in *Drosophila melanogaster*. Evolution 38: 1004-1010.

Rose, M. R., 1989. Genetics of increased lifespan in *Drosophila*. Bioessays 11: 132-135.

Rose, M. R., 1991. Evolutionary biology of aging. Oxford University Press, New York.

Rose, M. R. & B. Charlesworth, 1981. Genetics of life-history in *Drosophila melanogaster*. II. Exploratory selection experiments. Genetics 97: 187-196.

Rose, M. R., M. L. Dorey, A. M. Coyle & P. M. Service, 1984. The morphology of postponed senescence in *Drosophila melanogaster*. Can. J. Zool. 62: 1576-1580.

Sanderson, M. J., 1989. Confidence limits on phylogenies: the bootstrap revisited. Cladistics 5: 113-129.

Service, P. M., E. W. Hutchinson, M. D. MacKinley & M. R. Rose, 1985. Resistance to environmental stress in *Drosophila melanogaster* selected for postponed senescence. Physiol. Zool. 58: 380-389.

Service, P. M., 1987. Physiological mechanisms of increased stress resistance in *Drosophila melanogaster* selected for postponed senescence. Physiol. Zool. 60: 321-326.

Sneath, P. H. A. & R. R. Sokal, 1973. Numerical taxonomy. W. H. Freeman and Co., San Francisco.

Sokal R. R. & P. H. A. Sneath, 1963. Principles of numerical taxonomy. W. H. Freeman and Co., San Francisco.

Swofford, D. L., 1990. PAUP. Phylogenetic analysis using parsimony (ver. 3. On). Ill. Nat. Hist. Surv., Champaign, Illinois.

Swofford, D. L. & W. P. Maddison, 1987. Reconstructing ancestral character states under wagner parsimony. Math. Biosci. 87: 199-229.

PART FOUR
Aging in mammals

Introduction

How many biologists have stared into the mirror on their sixtieth birthdays and decided that they had better start doing research on aging? This decision may be renounced by the end of the day's celebrations, but the facts of mammalian aging are more intimately known to biologists, and others, than those of any other group of animals. The overwhelming majority of funding for aging research arises, directly or indirectly, from a concern about human aging. And thus the vast majority of publications in the aging field concern mammals, or at least their cells.

One of the most promising of all research strategies used with mammalian aging has been dietary, or caloric, restriction: giving rodents less food, but sufficient vitamins and the like. This line of work is represented by Holliday, in passing, and at greater length by Richardson and Pahlavani. Of course, one of the biggest problems with 'DR' research is what it all means, and both of the articles here address this question. Holliday also gives a somewhat cinematic view of the evolution of mammalian life-histories, including aging.

The remaining articles attempt various genetic dissections of rodent and human aging. Pleasingly closing the circle, the articles by Martin and Albin return to broad evolutionary issues first raised in the initial articles of this volume. Although the mechanisms of antagonistic pleiotropy and mutation-accumulation arise sporadically throughout this collection, having them surface finally and dramatically with Albin's essay on human genetic variation indicates just how interconnected, sometimes even unified, aging research has become. Perhaps Janus would be pleased.

M. R. Rose and C. E. Finch (eds), Genetics and Evolution of Aging, 217–225, 1994.
© 1994 *Kluwer Academic Publishers. Printed in the Netherlands.*

Longevity and fecundity in eutherian mammals

Robin Holliday
CSIRO Division of Biomolecular Engineering, Sydney Laboratory, PO Box 184, North Ryde, NSW 2113, Australia

Key words: longevity, fecundity, eutherian mammals

Abstract

The disposable soma theory of the evolution of aging and longevity proposes that there is a trade-off between investment of metabolic resources into reproduction and into the maintenance and lifespan of the soma. Thus, within major taxonomic groups fecundity should be inversely related to longevity. This was examined for 47 representative genera of eutherian mammals, for which satisfactory data is available. The prediction of the theory is strongly confirmed. Most short lived mammals have a reproductive potential of 40 or more offspring, whereas the longest lived genera can produce only about 10 offspring, if infant mortality is disregarded. However, some primates which breed slowly have relatively short lifespans, which in part can be attributed to low mortality of offspring. During the evolution of primates it seems that the rate of development to adulthood gradually decreased and this is correlated with increased longevity. Some environmental influences which reduce fecundity, such as low calorie intake, also increase longevity. This may be an adaptation to maximise reproduction when food is available in a natural environment, and to preserve the soma during periods of partial starvation.

Introduction

The early evolution of metazoa gave rise to plants and animals in which the distinction between germ-line and somatic cells was not clear cut. This is true for many plants today, since vegetative cells can give rise to a whole new plant with the circumvention of sexual reproduction. Clonal vegetative proliferation produces cell lineages which apparently do not age. Similarly, simple animals such as flatworms have pools of totipotent cells which can replace all differentiated cells. This regenerative ability allows animals to survive indefinitely without sexual reproduction (Child, 1915; Sonneborn, 1930). Ageing in metazoan animals probably arose when post-mitotic cells could no longer be replaced by totipotent cells. This could have occurred quite early in evolution, as simple animals such as nematodes have a finite lifespan, and can only survive by sexual reproduction. In such organisms, the distinction between germ-line and soma becomes very clear cut. Already, at this early stage of evolution, there must be some balance between the survival time of the soma, and the production of offspring which will perpetuate the species. A crucial question is therefore the relationship between longevity and fecundity. The 'disposable soma' theory of the evolution of ageing proposes that there is a trade off between investment of metabolic resources in maintaining the integrity of the soma, and the investment in reproduction (Kirkwood, 1977, 1981; Kirkwood & Holliday, 1979, 1986; Holliday, 1994).

The factors which determine this balance or trade off are obviously highly complex, and the ecological niche which the organism inhabits is a crucial component. There is clearly no simple relationship between survival of the soma and reproduction. For example, many fish are long lived, and produce very large numbers of potential offspring. In this case, an extremely small proportion of the fertilised eggs survive to adulthood. The reverse is seen in other cases. For example, some parasitic wasps lay few eggs on their paralysed prey, because the food resource is limited and allows only a few larvae to develop to adulthood. Yet among insects, these wasps are not especially long-lived. In this example, the larvae are provided with a source of food secured in a burrow, and there is a high probability of development to adults. In view of the variety of life styles and environments, it might be thought that generalisations about fecundity and longevity in animals are not feasible. Yet this is not necessarily the case, if species of one taxonomic group are considered. Birds and mammals have a considerable range of longevities, and homothermy to some extent standardises their metabolism in a variety of environments. The disposable soma theory of the evolution of ageing predicts that in these groups there should be an inverse relationship between fecundity and longevity. Although this analysis could be attempted with avian species, the existing data, especially longevity data, are more reliable for eutherian mammalian species. (Marsupial mammals are excluded, because their reproduction is very different.) Even so, there are a number of uncertainties in estimating the real reproductive potential of various species, and longevities are often based on very small numbers of captive individuals. In this review, I select those species or genera for which the data seem reasonably certain. The major sources of information are Adsell's Patterns of Mammalian Reproduction (Hayssen *et al.*, 1993), Walker's Mammals of the World (5th ed. Novak, 1991) and M.L. Jones (1982 and personal communication).

Reproductive potential of eutherian mammals

The major parameters which are considered are (1) the time from birth of females to reproductive maturity; (2) the gestation period; (3) the litter size; (4) the interval between litters. One major uncertainty is the effect of the loss of offspring on the time interval to the subsequent birth(s). It is common for females who are suckling young to be non-fertile until the offspring have reached a certain age. If the offspring die from disease or predators, then fertility is brought forward. Thus, infant mortality is related to fecundity or reproductive potential. In the context of this discussion, the important parameter is the reproductive potential which could give rise to breeding adults of the next generation. Therefore infant mortality is ignored, and the interlitter interval is based on the supposition that in each case offspring are reared. Another uncertainty relates to climatic variability. There are several examples of mammals breeding once a year in temperate or cold climates, whereas in warm or tropical climates there may be two breeding cycles. A major problem in estimating reproductive potential is the decline of fertility of females with age. In human females, the menopause is the end of fertility, but in most species there is simply a gradual decline in fertility with age. The rate of this decline almost certainly relates to the survival of young offspring. Females which persistently lose offspring are more likely to breed for a considerable proportion of their lifespan, than females which successfully raise many litters. It is probable that rapidly breeding species, including rodents, lose fertility more quickly, in relative terms, than do species which breed slowly. In any case, the data relating fertility to age are quite sparse, especially for animals in their natural environments. Therefore, a rule of thumb assumption is made that females breed from sexual maturity for two thirds of their lifespan. It goes without saying that females invest more in reproduction than males, so their fecundity is the measure of the reproductive potential of the species in question.

In the survey an attempt is made to choose representative species or genera. The sample of 47 includes 10 rodents, 10 herbivores, 10 carnivores, 7 primates, 3 Chiroptera (bats) and representatives from some other orders. Usually only one representative of closely related species is chosen (e.g. large cats, great apes, sheep and goats). Although there is much data for domestic species, it is preferable to choose their natural ancestors, where possible. For example, the domestic pig has been selected for high fertility and the wild pig has a smaller litter size. In several cases, fertility or longevity data came from more than one species in a genus, therefore 47 genera rather than species have been selected. The data are listed in Tables 1–3, with the

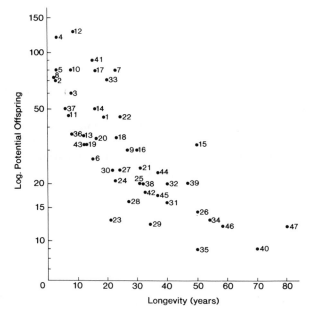

Fig. 1. The relationship between the reproductive potential and longevity of 47 genera of eutherian mammals, based on the data in Tables 1–3. The numbers indicate the genus listed in column 2 of Tables 1–3. Reproductive potential is the maximum number of offspring which might be produced under ideal conditions. Longevity is based primarily on the maximum lifespan of limited numbers of individuals in captivity (see text).

numbering for Figure 1. In estimating the maximum reproductive potential, the important parameters are the period of fertility (67% of female adult lifespan), the litter size and the litter intervals.

Maximum longevities

Estimating the longevity of mammals is fraught with difficulties. In their natural environments most individuals rarely achieve their maximum lifespan, whereas they may do so in a protected environment. Therefore captive zoo animals provide the best data, and the most up to date and reliable records have been compiled by M.L. Jones (1982 and personal communication). In general, the early zoo records are unreliable, because the age at which the particular animal has been placed in a zoo may not be known, and food, the environment and veterinary care was less good in the past then it is now. Recorded longevities have therefore tended to increase

throughout the century. A more serious problem is statistical. In most cases, very few individuals have lived out their natural lifespan in zoos. For instance, a handful of great apes have lifespans in the range 48–59 years. One might ask what the corresponding figure would be for a small number of humans kept in the same conditions? A reasonable estimate might be 70–80 years. However, we know from the millions of records of human lifespan that the actual maximum is 120 years. Not only is the sample size enormous, but there also may be rare long-lived variants. Equivalent numbers of gorillas or chimpanzees might yield a maximum lifespan of 80 or 90 years. The same problem applies to domestic animals, especially those kept as pets. There are records of domestic cats living well above 20 years, and also exceptional longevities for dogs, rabbits and so on. In the case of experimental animals kept under good laboratory conditions, it is common to determine the lifespan of cohorts of 30–100 rats or mice of each sex. A reasonable measure of longevity is then the last 5 or 10% of surviving individuals. (Note that even genetically uniform inbred mice kept under identical conditions have very variable lifespans (Zurcher *et al.*, 1982), suggesting that stochastic features of the ageing process are important determinants.) For all the reasons just elaborated, estimates of 'maximum lifespans' are necessarily approximate. The values in Tables 1–3 and Figure 1 are based primarily on the zoo records, which in turn are based on small sample sizes.

Relationship between fecundity and lifespan

The data in Tables 1–3 are plotted in Figure 1 which is a linear-log plot relating lifespan to reproductive potential. In spite of the variability in anatomy and biological life style of the 47 genera, the Figure shows that there is quite a clear relationship between longevity and fecundity in eutherian mammals. The overall trend which is seen confirms a major prediction of the disposable theory of ageing. For a species to maintain its numbers, at least two offspring must survive to adulthood and themselves reproduce. The figure shows that for short lived species the excess number of offspring is about forty-fold, whereas for long lived species the excess is about five-fold.

There are, however, some significant exceptions to the inverse relationship between longevity and fecundity. Amongst rapidly breeding rodents, there are some which live 2–3 years, and others which live significantly longer. Notable examples are the similar sized mice *Mus* and *Peromyscus*, which have provided material for several important experimental studies on ageing (see, for example, Sacher and Hart, 1978). Squirrels, rabbits and bears seem to produce more offspring than would be expected for their longevities (7, 12 and 15 in Fig. 1). Although zoo specimens of the grey squirrel have lived for 23 years, their survival in the wild is surprisingly brief. Less than 10% of animals survive for 4 years in a natural environment (Gurnell, 1987). Animals which have relatively low fecundity in relation to longevity are some bats, herbivores and some primates (e.g. 23, 29, 35 in Fig. 1). This can probably be related to their evolved life style. Small ground living mammals are very frequently killed by predators, but bats of similar size escape predators by flight and by roosting on the roof of caves. Reproduction is obviously more difficult and many species produce only one offspring per year. The evolution of this life style is associated with a considerable increase in longevity. Herbivores (especially Families *Bovidae* and *Cervidae*) usually have a single offspring which is extremely well developed at birth, grows rapidly and is able to then escape from predators. Also, the rapid growth results in early sexual maturity, so these animals start breeding at a younger age than many other mammals of comparable size.

Evolution of primates

Primates have particular features which tend to separate them from other groups of mammals. During their evolution there was both an increase in size and also in longevity, but the relationship between longevity and fecundity is not straightforward. Some small primates such as the mouse lemur (*Microcebus*) and the

Table 1. Reproduction and longevity of eutherian mammals: Rodents and carnivores.

Common name	Genus (and no. in Figure 1)	Time to reproduction	Gestation (days)	Litter size (range or average)	Interlitter interval	Longevity (years)	Reproductive potential (max. no. offspring)
Mouse	*Mus* (4)	5–7 w	19–21	5–6	36 d	3	120
Rat	*Rattus* (5)	3–5 m	21–22	8	70 d	3	80
White-footed mouse	*Peromyscus* (3)	30–49 d	12–27	3–4	73–90 d	8	60
Hamster	*Cricetus* (2)	43 d	18–20	6–8	30 d	2.5	70
Guinea pig	*Cavea* (10)	60 d	56–74	4	90–120 d	8	80
Lemming	*Lemmus* (8)	16–49 d	16–21	7.3	30–35 d	1.5.2	73
Grey Squirrel	*Sciurus* (7)	1 y	44	3	1–2 y	23	60
Chinchilla	*Lagidium* (1)	6 m	111	1–2	6 m	19	60
Beaver	*Casta* (6)	1.5 y	120	2–4	1 y	15	27
Porcupine	*Hystrix* (9)	9–16 m	120	1–2	1 y	27	22
Fox	*Vulpes* (13)	10 m	49–56	5	1 y	12	36
Dog	*Canis* (14)	10–24 m	63	3–10	1 y	16	50
Cat	*Felis* (17)	10–12 m	65	3–5	6 m	16	80
Otter	*Lutra* (18)	2 y	60–63	2–3	1 y	23	35
Badger	*Meles* (20)	1 y	365*	3–6	1 y	16	35
Hyaena	*Hyaena* (21)	2–3 y	88–92	2–4	1 y	31	24
Bear	*Ursus* (15)	4–6 y	180–266	1–4	2 y	50	32
Racoon	*Procyon* (22)	1–2 y	63	3–4	1 y	21	45
Skunk	*Mephitis* (19)	1 y	59–77	4–5	1 y	13	32
Lion	*Panthera* (16)	3–4 y	100–119	3–4	18–26 m	30	30

Abbreviations: d, days; m, months; w, weeks; y, years.

* Including a long period of delayed implantation.

marmoset (*Calithrix*) have fairly short lifespans and breed fairly rapidly. There may be 2–3 offspring per litter and more than one litter per year. Other small primates such as lorises (*Nyctocebus* and *Loris*) and the squirrel monkey (*Saimiri*) have only one offspring per year and relatively short lifespans. It therefore seems that the probability of survival in a natural environment is high. It is likely that the adoption of an arboreal habitat, the evolution of

intelligence and memory, and especially the protection of single young offspring by the mother have all resulted in a significant decline in mortality. This would mean that the probability of an infant surviving to adulthood and itself reproducing would be higher than in most other mammals with comparable longevity.

The evolution of man does not seem to have been associated with a decline in fecundity. The

Table 2. Reproduction and longevity of eutherian mammals: Herbivores, and primates.

Common name	Genus (and no. in Figure 1)	Time to reproduction	Gestation (days)	Litter size (range or average)	Interlitter interval	Longevity (years)	Reproductive potential (mx. no. offspring)
Vicuna	*Vicugnia* (27)	2 y	330–350	1	1 y	24	19
Camel	*Camelus* (26)	5 y	370–440	1	2 y	50	15
Deer	*Cervus* (28)	28 m	235	1	1 y	27	16
Giraffe	*Giraffa* (29)	3.5 y	437	1	20 m	36	12
Sheep	*Ovis* (30)	1–2 y	150–180	1–2	1 y	20	22
Ox or Bison	*Bos* or *Bison* (31)	2–3 y	285	1	1 y	40	16
Horse	*Equus* (32)	2 y	332–342	1	1 y	40	20
Hippopotamus	*Hippopotamus* (34)	10 y	227–240	1	2 y	54	13
Rhinoceros	*Rhinoceros* (35)	5–7 y	462–491	1	3 y	47	9
Indian Elephant	*Elephus* (40)	15–16 y	644	1	4 y	70	9
Mouse lemur	*Microcebus* (41)	1 y	54–68	2–3	71–75 d	15	90
Brown lemur	*Lemur* (42)	18 m	117	1	1 y	30	18
Marmoset	*Calithrix* (43)	12–15 m	133–143	2–3	6 m	16	32
Macaque	*Macaca* (44)	3–5 y	160	1	1 y	37	23
Baboon	*Papio* (45)	5 y	170–173	1	15 m	37	17
Orangutan	*Pongo* (46)	10 y	233–265	1	2.5–3 y	58	12
Man	*Homo* (47)	16 y	275	1	27 m	80	12

Abbreviations: d, days; m, months; y, years.

Table 3. Reproduction and longevity in eutherian mammals: Lagomorpha, Chiroptera and other taxonomic groups.

Common name	Genus (and no. in Figure 1)	Time to reproduction	Gestation (days)	Litter size (range or average)	Interlitter interval	Longevity (years)	Reproductive potential (max. no. offspring)
Rabbit	*Oryctolagus* (12)	6 m	28–33	5–6	50–70 d	9	165
Hare	*Lepus* (12)	7–8 m	41–47	3.8	90–120 d	7	46
Fruit bat	*Eidolan* (23)	1 y	270*	1	1 y	21	13
Collared fruit bat	*Roussettus* (24)	1 y	120	1–2	1 y	23	22
Flying fox	*Pteropus* (25)	6 m	140–150	1	1 y	31	20
Pig	*Sus* (33)	1.5 y	100–140	4–8	1 y	20	72
Hedgehog	*Ericaceris* (37)	1 y	35–40	4–6	1–2 y	6	50
Elephant shrew	*Elephentulus* (36)	6 m	49–57	1–2	64–70 d	8	37
Sloth	*Choleopus* (38)	3.5 y	410	1	14–16 m	32	20
Minke whale	*Balaenoptera* (39)	7–8 y	300–320	1	18–24 m	47	20

Abbreviations: d, days; m, months; y, years.
* Implantation is delayed.

interbirth interval is 2–3 years in the chimpanzee, gorilla, orangutan and the human female (Hayssen *et al.*, 1993), which means that the maximum number of offspring per female is similar. In the case of primates, it seems that longevity relates far more strongly to the time taken to reach sexual maturity, that is, the rate of growth to adulthood. This is illustrated in Figure 2 for 13 representative species. It seems that the increase in longevity is associated with a declining rate of development, rather than a reduction in fecundity. A strong argument can be made that human evolution occurred by a process known as neotony (de Beer, 1958). This is a process whereby development is slowed down so that the adult of a newly evolved species has characteristics of a young individual of an ancestor species. On anatomical grounds, adult humans more closely resemble young apes than adults, but sexual reproduction must be brought forward relative to overall body development, although it occurs later than in the ancestor species. In the evolution of primates as a whole there is clearly a slowing down of development, an increase in the time to reach sexual maturity and an increase in longevity. These in turn very likely relate to the evolution of learning ability and intelligent behaviour, which better adapts the species to a hostile environment. This adaptation is also associated with a decline in mortality and a lower rate of reproduction. There is also a strong argument made by Wilson (1991), that rates of evolution correlate with brain size in vertebrates. He suggested that the better an animal relates its *behaviour* to the environment it finds itself in, the faster it will adapt in Darwinian terms to this environment. Intelligent awareness of the environment would be correlated with brain size, and therefore an increase in brain size would be related to the rate of evolutionary change. This could apply strongly to higher primates and help to explain the rapid evolution of brain size. Furthermore, the evolution of language in man as an extension of behaviour would further accelerate successful adaptation to the environment. The acquisition of language and other skills are favoured if development to adulthood is slowed down. Thus, we would expect puberty to occur later in man than the great apes, and this in turn relates to delayed reproduction and a lengthened lifespan.

Whether this increase in longevity evolved rapidly is a matter for debate. A claim has been made that there was a particularly rapid increase in lifespan during the evolution of man, which lead to the conclusion that rather few genes may be involved in determining lifespan (Cutler, 1975). This claim and conclusion demand critical examination. In particular, the estimation of maximum lifespans of the larger primates is difficult, as there are so few specimens kept in zoos. As was explained previously, it is likely that zoo lifespans of chimpanzees and gorillas (about 50 years) correspond to human lifespans of 70–80 years, not 120 years, which is a frequently quoted figure. Thus, instead of a two-fold difference in longevity, there is more likely to be a difference of 20–30 years. This somewhat weakens the view that an increase in longevity evolved rapidly.

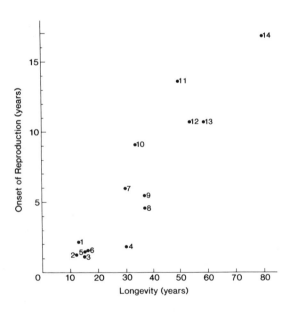

Fig. 2. The relationship between the age of onset of reproduction of female primates, including the period from fertilisation to birth, and the longevity of the species. 1, Slow loris (*Nyctocebus*); 2, Slender loris (*Loris*); 3, Mouse lemur (*Microcebus*); 4, Brown lemur (*Lemur*); 5, Squirrel monkey (*Saimiri*); 6, Marmoset (*Calithrix*); 7, Spider monkey (*Ateles*), 8, Macaque (*Macaca*); 9, Baboon (*Papio*); 10, Gibbon (*Hylobates*); 11, Chimpanzee (*Pan*); 12, Gorilla (*Gorilla*); 13, Orangutan (*Pongo*); 14, Human (*Homo*). See legend to Fig. 7.5 for sources of data.

Human demography and longevity

It is generally agreed that when humans evolved, they adopted a hunter-gatherer lifestyle. They probably lived in small groups of interrelated individuals, just as do some of the great apes today. It is important to consider the demography of such human populations in relation to reproduction and longevity. The early aboriginal population of Australia presumably comprised a reasonably homogeneous population which subsequently broke up into many subgroups or tribes. Archaeological evidence suggests that the aborigines arrived about fifty thousand years ago, and at the time the white settlers arrived in 1788, it is estimated that the total population was about 500,000. If we assume that the founder population was no more than a few hundred individuals, say 1,000, and that there are on average 4–5 generations per century, then it can be calculated that the average increase in population size per generation is extremely small. For a time span of 50,000 years at 4.5 generations per century there are 2250 generations. If the increase in the number of individuals is 500-fold , then the average increment in population size x per generation is given by the formula

$$x^{2250} \approx 500$$

$x \approx 1.0028$ or 0.28% per generation.

This value will be slightly lower if the founder population was greater, or the number of generations was greater. Obviously, this is an overall estimate of growth, and there would have been major fluctuations in this rate, depending on food supply and other environmental conditions. Nevertheless, the demographic situation is scarcely different from a steady state population. To maintain a population in a steady state each reproducing female must herself produce, on average, two reproducing adults.

A study in East Africa shows that the average interbirth interval for lactating women is 27 months, and 15 months for non-lactating women (Saxton & Servadda, 1969). Thus, the number of offspring produced is in part determined by the survival or death of infants. In chimpanzees and gorillas in their natural environments, infant mortality is about 25% (Harcourt et al., 1981;

Courtenay & Santow, 1989). If we assume the same value for human infants in a hunter-gatherer community, then females will produce on average one offspring every 2 years. One also has to take into account the survival of female children to breeding age, and the mortality of breeding females, including death in childbirth. This introduces several imponderables, but assuming that the population renews itself without increase, then one can estimate (1) the average number of offspring per female after they reach reproductive age, and (2) the expectation of life. These values are about 6 offspring, and the expectation of life of females who reach 16 years of age is 28 years. The expectation of life at birth is only 15 years. These estimates are based on an annual mortality rate of about 7% after the age of one year. This is an over-simplification, since it assumes an exponential survival curve, which is probably unrealistic. However, from the analysis of skeletons in graveyards in ancient Greece (Angel, 1947), it has been estimated that survival was not far removed from an exponential loss of life, whereas one might have expected otherwise in a well organised society.

The evolution of human longevity would be the result of an interplay between many phenotypic characteristics, including the ability to obtain a food supply from the environment, which depends in turn on intelligent behaviour; the extent of mortality in infancy and childhood; the likelihood of adults producing sufficient offspring to maintain the population, and the social organisation of hunter-gatherer groups. In developed nations today the maximum lifespan is about 25–30 years greater than the expectation of life. In early hunter-gatherer communities the difference would have been very much greater, perhaps as much as 60–70 years. Nevertheless, in such communities, some adults would survive to old age. In small communities of very mixed age, kin selection and the evolution of altruistic behaviour is likely to become very important. To ensure survival of offspring, it would be counterproductive for females to keep reproducing, rather than successfully rearing existing children. The evolution of the menopause was in all likelihood an adaptation to ensure survival of younger individuals in the society. With 25% infant mortality and a 7%

annual mortality rate, about 3% of the female population would reach the age of 45, but it should be noted that these women would have probably produced the most offspring in the community. Therefore the kin-selective force for non-reproduction would not be trivial, given the possible number of children and grandchildren which might benefit from this trait. It is probable that non-reproducing females cared not only for their own children, but also for grandchildren or other young relatives, just as they often do today. The existence of the menopause indicates that even in primitive hunter-gatherer societies, some individuals achieved a longevity which was a considerable proportion of the maximum lifespan.

Environmental stress, reproduction and longevity

In animals which reproduce rapidly, a glut in the availability of food can lead to a population explosion. This can be followed by mass migration in the search of new habitats and food. Alternatively, if the expanded population is confined to the same habitat, then competition for food supply and living space is intense. Under these conditions, reproduction slows down and may cease (Pennycuick *et al.*, 1986, 1987), which in turn may affect subsequent longevity. Indeed, the common life style for small ground animals such as mice and rats is likely to be the alternation of periods when the food supply is adequate for breeding, with periods of limited food supply which limits breeding. It is known from many observations of laboratory mice and rats, that calorie deprivation (usually 50–60% of an *ad libitum* diet) results in partial or complete sterility. It is well established that animals kept on such a diet live for a significantly longer period than animals fed *ad libitum*. This may well be an adaptive mechanism which allows animals to survive through a period of limited food, without breeding, until food again becomes available, when breeding will resume (Holliday, 1989). Indeed, the experiments show that animals which are held on a calorie restricted diet can subsequently *breed at a later age* than animals which are fed throughout *ad libitum*. The whole adaptive response in these fluctuating environments is

therefore to breed and store food in the form of fat when food is plentiful, then to invest resources, including stored fat, into survival when food is scarce. These resources are best used by diverting them from normal breeding, which would be a futile exercise when food is unavailable, into maintenance of the soma, which stretches out the normal lifespan. It is not yet known if a similar physiological response occurs in larger slow breeding animals. Anorexic human females become infertile, and most long lived individuals are lightly built, but a direct connection between dietary intake and longevity has not been established in man, or other large mammalian species.

Conclusions

In the adaptive radiation of mammals from their early ancestors, we can see the evolution of various strategies for survival. Small omnivorous ground living mammals are subject to heavy predation from other mammals and birds, and also to a fluctuating food and water supply. When food is plentiful, resources are channelled into rapid breeding, and longevity is short. Of the many offspring produced, only a small proportion are likely to breed. Other small mammals evolved a more specialised life style which was less hazardous. In particular, the bats are able to escape predators by flight and by roosting on the roof of caves. The much reduced mortality is accompanied by a far slower rate of reproduction, and also a very significantly longer lifespan. Many herbivores escape predators by alertness and flight. They breed fairly slowly (usually one offspring per year), but the newborn are highly developed and grow rapidly to adulthood. Low mortality and slow reproduction are associated with a fairly long lifespan. Carnivores are in a somewhat intermediate situation. Normally several offspring are produced in a litter, and their survival largely depends on the ability of parents to find prey, and also on the hunting successes of the offspring as they mature. Longevity is greater in the slower breeding large carnivores than in the faster breeding small species. Indeed, in this group it is believed that the smallest species, such as weasels and stoats, evolved larger ones. These highly specialised small species are dependent on a continual source of food; their litters are large, and their lifespans are quite short. In the primates, extended parental care reduces the mortality of offspring. Also, learning and intelligence came to play a more important role in survival. In this group we see the evolution of an extended period of development prior to sexual maturity, which is also associated with increased longevity. The largest land mammals, the rhinoceros, hippopotamus and elephant, are relatively free from predators. The single offspring are well developed and are protected by their parents. These animals breed very slowly, and have long lifespans.

All these examples of adaptation illustrate the strong inverse relationship between fecundity and longevity. This confirms a major prediction of the disposable soma theory of ageing, namely, that organisms maintain their soma by a variety of mechanisms, all of which require metabolic resources. The diversion of some of these resources into reproduction reduces the efficiency of maintenance and therefore the lifespan. When mammals become very well adapted to their environment, mortality is reduced and the chance of offspring surviving to adulthood is increased. To maintain the species, less is invested in reproduction, more in maintenance of the soma, and longevity is increased. Many previous studies have demonstrated that long lived mammalian species have more efficient somatic cell maintenance mechanisms than short lived ones (reviewed by Holliday, 1994).

References

Angel, J.L., 1947. The length of life in ancient Greece. J. Gerontol. II, 1: 18–24.

Child, C.M., 1915. Senescence and Rejuvenescence. Chicago University Press, Chicago.

Courtenay, J. & G. Santow, 1989. Mortality of wild and captive chimpanzees. Folia Primatol. 52: 167–177.

Cutler, R.G., 1975. Evolution of human longevity and the genetic complexity governing aging rate. Proc. Natl. Acad. Sci. USA 72: 4664–4668.

de Beer, G., 1958. Embryos and Ancestors, 3rd. edn. Oxford University Press, Oxford.

Gunrell, J., 1987. The Natural History of Squirrels. Christopher Helm, London.

Harcourt, A.H., D. Fossey & J. Sabater-Pi, 1981. Demography of Gorilla gorilla. J. Zool. Lond. 195: 215–233.

Hayssen, V., A. van Tienhoven & A. van Tienhoven, 1993. Asdell's Patterns of Mammalian Reproduction. Cornstock, Cornell University Press, Ithaca.

Holliday, R., 1989. Food, reproduction and longevity: is the extended lifespan of calorie restricted animals an evolutionary adaptation? BioEssays 10: 125–127.

Holliday, R., 1994. Understanding Ageing. Cambridge University Press.

Jones, M.L., 1982. Longevity of captive mammals. Zool. Gart. N.F. Jena 52: 113–128.

Kirkwood, T.B.L., 1977. Evolution of ageing. Nature 270: 301–304.

Kirkwood, T.B.L., 1981. Repair and its evolution: survival versus reproduction. In Physiological Ecology: An Evolutionary Approach to Resource Use, ed. C.R. Townsend & P. Calow pp. 165–189, Blackwell Scientific, Oxford.

Kirkwood, T.B.L. & R. Holliday, 1979. The evolution of ageing and longevity. Proc. Roy. Soc. B 205: 532–546.

Kirkwood, T.B.L. & R. Holliday, 1986. Ageing as a consequence of natural selection. In The Biology of Human Ageing, ed. K.J. Collins & A.H. Bittles, pp. 1–16. Cambridge University Press, Cambridge.

Nowak, R.M., 1991. Walker's Mammals of the World, 5th edn. Johns Hopkins Press, Baltimore.

Pennycuik, P.R., P.G. Johnston, N.H. Westwood & A.H. Reisner, 1986. Variation in numbers in a house mouse population housed in a large outdoor enclosure: seasonal fluctuations. J. Anim. Ecol. 55: 371–391.

Pennycuik, P.R., A.H. Reisner & N.H. Westwood, 1987. Effects of variations in the availability of food and home sites and culling on populations of house mice housed in outdoor pens. Oikos 50: 33–41.

Sacher, G.A. & R.W. Hart, 1978. Longevity, aging and comparative cellular and molecular biology of the house mouse, Mus musculus, and the white-footed mouse, Peromyscus leucopus. In Genetic Effects on Aging, ed. D. Bergsma & D.E. Harrison, pp. 71–96. Alan Liss, New York.

Sonneborn, T.M., 1930. Genetic studies on Stenostomum incaudatum (nov. sp.). I. The nature and origin of differences among individuals formed during vegetative reproduction. J. Exp. Zool. 57: 57–108.

Zurcher, C., M.J. van Zwieten, H.A. Solleveld & C.F. Hollander, 1982. Ageing research. In The Mouse in Biomedical Research, vol. IV, ed. H.L. Foster, J.D. Small & J.G. Fox, pp. 11–35. Academic Press, New York.

M. R. Rose and C. E. Finch (eds), Genetics and Evolution of Aging, 226–231, 1994.
© 1994 *Kluwer Academic Publishers. Printed in the Netherlands.*

Thoughts on the evolutionary basis of dietary restriction

Arlan Richardson & Mohammad A. Pahlavani
*Geriatric Research, Education and Clinical Center, Audie L. Murphy Memorial Veterans Hospital,
and Department of Medicine, University of Texas Health Science Center,
San Antonio, TX 78284, USA*

The dramatic effect of nutrition on the life span of rodents was shown initially by McCay's laboratory in the 1930's (McCay *et al.*, 1935, McCay *et al.*, 1939). They found that both the median and maximum survival of rats were increased significantly when the diet of weanling rats was restricted severely and growth retarded. This phenomenon became known as dietary or calorie restriction, and subsequent studies have demonstrated that the life span of rats could be extended significantly using less severe restriction regimens; e.g., a 30 to 50% restriction of calories generally results in a 20 to 50% increase in life span (mean and maximum survival) (Masoro, 1985, Masoro, 1988, Masoro, 1992a, Weindruch & Walford, 1988). At the present time, it is generally accepted that dietary restriction extends the life span of laboratory rodents by retarding the aging process because dietary restriction increases the maximum survival of rodents and alters most physiological and pathological processes that change with increasing age (Richardson & McCarter, 1991; Masoro, 1992a). Therefore, dietary restriction offers gerontologists a unique system for studying the aging process.

Although reduced calorie consumption is the primary nutritional component responsible for the increased life span of laboratory rodents (Masoro, 1992b), the biological mechanism whereby reduced calorie consumption increases the survival of rodents and retards aging is currently unknown. At the present time, studies on the biological mechanism underlying dietary restriction have focused on defining the physiological and biochemical changes that occur with dietary restriction (Masoro *et al.*, 1991, Yu *et al.*, 1991, Richardson, 1985, Heydari & Richardson, 1992). In this paper, we will consider the possible evolutionary origins of dietary restriction in an attempt to gain greater insight into the underlying mechanism.

If the phenomenon of dietary restriction evolved, what were the evolutionary advantages of dietary restriction? In 1988, Harrison and Archer hypothesized that the selective advantage of dietary restriction was the extension of reproductive life span, which allowed an organism (e.g., rodent) to extend its reproductive life beyond periods of food shortage in the wild. In 1989, Holliday advanced a similar concept and proposed that dietary restriction changed the strategy of survival from rapid reproduction over a short life span to reduced reproduction over a longer life span. Holliday (1989) suggested that this change in survival strategy occurred because an organism placed more resources in repair/maintenance and fewer resources in reproduction. However, Phelan and Austad (1989) have argued against the view that dietary restriction evolved because it increased reproductive longevity since reproductive senescence is largely irrelevant in nature.

We propose that dietary restriction evolved as a system that gave an organism the ability to maximize reproductive effectiveness over a range of nutritional conditions by adjusting the investment of resources that an organism makes in reproduction and repair/maintenance. We define reproductive effectiveness as the ability of an animal to produce the maximum number of strong, healthy offspring that have a high probability of reproducing. A schematic representation of our model is given in Figure 1. When food is plentiful, the organism places more of its resources into reproduction resulting in more frequent breeding and a larger litter size. However, in times of limited food supply, the organism reduces the resources it invests in reproductive functions and increases the

- ▷ Repair/Maintenance
- ◆ Reproduction
- ▷ Basic/Essential Functions

Unrestricted   Restricted

Fig. 1. A model for the evolutionary basis of dietary restriction. The two graphs show the proportion of resources an organism uses for three categories of function (repair/maintenance, reproduction, and basic essential functions, e.g., breathing, heat production, movement, basal metabolism, etc.) under conditions of high food availability (unrestricted) and limited food availability (restricted). The assignment of resources for the three categories of function is arbitrary; however, the graphs show how the distribution of resources in the three categories change with changes in food availability. (From: *Thoughts on the Evolutionary Basis of Dietary Restriction* by Richardson & Pahlavani)

resources it invests in the repair/maintenance of cells and tissues. Such an adjustment in resources would result in reduced breeding and smaller litter size, which will maximize the chance that an organism can carry a fetus to term and feed the offspring.

According to our model, the evolutionary advantage of dietary restriction would arise because it allows an organism to maximize its reproductive effectiveness in response to changes in the food supply, not because of its ability to extend reproductive life span. Therefore, we believe that the extension of life span by dietary restriction is fortuitous and arises secondarily from an organism shifting its resources from reproduction to repair/maintenance in response to reduced calorie consumption. Our model is consistent with the disposable soma theory of aging proposed by Kirkwood (1989, 1977). This theory argues that a trade-off occurs between the investment of resources in repair/maintenance, which will result in a potentially longer life, and the immediate investment of resources in reproduction. This theory also argues that the trade off will take place in such a way that the investment of resources in repair/maintenance is always less than that required for indefinite somatic longevity. Thus, one would predict from the disposable soma theory of aging that the life span of an organism would be increased significantly when resources are shifted from reproduction to repair/maintenance, i.e., a shift

that we believe occurs when an organism adjusts its reproductive strategy in response to reduced calorie consumption.

Data from several areas support the hypothesis that dietary restriction results in a shift in resources from reproduction to repair/maintenance. First, several studies have shown that dietary restriction alters the reproductive status of rodents. Holehan and Merry (1985a, 1985b) reported that dietary restriction initiated in weanling rats reduced both litter size and overall fertility in female rats in addition to delaying the onset of puberty. Breeding capacity is also reduced when calorie restriction is imposed in adult mice (Visscher *et al.*, 1952). In addition, Chapin *et al.* (1993) showed that a calorie restriction resulting in only a 10% decrease in body weight reduced litter size significantly in female mice. Thus, even a small change in calorie consumption appears to alter the reproductive status of females. Fertility is also reduced by dietary restriction in male mice; however, the effect is variable, and the effect of dietary restriction on the male reproductive system is less pronounced than for the female reproductive system (Chapin *et al.*, 1993). Dietary restriction has also been shown to increase life span and reduce fecundity in invertebrates, e.g. spiders (Austad, 1989) and nematodes (Klass, 1977). Thus, all of the current data support the view that the ability of animals to reproduce is reduced when calorie consumption is reduced.

Our model is also supported by the studies on the effect of dietary restriction on processes involved in repair/maintenance. In general, dietary restriction improves these processes. For example, it is well established that dietary restriction enhances the immunological defense system, which plays a critical role in the response of an organism to a variety environmental challenges (e.g., infectious agents, cancerous cells, etc.). Antigen presentation, antigen-specific lymphocyte proliferation, and antibody production in response to environmental challenges are all elevated in calorie restricted rodents (Tada, 1992, Weindruch *et al.*, 1986). In addition, tissues and cells of calorie restricted rodents also seem to have a more active antioxidant defense system, which appears to lead to a lower level of free radical damage in cells and tissues (Lee & Yu, 1990, Yu *et al.*, 1991).

Because free radicals are believed to play a fundamental role in a variety of pathological processes (Harman, 1991, Gutteridge, 1993), a more active antioxidant defense system would be predicted to result in reduced pathology and increased survival. Two other processes involved in cellular repair are enhanced by dietary restriction: DNA repair (Haley-Zitlin & Richardson, 1994), and protein turnover, which removes damaged proteins from cells (Richardson & Ward, 1994). Recently, our laboratory also showed that dietary restriction enhanced the ability of cells to respond to stress. The expression of the heat shock protein hsp70 is significantly higher in cells isolated from calorie restricted rats after a heat shock (Heydari et al., 1993). Hsp70 plays a critical role in protecting cells and tissues from a variety of stresses including hyperthermia. We also found that calorie restriction provided the rats with protection against the adverse effects of hyperthermia. For example, the data in Figure 2 show that only 16% of rats fed ad libitum survived a period of prolonged heat stress. In contrast, 75% of the rats fed the calorie restricted diet survived the heat stress. Thus, the data currently show that a variety of physiological processes involved in repair/maintenance are enhanced by dietary restriction and that calorie restricted rodents are able to survive a variety of insults/stresses better than rodents fed ad libitum.

How does an organism respond to changes in food availability and adjust its reproductive strategy, i.e., shift resources from reproduction to repair/maintenance or visa versa? We believe that the endocrine system is the physiological system most likely to be involved in adjusting reproductive strategy in response to diet because reproduction and a variety of other cellular/biochemical processes are regulated through changes in hormone levels. Over the past decade, several laboratories have compared the hormonal status of rodents fed ad libitum and rodents fed a calorie restricted diet, and the results of these studies are summarized in Table 1. It is quite obvious from Table 1 that dietary restriction has a major affect on the endocrine system. While dietary restriction reduces the plasma levels of most hormones, the levels of a few hormones are elevated in response to dietary restriction. The decrease in the levels of growth hormone and the hormones involved in sexual development is consistent with our model that dietary restriction shifts resources from

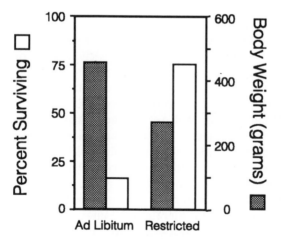

Fig. 2. The effect of dietary restriction on survival after heat stress. Male 20-month-old rats fed *ad libitum* or a calorie restricted diet were exposed accidentally to a high temperature. The figure gives the body weights of the rats and percent of the rats that survived the heat stress. The data were taken from Heydari *et al.* (1993). (From: *Thoughts on the Evolutionary Basis of Dietary Restriction* by Richardson & Pahlavani)

Table 1. Effect of dietary restriction on hormone levels.

Hormone	Effect of dietary restriction	Reference
Insulin	Decrease	Masoro *et al.* (1992) Chu *et al.* (1991)
Glucocorticoides	Increase	Sabatino *et al.* (1991)
Thyroid Hormone (T₃)	Decrease	Holehan & Merry (1985) Herlihy *et al.* (1990)
Growth Hormone (IGF-I)	Decrease	Sorrentino *et al.* (1971) Chan *et al.* (1993) Armario *et al.* (1987) Sonntag *et al.* (1992)
Ovarian/Testicular Hormones	Decrease	Holehan & Merry (1981, 1985c) Chapin *et al.* (1993)
Catacholeamines	Increase (?)	Strong *et al.* (1990)
Calcitonin	Decrease	Salih *et al.* (1993)
Parathyroid Hormone	Decrease	Armbrecht *et al.* (1988) Salih *et al.* (1993)

reproduction. In addition, it is interesting to note that the effect of dietary restriction on several hormones (e.g., insulin, growth hormone, glucocorticoids, and T_3) occurs very rapidly in response to calorie restriction; the changes in the levels of these hormones are observed at the earliest age studied after implementation of the dietary regimen. For example, the changes in plasma levels of growth hormone (Chan *et al.*, 1993) and insulin (Chu *et al.*, 1991) are observed within weeks of the calorie restriction. Such a response would be expected if an animal were to respond quickly to the availability of resources in the environment and adjust its reproductive strategy.

In summary, we propose that the anti-aging action of dietary restriction evolved because it gave an organism the ability to adjust its reproductive strategy so as to maximize its reproductive effectiveness over a range of nutritional conditions. A major advantage of our model is that it does not rely on the increase in reproductive life span as the evolutionary driving force behind the selective advantage of dietary restriction. Phelan and Austad (1989) have argued that the selective pressure for extending reproductive life span through natural selection would be very weak because reproductive senescence is largely irrelevant in nature, e.g., very few animals survive beyond their reproductive life span in the wild. According to our model, the extension of life span by dietary restriction is secondary to an organism adjusting its reproductive strategy to decreased calories and shifting its resources from reproduction to repair/maintenance. Based on the disposable soma theory of aging (Kirkwood, 1989, Kirkwood, 1977), we propose that the shift in resources to repair/maintenance results in longer life span of laboratory rodents in the laboratory situation where predation is eliminated.

Our model for the evolutionary basis of dietary restriction leads to at least two predictions. The first prediction is that dietary restriction would have to be implemented throughout most of the life span of an animal to have the maximal effect on survival and longevity. According to our model, changing an animal's diet form a low calorie regimen to a high calorie regimen would trigger a shift in resources from repair/maintenance to reproduction. This shift would be expected to result in a decreased potential for increased longevity. The limited studies in this area are consistent with this prediction. In an early study, Ross (1972) reported that dietary restriction implemented over a relatively short period of time was much less effective in increasing survival than dietary restriction implemented over the life span of rats. More recently, Yu *et al.* (1985) compared the survival of rodents fed a calorie restricted diet for various periods of time. They found that life-long dietary restriction increased the median survival of rats 50%; however, when rats were fed a calorie restricted diet only for the first 6 months of life and then fed *ad libitum*, the median survival was increased only 15%. In addition, Beauchene *et al.* (1986) found that calorie-restriction during either the first or second year of life was approximately one-half as effective at increasing life span as life long dietary restriction. A second prediction that would arise from our model is that only organisms that show a shift in their reproductive strategy in response to changes in nutrition (i.e., a decrease in reproduction in response to decreased calories) would show an increased life span in response to dietary restriction. To our knowledge, there is no information in this area at the current time.

Acknowledgments

The authors were supported by research grants AG 01548 and AG 0118 awarded by the National Institute on Aging and Medical Research funds from the Veterans Administration.

References

Armario, A., J.L. Montero & T. Jolin, 1987. Chronic food restriction and the circadian rhythms of pituitary-adrenal hormones, growth hormone and thyroid-stimulating hormone. Ann. Nutr. Metab. 31: 87–87.

Armbrecht, H.J., R. Strong, M. Boltz, D. Rocco, W.G. Wood & A. Richardson, 1988. Modulation of age-related changes in serum 1,25-dihydroxyvitamin D and parathyroid hormone by dietary restriction of Fischer 344 rats. J. Nutr. 118: 1360–1365.

Austad, S.N., 1989. Life extension by dietary restriction in the bowl and doily spider, *frontinella pyramitela*. Exp. Gerontol. 24: 83–92.

Beauchene, R.E., C.W. Bales, C.S. Bragg, S.T. Hawkins & R.L. Mason, 1986. Effect of age on initiation of feed restriction on growth, body composition, and longevity of rats. J. Gerontol. 41: 13–19.

Chan, W., R.J. Krieg, Jr., T.E. Sayles & D.W. Matt, 1993. Caloric restriction and expression of liver insulin-like growth factor-I, growth hormone receptor and pituitary growth hormone. Nutr. Res. 13: 1343–1350.

Chapin, R.E., D.K. Gulati, P.A. Fail, E. Hope, S.R. Russell, J.J. Heindel, J.D. George, T.B. Grizzle & J.L. Teague, 1993. The effects of feed restriction on reproductive function in Swiss CD-1 mice. Fundam. Appl. Toxicol. 20: 15–22.

Chu, K.U., J. Ishizuka, G.J. Poston, C.M.J. Townsend, C.H.J. Greeley, B.P. Yu & J.C. Thompson, 1991. Change in endocrine pancreatic function in short-term dietary restriction. Nutrition 7: 425–429.

Gutteridge, J.M.C., 1993. Free radicals in disease processes: A compilation of cause and consequence. Free Radic. Res. Commun. 19: 141–158.

Haley-Zitlin, V. & A. Richardson, 1994. Effect of dietary restriction on DNA repair and DNA damage. Mutation Res. (in press).

Harman, D., 1991. The aging process: Major risk factor for disease and death. Proc. Natl. Acad. Sci. USA 88: 5360–5363.

Harrison, D.E. & J.R. Archer, 1988. Natural selection for extended longevity from food restriction. Growth. Dev. Aging 52: 65.

Herlihy, J.T., C. Stacy & H.A. Bertrand, 1990. Long-term food restriction depresses serum thyroid hormone concentrations in the rat. Mech. Ageing Dev. 53: 9–16.

Heydari, A.R., B. Wu, R. Takahashi, R. Strong & A. Richardson, 1993. Expression of heat shock protein 70 is altered by age and diet at the level of transcription. Mol. Cell. Biol. 13: 2909–2918.

Heydari, A.R. & A. Richardson, 1992. Does gene expression play any role in the mechanism of the antiaging effect of dietary restriction. Ann. NY Acad. Sci. 663: 384–395.

Holehan, A.M. & B.J. Merry, 1985a. The control of puberty in the dietary restricted female rat. Mech. Ageing. Dev. 32: 179–191.

Holehan, A.M. & B.J. Merry, 1985b. Lifetime breeding studies in fully fed and dietary restricted female CFY Sprague-Dawley rats. 1. Effect of age, housing conditions and diet on fecundity. Mech. Ageing. Dev. 33: 19–28.

Holehan, A.M. & B.J. Merry, 1985c. Modification of the oestrous cycle hormone profile by dietary restriction. Mech. Ageing Dev. 32: 63–76.

Holliday, R., 1989. Food, reproduction and longevity: Is the extended lifespan of calorie-restricted animals an evolutionary adaptation? BioEssays 10: 125–172.

Kirkwood, T.B., 1977. Evolution of ageing. Nature 270: 301–304.

Kirkwood, T.B., 1989. DNA, mutations and aging. Mutation Res. 219: 1–7.

Klass, M.R., 1977. Aging in the nematode *Caenorhabditis elegans*: major biological and environmental factors influencing life span. Mech. Ageing Dev. 6: 413–429.

Lee, D.W. & B.P. Yu, 1990. Modulation of free radicals and superoxide dismutase by age and dietary restriction. Aging Clin. Exp. Res. 2: 357–362.

Masoro, E.J., 1985. Nutrition and aging: a current assessment. J. Nutr. 115: 842–848.

Masoro, E.J., 1988. Food restriction in rodents: An evaluation of its role in the study of aging. J. Gerontol. 43: B59–B64.

Masoro, E.J., I. Shimokawa & B.P. Yu, 1991. Retardation of the aging processes in rats by food restriction. Ann. NY Acad. Sci. 621: 337–352.

Masoro, E.J., 1992a. Potential role of the modulation of fuel use in the antiaging action of dietary restriction. Ann. NY Acad. Sci. 663: 403–411.

Masoro, E.J., 1992b. Retardation of aging processes by food restriction: An experimental tool. Am. J. Clin. Nutr. 55: 1250S–1252S.

Masoro, E.J., R.J. McCarter, M.S. Katz & C.A. McMahan, 1992. Dietary restriction alters characteristics of glucose fuel use. J. Gerontol. 47: B202–B208.

McCay, C.M., M.F. Crowell & L.A. Maynard, 1935. The effect of retarded growth upon the length of life span and upon the ultimate body size. J. Nutr. 10: 63–79.

McCay, C.M., L.A. Maynard, G. Sperling & L.L. Barnes, 1939. Retarded growth, life span, ultimate body size and age changes in the albino rat after feeding diets restricted in calories. J. Nutr. 18: 1–13.

Merry, B.J. & A.M. Holehan, 1981. Serum profiles of LH, FSH, testosterone and 5-alpha-DHT from 21 to 1000 days of age in *ad libitum* fed and dietary restricted rats. Exp. Gerontol. 16: 431–444.

Merry, B.J. & A.M. Holehan, 1985. The endocrine response to dietary restriction in the rat, pp. 117–141 in Molecular Biology of Aging, edited by A.D. Woodhead, A.D. Blackett and A. Holleander. Plenum, New York.

Phelan, J.P. & S.N. Austad, 1989. Natural selection, dietary restriction, and extended longevity. Growth. Dev. Aging 53: 4–5.

Richardson, A., 1985. The effect of age and nutrition on protein synthesis by cells and tissues from mammals, pp. 31–48 in Handbook of Nutrition and Aging, edited by R.R. Watson. CRC Press, Boca Raton, Florida.

Richardson, A. & R. McCarter, 1991. Mechanisms of food restriction: change of rate or change of set point? pp. 177–272 in The potential for nutritional modulation of the aging processes, edited by D.K. Ingram, G.T. Baker and N.W. Shock. Food & Nutrition Press Inc., Trumbull, CT.

Richardson, A. & W.F. Ward, 1994. Changes in protein turnover as a function of age and nutritional status, pp. 309–315 in Handbook of nutrition in the aged, edited by R.R. Watson. CRC press, Boca Raton, Florida.

Ross, M.H., 1972. Length of life and caloric intake. Am. J. Clin. Nutr. 25: 834–838.

Sabatino, F., E.J. Masoro, C.A. McMahan & R.W. Kuhn, 1991. Assessment of the role of glucocorticoid system in aging processes and in the action of food restriction. J. Gerontol. 46: B171–B179.

Salih, M.A., D.C. Herbert & D.N. Kalu, 1993. Evaluation of the molecular and cellular basis for modulation of thyroid c-cell hormones by aging and food restriction. Mech. Ageing Dev. 70: 1–21.

Sonntag, W.E., J.E. Lenham & R.L. Ingram, 1992. Effects of aging and dietary restriction on tissue protein synthesis: relationship to plasma insulin-like growth factor-1. J. Gerontol. 47: B159–B163.

Sorrentino, S., R.J. Reiter & D.S. Schalch, 1971. Interactions of the pineal gland, blinding, and underfeeding on reproductive organ size and radioimmunoassayable growth hormone. Neuroendocrinology 7: 105–115.

Strong, R., M.A. Moore, C. Hale, W.J. Burke, H.J. Armbecht & A. Richardson, 1990. Age-related changes in adrenal catecholamine content and tyrosine hydroxylase gene expression: effects of dietary restriction, pp. 218–227 in Endocrine function and aging, edited by J.A. Armbrecht, R. Coe and N. Wongsurawat. Springer-Verlag, Berlin.

Tada, T., 1992. Nutrition and the immune system in aging: An overview. Nutr. Rev. 50: 360.

Visscher, M.B., J.T. King & Y.C.P. Lee, 1952. Further studies on influence of age and diet upon reproductive senescence in strain A female mice. Am. J. Physiol. 170: 72–76.

Weindruch, R., R.L. Walford, S. Fligiel & D. Guthrie, 1986. The retardation of aging in mice by dietary restriction: longevity, cancer, immunity and lifetime energy intake. J. Nutr. 116: 641–654.

Weindruch, R. & R.L. Walford, 1988. The retardation of aging and disease by dietary restriction. Thomas, Springfield, Illinois.

Yu, B.P., E.J. Masoro & C.A. McMahan, 1985. Nutritional influences on aging of Fischer 344 rats: I. Physical, metabolic, and longevity characteristics. J. Gerontol. 40: 657–670.

Yu, B.P., D.-W. Lee & J.-H. Choi, 1991. Prevention of free radical damage by food restriction, pp. 191–197 in Biological effects of dietary restriction, edited by L. Fishbein. Springer-Verlag, New York.

M.R. Rose and C.E. Finch (eds.), Genetics and Evolution of Aging, 232–242, 1994.
© 1994 *Kluwer Academic Publishers. Printed in the Netherlands.*

Genetic control of retroviral disease in aging wild mice

Murray B. Gardner

Department of Pathology, University of California, Davis, School of Medicine, MS1-A, Room 3453, Davis, CA 95616, USA

Received 22 June 1993   Accepted 22 June 1993

Key words: aging wild mice, retroviruses, murine leukemia virus, murine mammary tumor virus, lymphoma, lower motor neuron disease

Abstract

Different populations of wild mice (*Mus musculus domesticus*) in Los Angeles and Ventura Counties were observed over their lifespan in captivity for expression of infectious murine leukemia virus (MuLV) and murine mammary tumor virus (MMTV) and for the occurrence of cancer and other diseases. In most populations of feral mice these indigenous retroviruses were infrequently expressed and cancer seldom occurred until later in life (> 2 years old). MMTV was found in the milk of about 50% of wild mice, but was associated with only a low incidence (> 1%) of breast cancer after one year of age. By contrast, in several populations, most notably at a squab farm near Lake Casitas (LC), infectious MuLV acquired at birth via milk was highly prevalent, and the infected mice were prone to leukemia and a lower motor neuron paralytic disease after one year of age. These two diseases were both caused by the same infectious (ecotropic)strain of MuLV and were the principal cause of premature death in these aging LC mice. A dominant gene called FV-4^R restricting the infection with ecotropic MuLV was found segregating in LC mice. Mice inheriting this FV-4^R allele were resistant to the ecotropic MuLV associated lymphoma and paralysis. The FV-4^R allele represents a defective endogenous MuLV provirus DNA segment that expresses an ecotropic MuLV envelope-related glycoprotein (gp70) on the cell surface. This FV-4^R encoded gp70 presumably occupies the receptor for ecotropic MuLV and blocks entry of the virus. The FV-4^R gene was probably acquired by the naturally occurring crossbreeding of LC feral mice with another species of feral mice (*Mus castaneus*) from Southeast Asia. The FV-4^R gp70 does not block entry of the amphotropic MuLV that uses a separate cell surface receptor. Therefore LC mice continued to be susceptible to the highly prevalent but weakly lymphogenic and nonparalytogenic amphotropic strain of MuLV. The study points out the potential of feral populations to reveal genes associated with specific disease resistance.

Introduction

Knowledge about the genetic influence on aging in mammals has been derived in part from inbred mouse genotypes that shorten lifespan by causing an early onset of specific disease rather than generally accelerating senescence (Finch, 1990). An understanding of the molecular genetics of these diseases of short-lived inbred mice often leads to the discovery of other genes or alleles that are associated with longer lived genotypes. On the other hand, other genetic experiments on mammalian ag-

ing have employed inbred and wild mouse strains that do not have inordinantly short lifespans (Sacher & Hart, 1978). We have studied several aging populations of long-lived wild mice (*Mus musculus domesticus*) in southern California and have observed one particular population that habitates a squab farm near Lake Casitas (LC) in Ventura County, and in which a highly prevalent infection with murine leukemia virus (MuLV) causes premature death from lymphoma and/or lower motor neuron disease. (For summary, see Gardner, 1978; Gardner & Rasheed, 1982) By classical genetic breeding experiments between LC wild mice and inbred, lymphoma-prone AKR mice, it was shown that susceptibility or resistance to the MuLV related diseases in individual LC mice depended on the segregation of a polymorphic MuLV resistance gene called FV-4^R (Gardner *et al.*, 1980). This paper reviews the natural history and biology of this particular model and points out the lessons learned and possible relevance to humans.

Rationale for studying retroviruses in aging wild mice

Retroviruses encompass a large family of infectious agents unified by a common structure and mode of replication. Although often noncytopathic and nonpathogenic, they are occasionally associated in animals with a variety of malignancies, immunodeficiencies, neurological degeneration and other effects (Coffin, 1992). Like all retroviruses, the MuLV and MMTV are enveloped particles of about 100 nm diameter with an internal spherical or conical core and a dimeric genome of polyadenylated RNA 7-10 kilobases (Kb) in length. The morphology of the MuLV and MMTV particles was called Type C and Type B, respectively. The virus genes encode a capsid protein (*gag*), the reverse transcriptase enzyme (*pol*) and envelope glycoprotein (*env*). After entry into the cell the RNA genome is copied into a double-stranded DNA molecule by the reverse transcriptase and subsequently this DNA is covalently joined to the genomic DNA of the host cell to form the integrated proviruses. The integrated proviruses are stable and serve as a template for viral mRNA and protein synthesis under the regulation of DNA elements within the long terminal repeats (LTR) flanking the proviral DNA at each end. The promoter sequences within the LTR occasionally activate cellular protooncogenes and thereby trigger lymphogenesis or mammary carcinogenesis. Rarely, some of the MuLV may also transduce cellular protooncogenes to form highly oncogenic viruses. Other MuLV strains (e.g. amphotropic MuLV) are now used for gene therapy.

In the late 1960s, an intense search for human retroviruses was launched in the U.S.A. under the auspices of the National Cancer Institute's War on Cancer. The discovery of the human retroviruses, first the human T-cell lymphotropic virus (HTLV), and then the human immunodeficiency virus (HIV), was still over a decade away. The principal models at that time (1960s) for understanding the relationship of retroviruses to cancer were certain inbred mouse strains (e.g. GR, AKR) that were highly susceptible to breast cancer or leukemia. Most remarkably, for both the Type B breast cancer-causing retrovirus (MMTV) and the Type C leukemia viruses (MuLV), the major route of transmission in these inbred mice was from one generation to the next by genetic inheritance. Horizontal transmission of exogenous Type B or Type C virus in these particular mouse strains played a lesser role in virus spread. The discovery of endogenous (inherited) infectious retroviral genes in inbred mice, and also at about the same time, in domestic chickens, led to the formulation of the Virus Oncogene hypothesis (Huebner & Todaro, 1969) that guided much of cancer virus research in the 1970s. As part of this effort, it was important to determine if similar infectious endogenous retroviruses occurred in the outbred feral progenitors of the inbred lab mice or whether these inherited cancer virus genes were an artifact of laboratory inbreeding and purposeful selection for high cancer incidence strains. With this goal in mind, a number of different populations of feral mice in southern California were studied for their natural history of Type C and Type B retroviruses and associated cancers.

Natural history of type C (MuLV) retroviruses in aging wild mice

Wild mice from about fifteen different and widely separated trapping areas in southern California were allowed to age in the laboratory. Over a decade (1968-1978), about 10,000 mice were housed singly in mason jars and allowed to live out their 'natural' lifespan while under observation for spontaneous tumors or other diseases (Gardner *et al.*, 1973, 1971a, 1971b, 1976; Rongey *et al.*, 1973, 1975; Gardner, Lund & Cardiff, 1980). The mice were autopsied when sick or moribund. Tissues were examined microscopically and samples were collected for Type C and Type B retrovirus assays. The mice were also studied in their natural habitation, bred in the laboratory, and subjected to various experimental procedures to induce latent retrovirus expression. The details of how these mice were housed, fed, screened for indigenous pathogens and examined for Type B and Type C retrovirus activity are covered elsewhere (Gardner et al., 1973, 1971a, 1971b, 1976; Rongey et al., 1973, 1975; Gardner, Lund & Cardiff, 1980). The rationale for studying aging wild mice proved very valid and led to the realization of a new biology of retroviruses, different in some major respects from that observed in the inbred mice and actually more predictive of the natural history of the Type C human T cell leukemia virus (HTLV) discovered a decade later in humans (Gardner, 1987). By contrast, a mammary tumor virus has never been found in humans (Cardiff & Gardner, 1985).

Spontaneous cancer, lymphoma, motor neuron disease and type C oncovirus (MuLV) infection in aging wild mice

In most (12 of 15) of the aging populations of wild mice, only a few tumors occurred late in life. The total tumor prevalence was 9.5% mostly after two years of age (Gardner et al., 1973). Most of the tumors were lymphomas localized to spleen or lymph node. The only other tumor types observed were pulmonary adenomas, hepatomas, fibrosarcomas and myeloid leukemia. Typical of these wild mouse populations in Los Angeles County were those located at a squab farm in Bouquet Canyon, a birdseed plant (Hartz Mountain) in downtown Los Angeles, an egg ranch at Munneke and a squab farm in Soledad Canyon. The cumulative total mortality and cumulative specific tumor mortality as calculated by life table methods for these three representative aging populations of wild mice in comparison with the lymphoma-paralysis prone Casitas population of wild mice, described in the next paragraph, is shown in Figure 1 and described in detail in Gardner et al., (1976). Other common pathology found in the aging mice included inflammation of the liver portal tracts (cholangitis) due to biliary obstruction from the dwarf tapeworm (*Hymenolepis nana*) and glomerulosclerosis of the kidneys. Neurologic disease did not occur in these particular mice. No amyloidosis was found. No microscopic abnormalities were found in about 25% of the moribund or dead aging mice. In most of the lymphomas, MuLV core (p30) antigen was detected by complement fixation and Type C particles were observed by electron microscopy (EM) (Gardner et al., 1973, 1971a). Virus was not demonstrable in the other tumor types or in spleens of normal aging mice. MuLV p30 antigen and Type C particles were detected in only a few sarcomas induced in older mice by 3-methyl-cholanthrene (Gardner et al., 1971a). The mice were also very resistant to x-irradiation induced tumors. Indigenous polyoma virus infection did not cause any increase in spontaneous tumors in the wild mice (Gardner et al., 1972b). Treatment with antilymphocytic sera caused an activation of latent cytomegalovirus and death but no appreciable activation of latent MuLV (Gardner et al., 1974). These findings indicated that the feral mice were generally very resistant to spontaneous tumor development and that Type C MuLV was strongly repressed in most feral populations and became detectable only infrequently late in life during spontaneous lymphogenesis or chemical sarcomagenesis. The MuLV that was isolated from lymphomas of these resistant mice was shown to be infectious for mouse embryo and other species' cells in tissue culture but only weakly lymphogenic. The wide *in vitro* tropism of this MuLV was designated 'amphotropic' (Rasheed, Gardner & Chan, 1976), in contrast to the infectious ('ecotropic') MuLV of laboratory mice which only grew in rodent cells. Under natural conditions, the amphotropic MuLV was apparently transmitted as a congenital milk-borne infection and the infected mice were specifically immune-tolerant to this virus because of the newborn age of exposure (see below). Following inoculation of susceptible newborn lab mice and wild mice, the amphotropic MuLV induced less than 30% incidence of lymphoma after a latent period of one year or more (Gardner et al., 1978). In retrospect, if we had been forced to work only with these low-tumor incidence, low MuLV expressor populations of wild mice, progress would indeed have been very slow in understanding the biology of Type C retroviruses in wild mice. We could conclude that wild mice, in general, if spared death from predators, fighting, starvation or infectious disease, were long lived (> 2 years) and cancer resistant. However, a few wild mice that had apparently acquired a congenital infection with amphotropic MuLV were prone to develop lymphoma in later life.

Fortunately, we discovered three populations of wild mice situated in widely separate locations – a duck farm at La Puente and a grain mill in Norwalk, both in Los Angeles County, and a squab farm near Lake Casitas (LC) in Ventura County – characterized by a high prevalence and level of amphotropic MuLV infection and the more frequent occurrence of lymphomas at a younger age (Gardner *et al.*, 1973, 1976). The LC population was most thoroughly studied. The lymphomas arose in the spleen and were of pre-B or null cell origin (Bryant *et al.*, 1981). These MuLV-infected mice were also prone to a fatal lower motor neuron disease with hind leg paralysis (for summary, see Gardner, 1985). The neurologic disease was characterized by high levels of MuLV in the CNS, and development of a spon-

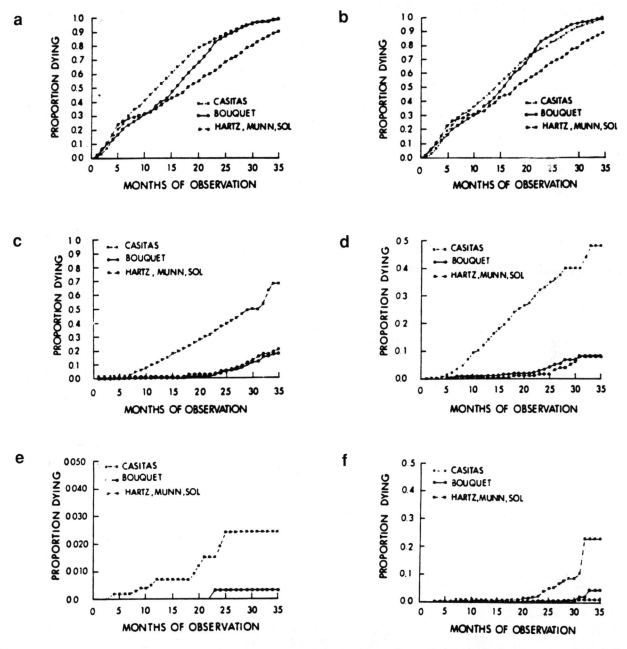

Fig. 1. a. cumulative total mortality from all causes; b. cumulative total mortality excluding deaths from tumors and paralysis; c. cumulative incidence rate for all tumors; d. cumulative incidence rate for lymphoma; e. cumulative incidence rate for carcinoma, excluding hepatoma; f. cumulative incidence rate for hepatoma; g. cumulative incidence rate for lung adenoma; h. cumulative incidence rate for sarcoma; i. cumulative incidence rate for paralysis.

236

giform, non-inflammatory pathology with gliosis and loss of anterior horn motor neurons, mainly in the lumbar spinal cord(Gardner *et al.*, 1973; Andrews & Gardner, 1974). The incidence of lymphomas in these high MuLV expressor mice was ten times greater than that observed in the more common, low MuLV expressor mice (Gardner *et al.*, 1976) (Fig. 1). After about one year of age, 15% of aging MuLV-infected LC mice eventually developed lymphoma, 10% developed paralysis, 2% had both diseases and 5% had epithelial tumors (breast and liver carcinomas, lung adenoma). Virologic studies showed that an ecotropic MuLV, intermixed with the amphotropic MuLV, was uniquely associated with the paralytic disease. Experimental transmission results in laboratory mice proved that the ecotropic MuLV (Officer *et al.*, 1973), including a molecular clone (Jolicoeur *et al.*, 1983), in-

duced both paralysis and lymphoma whereas the amphotropic virus induced only lymphoma (Gardner *et al.*, 1978). In these high MuLV expressor wild mice, both amphotropic and ecotropic viruses were transmitted congenitally via milk leading to a lifelong systemic infection with high titered viremia (10^4 - 10^6 infectious units/ml) and resultant specific immune tolerance (Gardner *et al.*, 1979; Klement *et al.*, 1976). However, general immunity, vigor, and reproductivity were not impaired in the viremic mice. The infectious MuLV was present in many tissues but the major sites of initial virus replication were B cell areas of the spleen (Gardner *et al.*, 1976).

Interestingly, about 15% of LC mice escaped congenital infection with either amphotropic or ecotropic MuLV and remained free of infectious MuLV and related lymphomas or paralysis

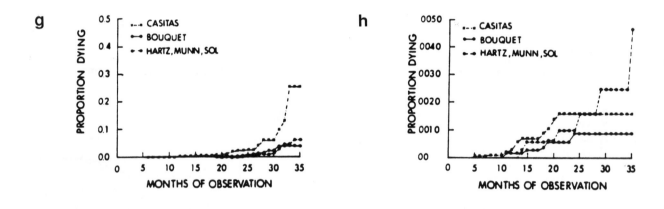

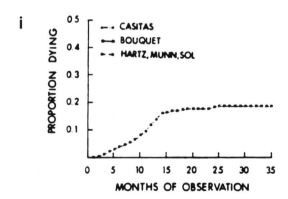

Fig. 1. Continued.

throughout their lifetime (Gardner, 1980). In this respect, they closely resembled the more common cancer resistant, low MuLV populations of wild mice. It was also possible through foster nursing on MuLV-free lab mice to largely eliminate infectious MuLV from the infected LC mice and, thereby, convert them into a long-lived disease resistant population. As with all laboratory mice, all feral mice also contain numerous copies of non-infectious, defective, endogenous MuLV-related proviral DNA in their genome. These endogenous MuLVs are called xenotropic or MCF (mink cell focus forming), and they represent separate envelope classes (Kozak & O'Neill, 1987). These defective proviral genes are thought to be the evolutionary relic of ancient infections with exogenous MuLV. The patterns of these proviral DNA sequences can be used to draw evolutionary lineages between different mouse subpopulations (Kozak & O'Neill, 1987), as will be illustrated later. In certain inbred mouse strains such as AKR, these endogenous, non-infectious MuLV genes recombine with endogenous infectious (ecotropic) MuLV to give rise to highly oncogenic MuLV (Hartley et al., 1977). A similar phenomenon may occur when amphotropic or ecotropic viruses from feral mice are passaged through inbred mice (Rasheed, Pal & Gardner, 1982; Rasheed, Gardner & Lai, 1983). However, this event seldom, if ever, occurs in wild mice (Gardner & Rasheed, 1982). In feral mice, the amphotropic and ecotropic MuLV strains are entirely exogenous (Barbacid, Robbins & Aaronson, 1979; Rassart, Nelbach & Jolicoeur, 1986; O'Neill et al., 1987), genetically stable, do not recombine with endogenous MuLV DNA in the feral mouse genome and remain only weakly oncogenic. Although most of the inherited MuLV proviral genes in laboratory and feral mice have no known biologic function, we shall see shortly a remarkable example in LC mice of a useful proviral function (i.e., the FV-4^R gene).

Natural history of Type B viruses in wild mice

The prevalence of Type B virus (mammary tumor virus) and of spontaneous breast tumors was uniformly low in all of the populations of wild mice, regardless of the level of MuLV infection. Less than 1% of wild mice developed breast tumors after one year of age. Type B particles and MMTV antigen were detectable in about 50% of the spontaneous breast tumors and in about 50% of normal lactating breast tissue from these mice (Rongey et al., 1973; Gardner, Lund & Cardiff, 1980). Type B particles were also detected by EM in seminal vesicles and salivary glands of normal wild mice (Rongey et al., 1975). MMTV in normal wild mouse milk was only weakly tumorigenic in foster nursed laboratory mice. In wild mice, the MMTV, like MuLV, was transmitted by milk and not genetically. Interestingly, some LC wild mice were discovered that completely lacked any MMTV, either in the form of infectious viruses or proviral DNA (Cohen & Varmus, 1979). Breast development and lactation were normal in these mice, although a few did develop hyperplastic mammary lesions (Faulkin et al., 1984). Thus, among wild mice, MMTV exists in low prevalence and causes a few breast tumors, but it plays an insignificant role in the overall biology or causes of death (Cardiff & Gardner, 1985).

Non-genetic control of Type C virus and associated disease in wild mice

Because the Type C virus (MuLV) was transmitted epigenetically in wild mice, it was possible to largely eliminate the virus by immunologic or animal husbandry methods. Passive immunization of newborn LC mice with goat antisera against the LC ecotropic MuLV markedly reduced the titer of ecotropic virus in the progeny of viremic mothers and completely prevented the development of paralysis (Gardner et al., 1980). However, since the amphotropic MuLV of these mice was not neutralized by this antisera, the progeny did develop some lymphomas later in life. Insofar as adult LC mice were already highly viremic, it was not possible to actively immunize them with inactivated whole amphotropic or ecotropic MuLV vaccines (Klement et al., 1976). In LC mice the ecotropic and amphotropic MuLV both replicated early in life primarily in the spleen. Therefore, splenectomy at six weeks of age significantly reduced the serum titer of both classes of MuLV. Since development of the neurologic disease depended on obtaining a high level of ecotropic virus in the CNS early in life, this disease was prevented by the splenectomy

(Gardner *et al.*, 1978). The reduction of viremia also reduced the incidence but did not eliminate the later occurrence of lymphomas in the splenectomized LC mice.

Perhaps the most dramatic means of non-genetic control of MuLV in these mice was by foster nursing on nonviremic laboratory mice (Gardner *et al.*, 1979). Because the amphotropic and ecotropic MuLV were transmitted almost, if not entirely, by milk, it was possible to virtually eliminate these viruses by foster nursing. Conversely, it was possible to introduce the LC MuLV into uninfected lab mice by foster nursing on viremic LC females (Gardner *et al.*, 1979). These infected lab mice then developed the lymphoma and paralytic diseases that were typical of naturally infected LC mice.

Selective breeding of nonviremic LC mice, which constituted about 15% of the total population, also was a very effective measure in that these mice remained free of infectious MuLV throughout their lifespan and did not develop the associated diseases (Gardner *et al.*, 1980). These nonviremic mice were susceptible, however, to the ecotropic and amphotropic MuLV when it was introduced by nursing on viremic LC mothers (Gardner *et al.*, 1979). The ability of this noninfected minority of LC mice to remain uninfected while living among the majority of infected LC mice is further evidence against the horizontal transmission of MuLV among unrelated mice and supports the contention that milk-borne virus is the major route of transmission of MuLV in nature. These findings also support the absence of activated endogenous MuLV as infectious agents in the aging uninfected wild mice.

Genetic control of Type C virus in wild mice by the introduction of the FV-1^b MuLV-resistance allele from inbred mice

Before Type C virus (MuLV) was discovered in wild mice, it had been shown that this type of virus was subject to genetic control in inbred laboratory mice. The gene in lab mice primarily responsible for this control was called FV-1 because it restricted the growth of Friend MuLV (Lilly & Pincus, 1973). Two alleles, FV-1^n and FV-1^b, present in different strains of inbred mice, restricted B-tropic or N-tropic MuLV growth. B-tropic and N-tropic MuLV were defined by their relative suscep-

tibility to growth in Balb/c or NIH Swiss cells, respectively. This dominant gene effect was exerted after viral entry during the process of reverse transcription. All of the wild mouse ecotropic and amphotropic viruses were found to be N-tropic and all of the LC wild mice tested were monomorphic for the FV-1^n genotype. Therefore, the FV-1 locus was fully permissive for the infectious MuLV present naturally in these wild mice and differences in prevalence of ecotropic or amphotropic MuLV in the LC wild mice could not be accounted for by segregation of the FV-1 alleles, as seen in lab mice. However, it was possible to show by cross breeding that the FV-1^b allele from the C57Bl inbred mouse strain could block the growth of the N-tropic LC-MuLV, because the F1 progeny of crosses between viremic LC wild mice and uninfected C57Bl mice were completely free of any infectious Type C virus, even after nursing on the infected LC mothers (Gardner *et al.*, 1976). Backcrosses of the F1 hybrids to the LC parental strain showed that this virus resistance segregated with the FV-1^b allele from the C57Bl parental strain.

Genetic control of Type C virus in wild mice by natural segregation of the FV-4 MuLV-resistance allele

As noted above, the LC wild mice are monophoric for the FV-1^n genotype and their amphotropic and ecotropic MuLV are N-tropic. Therefore, this locus could not account for the control of ecotropic MuLV in these animals. Surprisingly, another dominant gene, called FV-4 was found to control replication of ecotropic MuLV in these mice (Gardner *et al.*, 1980). This gene is distinct from other MuLV restriction genes (e.g., FV-2 and FV-3) described in laboratory mice (Lilly & Pincus, 1973), that have not been described in LC or other wild mice. When FV-4 was first discovered in LC mice it was called Akvr-1 because it restricted replication of the AKR endogenous ecotropic MuLV and prevented lymphoma in AKR × LC F1 hybrids (Gardner *et al.*, 1980). Later, it became clear that Akvr-1 was allelic with and identical in sequence to the FV-4 dominant resistance gene which was first described as preventing exogenous infection by N- and NB-tropic Friend MuLV in Japanese wild mice (*M. m. molossinus*) and the derivative G inbred line

Table 1. Properties of FV-4 restriction gene.

1. Segregates as a dominant gene in LC mice.
   Allele frequency 0.56.
2. Present also on chromosome 12 in Asian wild mice –
   M.m. castaneus and *M.m. molossinus.*
3. Not present in laboratory mice (except G strain derived
   from *M.m. molossinus*). Acquired in LC mice by
   interbreeding with *M.m. castaneus.*
4. Represents a defective endogenous MuLV provirus
   expressing an ecotropic envelope glycoprotein (gp70)
   on the surface of uninfected cells.
5. Interferes with infection by all ecotropic MuLV by
   blocking of the receptors.
6. Determines susceptibility or resistance to exotropic
   MuLV-caused paralysis or lymphoma in individual
   LC mice.

(Suzuki, 1974; Odaka *et al.*, 1981). Genetic studies then confirmed that the ecotropic MuLV resistance gene of G mice, *M. m. molossinus* and LC mice were alleles of a single locus on chromosome 12 (Odaka *et al.*, 1981; O'Brien *et al.*, 1983).

The FV-4 resistance gene was discovered serendipitously while cross-breeding LC mice with AKR inbred mice (Gardner *et al.*, 1980). Three patterns of MuLV viremia were observed in the F1 progeny of the individual AKR × LC crosses; at two months of age, either all of the F1 progeny were viremic, all were nonviremic, or about 50% were viremic. The viremia was entirely due to AKR ecotropic MuLV because only LC males or nonviremic LC females were bred to the AKR mice and we had already determined that infectious MuLV is transmitted only by LC females (Gardner *et al.*, 1979). In F1 backcrosses to AKR mice and in F2 progeny, viremia was present in about 50% and 25%, respectively, thus indicating the segregation in LC mice of a dominant gene capable of strongly blocking the expression of the endogenous ecotropic MuLV inherited from the AKR parent. The dominant MuLV-restrictive allele was called FV-4[R], and the recessive allele FV-4[S]. The FV-4[R] restriction effect was long lasting (> 18 mo) and associated with prevention of lymphoma in the AKR x LC F1 progeny. The properties of this MuLV restriction gene are summarized in Table 1. FV-4 (Akvr-1) strongly blocks cell to cell spread of all exogenous and endogenous infectious ecotropic MuLVs (N-tropic, B-tropic or NB-tropic) both *in vivo* and *in vitro* (Odaka *et al.*, 1981; Rasheed &

Gardner, 1983). All ecotropic MuLVs use the same receptor encoded by the Rec-1 locus on mouse chromosome 5 (Sarma, 1967) and thus would be interfered with by binding of the FV-4 encoded gp70 to this receptor. This ecotropic MuLV receptor has recently been cloned and sequenced (Albritton *et al.*, 1989) and its normal function shown to be that of a basic amino acid transporter (Kim, *et al.*, 1991; Wang *et al.*, 1991). The homologous human gene has also been cloned (Yoshimoto, Yoshimoto & Meruelo, 1991), and both mouse and human genes are activated in rapidly proliferating cells (Yoshimoto, Yoshimoto & Meruelo, 1992). FV-4[R] does not, however, block amphotropic MuLV, which uses a different receptor encoded by a gene on chromosome 8 (Gazdar *et al.*, 1977). Restriction is generally stronger *in vivo* than *in vitro* and stronger in hematopoietic cells than in fibroblasts.

Expression of the FV-4[R] locus in uninfected cells is associated with the presence of an ecotropic MuLV-related envelope glycoprotein of about 70,000 kD (gp70) on the cell surface (Yoshikura & Odaka, 1982; Ikeda & Odaka, 1983; Dandekar *et al.*, 1987). Hirt analysis of FV-4[R] resistant cells shows no proviral DNA after ecotropic virus challenge, whereas the same cells do show proviral DNA after challenge with amphotropic virus (Dandekar *et al.*, 1987). The block to infection at the cell surface level presumably occurs by receptor interference. Using an AKR ecotropic env probe, the FV-4[R] gene was cloned and shown to be a truncated proviral genome containing a small segment of *pol*, the entire *env* gene (gp70) and a 3' LTR (Dandekar, S. 1987; Kozak, C.A. *et al.*, 1984; Ikeda, H. *et al.*, 1985). Sequence analysis showed that the FV-4[R] gp70 was 70% related to the AKR ecotropic gp70, 58% similar to endogenous xenotropic and MCF envelope sequences, but 90% related to the LC ecotropic gp70. The protein encoding sequence of the FV-4[R] allele, cloned from LC mice, was identical to that of the FV-4[R] gene from *M. m. molossinus* (Dandekar *et al.*, 1987). Southern blot hybridization showed that *M. m. molossinus, M. m. castaneus* and LC mice each contained the same FV-4[R] provirus (Kozak & O'Neill, 1987). *M. m. mollosinus* contains a single FV-4[R] provirus whereas LC mice and *M. m. castaneus* each carry several additional copies of the FV-4[R] gene. Moreover, *M. m. castaneus* and LC mice share two specific FV-4 proviral integrations. Therefore, the FV-4 resistance gene was probably introduced into *M. m. molossinus* and LC mice by natural interbreeding with *M. m. castaneus* during this past century. LC feral mice are thus hybrids between *M. m. domesticus* from Europe and *M. m. castaneus* from Asia (Kozak & O'Neill, 1987). Opportunities for such interbreeding would have occurred via the shipping trade or immigration from Asian countries.

The observed frequency of the FV-4 resistance allele in LC mice randomly trapped at the squab farm was 56% which does not vary from expectation of the Hardy-Weinberg equilibrium (Gardner *et al.*, 1980). The probable frequency of LC mice that contain at least one FV-4 restriction allele is 80% and all of these mice would therefore be resistant to the LC ecotropic MuLV and associated paralysis and lymphoma. They would not, however, be totally resistant to lymphoma, because they may still be congenitally infected with amphotropic MuLV. Compared to the ecotropic MuLV, however, the amphotropic MuLV is less lymphogenic. The 20% of LC mice not inheriting this resistance allele would be considered homozygous for the susceptibility allele (i.e. FV-4^{ss}) and would be vulnerable to ecotropic virus congenital infection and the associated diseases later in life. Inheritance of FV-4^{R} would thus have a protective value on the survival of individual LC mice. Although manifested after the onset of breeding, evolutionary conservation of this locus might be expected because of its beneficial effect on survival and reproductive life span. In summary, this fascinating wild mouse model (summarized recently in Gardner, Kozak & O'Brien, 1991) has taught us that susceptibility or resistance to lymphoma and a slow neurologic disease, both developing after midlife (> 1 year of age) is determined, not only by exposure at birth to maternally transmitted ecotropic-MuLV, but ultimately by inheritance of an ecotropic-MuLV resistance gene (FV-4^{R}) segregating in this outbred population.

Lessons learned and possible relevance to humans

A major lesson learned from this study is that feral mice are indeed a valuable model for studying the biology of retroviruses in relation to etiologically-associated tumors and non-oncogenic diseases. The natural history of MuLV in wild mice, discovered in the 1970s, resembled the natural history of HTLV discovered in humans a decade later (Gardner, 1987). Remarkable similarities between the Type C leukemia viruses of wild mice and humans include the primary transmission via maternal milk early in life, the long latent period before onset of lymphoma, the genetic stability of the virus, and the ability of the same virus to induce both lymphoma and a non-neoplastic, degenerative neurologic disease. As shown with MuLV in this wild mouse model, avoidance of breast feeding appears effective in preventing transmission of HTLV infection from carrier mothers to their children in endemic regions of southern Japan (Hino & Doi, 1989).

In respect to human retrovirology, it is possible that natural resistance to HTLV and HIV infection may someday be found attributable to resistance genes such as FV-4^{R}. The human genome certainly contains a large complement of endogenous defective HTLV-related proviral DNA (Shih, Misra & Rush, 1989). Introduction of such cloned resistance genes into bone marrow progenitor cells might become a future application of gene therapy directed at control of human retroviral diseases. A counterpart to the mammary tumor virus of mice has not been found in humans, and this virus has only a minimal oncogenic potential in feral mice. In respect to aging, these studies point out the potential of feral populations to uncover other loci associated with increased longevity by selection for specific disease resistance. However, it remains unclear if FV-4 and other virus resistance genes have any effect on general senescence.

References

Albritton, L.M., L. Tseng, D. Scadden & J.M. Cunningham, 1989. A putative murine ecotropic retrovirus receptor gene encodes a multiple membrane-spanning protein and confers susceptibility to virus infection. Cell 57: 659-666.

Andrews, J.M. & M.B. Gardner, 1974. Lower motor neuron degeneration associated with type C RNA virus infection in mice: neuropathological features. J. Neuropath. Exp. Neuro. 33: 285-307.

Barbacid, M., K.C. Robbins & S.A. Aaronson, 1979. Wild mouse RNA tumor viruses. A nongenetically transmitted virus group closely related to exogenous leukemia viruses of laboratory mouse strains. J. Exp. Med. 149: 254-266.

Bryant, M.L., J.L. Scott, B.K. Pak, J.D. Estes & M.B. Gardner, 1981. Immunopathology of natural and experimental lymphomas induced by wild mouse leukemia virus. Am. J. Pathol. 104: 272-282.

Cardiff, R.D. & M.B. Gardner, 1985. The human-MuMTV interrelationship: does it really matter? In: Understanding Breast Cancer: Clinical and Laboratory Concepts, edited by M. Rich, J. Hager, and P. Furmanski, New York, Marcel Dekker, p. 317-333.

Coffin, J., 1992. Structure and Classification of Retroviruses. In: The Retroviridae, Vol. 1, edited by J.A. Levy, New York, Plenum Press, p. 19-49.

Cohen, J.C. & H.E. Varmus, 1979. Endogenous mammary tumour virus DNA varies among wild mice and segregates during inbreeding. Nature 278: 418-423.

Dandekar, S., P. Rossitto, S. Picket, G. Mockli, H. Bradshaw, R. Cardiff & M. Gardner, 1987. Molecular characterization of Akvr-1 restriction gene: a defective endogenous retrovirus identical to Fv-4R. J. Virol. 61: 308-314.

Faulkin, L.J., D.J. Mitchell, L.J.T. Young, D.W. Morris, R.W. Malone, R.D. Cardiff & M.B. Gardner, 1984. Hyperplastic and neoplastic changes in the mammary glands of feral mice free of endogenous MuMTV provirs. J. Natl. Cancer Inst. Monog. 73: 971-982.

Finch, C.E., 1990. Longevity, Senescence and the Genome, Chicago, IL, University of Chicago Press, pp. 317-339.

Gardner, M.B., 1978. Type C viruses of wild mice: characterization and natural history of amphotropic, ecotropic, and xenotropic MuLV. In: Current Topics in Microbiology and Immunology, edited by W. Arber, W. Henle, P.H. Hofschneider et al., New York, Springer-Verlag, p. 215-260.

Gardner, M.B., 1985. Retroviral spongiform polioencephalomyelopathy. Rev. Infect. Dis. 7: 99-110.

Gardner, M.B., 1978. Naturally occurring leukaemia viruses in wild mice: how good a model for humans. In: Cancer Surveys, edited by L.M. Franks, p. 55-71.

Gardner, M.B., A. Chiri, M.F. Dougherty, J. Casagrande & J.D.Estes, 1979. Congenital transmission of murine leukemia virus from wild mice prone to the development of lymphoma and paralysis. J. Natl. Cancer Inst. 62: 63-70.

Gardner, M.B., J.D. Estes, J. Casagrande & S. Rasheed, 1980. Prevention of paralysis and suppression of lymphoma in wild mice by passive immunization to congenitally transmitted murine leukemia virus. J. Natl. Cancer Inst. 64: 359-364.

Gardner, M.B., B.E. Henderson, J.D. Estes, H. Menck, J.C. Parker & R.J. Huebner, 1973. Unusually high incidence of spontaneous lymphomas in wild house mice. J. Natl. Cancer Inst. 50: 1571-1579.

Gardner, M.B., B.E. Henderson, J.D. Estes, R.W. Rongey, J. Casagrande, M. Pike & R.J. Huebner, 1971. Spontaneous tumor occurrence and C-type virus expression in polyoma infected aging wild mice. J.N.C.I. 52: 979-981.

Gardner, M.B., B.E. Henderson, J.D. Estes, R.W. Rongey, J. Casagrande, M. Pike & R.J. Huebner, 1976. The epidemiology and virology of C-type virus-associated hematological cancers and related diseases in wild mice. Cancer Res. 36: 574-581.

Gardner, M.B., B.E. Henderson, J.E. Officer, R.W. Rongey, J.C. Parker, C. Oliver, J.D. Estes & R.J. Huebner, 1973. A spontaneous lower motor neuron disease apparently caused by indigenous type-C RNA virus in wild mice. J. Natl. Cancer Inst. 51: 1243-1254.

Gardner, M.B., B.E. Henderson, R.W. Rongey, J.D. Estes & R.J. Huebner, 1973. Spontaneous tumors of aging wild house mice. Incidence, pathology, and C-type virus expression. J Natl. Cancer Inst 50: 719-734.

Gardner, M.B., Klement, V., Henderson, B. et al., 1978. Lymphoma, Paralysis and Oncornaviruses of Wild Mice. In: Tumours of Early Life in Man and Animals, edited by L. Severi, Monteluc, Italy: Perugia Quadrennial Conferences, p. 343-356.

Gardner, M.B., V. Klement, B.E. Henderson, H. Meier, J.D. Estes & R.J. Huebner, 1976. Genetic control of type C virus of wild mice. Nature 259: 143-145.

Gardner, M.B., V. Klement, R.R. Rongey, P. McConahey, J.D. Este & R.J. Huebner, 1976. Type C virus expression in lymphoma-paralysis-prone wild mice. J. Natl. Cancer Inst. 57: 585-590.

Gardner, M.B., C.A. Kozak & S.J. O'Brien, 1991. The Lake Casitas wild mouse: evolving genetic resistance to retroviral disease. Trends in Genetics 7: 22-27.

Gardner, M.B., J.K. Lund & R.D. Cardiff, 1980. Prevalence and distribution of murine mammary tumor virus antigen detectable by immuocytochemistry in spontaneous breast tumors of wild mice. J. Natl. Cancer Inst. Monog. 64: 1251-1257.

Gardner, M.B., J.E. Officer, J. Parker, J.D. Estes & R.W. Rongey, 1974. Induction of disseminated virulent cytomegalovirus infection by immunosuppression of naturally chronically infected wild mice. Infect. Immun. 10: 966-969.

Gardner, M.B., J.E. Officer, R.W. Rongey, J.D. Estes, H.C. Turner & R.J. Huebner, 1971. C-type RNA tumour virus genome expression in wild house mice. Nature 232: 617-620.

Gardner, M.B. & S. Rasheed, 1982. Retroviruses in feral mice. Int. Rev. Exp. Path. 23: 209-267.

Gardner, M.B., S. Rasheed, J.D. Estes & J. Casagrande, 1980. The history of viruses and cancer in wild mice. In: Viruses in Naturally Occurring Cancer, Cold Spring Harbor Conferences on Cell Proliferation, Vol. 7, Cold Spring Harbor, Cold Spring Harbor Laboratory, p. 971-987.

Gardner, M.B., S. Rasheed, B.K. Pal, J.D. Estes & S.J. O'Brien, 1980. Akvr-1, a dominant murine leukemia virus restriction gene, is polymorphic in leukemia-prone wild mice. Proc. Natl. Acad. Sci. U.S.A. 77: 531-535.

Gazdar, A.F., H. Oie, P. Lalley, W.W. Moss & J.D. Minna, 1977. Identification of mouse chromosomes required for murine leukemia virus replication. Cell 11: 949-956.

Hartley, J.W., N.K. Wolford, L.J. Old & W.P. Rowe, 1977. A new class of murine leukemia virus associated with development of spontaneous lymphomas. Proc. Natl. Acad. Sci. U.S.A. 74: 789-792.

Hino, S. & H. Doi, 1989. Mechanisms of HTLV-1 Transmission. In: HTLV-1 and the Nervous System, edited by G.C. Roman, J.-C. Vernant & M. Osame, New York, Alan R. Liss, Inc., p. 495-501.

Huebner, R.J. & G.J. Todaro, 1969. Oncogenes of RNA tumor viruses as determinants of cancer. Proc. Natl. Acad. Sci. U.S.A. 64: 1087-1094.

Ikeda, H., F. Laigret, M.A. Martin & R. Repaske, 1985. Characterization of a molecularly cloned retroviral sequence associ-

242

ated with Fv-4 resistance. J. Virol. 55: 768-777.

Ikeda, H. & T. Odaka, 1983. Cellular expression of murine leukemia virus gp70-related antigen on thymocytes of uninfected mice correlates with Fv-4 gene-controlled resistance to Friend leukemia virus infection. Virol. 128: 127-139.

Jolicoeur, P., N. Nicolaiew, L. DesGroseillers & E. Rassart, 1983. Molecular cloning of infectious viral DNA from ecotropic neurotropic wild mouse retrovirus. J. Virol. 56: 639-643.

Kim, J.W., E.I. Closs, L.M. Albritton & J.M. Cunningham, 1991. Transport of cationic amino acids by the mouse ecotropic retrovirus receptor. Nature 352: 725-728.

Klement, V., M.B. Gardner, B.E. Henderson & et al., 1976. Inefficient humoral immune response of lymphoma-prone wild mice to persistent leukemia virus infection. J. Natl. Cancer Inst. 57: 1169-1173.

Kozak, C.A., N.J. Gromet, H. Ikeda & C.E. Buckler, 1984. A unique sequence related to the ecotropic murine leukemia virus is associated with the Fv-4 resistance gene. Proc. Natl. Acad. Sci. U.S.A. 81: 834-837.

Kozak, C.A. & R.R. O'Neill, 1987. Diverse wild mouse origins of xenotropic, mink cell focus-forming, and two types of ecotropic proviral genes. J. Virol. 61: 3082-3088.

Lilly, F. & T. Pincus, 1973. Genetic control of murine viral leukemogenesis. Adv. Cancer Res. 17: 231-277.

O'Brien, S.J., E.J. Berman, J.D. Estes & M.B. Gardner, 1983. Murine retroviral restriction genes Fv-4 and Akvr-1 are alleles of a single locus. J. Virol. 47: 649-651.

O'Neill, R.R., J.W. Hartley, R. Repaske & C.A. Kozak, 1987. Amphotropic proviral envelope sequences are absent from the Mus germ line. J. Virol. 61: 2225-2231.

Odaka, T., H. Ikeda, H. Yoshikura, K. Moriwaki & S. Suzuki, 1981. FV-4: gene controlling resistance to NB-tropic Friend murine leukemia virus. Distribution in wild mice, introduction into genetic background of BALB/c mice, and mapping of chromosomes. J. Natl. Cancer. Inst. 67: 1123-1127.

Officer, J.E., N. Tecson, J.D. Estes, E. Fontanilla, R.W. Rongey & M.B. Gardner, 1973. Isolation of a neurotropic type C virus. Science 181: 945-947.

Rasheed, S. & M.B. Gardner, 1983. Resistance of fibroblasts and hematopoietic cells to ecotropic murine leukemia virus infections: An Akvr-1R gene effect. Int. J. Cancer 31: 491-496.

Rasheed, S., M.B. Gardner & E. Chan, 1976. Amphotropic host range of naturally occurring wild mouse leukemia viruses. J. Virol. 19: 13-18.

Rasheed, S., M.B. Gardner & M.M.C. Lai, 1983. Isolation and characterization of new ecotropic murine leukemia viruses

after passage of an amphotropic virus in NIH Swiss mice. Virol. 130: 439-451.

Rasheed, S., B.K. Pal & M.B. Gardner, 1982. Characterization of a highly oncogenic murine leukemia virus from wild mice. Int. J. Cancer 29: 345-350.

Rassart, E., L. Nelbach & P. Jolicoeur, 1986. Cas-Br-E murine leukemia virus: sequencing of the paralytogenic region of its genome and derivation of specific probes to study its origin and the structure of its recombinant genomes in leukemic tissues. J. Virol. 60: 910-919.

Rongey, R.W., A.H. Abtin, J.D. Estes & M.B. Gardner, 1975. Mammary tumor virus particles in the submaxillary gland, seminal vesicle, and nonmammary tumors of wild mice. J. Natl. Cancer Inst. 54: 1149-1156.

Rongey, R.W., A. Hlavackova, S. Lara, J. Estes & M.B. Gardner, 1973. Types B and C RNA virus in breast tissue and milk of wild mice. J. Natl. Cancer Inst. 50: 1581-1589.

Sacher, G.A. & R.W. Hart, 1978. Longevity, aging and comparative cellular and molecular biology of the house mouse, mus musculus, and the white-footed mouse, peromyscus leucopus. In: Genetic Effects on Aging, Birth Defects Original Article Series, Vol. 14, No. 1, edited by D. Bergsma & D.E. Harrison, New York, A.R. Liss, p. 71-96.

Sarma, P.S., M.P. Cheong, J.W. Hartley & R.J. Huebner, 1967. A viral interference test for mouse leukemia viruses. Virol. 33: 180-184.

Shih, A., R. Misra & M.G. Rush, 1989. Detection of multiple, novel reverse transcriptase coding sequences in human nucleic acids: relation to primate retroviruses. J. Virol. 63: 64-75.

Suzuki, S., 1974. FV-4: A new gene affecting the splenomegaly-induction by Friend leukemia virus. Jpn. J. Exp. Med. 45: 473-478.

Wang, H., M.P. Kavanaugh, R.A. North & D. Kabat, 1991. Cell-surface receptor for ecotropic murine retroviruses is a basic amino-acid transporter. Nature 352: 729-731.

Yoshikura, H. & T. Odaka, 1982. Surface antigen expressed in hematopoietic cells derived from Fv-4r mouse strains. J. Natl. Cancer Inst. 68: 1005-1009.

Yoshimoto, T., E. Yoshimoto & D. Meruelo, 1991. Molecular cloning and characterization of a novel human gene homologous to the murine ecotropic retroviral receptor. Virol. 185: 10-17.

Yoshimoto, T., E. Yoshimoto & D. Meruelo, 1992. Enhanced gene expression of the murine ecotropic retroviral receptor and its human homolog in proliferating cells. J. Virol. 66: 4377-4381.

M.R. Rose and C.E. Finch (eds.), Genetics and Evolution of Aging, 243–255, 1994.
© 1994 *Kluwer Academic Publishers. Printed in the Netherlands.*

Genetics of life span in mice

E. J. Yunis[1] & M. Salazar[2]
[1] *Dana Farber Cancer Institute. Boston, MA 02115, USA*
[2] *Department of Pathology Harvard Medical School. Boston, MA 02115, USA*

Received and accepted 22 June 1993

Key words: MHC (Major Histocompatibility complex), TCR (T cell receptors), thymus, viral infection

Abstract

Thymic involution that occurs earlier in some individuals than others may be the result of complex interactions between genetic factors and the environment. Such interactions may produce defects of thymus-dependent immune regulation associated with susceptibility to developing autoimmune diseases, malignancy, and an increased number of infections associated with aging.

The major histocompatibility complex may be important in determining profiles of cause of death and length of life in mice. Genetic influences on life span involve interactions between loci and allelic interactions during life which may change following viral infections or exposure to other environmental factors. We have used different experimental protocols to study the influence of H-2 on life span and found that interactions between genetic regions, are inconsistent, particularly when comparing mice infected or not infected with Sendai virus.

Genes important for life span need to be studied against many genetic backgrounds and under differing environmental conditions because of the complexity of the genetics of life span. Several genetic models were used to demonstrate that the MHC is a marker of life span in backcross and intercross male mice of the H-2^d and H-2^b genotypes in B10 congenic mice. Females lived longer than males in backcross and intercross mice, while males lived longer than females in B10 congenics. H-2^d was at a disadvantage for life span in backcross mice of the dilute brown and brown males exposed to Sendai infection, but intercross mice not exposed to Sendai virus of the same genotype were not at a disadvantage. H-2^d mice were not disadvantaged when compared to H-2^b in B10 congenics that had not been exposed to Sendai virus infection but the reverse was true when they were exposed. Overall, all our studies suggest that genetic influences in life span may involve interactions between loci and many allelic interactions in growing animals or humans. These genetic influences on life span may vary after they are exposed to infections or other environmental conditions. This paper emphasizes the need to use several genetic models, especially animals that have been monitored for infections, to study the genetics of life span.

Introduction

The potential life span of a species, defined as the duration of life of the longest survivors, is determined by genetic factors which control the rate of cellular and organ development and involution. It has been difficult to study the cellular and genetic factors influencing life span because the incidence of disease increases exponentially with age (Simms, 1946; Jucket & Rosenberg, 1988) and it is difficult to determine if the same mechanisms that mediate some diseases also control aging. Longevity of individuals within a species can also be determined by an absence of genetic susceptibility to diseases and the maintenance of the vigor required to live a normal life. Among the critical functions, a well-balanced immune system is necessary to cope with environmental stimuli and internal degenerative changes during life. The finding that life spans vary in different strains of mice suggested

244

that genetic markers for life span may be identified to explain such differences. Among the biomarkers of life span, the vigor of the immune system, which is under genetic control, has been thought to be important.

Theories concerning immunological dysfunction as a factor in aging have been stated by Walford (1969) and promulgated and developed by Burnet (1971). One theory holds that the immune system is essential for maintenance of health, and that its integrity determines survival advantage. Walford suggested that the breakdown in self recognition among cells by the immune system is secondary to genetic changes in somatic cells controlling immunity, resulting in the production of autoimmunity, malignancy, or increased susceptibility to developing infections.

The decline of vigorous immune function in aging humans or mice was found to be associated with the same diseases and immunological abnormalities found in individuals lacking T cells (Good & Yunis, 1974). Observations of this type led Burnet (1958) to suggest that clones of cells, ordinarily eliminated by the immune system, persist as immune functions become deficient with age. Immunodeficiencies, together with the persistence of such cells, may explain the autoimmunity which is frequently observed in both aging and immunodeficient individuals (Burnet, 1970).

Alternatively, others (Good & Yunis, 1974) interpreted these abnormalities in terms of a 'forbidden antigen' theory of autoimmunity, arguing that under immunodeficient circumstances, antigens otherwise excluded from the body or promptly eliminated are permitted to enter, to remain and to generate cross-reacting antibodies. They believe that a possible basis for an immunological theory of aging is based on thymic involution which can occur at varying ages (Yunis et al., 1972). It was suggested that such a decline may be a genetically controlled 'clock' which operates at a rate consistent with the median life span of the species. Therefore, for the longest-lived member of a species, the 'clock' adheres to the limits imposed by the postulate of Hayflick (1965). It is unknown what mechanism controls the involution of the thymus. The genetic control of thymus involution is unknown but it may involve the CNS-endocrine system and not immune mechanisms *per se*. However, thymus involution could produce immune alterations and

Table 1. Influence of age of donor of thymus and spleen and reconstitution of neonatally thymectomized mice.

Strain	Age of donor	8 month survivors	
		Thymus graft	Spleen cells
A	4 months	46% (28)	85% (34)
	12 months	52% (23)	87% (30)
	24 months	23% (13)	35% (26)
CBA	4 months	91% (11)	100% (15)
	12 months	92% (12)	91% (22)
	24 months	76% (17)	86% (21)

Total number of mice studies per group in parentheses.
Mice were treated intraperitoneally at 2 weeks of age with thymus grafts and spleen cells from mice obtained at different ages. From Yunis *et al.*, 1972.

diseases that curtail life span, such as infections, neoplastic and autoimmune diseases, which could occur in immunodeficiency states as well as during aging (Roberts-Thomsen *et al.*, 1974; Hallgren *et al.*, 1978). Consistent with this explanation was the finding that some aged humans have an increased number of immunoglobulin-secreting cells and a deficiency of a regulatory T cell (Strelkauskas, Andrew & Yunis, 1981). This regulatory cell with the CD45 surface marker may be the cell necessary to provide the signal to increase the transduction defect of lymphocytes from aged mice and humans.

Involution of the thymus and ensuing deficiency in cell mediated immune function occur earlier in certain individuals and inbred strains than others. For example, immunodeficiency observed in strain A mice during the second year of life is strikingly similar to that produced by neonatal thymectomy (Yunis, Fernandes & Stutman, 1971; Yunis *et al.*, 1972; Good & Yunis, 1974). Conversely, CBA mice, in which thymic involution is relatively late, tend to be long-lived and to maintain immunologic function longer than others. An experiment demonstrating that thymus involution occurs at earlier age in autoimmune susceptible strain A than a long lived strain (CBA) is shown in Table 1. Thymus grafts or spleen cells from CBA mice prevented death from wasting disease in neonatally thymectomized mice in a significant number of animals treated with thymus grafts from 4, 12 or 24 month old donors. The capacity of spleen cells obtained from A mice (known to be autoimmune susceptible and to have a short life span) prevented wasting disease and death when the donors were 4 and 12 months of age but prevented wasting disease and death in only 35% when donors were 24 months of age. Thymus grafts from A mice prevented death from wasting less frequently, especially with donors at 24 months of age, than those grafted with thymus from CBA mice. These results showed that thymus involution and cells from the thymus-dependent system occur earlier in A stain than in CBA mice.

Neonatal thymectomy shortened the incubation period for development of antinuclear antibodies and immune complex glomerular disease in both the NZB and its (NZB × NZW)F1 hybrids. The spontaneous development of antinuclear antibodies in NZB and A mice (Teague *et al.*, 1970; Good & Yunis, 1974; Fernandes, Good & Yunis, 1977) was likewise facilitated by neonatal thymectomy. Such mice produced cell proliferation in the lymphoid tissues similar to that observed in aging mice of these autoimmune-susceptible strains. In contrast, CBA/H and C3H mice do not develop spontaneous autoimmunity during aging and showed decreased production of autoantibodies following neonatal thymectomy.

Neonatal thymectomy in autoimmune-susceptible mice produces several alterations: immunologic deficiencies, autoimmune hemolytic anemia, antinuclear and anti-DNA antibodies, and the hematologic, hepatic, splenic and renal lesions which appear early in life. These changes represent an acceleration of the processes associated with cellular immunodeficiency during aging in these strains (Yunis *et al.*, 1972). Table 2 summarizes one of these experiments. It shows that anti-DNA antibodies are found with higher frequency at 12 months of age in (NZB × NZW)F1 or NZB strains which are genetically susceptible to developing autoimmune diseases; the A, NZW and (A × NZW)F1 strains developed these antibodies later and with lower frequency. Further, C3H and CBA/H were strains resistant to developing autoimmunity and did not develop anti-DNA antibodies even at 18 months of age.

Attempts to correct immunodeficiency in thymectomized animals provided another line of evidence linking immunodeficiency to autoimmunity. In neonatally thymectomized mice both immunodeficiency and autoimmunity could be prevented by transplantation of thymus or spleen lymphocytes from syngeneic, semiallogeneic or allogeneic young donors (Good & Yunis, 1974; Fernandes, Good & Yunis, 1977). Even after wasting disease and autoimmune processes have appeared, effects in neonatally thymectomized mice were reversed by treatment with multiple thymus grafts or by injections of large number of thymocytes or peripheral lymphoid cells (Fernandes, Good & Yunis, 1977; Yunis *et al.*, 1972).

In summary, thymic involution is associated with the progressive decline of immune functions (Fernandes, Good & Yunis, 1977; Hirokawa, 1977). Although immune functions are important in the aging process, there are other functions that are equally important. Autoimmune diseases, infections and neoplastic diseases increase dramatically as the immune functions decline; such diseases are hallmarks of the aging process. As will be dis-

Table 2. DNA antibodies in different strains of mice.

Strain	6 months		12 months		18 months	
	%	Nº mice	%	Nº mice	%	Nº mice
(NZB × NZW) F1	41%	17	59%	17	*	
NZB	37%	32	57%	23	*	
Af	5%	20	19%	21	21%	19
NZW	5%	22	28%	18	*	
(NZB × A) F1	7%	13	25%	15	*	
(NZB X C3H) F1	0%	23	4%	23	*	
C3H	0%	14	0%	14	0%	9
CBA/H	0%	24	0%	22	0%	15

* All animals died before 18 months of age; % indicates percentage of mice with DNA antibodies.
From Yunis *et al.*, 1972.

cussed in this chapter, other genetic systems may be important markers of life span. Environmental influences can also accelerate or delay this natural process. For example, immune abnormalities of aging can be restored by caloric restriction (Fernandes, Good & Yunis, 1977). Viral infections can also alter the genetic interactions which probably influence the immune changes that occur later in the life of some strains of mice.

Genes that affect life span in relation to the immune system

Variable life spans of pure inbred strains suggest that there are genetic factors that may be responsible for this variation. In regard to the role of genes in life span, many reports have shown that females live longer that males and that genetic defects that cause immunodeficiency shorten the life span of humans. Immune dysfunctions can be found in mice with mutations of genes in chromosomes X, Y,1, 2, 3, 4, 5, 6, 10, 11, 13, 14, 15, 16 and 19. These mice are commercially available but life span studies in them have not been performed (Shultz, 1993). Also, MHC alleles are important in immune responses and certain alleles may be involved in the control of autoimmunity and susceptibility for malignancies, which may indicate that certain MHC alleles may be associated with longer life span. We believe that certain MHC profiles are associated with shorter life span and others with longer life span, and that the genetic interactions involved differ when the animals are raised under different environmental conditions. The involvement of the MHC in such interactions which influence the development of malignancy and autoimmune disseases in aged individuals warrants further discussion.

MHC and malignancy

The first experimental evidence implicating the MHC in disease came from experimental work in mice which suggested that certain H-2 types mediated susceptibility to developing leukemia. In the case of Gross leukemia, H-2^b strains are resistant (Lilly, Boyse & Old, 1964). These results stimulated the studies of MHC associations with malignancy (Amiel, 1967). In other studies, increased tumor resistance of F1 hybrids to parental strain

tumors was a manifestation of MHC linked genes, but such genes were not exclusively in the I region of the mouse H-2 complex and thus included genes that are distinct from the presently identified H-2 linked immune response genes. HLA associations to particular malignancies may represent survival differences rather than tumor incidence differences. This is a particular problem because of the necessary retrospective character of the HLA and malignancy studies. The most striking examples to be noted are an excess of HLA-A2 in long term survivors of acute lymphatic leukemia and the excess of HLA-A19 and HLA-B5 among short-term survivors of Hodgkin's disease (Simons & Amiel, 1977; Falk & Osoba, 1977).

Studies in mice suggested an influence of the MHC on natural killer cell activity as well as the possibility that natural killer cell activity influences the *in vivo* behavior of transplanted tumors (Williams & Yunis, 1978). Natural killer cell activity was elevated in F1 hybrids and controlled by MHC genes, the Hh genes (Petranyi *et al.*, 1976). HLA-B12 individuals had relatively high levels of natural killer cell activity (Simons & Amiel, 1977). Also, patients with acute myelogenous leukemia responded better to chemotherapy when they had the HLA-B12 phenotype (Parrish, Heise & Cooper, 1977; Pross & Baines, 1976). Thus, MHC genes as well as Hh genes may control the levels of NK killing and may explain HLA associations with survival of patients with AML.

The cell-mediated defect in mice or patients carrying tumors may not only involve NK cells but also T cells or soluble factors produced by them. Cancer patients and tumor-bearing mice demonstrated deficient immune functions manifested by decreased cell-mediated-immune assays and decreased lymphocyte cytotoxicity (Broder & Waldman, 1978; Hersch & Openheim, 1965; Takasugi, Ranaseyer & Takasugi, 1977; Young *et al.*, 1972). Furthermore, several mechanisms have been proposed to explain the immune impairment in cancer-carrying animals. These included production of suppressor cells, suppressor factors by tumor cells, deletion of tumor-specific clones and diminished production of cytokines (North & Bursuker, 1984; Webb, Morris & Sprent, 1990; Fearon *et al.*, 1990).

Association of MHC and autoimmune disease

HLA antigens have been found associated with

many diseases, a significant number of which are autoimmune in nature. In Caucasians a large number of diseases such as Insulin Dependent Diabetes, Rheumatoid Arthritis, several endocrinopathies, and Pemphigus Vulgaris (PV) were associated with class II MHC gene products carried by haplotypes marked by either HLA-DR3 or HLA-DR4. In autoimmune diseases characterized by autoantibodies, several genetic factors including the MHC, in addition to environmental factors, may be important in pathogenesis. An example of the role of the MHC and environmental factors in autoantibody production has been described in Pemphigus Vulgaris. This autoimmune disease caused by high concentrations of antibody to epidermal cadherin was associated with two kinds of HLA-DR4, DQ8 haplotypes dominantly inherited among Jewish patients, and these haplotypes plus DR6, DQ5 haplotypes in non-Jewish patients. Low levels of the PV antibody were found in asymptomatic family members. The inheritance of low levels of antibody in asymptomatic relatives has been linked to the MHC. Therefore, disease appears to occur in susceptible individuals with low levels of antibody when a second factor, either environmental or genetic, induces high levels of autoantibody sufficient to produce the disease (Ahmed *et al.*, 1993).

Association of MHC and life span

The MHC should influence life span, since it represents the main genetic factor regulating the immune system (Benacerraf, 1981). Evidence in support of the role played by MHC in life span comes from studies of congenic mice with three different strain backgrounds. Despite definite background-dependent differences in longevity when the H-2 allele was the same, distinct differences appeared among different H-2 congenics with the same background genome (Smith & Walford, 1977; 1978). The longest-lived strain with the C57BL/10 background, B10.R111, displayed the highest response to mitogens throughout most of life. The shortest-lived strain, B10.AKM, had the lowest proliferative response (Meredith & Walford, 1977). B10.F mice (H-2^n) demonstrated immunodeficiency and were shown to be short-lived (Popp, 1978).

Associations of immune functions with genotypes and life span in humans have not been studied systematically, but it has been shown that women older than 70 years of age had a lower frequency of HLA-B8 than younger women or men (the work was performed at a time when anti-DR reagents were not available). Women of this phenotype had lower T cell proliferative responses beyond 70 years of age than young controls or men carrying that phenotype (Greenberg & Yunis, 1978). Evidence presented at the Eighth International Histocompatibility Workshop Conference demonstrated association of HLA-B7 with Alzheimer's disease (Walford & Hodge, 1980) and association of HLA-DR1 with Xeroderma pigmentosum (Hodge, Degos & Walford, 1980).

MHC alleles are markers of long life span

Genes that may be markers for shortened life span have been identified. It is more difficult to identify genetic markers of long life. One example in mice was the association of longer life span in H-2^b congenic strains when they were compared with congenic mice of other H-2 haplotypes. Within congenic mice, certain H-2 alleles significantly influenced life span but the same alleles may have comparatively different effects on life span on different backgrounds. On C57BL/10 and C3H backgrounds H-2^b tended to mark long survival, on A strain to short survival. (Smith & Walford, 1978). Therefore, the use of congenic strains for genetics of life span may not be useful in understanding the genetic interactions that may be involved in the many aspects of life span in outbred mice. Analysis of HLA distribution in aged normal individuals showed an increased HLA-DR heterozygosity (Hodge & Walford, 1980). More importantly, subjects over 90 years of age had a low frequency of HLA-DR9 and increased frequency of HLA-DR1. The authors argued that a high frequency of HLA-DR9 and a low frequency of HLA-DR1 are associated with autoimmune or immunodeficiency diseases, which indicated that genetic protection against them may contribute to longevity (Takata *et al.*, 1987).

Experimental models to study genes of life span

The primary focus of this paper is the analysis of murine genetic research using four different genetic models to show possible gene variants influencing life span and the emphasis of the importance of the genetic interactions affecting life span in the exposure of mice to infections.

Table 3. Proportional hazards regression models and linear regression models of genetic markers.

Chromosome	Statistically indistinguishable Markers	Allele associated with longer survival	Significance level in final proportional hazards model	Significance level in final linear model
7	P450			
	Coh	B	0.0011	N/A
	Xmmv–35			
2	Ly. 24	B	<0.0001	0.0006
2	B2m			
	H–3	D	0.0006	0.0012
1	Lamb–2	D	<0.0001	0.0006
1	Ltw–4	B	<0.0001	<0.0001
12	Igh–Sa4			
	Igh–Sa2			
	Igh–Bg1			
	Igh–Nbp	D	0.0018	<0.0001
	Igh–Npa			
	Igh–Gte			
	Odc–8			
	Ox–1			
	Npid			
12	D12Nyu1	D	N/A	0.0011

Markers are indistinguishable if their genotypes were not known to differ in any of the twenty strains in the model.
From Gelman *et al.*, 1988.

Experiments using Recombinant Inbred strains (RI)
Recombinant Inbred strains have been shown to be useful in the analysis of segregation and linkage of multifactorial traits (Blizard, 1992).

Twenty strains of B × D RI mice (Taylor, 1989), composed of 395 females obtained from the Jackson Laboratory, were studied. No male mice were included in these studies (Gelman *et al.*, 1988). At the time of receipt at the Michael Redstone Animal Facility of the Dana-Farber Cancer Institute, the mice were between five and ten weeks old. They were representative of the 23 B × D RI strains available in 1982. Fifteen of the strains analyzed included 19 to 21 mice and five included nine or ten mice. Care was taken to minimize environmental influences such as temperature, noise, humidity, cage location, etc.

Published strain distribution patterns were available for up to 141 genes (actually, markers of an area of a chromosome) which were identified as being from the long-lived B or short-lived D parent. Most of these strains did not have data on several genes (For example: *Ly-22* on chromosome 4, *Saac* on Chromosome 9, *Igh-Bgl, Igh-Npa* on chromosome 12, and *Lyb-7*). Strains 2 and 14 were asso-

ciated with significantly shorter life span, and strain 19 was significantly associated with longer life span. Four additional strains were associated with longer life span than the other 13 strains studied using a proportional hazards model of survival. The mean survival of the shortest lived strain was 479 days and the mean survival of the longest lived strain was almost double (904 days). Ranges of survival within strains were large (average of 642 days) and strains accounted for only 29% of the variation of survival; there were important environmental effects on life span, even in a colony housed in a single room. 101 markers of 15 chromosomes had distinguishable distributions on the 20 strains. The single region most significantly correlated with survival (marked by *Coh, Xmmv-35* on chromosome 7) divided the mice into two groups with survival medians which differed by 153 days (755 days for mice with the B genotype and 602 days for mice with the D genotype). It was expected that the genes of the B parent (using a proportional hazards or a linear regression model analysis) were better predictors of longer life span, but in general this was not corroborated.

Two different statistical models were used to

identify genetic regions of life span as shown in Table 3. The significance of the linkage between different genes in our studies was based on the use of proportional hazards and a linear model (Peto & Peto, 1972; Cox, 1972) using least squares and analyses of variance (Draper & Smith, 1966) and the R-squared and Cp with algorithm (Furnival & Wilson, 1974) with p values ranging between <0.001 and .001. Others consider that statistical evaluation of RI strains is based on Bonferroni's corrections for multiple testing (Neuman, 1992). It is quite possible that these genetic regions will not be as important predictors of life span as other genetic polymorphisms which were unknown at the time of the analysis (Gelman *et al.*, 1988). Future experiments using other RI mice are necessary to determine if these findings are generalizable. Also, male RI mice may show different genetic profiles for life span than the profile described for females.

Backcross study

Backcross mice [(C57/6 × DBA/2)F1 × DBA/2] demonstrated that several chromosomal regions and the environment act together to influence life span in mice (Yunis *et al.*, 1984). The Cox model, used to test for interactions between the different variables, revealed two significant three-way interactions. (H-2^b × H-2^d) F1 lived longer than H-2^d homozygous in animals heterozygous for the

Brown locus b (B/b) or homozygous (b/b). It was found that backcross mice produce different genetic life span profiles in relation to coat color markers and H-2 genotype. These backross mice showed primarily two genetic effects, the major histocompatibility complex in males and the coat color markers in females. Analyses of the *b* locus in relation to life span showed that the Bb mice lived longer than bb females, but the *dilute* locus (d) on chromosome 9 did not influence life span. The genetic interactions demonstrated that the *dilute* locus and the brown coat color have shorter life span in females. More importantly, there was a strong heterozygosity effect influencing life span. The longest-lived mice were females heterozygous at the *H-2* and *Brown* (b) loci. The shortest-lived mice were males homozygous at the *H-2* and *Brown* loci.

Intercross study

Analysis of genetic interactions in the F2 in an intercross of (C57BL/6 × DBA/2) F1 revealed influences of genetic factors on life span (Table 4). Females lived longer than males. *Dilute* brown females died sooner than females of other colors. H-2^b/H-2^b males died sooner than H-2^d/H-2^d or H-2^b/H-2^d males, except that among *dilute* brown males those of type H-2^b/H-2^d died sooner. Cluster analysis suggested that male and female genotypes each fall into two groups. Female *dilute* brown

Table 4. Group characteristics (life span in months).

Genotype		N° of mice		Ranks of medians		Median life span (95% confidence interval)		Log rank P-value
H-2	Coat color	F	M	F	M	Females	Males	
bb	*[B.][D.]*	152	157	6	9	27.3 (26.3, 27.6)	25.6 (24.8, 26.0)	0.003
bb	*[B.][dd*	46	58	8	11	26.5 (25.8, 28.0)	24.7 (24.2, 25.0)	0.157
bb	*bb[D.]*	65	36	7	10	26.7 (26.2, 27.2)	25.4 (23.9, 26.2)	0.0071
bb	*bbdd*	19	15	12	5	24.1 (23.9, 24.6)	26.8 (23.3, 28.5)	0.279
bd	*[B.][D.]*	301	294	2	4	28.4 (27.9, 28.7)	26.8 (26.4, 26.9)	0.0149
bd	*[B.][dd*	114	93	9	6	26.3 (25.7, 27.3)	26.4 (26.0, 27.3)	0.0923
bd	*bb[D.]*	90	114	3	2	28.1 (26.9, 28.5)	27.2 (25.8, 28.2)	0.0241
bd	*bbdd*	37	27	10	12	25.9 (25.3, 26.5)	24.6 (23.3, 26.2)	0.379
dd	*[B.][D.]*	158	162	4	3	27.9 (27.5, 28.4)	26.9 (26.6, 27.4)	0.113
dd	*[B.][dd*	50	53	1	7	29.3 (28.1, 30.2)	26.2 (25.3, 27.9)	0.021
dd	*bb[D.]*	47	43	5	8	27.8 (27.2, 28,3)	26.1 (24.6, 27.9)	0.0658
dd	*bbdd*	16	11	11	1	24.7 (23.2, 27.1)	27.6 (26.8, 28.9)	0.0318

Coat colors: [B.][D.] = black, [B.]dd = dilute black, bb[D.] = brown, bbdd = dilute brown. Brackets indicate phenotypes. Values are for comparisons of sex in each of the different genotypes. From Dear *et al.*, 1992.

mice have shorter lives than other females and male (H-2^b/H-2^b)F1 mice. *Dilute* brown females lived shorter life span than females of other genotypes regardless the H-2 genotype. *Dilute* brown (H-2^b/H-2^d) and H-2^d homozygotes lived longer than the H-2^b × H-2^d)F1. The remaining H-2^d homozygotes and (H-2^d × H-2^b)F1 males lived longer lives than H-2^b homozygous mice. The association of heterozygosity with life span was clearer in females than in males, yet the longest-lived female genotype was homozygous H-2^d/H-2^d of dominant black phenotype at the *Brown* locus of chromosome 4 and homozygous dd at the *Dilute* locus of chromosome 9. The shortest-lived females were *Dilute* brown H-2^b/H-2^b. The longest and shortest-lived male genotypes were *Dilute* brown H-2^d/H-2^d and *Dilute* brown H-2^b/H-2^d, respectively. (Dear *et al.*, 1992).

H-2 congenic study

Ten congenic strains of mice were studied for life span. It was possible to compare the K-end and D-end of *H-2* haplotypes. B10, B10.2R and B10.4R are H-2^b at the D-end; B10.A, B10.5R, B10.HTT, B10.T6R and B10.D2N are H-2^d at the D-end. B10, and B10.5R are H-2^b at the K-end; B10.D2^n H-2^d at the K end. H-2^k and H-2^s haplotypes were also examined on the B10 background. There was no influence of the K-end of H-2 in life span, but the differences at the E alpha in six of eight strains and at the D-end in eight of eight strains influenced life span. Another important finding was that the H-2^d haplotype was associated with shorter survival when compared to the other three H-2 haplotypes studied which included H-2^b (Gelman *et al.*, 1990).

New experiments supporting the association of MHC with immune responses, life span and development of malignancy

Analysis of *H-2* (H-2^b, H-2^k and H-2^d) haplotypes in congenic strains of mice and their hybrids revealed the unexpected finding that males lived longer than females in H-2^k and H-2^b homozygous and the three heterozygous combinations. H-2 interactions with gender demonstrated that the influence of H-2 in life span was found primarily in males of the H-2^d and H-2^k haplotypes. The association of heterozygosity with longer survival was evident only when comparing mice carrying the H-2^b haplotype with the (H-2^b × H-2^d)F1 hybrids. Males and female mice homozygous or heterozygous for H-2^d or H-2^k haplotypes lived longer than homozygous H-2^b or (H-2^k × H-2^b)F1 mice (Table 5, Salazar *et al.*, unpublished observations). H-2^d homozygotes and (H-2^b × H-2^d)F1 lived longer than H-2^b homozygotes (p = .001 in males or females). The incidence of lymphomas was lower in H-2^d homozygotes than H-2^b homozygotes (23% and 50% respectively, p = .001), or than (H-2^b × H-2^d)F1 (34%. p = .05). The incidence of lymphomas of H-2^b homozygotes was higher than that of the F1 hybrid (p = .009).

Role of viral infection in genetic interactions involved in life span

Mouse hepatitis virus infection is latent in many strains of mice. Neonatally thymectomized mice developed acute hepatitis, and this disease was

Table 5. H–2 congenics. Life span and group characteristics.

Genotype H–2	N° of mice		Ranks of median		Median life span 95% Cl		Log rank P-value
	F	M	F	M	Females	Males	
b/b	104	102	6	5	26.0 (25.0, 26.8)	28.8 (27.6, 29.4)	0.0098
k/k	118	106	2	3	29.1 (27.3, 29.9)	30.0 (28.1, 31.6)	0.0049
d/d	45	57	1	2	29.5 (26.9, 31.5)	30.2 (28.5, 33.4)	0.84
bk	161	164	5	6	27.7 (26.0, 28.1)	28.1 (26.8, 29.1)	0.0078
bd	124	133	3	4	29.0 (27.6, 29.9)	29.5 (27.8, 31.2)	0.0044
dk	241	197	4	1	28.3 (27.5, 29.4)	31.7 (30.2, 32.7)	< 0.001

The median life spans are ranked separately for each sex. Confidence intervals are wider for smaller groups of mice. The log rank tests compare entire survival curves, not just the medians. P-values are for comparisons between males and females.

Table 6. Comparison between backcross, intercross and congenic life span studies.

Parameter	Backcross (C57BL/6 × DBA/2) F1 × DBA/2	Intercross (C57BL/6 × DBA/2) F1 × F1	B10 Congenic Exp 1	B10 congenic Exp 2
H–2 locus influence	Males	Males	Males and females	Males and females
H–2d short life in males	Yes	No	Yes	No
Sendai infection	Yes	No	Yes	No
b locus on chromosome 4: Bb better than bb in females	Yes	Not studied	N/A	N/A
d locus on chromosome 9: Dd not better than dd	Yes	Not studied	N/A	N/A
Longer life span	Females	Females	Males	Males
Genetic interactions	Dilute brown/brown have shorter life span in females	Dilute brown shorter life span in females	N/A	N/A
Heterozygosity effect	Strong	Relative	Relative	Relative
H–2d/H–2d	Disadvantaged in dilute brown males and females	Not disadvantaged	Disadvantaged	Not disadvantaged

primarily found in animals developing wasting disease (Stutman, Yunis & Good, 1972). Such studies suggested that T cells are required to prevent activation of latent viral infections.

In the case of Sendai virus infection, T cells and the major histocompatibility complex were involved in susceptibility to acute infection and early death (Stewart *et al.*, 1978), with highest incidence in H-2d/H-2d mice (DBA/2) (Parker, Whiteman & Richter, 1978). However, we believe that mice that were exposed to Sendai virus and survived the exposure to infection developed T cell defects detectable later in life. In the absence of virus infection or exposure to the virus, the H-2d conferred longer life span than mice of H-2k or H-2d genotyopes.

In several studies summarized in Table 6, using different experimental models, significant genetic interactions between genes on chromosomes 4, 9, 17 and gender were demonstrated (Yunis *et al.*, 1984; Gelman *et al.*, 1988; Gelman *et al.*, 1990; Dear *et al.*, 1992). In experiments where the mice had been exposed to Sendai virus infection it was found that the H-2b phenotype was associated with longer life span than was H-2d or H-2k (Smith & Walford, 1977; Gelman *et al.*, 1990). However, in

other experiments using a different experimental model, it was demonstrated that mice H-2d homozygous or (H-2d × H-2b)F1 lived longer than the H-2b mice when they were not exposed to the same infection. The only important difference between the studies reported was that the mice had been exposed to and possibly infected with Sendai virus. These experiments suggested that exposure to infection can change the profile of life span (Dear *et al.*, 1992). Earlier findings that H-2 significantly influences life-span primarily in males (Yunis *et al.*, 1984) were corroborated. While the same loci remain important, it appears that environmental changes such as exposure to Sendai virus and the segregation of other genetic systems in different experimental models affected the details of genetic interactions. In the backcross experiments H-2b/H-2b animals lived longer than the H-2d/H-2d animals, suggesting an influence of heterozygosity on life span, but in the F2 experimental model the H-2d haplotype conferred long life span. Furthermore, the heterozygosity index was associated significantly with life span in animals that had been exposed to Sendai. In the F2 experimental model heterozygosity *per se* did not increase life span in mice. Although the association of life span with heterozygosity was significant in females, the longest-lived mice were females predominantly heterozygous for the *Brown* locus of chromosome 4, but homozygous for H-2^d/H-2^d and for the *Dilute* locus of chromosome 9.

It is noteworthy that Sendai infected congenic mice, which differed only at H-2, demonstrated that H-2^b mice lived longer than H-2^d mice. This finding was confirmed in backcross mice. In contrast, in the absence of Sendai infection H-2d was a better predictor of long life span than H-2b as demonstrated in intercross mice (Dear *et al.*, 1992). Unpublished experiments using congenic mice confirmed that in the absence of Sendai virus infection, H-2d mice live longer than H-2^b mice (Salazar *et al.*, unpublished). The studies also demonstrated that T cell functions were diminished in H-2^b homozygotes as compared to H-2^d homozygous or (H-2^b × H-2^d)F1 in males and females. It is not unusual that proliferative responses to lectins such as PHA are decreased in aged mice. What is different is that the defect was found more significantly in mice of H-2^b genotype.

In previous studies, histopathological findings failed to demonstrate a relationship between cause of death, genotype and life span (Salazar *et al.*, unpublished). This may be due to the fact that previous studies included autopsies at death in less than 25% of the animals as compared to more than 75% autopsies performed in our unpublished experiments using H-2 congenic mice and their hybrids. A higher incidence of lymphomas was correlated to H-2^b homozygous and (H-2^b × H-2^k)F1 when compared to a lower incidence in H-2^d homozygotes and (H-2^d × H-2^k)F1 mice.

Discussion

It has not been possible to determine a single gene that can predict long life in outbred populations of mammals. Different life spans in various inbred strains of mice suggested genetic influences which have been associated with specific diseases. In many experiments females lived longer than males and F1 hybrids lived longer than parental strains (Russell, 1975; Yunis *et al.*, 1984). We have summarized evidence in favor of genetic factors that influence immune responses during life span. It is

remarkable that MHC influenced life span in several experimental models in the absence or presence of multiple genetic interactions. Nevertheless, studies also suggested that life span results from environmental factors interacting with several genes, which include the genes of the major histocompatibility complex.

With regard to the influence of H-2 alleles in the frequency of lymphomas in H-2 congenic mice, H-2^b was more susceptible to lymphomas in older animals than H-2^d mice of the same age. However, other studies had shown that H-2 alleles alone could not alter susceptibility to the genesis of lymphomas. Our studies demonstrated that H-2^d was associated with longer life span, maintained T cell immune vigor in old age, and had lower incidence of lymphomas. Current models for lectin-induced T cell proliferation suggest that activation of protein kinase C (PK-C) and elevation of cytoplasmic CA$^+$ may both play important roles in the earliest phases of signal transduction. T cells from young mice responded as well to optimal combinations of these agents as they did to the strong polyclonal activator Con A, but T cells from old mice responded better to optimal combinations of PMA (phorbol myristate acetate, a PK-C activator) plus calcium ionophore ionomycin than they did to Con A. This suggests that an inability to transduce the signal supplied by extracellular ligands into the intracellular signals represented by PK-C and CA$^+$ activators could be the underlying mechanism for age-associated loss of T cell reactivity. Furthermore, T cells from old mice required higher levels of ionomycin for maximal proliferation than cells from young animals (Miller, 1986).

Recently it has been reported that animals bearing a tumor longer than 26 days develop CD8$^+$ T cells with impaired cytotoxic function, decreased expression of the tumor necrosis factor-α and ganzyme β genes, and decreased ability to mediate an antitumor response *in vivo*. The structure of the TCR was evaluated by surface iodination of the purified T cells from normal and tumor-bearing mice. T lymphocytes from tumor-bearing mice expressed T cell antigen receptors that contained low amounts of CD3γ and completely lacked CD3ζ which was replaced by the Fcϵ γ-chain. Expression of the tyrosine kinases p56^{lck} and p59^{fyn} was also reduced. These changes were interpreted to explain the basis of the immune defects in tumor-bearing animals. These authors also reported that the same findings occur in human cancer patients whose peripheral blood T cells lacked expression of lck and Fyn protein. Thus, structural and functional signal transduction molecules from tumor-bearing mice and humans may explain T cell abnormalities in cancer bearing animals and humans (Mizoguchi *et al.*, 1992). They demonstrated that these clones are manifested clinically only in old age. Other authors showed that the immune defects found in murine hosts bearing tumors have a decrease in cytotoxic function of CD8-cells, suggesting one mechanism responsible for immunologic defects of CD8 effector T cells in patients with tumors (Loeffler *et al.*, 1992; Mizoguchi *et al.*, 1992). It seems that tumors can produce a factor that changes the functions of T cells. It is unknown if premalignant clones (of spontaneous tumors) *per se* can produce some of these alterations, or if other factors (environmental, such as infections) can produce similar changes, which would predispose animals or humans to develop tumors during aging (Newell, Spirtz & Sider, 1989). It is not known if the defect of T cell function found in aged mice and humans is identical to that found in T cells of animals carrying tumors, defective expression of p56^{lck} and p59^{fyn} and lack of expression of CD3γ in association with TCR. For example, changes in T cells of tumor-bearing mice may be found in aging mice, which might explain the immune defects associated with aging. Low T cell responses during aging may represent evidence that they have incipient malignancy or predisposition for malignant disease. It is possible that lymphomas seen late in life arise from premalignant clones of cells that have become committed to neoplasia in young animals (Van Houten *et al.*, 1989). The progressive growth of a subcutaneous implant of a tumor in mice resulted in decreased lytic function of the CD8$^+$T lymphocytes that was associated with decreased expression of mRNA for tumor necrosis factor-α and granzyme β, and the complete loss of the ability of adoptively transferred cells to mediate an antitumor effect *in vivo* (Loeffler *et al.*, 1992). In these experiments there was no detection of suppressor function. These results contrasted with early studies in the development of spontaneous tumors due to mammary tumor virus, where there was a proven role of suppressor cells in the tumor development (Fernandes, Yunis & Good, 1976).

The genetics of life span are complex and we anticipate that genes important in life span will need to be studied in many genetic backgrounds and environmental conditions. It is remarkable that even after minimizing the genetic and environmental variations we have found complex patterns of disease and life span, which raises doubts as to the feasibility of studying genetic effects of life span in outbred populations. At the very least, it would be necessary to identify genes that under all environmental conditions influence life span, a task difficult to undertake in the near future. Since the National Institute of Aging has recently undertaken an initiative to identify genes of life span in outbred populations of mammals, careful attention must be paid to the problem of identifying these genes for this initiative to be productive. We emphasized the importance of the thymus and T cells during the life of an animal and that thymus-dependent immunity prevents development of diseases common in aged individuals. MHC alleles may also determine susceptibility to disease but it is not known if its genetic influence can direct the involution of the thymus or the thymus-dependent immune functions. New research is needed to invesitage the role of environmental factors at different times in the life of an animal and the possible production of latency to developing diseases later in life. Areas of research that need intensive investigation include at least two:

a) Identification of premalignant clones of cells latent in animals susceptible to spontaneous malignancies.

b) Investigation of mechanisms of immune regulation that may occur following infections during the neonatal period, for example the role of endogenous retroviruses following exogenous viral infections or other infections (Talal, Flesher & Dang, 1992). These environmental factors may activate retroviruses or other molecules which could be involved in the development of immune dysfunctions and lymphomas, or other malignancies that develop later in life.

In other words, experiments are required to prove that the environment may be playing a more significant role in the immune or other functions of the growing animal than had been previously thought. In that case, genetic influences on life span would involve interactions between loci and many allelic interactions in the growing individual which may vary after the individual is exposed to different infections or other environmental factors.

Acknowledgement

This work was supported by National Institutes of Health grants RO1-AG-02329 and CA-06516. The authors wish to thank Borghild Yunis and Dr. Jai Dev Dasgupta for editorial assistance.

References

Ahmed, A. R., A. Mohinem, E. J. Yunis, N. Mirza, V. Kumar, E. H. Beutner & C. A. Alpe, 1993. Linkage of Pemphigus Vulgaris Antibody to the Major Histocompatibility Complex in Healthy Relatives of patients. J. Exp. Med. 177: 419-424.

Amiel, J. L., 1967. Study of the leukocyte phenotypes in Hodgkin's disease, p. 79 in Histocompatibility Testing, edited by E. S. Curtoni and R. M. Tosi. Munksgaard, Copenhagen.

Benacerraf, B., 1981. A hypothesis to relate the specificity of T lymphocytes and the activity of I region specific in genes in macrophages and B lymphocytes. J. Immunol. 128: 1809-1812.

Blizard, D. A., 1992. Recombinant Inbred strains: General methodological considerations relevant to the study of complex characters. Behavior genetics 22: 621-623.

Broder, S. & T. A. Waldman, 1978. The Suppressor-Cell Network in Cancer. N. Eng. J. Med. 299: 1281-1283.

Burnet, F. M., 1958. The clonal selection theory of acquired immunity. Vanderbilt and Cambridge Presses, Nashville, Tennessee.

Burnet, F. M., 1970. Immunological surveillance. Pergamon Press, Oxford, New York.

Burnet, F. M., 1971. An immunological approach to aging. Lancet 2: 358-360.

Cox, D. R., 1972. Regression models and life tables. J. R. Stat. Soc. (Ser. B). 34: 187-202.

Dear, K. B. G., M. Salazar, A. L. M. Watson, R. S. Gelman, R. Bronson & E. J. Yunis, 1992. Traits that influence longevity in mice: A second look. Genetics 132: 229-239.

Draper, N. R. & H. Smith, 1966. Applied Regression Analysis. John Wiley & Sons. New York.

Falk, J. & d. Osoba, 1977. The HLA system and survival in malignant disease: Hodgkin's disease and Ca of the breast, pp. 205-216 in HLA and malignancy, edited by Gerald P. Murphy. Alan R. Liss, New York.

Fearon, E. R., D. M. Pardoll, T. Itaya, P. Golumbek, H. I. Levitsky, J. W. Simons, H. Karasuyama, B. Vogelstein & P. Frost, 1990. Interleukin-2 production by tumor cells bypasses T helper function in the generation of an antitumor response. Cell 60: 397-403.

Fernandes, G., R. A. Good & E. J. Yunis, 1977. Attempts to correct age-related immunodeficiency and autoimmunity by cellular and dietary manipulation in inbred mice, pp. 111-133 in Immunology and Aging, edited by T. Makinodan and E. J. Yunis. New York, Plenum.

Fernandes, G., E. J. Yunis & R. A. Good, 1976. Suppression of adenocarcinoma by the immunological consequences of calorie restriction. Nature 263: 504-507.

Furnival, G. M. & R. W. Wilson, Jr., 1974. Regression by leaps and bounds. Technometrics 16: 499-511.

Gelman, R., A. Watson, R. Bronson & E. J. Yunis, 1988. Murine chromosomal regions correlated with longevity. Genetics 125: 167-174.

Gelman, R., A. Watson, E. J. Yunis & R. M. Williams, 1990. Genetics of survival in mice: subregions of the major histocompatibility complex. Genetics 25: 167-174.

Good, R. A. & E. J. Yunis, 1974. Association of autoimmunity, immunodeficiency and aging in man, rabbits and mice. Fed. Proc. 33: 2040-2050.

Greenberg, L. J. & E. J. Yunis, 1978. Genetic control of autoimmune disease and immune responsiveness and the relationship to aging, pp. 249-260 in Genetic effects of aging, edited by D. Bergsma and D. Harrison. Allan R. Liss, Pub., New York.

Hallgren, H. M., J. H. Kersey, D. P. Dubey & E. J. Yunis, 1978. Lymphocyte subsets and integrated immune function in aging humans. Clin. immunol. Immunopathol. 10: 65-68.

Hayflick, L., 1965. The limited *in vitro* lifetime of human diploid cell strains. Exp. Cell. Res. 37: 614-636.

Hersch, E. M. & J. J. Openheim, 1965. Impaired *in vitro* lymphocyte transformation in Hodgkin's disease. N. Eng. J. Med. 273: 1006-1012.

Hirokawa, K., 1977. The thymus and aging, pp. 51-72 in Immunology and aging. Edited by T. Makinodan and E. J. Yunis. Plenum Press. New York.

Hodge, S. E., L. Degos & R. L. Walford, 1980. Four chromosomal instability syndromes: Bloom's syndrome, Fanconi's anemia, Werner's syndrome, and xeroderma pigmentosum pp. 730-733 in Histocompatibility testing 1980, edited by P. I. Terasaki. UCLA tissue typing laboratory, Los Angeles, CA.

Hodge, S. E. & R. L. Walford, 1980. HLA distribution in aged normals, pp. 722-726 in Histocompatibility testing 1980, edited by P. I. Terasaki. UCLA tissue typing laboratory, Los Angeles, CA.

Juckettt, D. A. & B. Rosenberg, 1988. Integral differences among human survival distributions as a function of disease. Mechanisms of aging and development 43: 239-257.

Lilly, F., E. A. Boyse & L. J. Old, 1964. Genetic basis of susceptibility to viral leukaemogeneis. Lancet 11: 31-35.

Loeffler, C. M., M. J. Smyth, D. L. Longo, W. C. Kopp, L. K. Harvey, H. R. Tribble, J. E. Tase, W. J. Urba, A. S. Leonard, H. A. Young & A. C. Ochoa, 1992. Immunoregulation in cancer-bearing hosts: Down-regulation of gene expression and cytotoxic function in CD8[+] T cells. J. of Immunol. 149: 949-956.

Meredith, P. J. & R. L. Walford, 1977. Effect of age on response to T and B cell mitogens in mice congenic at the H-2 locus. Immunogenetics 5: 109-128.

Meyer, T. E., M. J. Armstrong & C. M. Warner, 1989. Effects of H-2 haplotype and gender on the life span of A and C57BL/6 mice and their F1, F2 and backcross offspring. Growth Dev. Aging. 53: 75-183.

Miller, R. A., 1986. Immunodeficiency of aging: restorative effects of phorbol ester combined with calcium ionophore. J. Immunol. 137: 805-808.

Mizoguchi, H., J. J. O'Shea, D. L. Longo, C. M. Loeffler, D. M. McVicar & A. C. Ochoa, 1992. Alterations in signal transduction molecules in T lymphocytes from tumor-bearing mice. Science 258: 1795-1998.

Newman, P. E., 1992. Inference in linkage analysis of multifactorial traits using Recombinant Inbred strains of mice. Behaviour genetics 22: 665-676.

Newell, G. R., M. R. Spitz & J. G. Sider, 1989. Cancer and Age. Seminars in Oncology 16: 3-9.

North, R. J. & I. Bursuker, 1984. Generation and decay of the immune response to a progressive fibrosarcoma. J. Exp. Med. 159: 1295-1311.

Parker, J. C., M. D. Whiteman & C. B. Richter, 1978. Susceptibility of inbred and outbred mouse strains to Sendai virus and prevalence of infection in laboratory rodents. Infect. Immun. 19: 123-130.

Parrish, E. J., E. R. Heise & M. R. Cooper, 1977. HLA association with acute myelogenous leukemia, pp. 82-89 in HLA and Malignancy, edited by G. P. Murphy. Alan R. Liss, New York.

Petranyi, G. G., R. Kiessling, S. Povey, G. Klein, L. Herzenberg & H. Wigzell, 1976. The genetic control of natural killer cell activity and its association with in vivo resistance against a Moloney lymphoma isograft. Immunogenetics 3: 15-28.

Peto, R. & J. Peto, 1972. Asymptotically efficient rank invariant test procedures. J. R. Stat. Soc (Ser. A). 135: 185-198.

Popp, D. M., 1978. Use of congenic mice to study the genetic basis of degenerative disease, pp. 261-279 in Genetic Effects of Aging. Vol. XIV, edited by D. Bergsma and D. Harrison. Alan R. Liss, New York.

Pross, H. F. & M. G. Baines, 1976. Spontaneous human lymphocyte mediated cytotoxicity against tumor target cells. I. The effect of malignant disease. Int. J. Cancer. 18: 593-604.

Roberts-Thomsen, I., S. Wittingham, U. Youngshaiud & I. R. MacKay, 1974. Aging, immune response and mortality. Lancet 2: 368-370.

Russell, E. S., 1975. Life span and aging patterns, pp. 511-519 in Biology of the Laboratory Mouse, Ed. 2. Dover publications, New York.

Schultz, L. D., 1993. Mouse models of immunodeficiency diseases. Immunology News. The Jackson Laboratory, Bar Harbor, Me.

Simms, H. S., 1946. Logarithmic increase in mortality as a manifestation of aging. J. Gerontol. 1: 13.

Simons, M. J. & J. L. Amiel, 1977. HLA and Malignant diseases, pp. 212-232 in HLA and disease, edited by J. Dausset and A. Vejgard. Munksgaard, Williams and Williams Co., Copenhagen.

Smith, G. W. & R. L. Walford, 1977. Influence of the main histocompatibility complex on aging in mice. Nature 270: 727-729.

Smith, G. S. & R. L. Walford, 1978. Influence of the H-2 and H-1 histocompatibility systems upon life span and spontaneous cancer incidences in congenic mice, pp. 281-312 in Genetic effects of gaging, edited by D. Bergsma and D. Harrison. Alan R. Liss, Pub, New York.

Stewart, R. B. & M. J. Tucker, 1977. Infection of inbred strains of mice with Sendai virus. Can. J. Microbiol. 42: 9-13.

Strelkauskas, A. J., J. A. Andrew & E. J. Yunis, 1981. Autoantibodies to a regulatory T cell subset in human aging. Clin. Exp. Immunol. 45: 308-315.

Stutman, O., E. J. Yunis & R. A. Good, 1972. Studies on Thymus function. III Duration of the thymic function. Journal of Experimental Medicine 135: 339-356.

Takasugi, M., A. Ramseyer & J. Takasugi, 1977. Decline of natural nonselective cell-mediated cytotoxicity in patients with tumor progression. Cancer Res. 37: 413-418.

Takata, H., M. Suzuki, T. Ishii, S. Sekilguchi & H. Iri, 1987. Influence of major histocompatibility complex region genes in human longevity among Okinawan Japanese centenarians and nonagarians. Lancet 2: 824-826.

Talal, N., E. Flesher & H. Dang, 1992. Are endogenous retroviruses involved in human autoimmune disease? Journal of Autoimmunity 5(Sup. 5): 61-66.

Taylor, B. A., 1989. Recombinant Inbred strains, pp. 773-789 in Genetic variants and strains of the laboratory mouse, 2nd ed., edited by M. F. Lyon and A. G. Searle. Oxford University Press.

Teague, P. O., E. J. Yunis, G. Rodey, C. Martinez & R. A. Good, 1970. Autoimmune phenomena and renal disease in mice. Role of thymectomy, aging, and evolution of immunologic capacity. Lab Invest. 22: 121-138.

Van Houten, N., P. B. Willoughby, L. W. Arnold & G. Haughton, 1989. Early commitment to neoplasia in murine B- and T-cell lymphomas arising late in life. Journal of the national Cancer Institute 81: 47-54.

Walford, R. L., 1969. The immunologic theory of aging. Munksgaard, Copenhagen.

Walford, R. L. & S. E. Hodge, 1980. HLA distribution in Alzheimer's disease, pp. 727-729 in Histocompatibility testing 1980, edited by P. I. Terasaki. UCLA tissue typing laboratory, Los Angeles, CA.

Webb, S., C. Morris & J. Sprent, 1990. Extrathymic tolerance of mature T cells: Clonal elimination as a consequence of immunity. Cell 63: 1249-1256.

Williams, R. M. & E. J. Yunis, 1978. Genetics of human immunity and its relation to disease, pp. 121-139 in Infection, Immunity, and genetics, edited by H. Friedman, T. J. Linna and J. E. Prier.

Young, R. C., M. P. Corder, H. A. Haynes & V. T. DeVita, 1972. Delayed hypersensitivity in Hodgkin's disease: A study of 103 untreated patients. Am. J. Med. 52: 63-72.

Yunis, E. J., G. Fernandes & O. Stutman, 1971. Susceptibility to involution of the thymus-dependent lymphoid system and autoimmunity. American Journal of Clinical Path. 56: 280-292.

Yunis, E. J., G. Fernandes, P. O. Teague, O. Stutman & R. A. Good, 1972. The thymus, autoimmunity and the involution of the lymphoid system, pp. 62-120 in Tolerance, atutoimmunity and aging, edited by M. Siegel and R. A. Good. Thomas, Sprigfield, II.

Yunis, E. J., A. Watson, R. S. Gelman, S. L. Sylvia, R. Bronson & D. E. Dorf, 1984. Traits that influence longevity in mice. Genetics 108: 999-1011.

M. R. Rose and C. E. Finch (eds.), Genetics and Evolution of Aging, 256–269, 1994.
© 1994 *Kluwer Academic Publishers. Printed in the Netherlands.*

Genes of the major histocompatibility complex and the evolutionary genetics of lifespan

Mark D. Crew
Department of Medicine, Division of Aging, University of Arkansas for Medical Sciences and the Geriatric Research Education and Clinical Center (GRECC), Department of Biochemistry and Molecular Biology, John L. McClellan Veterans Administration Hospital, Little Rock, AR 72205-5484, USA

Received and accepted 22 June 1993

Key words: aging, evolution, histocompatibility antigens, MHC

Abstract

Mice that presumably differ just in the major histocompatibility complex (MHC) chromosomal region provide the best evidence that MHC genes affect lifespan. Further evidence is that MHC region genes in some cases are known to influence reproduction, growth, and development. Moreover, MHC genetic associations with disease are well documented. This paper summarizes and defines aspects of the molecular biology, cellular function, and evolution of MHC genes (with special emphasis on the polymorphic MHC class I and II genes) which are important in aging, and attempts to integrate these into an evolutionary genetic perspective of senescence. It is suggested that MHC genes provide a mammalian paradigm for the genetics of lifespan because of their intra- and interspecies diversification, evolutionary selection, and age-specific effects.

Introduction

A decrease in the power of natural selection with advancing age is an evolutionary explanation for senescence that is substantiated in part by artificial selection with *Drosophilia* (Rose & Charlesworth, 1980; Rose, 1984; Charlesworth, 1990). Mutation accumulation and antagonistic pleiotropy are related theories which embody the notion of reduced selection for reproductive fitness at later ages (Charlesworth, this volume). Mutation accumulation suggests that heritable mutations may be neutral with respect to reproductive success but may be causative in age-associated pathologies. Antagonistic pleiotropy, on the other hand, asserts that some mutations which have late-age deleterious effects may actually be selected for because they impart increased fitness. The disposable soma theory is complementary to both mutation accumulation and antagonistic pleiotropy in that it describes the evolution of senescence in terms of the physiological costs of maintaining germ line versus somatic cells

(Kirkwood, 1977; Kirkwood & Rose, 1991). However, loci which fit predictions of these theories have not been well documented in mammals.

There is direct experimental support for an influence of major histocompatibility complex (MHC) genes on aging in mammals (Smith & Walford, 1977; reviewed in Finch, 1990, and Walford, 1990). The MHC is thus one of only a few chromosomal regions which are known to affect mammalian maximum lifespan disbarring loci involved in genetic diseases that drastically decrease lifespan (and reproductive fitness). In this paper, the biology and evolution of MHC genes, particularly class I genes, will be discussed in relationship to aging. The major contention of this treatise is that many features of MHC class I genes are in accord with evolutionary theories of aging, especially antagonistic pleiotropy. MHC class I genes evolve under the intense selective pressure of pathogens and therefore can significantly increase fitness, yet MHC class I genes are associated with diseases many of which have late-age onsets (i.e. the diseases minimally affect fecundity). Moreover, MHC class I genes influence reproductive schedules and developmental rates. MHC genes therefore may offer a real-life example of antogonistic pleiotropy and provide a model with which to compare other loci that alter the rate of aging in mammals once they are better defined.

A molecular and functional description of the MHC and MHC class I genes will precede a review of the evidence that MHC genes influence rates of development and aging. A case will be made for a role of MHC class I genes in these phenomena. Mechanisms by which MHC class I genes might influence senescence then will be described as quantitative or qualitative and putative examples of each will be given. Various aspects of the phylogeny and evolutionary history of MHC class I genes will then be related to the evolution of differential rates of aging. A summary will point to future directions and experimental approaches.

Molecular and functional description of MHCs

Structure and diversity of MHCs

The MHC of mice is called the *H-2*; in humans the MHC is refered to as HLA. The MHC may be broadly defined as an approximately 2 megabase chromosomal region roughly demarcated by *pgk-2* and *H-2K* in mice and PGK2 and DQA in humans (Klein, 1986). The organization of genes in the *H-2* (Fig. 1) is very similar to that of the HLA with a few notable exceptions. For one, all HLA class I genes are telomeric to class II loci. In fact, *Mus* and *Rattus* appear to be unique in having class I loci centromeric to the class II region (Klein & Figueroa, 1986; Koller *et al.*, 1989). In addition, there is a greater number of class I protein-encoding loci in the *H-2* versus HLA. Importantly, orthologous relationships between *H-2* and HLA class I genes are not discernible in contrast to the majority of other MHC loci in the two species.

Historically and functionally, MHC class I and II loci are the hallmarks of MHCs (Klein, 1986). Class I antigens are composed of the class I gene-encoded heavy chain in noncovalent association with β_2-microglobulin (β_2m) and present peptides derived from cytosolic pools most often to CD8+ (cytotoxic) T cells. Class II antigens are heterodimeric complexes consisting of MHC-encoded α and β chains. Peptides primarily from lysosomal or endosomal compartments are presented to CD4+ (helper) T cells by class II antigens (Yewdell & Bennink, 1990). Class I and II antigens differ in their tissue distribution as well – 'classical' class I antigens (H-2K, D, and L; HLA-A,B and C) are ubiquitously expressed in moderate to high levels in all somatic cells except for those in the brain, where expression is low but detectable (David-Watine, Israel & Kourilsky, 1990). In contrast, class II expression is restricted primary to cells of monocyte and lymphocyte lineages.

There are over 60 genes in the *H-2*. In addition to the structurally and functionally homologous class I and II genes, the MHCs of humans and mice contain a host of other genes. Those of known function residing between class I and II are commonly referred to as class III and include components of the complement pathway, steroid hydroxylase, tumor necrosis factor α and β, and heat shock proteins. Elsewhere in the MHC reside genes encoding a transcription factor (*Oct-3*; Uehara *et al.*, 1992), some components of the antigen processing pathway (discussed below), and a variety of transcribed loci with unknown function (Abe *et al.*, 1988; Spies *et al.*, 1989; Hanson & Trowsdale,

region:	K	class II	class III	D	Qa	T1	Hmt
genes:	*H-2K*	Aα, Aβ, Eα, Eβ, Ham1, Ham2 LMP2, LMP7	C2, C4, 21-OH hydroxylase, TNFα, β, hsp70	*H-2D*, D2-4, *H-2L*	Q1-10	T1-24	M1-7
				<--- expansion and contraction --->			

Fig. 1. Gross organization of the *H-2* complex. Commonly referred to regions of the H-2 are boxed and shown with the telomeric end at the right. Some of the genes in each region with known function are listed below the boxed regions and those displaying functionally significant polymorphism are underlined. The arrow below *D – Hmt* is meant to show the highly variable number of class I loci between haplotypes in these regions.

1991). Except for the transporters associated with antigen processing, evidence of functionally significant polymorphism in these nonclass I or II loci, a necessary requisite for loci to be implicated in the observed MHC effects on senescence, is sparse, so these other genes will be excluded from further discussion.

Extensive diversity among MHCs is exhibited both within and between species. Haplotypic (intraspecific) variation is apparent at two levels. First is the astounding variation between alleles of class I and II antigens. For example, to date over 50 serologically defined HLA-B alleles have been detected. The actual number of alleles is much greater, since in several cases more than five alleles differing in antigen binding region sequences share the same serological specificity. Haplotypic variation is rarely limited to a single locus – each haplotype is composed of assortments of alleles at several loci. Second, haplotypic variation entails, especially for class I genes, variation in the number of loci. In mice and to a lesser extent in humans, there is extensive polymorphism in nonclassical class I gene regions such that H-2 haplotypes may differ by two-fold in the number of nonclassical class I genes (Stroynowski, 1990; O'Neill $et\ al.$, 1986). Both levels of haplotypic variation may be involved in influencing senescence as discussed below.

Multiformity of MHCs among species is most evident in comparisons of class I genes and the organization of class I gene regions. Sequence variation between H-2 and HLA class I genes is so great that orthologous loci are difficult, if not impossible, to discern (Kindt & Singer, 1987: Lawlor $et\ al.$, 1990). This most likely reflects the role of chromosomal expansion and contraction in the diversification of class I genes (discussed below). Orthologous HLA and H-2 class II loci are more apparent because these genes, unlike class I genes, probably evolved through direct descent (Lundberg & McDevitt, 1992). As with haplotypic variation, there are large differences between species in the number of class I antigen-encoding loci ranging from fifty to sixty in $Peromyscus\ leucopus$ and $Rattus$ (Crew $et\ al.$, 1990; Jameson $et\ al.$, 1990) to seven in minature swine (Singer $et\ al.$, 1987). The proposition here is that MHC class I gene diversity between species mirrors the profound interspecific diversity of life history patterns.

$Functions\ of\ MHC\ genes$

As alluded to throughout the above discussion, MHC class I and II antigens have a primary role in the immune system via their presentation of antigens to T cells. In fact, the whole MHC has been likened to a bacterial operon because of the role of individual, nonhomologous loci in antigen presentation and overall immune function (Robertson, 1991). In this regard, the MHC appears to be unique among eukaryotic chromosomal regions. For instance, several proteins involved in the pathway leading to presentation of virally encoded antigens by MHC class I molecules are encoded within the MHC. The initial step in viral antigen presentation is proteolysis of viral peptides by large multifunctional proteosomes comprised in part of low mobility proteins, LMPs. LMP2 and LMP7 genes have been mapped within the class II region of mouse and human MHCs (Brown, Driscoll & Monaco, 1991; Glynne $et\ al.$, 1991). Presently, the exact role of LMP2 and LMP7 in the proteosome as well as the degree to which they might influence proteolytic specificity is uncertain. Further down the antigen presenting pathway, two subunits of a transporter protein which putatively translocate proteosome-derived peptides into the endoplasmic reticulum are likewise encoded within the MHC (Monaco, Cho & Attaya, 1990; Trowsdale $et\ al.$, 1990; Spies $et\ al.$, 1991; Deverson $et\ al.$, 1991). In this case, biologically relevant polymorphism with respect to peptide specificity has been recently described (Powis $et\ al.$, 1992). Finally, the highly polymorphic MHC class I proteins (which originally defined the H-2) bind peptides and β_2m in the ER and are transported to the cell-surface where T cells recognize the peptide in association with MHC class I antigen (Yewdell & Bennink, 1990).

The primary function of the MHC class I and II antigens as antigen presenting molecules is manifest in the localization of polymorphic residues to protein domains associated with antigen binding and T cell receptor (TcR) recognition as shown by crystallographic studies (Bjorkman $et\ al.$, 1987a, 1987b). This is exemplified in Figure 2 – the $\alpha1$ and $\alpha2$ domains contain over 75% of the amino acid variation between H-$2D$ genes of three haplotypes studied for lifespan (see below). The function of nonclassical class I genes (those in the Qa, Tl, and Hmt regions, Fig. 1) is not clear, though there are indications that some nonclassical class I

```
domain:      [        α1        ][         α2         ][α3][TM,CY]
consensus ATPVAGENT-VREGEINS-TRKIWSVG--CMQRRAR-RTRY--ARG-H-G-MMAA-
        b ---e-es-kyap---kqwsngalq-l-wlr----shykeh-kn---sry-mkr--r
        k va-ev-s--d--------d--e-----we--lkh--d----el-hvs-yvv--ttp
        r --r----d-v--ven--in-s---f-egq-t-----k--lhkne--a-h-m----q
          5223344556777889991111111111111111111111112222223333333
          0635284593490704700011222233466667788899990027881233456
          1473913781852589490173578151046835 6801
```

Fig. 2. Comparison of *H-2D* genes from three *H-2* haplotypes which have been studied for lifespan and fecundity. *H-2D* genes from the b, k and r haplotypes (GenBank) were translated and compared to a consensus derived from six *H-2D* genes. Out of 362 positions, 56 were variable between these three haplotypes. Only the variable positions are shown. The position number is at the bottom of the comparison (read vertically). The domains corresponding to the sequences shown are given at the top of the comparison (TM,CY denotes transmembrane and cytoplasmic domains).

proteins may present specific antigens and serve as restriction elements a la the classical transplantation antigens (Stroynowski, 1990; Ito *et al.*, 1990; Wang *et al.*, 1991). As such, polymorphic residues are less concentrated in the α1 and α2 domains.

The sphere of influence of MHC class I genes is not limited to the immune system. Reports that MHC class I proteins are co-immunoprecipitated with receptors for insulin and leutenizing hormone (LH) have led to the consideration that class I antigens act as modulatory subunits with peptide hormone receptors (Verland *et al.*, 1989; Solano *et al.*, 1988). Importantly, MHC class I polymorphism may be manifest in some of these interactions (LaFuse & Edidin, 1980). Judging from the three-dimensional structure of MHC class I molecules (Bjorkman *et al.*, 1987a, 1987b), interaction of MHC class I molecules with other integral membrane proteins on the same cell might occur through the α3 domain. Thus, polymorphism outside of the antigen and TcR binding regions may be relevant to peptide hormone receptor interactions with MHC class I molecules and therefore important in altering reproductive patterns as described below.

The role of MHC-encoded genes in longevity

Evidence from H-2 congenic and recombinant inbred mice

The most direct evidence that the MHC influences the rate of aging is that mice that presumably differ in just the *H-2* have significantly different lifespans (Smith & Walford, 1977). [The qualifier 'presumably' is necessary since it is possible that during the course of breeding other differences between regions outside of the MHC may have arisen]. These studies, originally described in 1977, were repeated ten years later with nearly identical results (R. Walford, pers. comm.), underscoring the robustness of the original observation and the phenotypic stability of the genetic differences. Genetic analyses by E. Yunis and colleagues (Yunis *et al.*, 1984; Watson *et al.*, 1990; Yunis & Salazar, this volume) also have suggested a role of MHC loci in aging. Meyer, Armstrong and Warner (1989) found only a trend toward *H-2* effects on lifespan. However, males of the haplotypes tested in that study also showed only weak differences between lifespan in the studies by Smith & Walford (1977).

Where examined, the *H-2*r haplotype is notable for its superiority with respect to biomarkers of aging (Walford, 1990; Table 1). Lifespan for B10.R111 (*H-2*r) males and females is longer than all other haplotypes on the B10 background. B10.RIII females have more delayed reproductive senescence when compared to b and k haplotypes (Lerner *et al.*, 1988; Lerner & Finch, 1991; summarized in Table 1). The concordance between lifespan and reproductive senescence (age at last litter) is striking. Furthermore, age-related declines in unscheduled DNA synthesis, removal of DNA adducts, mitogen-induced lymphocyte proliferation, and mixed function oxidase induction are lessened in B10.RIII relative to B10 and B10.BR mice (Hall, Bergmann & Walford, 1981; Koizumi, Walford & Hasegawa, 1987; Walford, 1990; and R. Walford, pers. comm.).

H-2D genes are the only orthologous loci among the haplotypes examined gerontologically for which DNA sequences are available. Variable residues in the protein sequences of *H-2D* antigens are

Table 1. Comparison of lifespan and fecundity of three *H–2* haplotypes on the B10 background.

| Haplotype | Lifespan[1] (weeks) | | Fecundity[2] | | | |
	Male	Female	# of litters	Age (weeks) at last litter	# of pups/ female	Pups/litter
H–2[k]	149 ± 1.1	161 ± 2.1	7.5 ± 0.5	44.4 ± 2.1	43 ± 3	5.8 ± 0.3
H–2[b]	155 ± 0.4	148 ± 1.2	5.4 ± 0.5	37.1 ± 2.4	34 ± 3	6.4 ± 0.3
H–2[r]	170 ± 0.8	165 ± 1.0	7.8 ± 0.5	45.3 ± 2.3	50 ± 3	6.7 ± 0.3

[1] Data from Smith & Walford (1977). Lifespan is the mean age of the last 10% surviving in that group which is a valid approximation of maximum lifespan. Differences between lifespans of each haploptype within each sex group are statistically significant ($p < 0.02$).
[2] Data from Lerner *et al.* (1988). Four measures of fecundity are given. *H–2*[k] and *H–2*[r] strains did not statistically differ in any of the measures. *H–2*[b] mice showed statistically significant differences from *H–2*[r] and *H–2*[k] mice in number of litters ($p < 0.05$), age at last litter ($p < 0.01$), and number of pups per female ($p < 0.01$).

shown in Figure 2 to illustrate the magnitude of polymorphism at classical class I loci and to show the complexity in pinpointing differences responsible for influencing the rate of aging (under the assumption the *H-2D* genes are responsible for the altered lifespan between *H-2* congenic lines). Note that polymorphism is more pronounced in the $\alpha 1$ and $\alpha 2$ domains, but there are four variable positions in the relatively well conserved $\alpha 3$ domain.

H-2 congenic mice were developed based upon variation at class I and II antigen encoding loci. Besides differences at these serologically defined loci there may be other tightly linked polymorphic loci. At present, the effects of these loci on MHC/ aging relationships can not be excluded. Figure 2 therefore should not be viewed as the only gerontologically or immunologically relevant differences between haplotypes – other differences in the class II, class III, or nonclassical class I gene regions may also be responsible for MHC influence on senescence. The characterization of gross and fine MHC structure among *H-2* haplotypes by gene mapping and DNA sequencing is therefore expected to be enlightening from not only an immunological perspective, but from a gerontological point of view as well.

MHC and disease associations
HLA associations with disease have been known for over twenty years, yet the molecular mechanisms remain elusive (Tiwari & Terasaki, 1985; Bell, Todd & Devitt, 1989). Several aspects, though, are germane to the topic of the evolution of senescence. First, linkage between HLA and disease is never complete. While individuals with the

HLA-B27 allele have at least an order of magnitude higher risk than non-HLA-B27 individuals for ankylosing spondylitis (AS), only 2% of HLA-B27 individuals are afflicted with AS (except in familial instances types of AS where the risk is up to 20%; Benjamin & Parham, 1990). Incomplete penetrance may be accounted for in two ways (Bell, Todd & McDevitt, 1989). 1) Serological definition of alleles does not usually resolve differences in the regions which bind peptide antigens and might be the most critical regions with regards to autoimmune disease susceptibility. In fact, some correlations approaching 100% can be identified using hypervariable regions common to several serologically different specificities (Bell, Todd & McDevitt, 1989). 2) More importantly, HLA-associated diseases are polygenic, multifactorial biological phenomena, as is senescence.

Particularly important with regards to the MHC and evolutionary theories of aging, there is often a late-age onset of HLA-associated diseases; individuals in post-reproductive years are predominantly affected (Tiwari & Terasaki, 1985; Klein, 1987). For example, increased mortality is evident in only a small portion of the AS patients (Diethelm & Schuler, 1991). Thus, fitness in an evolutionary sense may not compromised in individuals with disease susceptibility-associated HLA alleles. It therefore seems that MHC genes may be candidate loci for evolutionary theories of aging – namely, mutation accumulation and antagonistic pleitropy.

Most likely reflecting the role of MHC class I and II antigens in presenting virus encoded peptides to T cells, MHC associations are particularly evident in autoimmune diseases. For example, there are clearly *H-2* haplotypic susceptibilities to experimental allergic encephalomyelitis (a popular paradigm of multiple sclerosis). Insulin dependent diabetes, rheumatoid arthritis, and AS also are thought to have autoimmune components and have strong correlations with certain class II and I (for AS) alleles (Tiwari & Teraskai, 1985). Besides autoimmune diseases, MHC involvement in various stages of neoplasia are also reported (Goodenow, Vogel & Linsk, 1985; Tanaka *et al.*, 1988) and there is some evidence of allelic associations.

MHC involvement in growth and development

The vast majority of literature regarding effects of MHC genes on growth and development implicates class I region genes. However, at the outset it should be made clear that MHC class I genes are not necessary for normal growth and development, as gene 'knockout' experiments indicate (Koller *et al.*, 1990; Zijlstra *et al.*, 1990). Functional, cell-surface expression of MHC class I molecules requires association with β_2m. The gene encoding β_2m was disrupted by homologous recombination in embryonic stem cells and the mice derived from these cells lacked expression of β_2m and consequently surface expression of *H-2* class I antigens (classical and nonclassical alike). They develop normally, though they are extremely vulnerable to some parasites (e.g. *Trypanosoma cruzi*; Tarleton *et al.*, 1992).

However, there is precedence for genes of the MHC to affect various stages of growth and development; in at least three instances, MHC regions encoding nonclassical class I genes have been implicated. *RT1*, the rat MHC, has regions (*RT1.G* and *C*) which probably correspond to *H-2 Tl* and *Qa* regions. Gill and coworkers (Kunz *et al.*, 1980) found that the growth regulatory control locus (*grc*) maps to the RT1.G/C region and is actually composed of two genes which affect body size and testicular development (*dw-3* and *ft*, respectively). Small rats with abnormal testes are observed in the *RT1¹* haplotype (*grc⁻*), which contains an approximately 70 kb deletion encompassing the *grc*. Probes derived from the *grc* region of *grc⁺* mice identified a homologous sequence in *Mus musculus* which was mapped to the *Tl* region (Vincek *et al.*, 1990) and at least one *grc* region is related to *H-2 Tla* region genes (Kirisits *et al.*, 1992).

Perhaps related to the *grc* locus in the rat, ectopic expression of a *H-2 Tl* region class I gene, *T18*, has remarkable effects superficially resembling delayed immunosenescence (M. Kronenberg, *pers. comm.*). Several independently derived transgenic lines harboring the *T18* gene under control of heterologous promoters exhibit delayed thymus involution compared to nontransgenic littermates. Moreover, cultured splenic T cells from these *T18* transgenic mice have greater persistence and less stringent growth factor requirements than nontransgenic controls. It is not known whether the rat ortholog of *T18* is in the *RT1 grc* region, but it is an attractive hypothesis.

The cleavage rate of of preimplantation embryos is different in B10BR (*H-2ᵏ*) and B10 (*H-2ᵇ*) mice, suggesting that a gene designated *Ped* (preimplantation embryo development) is linked to the MHC (Warner, Brownell & Ewoldsen, 1988). Using congenic mice that differ just in the *Qa* region, it was found that *Ped* is closely linked to the *Qa-2* antigen (Warner, Brownell & Rothschild, 1991). These studies also showed that the *Ped* gene (or other genes in the *Qa* region) significantly affect birth weight and litter size, implying that reproductive fitness can be modulated by genes in this region of the *H-2*. Warner and colleagues (Ford *et al.*, 1988; Conley *et al.*, 1988) have shown that the MHC effects on embryo development are demonstrable in pigs as well as as mice. Since the MHC of these animals contains only seven class I genes (Singer *et al.*, 1987) it may offer a more tractable system for understanding MHC influences on early developmental rates.

The differentiation of rat myoblasts to multinucleate myotubes is inhibited by antibodies against rat MHC class I antigens (Honda & Rostami, 1989). These data provide yet another example of MHC class I involvement in growth and extend the range of tissues that may be developmentally influenced by MHC class I antigens.

MHC and reproduction

In addition to the aforementioned effects on testicular development by the *ft* locus in *RT1*, effects of MHC polymorphism are evident in the reproductive physiology of mice, pigs, and chickens (comprehensively reviewed in Lerner & Finch, 1991). Table 1 shows some of the significant features of *H-2* haplotypic variation on fecundity and reproductive senescence in relation to lifespan. Furthermore, the *H-2* (together with other non-*H-2* loci) influences the length of estrous cycles (Lerner *et al.*, 1988). Lerner and Finch (1991) posit that MHC-reproductive relationships reflect interaction of MHC class I antigens with peptide hormone receptors, especially LH receptors. This is a compelling scenario that deserves more experimental attention.

Implication of MHC class I genes in aging
Several features, when taken together, lead to the hypothesis that class I genes may mediate the *H-2* regulation of senescence. MHC class I genes are ubiquitously expressed and are central to immune function, yet are implicated in nonimmume interactions as well. There are disease susceptibility correlations with both class I and II genes and the diseases, like aging, are of a multifactorial and polygenic nature. Finally, the influence of MHC class I genes on developmental rates and reproductive schedules is compelling support for class I involvement. Because class I genes do not seem necessary for normal growth and development in sterile settings, they must be viewed as modulators of aging rates rather than primary determinants.

MHC and lifespan – mechanisms

Differential effects of MHC haplotypes on senescence can be described in mechanistic terms as either due to qualitative or quantitive differences between alleles. Phenotypic variations due to primary sequence differences between alleles are from qualitative mechanisms. Variation in the relative level of expression of each allele is a quantitative mechanism. Clearly, these two possibilities are not mutually exclusive and two alleles may impart phenotypic variation by quantitative as well as qualitative mechanisms.

Quantitative variation
That naturally occurring deletions in the *grc* region of *RT1* lead to altered growth and development implies that quantitative variation in MHC gene expression has phenotypic effects on growth and development. In this case the variation is an all-or-nothing situation. Another example of quantitave variation is the overexpression of *H-2T18* in transgenic mice. The strains used for microinjection express undetectable levels of *T18* in the thy-

mus. Thus, increasing the expression in thymocytes and cell-types that normally do not synthesize appreciable levels of *T18* has effects on a near-universal biomarker of aging in mammals, thymic involution. The effect of *Qa* region genes on pre-implantation embryo development might relate to quantitative variation at the all-or-nothing level as well, because there are large deletions in the *H-2*[k] (*Ped*[slow]) versus the *H-2*[b] (*Ped*[fast]) *Qa* region (O'Neill *et al.*, 1986).

Relevant to the issue of quantitative variation is the trend that innappropriate expression of MHC class I and II genes has dramatic physiological effects. For example, ectopic expression of *H-2K*[b] under control of the mylein basic protein gene promoter causes severe hypomyelination (Turnley *et al.*, 1991). Likewise, destruction of pancreatic β cells occurs when MHC class I antigens are over-expressed in these cells (Allison *et al.*, 1988). Regardless of the pathological mechanisms in these experimentally induced conditions, it is evident that such drastic changes in expression would decrease fecundity. However, a less severe increase in expression might be positively selected owing to better immune protection. Neurons express MHC class I antigens at very low levels and as an outcome viral infection is persistent (Joly, Mucke & Oldstone, 1991). It is reasonable to hypothesize that in some instances increased expression in neurons would be advantageous. Mutations leading to this situation might increase fitness but with detrimental consequences at later ages.

Also under the rubrik of quantitative variation should be mentioned the observed changes in MHC expression with age. Age-related increases in cell-surface *H-2* class I antigens have been reported (Sidman *et al.*, 1987; Janick-Buckner & Warner, 1990; Janick-Buckner *et al.*, 1991). Where tested, the increase in cell-surface expression parallels an increase in MHC class I encoding mRNA (Janick-Buckner *et al.*, 1991; M. Crew, unpublished data). It is not yet known whether there are differences between *H-2* haplotypes in the magnitude of the increased class I expression with age, but this aspect certainly merits consideration. In addition, studies at the mRNA level have utilized probes that cross-hybridize with all class I mRNAs. There may be drastic age-associated changes in the expression of nonclassical class I antigens that are masked by this method.

Qualitative (allelic) variation

Relationships of the MHC to various diseases are probably by and large due to qualitative (i.e. amino acid) differences between alleles rather than altered expression of an allele. For example, in HLA-B27 AS individuals versus healthy individuals there are no clear differences in HLA-B cell-surface levels (Benjamin & Parham, 1990). The same holds true for other HLA alleles associated with disease, probably without exception (Bell, Todd & McDevitt, 1989; Tiwari & Terasaki, 1985).

Likewise, if classical MHC class I antigens are involved in reproduction and developmental rates, then qualitative mechanisms may be invoked since the expression of the classical class I genes probably does not differ appreciably between haplotypes (O'Neill & McKenzie, 1980). However, as mentioned, haplotypic variation in age-related changes in MHC class I expression has yet to be thoroughly explored.

Phylogeny and evolutionary history of the MHC

Darwinian evolution entails natural selection of existing genetic variation. MHC evolution has inspired much debate, in part due to the difficulty in distinguishing the mechanisms giving rise to genetic diversity from the selection for and against variants. Pease *et al.* (1991) emphasized the importance in realizing that the mechanisms of creating genetic variation within the MHC are different from the mechanisms by which new alleles are selected and that the full range of selective forces acting on MHC evolution are not fully understood. However, the principle that unconventional mutations coupled with intense selection by pathogens are intrinsic to the evolution of MHC class I and II genes is widely supported.

Mutational mechanisms

DNA sequence comparisons have revealed that both conventional (single point) mutations and unconventional mutations such as gene conversion events (also referred to as template-directed exchange or nonreciprocal recombination) have contributed to the diversification of MHC class I and II genes (Parham *et al.*, 1989; Lawlor *et al.*, 1990; Pease *et al.*, 1991). Two types of gene conversion events are observed which are defined by the 'do-

nor' sequence (i.e. the sequences contributing to the observed mutation): 1) intraloci (between alleles) and 2) interloci. The former is a leitmotif in HLA evolutionary histories (Kuhner *et al.*, 1991), while both are observed in the phylogeny of *H-2* class I gene (Pease *et al.*, 1991). Putative examples of interloci conversion are not limited to MHC class I genes but are also discerned in other multigene families (*c.f.* Becker & Knight, 1990; Wines *et al.*, 1991).

Variability in class I genes is predominantly localized to the regions encoding antigen-binding and TcR interacting residues. Intra- and interloci sequence exchange clearly takes part in generating this diversity. Yet gene conversion-like events may also homogenize class I genes in regions outside of the hypervariable extracellular domains (Rada *et al.*, 1990). Homogenization is thought to engender species-specific residues which are evident in comparing α3 to cytoplasmic domain-encoding gene sequences of rat and mouse MHC class I genes (Rada *et al.*, 1990) and in comparisons of *Peromyscus leucopus*, *Rattus*, and *Mus* MHC class I transmembrane domain-encoding exon sequences (Crew *et al.*, 1991). Diversification of hypervariable regions versus homogenization of more conserved domains may have consequences pertinent to nonimmune functions of MHC class I genes. Mutations outside of the antigen-binding regions may be selectively neutral or even negative, but may be outweighed by the benefits of diversity in the antigen-binding region.

For MHC class I genes, an additional level of diversification is apparent which relates to the expansion and contraction via unequal crossover of class I loci. The classical class I antigens between species are not necessarily related by direct descent. Instead, over evolutionary time, class I gene loci vary in their capacity as antigen presenters (Hughes, 1991). Nonclassical class I loci may later be called upon to serve as a classical class I antigens (Wang *et al.*, 1991). Such is the inference from examination of class I genes in New World primates. The classical class I loci in cotton top tamarins appear to have descended from loci homologous to certain HLA nonclassical class I genes (Watkins *et al.*, 1990). Thus, keeping a battery of unused class I genes might be advantageous in future environmental settings though, as described above, the variability in nonclassical class I gene regions may affect developmental growth rates and reproductive schedules.

Lastly, there are conflicting views concerning the rate of mutation within the MHC. On the one hand, at least one allele, H-$2K^b$, apparently has a higher mutation rate than the rest of the genome (Nathenson et al., 1986). However, there is little evidence that this is the case in other MHC genes (Flaherty, 1988). The high degree of polymorphism shown by MHC class I and II genes may be accounted for by intense selection for variation rather than increased mutation rates. Further, intra- and interlocus exchange have similar advantages over single base substitutions. Purifying selection should remove single base substitutions more than gene conversion generated mutations, since sequences within coding regions are exchanged to equivalent positions in homologous genes. As Howard (1992) points out, because of this the frequency of gene conversion need not be higher than conventional single base substitutions.

Selective pressures

Since MHC class I and II antigens bind foreign peptides for presentation to T cells, resistance to pathogens has long been proposed as a selective force in maintaining diversity in MHC genes (Doherty & Zinkernagel, 1975). Hughes and Nei (1988) showed that in allelic comparisons of MHC class I genes, nonsynonymous substitutions prevailed in the antigen-binding regions of class I antigens but not elsewhere in the extra-cellular domain-encoding portions of the genes. This suggests that polymorphism is positively selected for and that MHC genes probably evolve by overdominant selection of new alleles.

Recently, the first example of increased resistance to an infectious agent conferred by HLA alleles was discovered (Hill et al., 1991), thereby substantiating previous intuitions. In a survey of HLA alleles in areas hyperendemic with malaria, usually rare class I (Bw53) and II (DRB*1302-DQB*0501) alleles were found to occur more frequently in these areas. Moreover, the rare alleles were highly associated with malaria resistance – the frequency of HLA-Bw53 was eight times higher in healthy individuals than in those with severe cases of malaria. These findings strongly support the role of pathogens in the evolution of MHC polymorphism.

Another side to positive selection for heterozygosity at MHC loci is found in the studies of Potts et al. (1991) where genotype frequencies were followed in an enclosed population of mice with known H-2 types. That female mice preferentially mated with MHC-disparate males was obvious in the observed versus expected frequency of H-2 heterozygotes. The H-2 influences urinary odor and presumably this is the basis for mating preferences (Boyse, Beauchamp & Yamazaki, 1987; Yamazaki et al., 1990). It seems plausible that avoidance of MHC homozygosity by odor type-determined mate selection evolved for enhanced pathogen resistance (Howard, 1991).

MHC in nonmammals

Functionally and structurally homologous molecules to mammalian MHC class I and II antigens can be discerned in all vertebrate classes. Molecular cloning and DNA sequence data have been reported for MHC class I genes of fish, reptiles, amphibians, and birds (Kaufman, Skjoedt & Salomonnsen, 1990; Hashimoto, Nakanishi & Kurosawa, 1990; Grossberger & Parham, 1992). The best studied nonmammalian MHCs (in terms of relative functional equivalence to mammalian MHCs and molecular characterization) are those from chickens and *Xenopus* (reviewed in Kaufman, Skjoedt & Salomonnsen, 1990).

The *Xenopus laevis* MHC (*Xela*) exemplifies features of nonmammalian MHCs relevant to this discussion (Flajnik & Pasquier, 1990). *Xela* encodes class I and II antigens (with each haplotype having only one detectable class I locus) and complement components as well, suggesting a genetic linkage that has persisted for over 300 million years. During amphibian development there are dramatic changes in the expression of MHC genes. Tadpoles express class II genes – adults express class I genes. Thus, while not causative of metamorphosis, developmental influences of MHC class I molecules may not be limited to mammals.

Perhaps the most relevant feature of interclass comparative analyses is that polymorphism is the rule rather than exception. This fact argues that the overall function of class I molecules is nearly identical across vertebrate classes and therefore the evolutionary selection is similar. The implication is that the evolution of aging and the MHC paradigm may be universally applied to vertebrates.

Evolution of the MHC and evolution of lifespan – comparison of rates and mechanisms

Evolutionary rates

Whether or not the mutation rate in the MHC is higher than in the rest of the genome, it seems clear that MHC class I genes evolve rapidly owing to the intense selection for variation. The recent identification of new HLA-B alleles (resulting from sequence exchange between existing alleles) in South American Indian tribes underscores this and suggests that new alleles can arise to appreciable frequencies in less than 40,000 years (Belich *et al.*, 1992; Watkins *et al.*, 1992). There are few clues as to the evolutionary rate at which mammalian senescence is altered. Extrapolating from studies in invertebrates, one can expect it also to be rapid – in *Drosophilia*, significant alteration in lifespan is observed after 15 generations with judicious selection for delayed reproduction (Rose, 1984; Rose, 1990).

Evolutionary mechanisms

Evolutionary mechanisms for altering aging rates are well developed at a theoretical level. Antagonistic pleiotropy predicts that some loci attain mutations which impart increased fitness but may be progressively more detrimental with advancing age. There is a striking similarity here to the evolution of MHC class I and II genes in that mutations may be strongly selected for by adding to protectiveness against infectious agents. However, along with this comes increased susceptibility to diseases which mainly occur in post-childbearing years. The trade-off between enhanced fitness early in life versus detrimental effects later may relate to the function of MHC class I and II genes, representing a balance between presentation of foreign peptides (imparting increased fitness) and self-peptides (resulting in old-age deleterious effects).

A requirement for evolutionary theories of aging is that there is sufficient variation for natural selection to act upon. Taking into account the large number of MHC class I and II alleles even in small populations, the MHC clearly satisfies this requirement. Furthermore, considering MHC-disparate mating preferences (Potts *et al.*, 1991), the MHC might be critical in maintaining diversity throughout the genome and therefore may indirectly influence the capacity for evolutionary changes in aging rates by maintaining variation at non-MHC loci.

In populations with high mortality rates, mutations with late-age-specific effects may equilibrate to a higher frequency than that observed in low mortality rate populations (Charlesworth, 1990). Perhaps the observed differential frequency of alleles among racial and ethnic populations reflects this theoretical prediction.

Molecular mechanisms

Molecular mechanisms that lead to alteration in life history patterns, including lifespan, are not well described. Here, molecular mechanisms by which the MHC may affect senescence have been classified as quantitative or qualitative. The latter is typified by gross deletions of MHC nonclassical class I genes. An allele at the *age-1* locus in *C. elegans* which increases lifespan by over 60% is a null allele (Johnson *et al.*, 1990), as would be genes that are deleted in some *H-2* haplotypes. Thus, quantitative mechanisms as described here for MHC genes may be broadly operative in senescence.

At the nucleotide level, the basis for qualitative variation among MHC class I genes is largely due to genetic exchange rather than substitution. Besides the aforementioned trade-off between presentation of self versus nonself peptide antigens, there may be additional physiological consequences which relate to this type of mutational mechanism. That is, while mutations in the antigen-binding regions may be positively selected for, concomitant mutations outside of the antigen-binding domains may also occur, for example, in the $\alpha 3$ domain. The $\alpha 3$ domain may interact with peptide hormone receptors, and thus if selection for antigen binding region mutations were strong enough, the frequency of a new allele which differentially associates with hormone receptors would increase significantly.

Summary and future directions

Evolutionary theories of aging are instructive, but loci which fit predictions of the theories are lacking. Accumulating evidence suggests that MHC region genes modulate developmental, reproductive, and aging rates. Especially implicated are the MHC class I genes, and much is known about their evolutionary history in terms of selection and mechanisms generating diversity. The premise put forward here is that MHC class I genes are an appropriate model to experimentally address evolutionary theories of senescence in vertebrates.

There are several ways to further substantiate and develop this hypothesis. Corroborative evidence might be found in further analyses of MHC genes in species of wide-ranging lifespans. Along these lines, examination of insular (low mortality rate) populations might be informative, especially if laboratory lifespan data are available. What kind and frequency of MHC alleles are observed relative to noninsular populations (e.g. island versus mainland populations), and how does this compare to the divergence at other loci?

Selection for delayed reproduction in mammals as has been done in *Drosophilia* (Rose, 1984) may be feasible (Johnson, 1988; Rose, 1988). If such studies are performed successfully, a comparison of the type and frequency of MHC alleles that exist in the young versus old strains would be highly recommended.

Causal effects of MHC genes on aging rates perhaps may be unequivocally proven by transgenic mice experiments. Two types of transgenic experiments could seperately address the issue of qualitative versus quantitative mechanisms defined above. Qualitative effects of MHC alleles may be addressed by site-specific recombination techniques. For example, replacing *H-2D*^{b} in C57B10 mice with the *H-2D*^{r} gene would directly test the hypothesis that *H-2D* genes are responsible for the observed differences in lifespan and biomarkers of aging between B10.RIII (*H-2*^{r}) and C57B110 (*H-2*^{b}) mice. The more routine transgenic techniques (random integration sites) are more applicable to quantitative mechanisms.

It is worth emphasizing that since MHC genes are distributed among all vertebrate classes, consideration of the effects of MHC genes on life history patterns should not be limited to mammals. The diversity of MHC genes among vertebrates mirrors the diversity of life history patterns, including aging rates.

Acknowledgements

I am grateful to Drs. Roy Walford and Mitchell Kronenberg for their communication of unpublished results. This work was supported in part by the Arkansas Experimental Program to Stimulate Competitive Research (funded by the National Science Foundation, the Arkansas Science and Technology Authority and University of Arkansas for Medical Sciences).

References

Abe, K., J. F. Wei, F. S. Wei, Y. C. Hsu, H. Uehara, K. Artzt & D. Bennett, 1988. Searching for coding sequences in the mammalian genome: the H-2K region of the mouse MHC is replete with genes expressed in embryos. EMBO J. 7: 3441-3449.

Allison, J., I. L. Campbell, G. Morahan, T. E. Mandel, L. C. Harrison & J. F. A. P. Miller, 1988. Diabetes in transgenic mice resulting from over-expression of class I histocompatibility molecules in pancreatic β cells. Nature 333: 529-533.

Becker, R. S. & K. L. Knight, 1990. Somatic diversification of immunoglobulin heavy chain VDJ genes: evidence for somatic gene conversion in rabbits. Cell 63: 987-997.

Belich, M. P., J. A. Madrigal, W. H. Hildebrand, J. Zemmour, R. C. Williams, R. Luz, M. L. Petzl-Erler & P. Parham, 1992. Unusual HLA-B alleles in two tribes of Brazilian Indians. Nature 357: 326-329.

Bell, J. I., J. A. Todd & H. O. McDevitt, 1989. The molecular basis of HLA-disease association, pp. 1-42 in Advances in Human Genetics, vol. 18, edited by H. Harris & K. Hirschhorn, Plenum Press, N.Y.

Benjamin, R. & P. Parham, 1990. Guilt by association: HLA-B27 and ankylosing spondylitis. Immunology Today 11: 137-142.

Bjorkman, P. J., M. A. Saper, B. Samraoui, W. S. Bennet, J. L. Strominger & D. C. Wiley, 1987a. Structure of the human class I histocompatibility antigen, HLA-A2. Nature 329: 506-511.

Bjorkman, P. J., M. A. Saper, B. Samraoui, W. S. Bennet, J. L. Strominger & D. C. Wiley, 1987b. The foreign antigen binding site and T cell recognition regions of class I histocompatibility antigens. Nature 329: 512-516.

Boyse, E. A., G. K. Beauchamp & K. Yamazaki, 1987. The genetics of body scent. Trends Genet. 3: 97.

Brown, M. G., J. Driscoll & J. Monaco, 1991. Structural and serological similarity of MHC-linked LMP and proteasome (multicatalytic proteinase) complexes. Nature 353: 355-357.

Charlesworth, B., 1990. Natural selection and life history patterns, pp. 21-40 in Genetic Effects on Aging II, edited by D. E. Harrison. Telford Press, Inc., Caldwell, N.J.

Conley, A. J., Y. C. Jung, N. K. Schwartz, C. M. Warner, M. F. Rotschild & S. P. Ford, 1988. Influence of SLA haplotypes on ovulation rate and litter size in miniature pigs. J. Reprod. Fertil. 82: 595-601.

Crew, M. D., M. E. Filipowsky, E. C. Zeller, G. S. Smith & R. L. Walford, 1990. Major histocompatibility complex class I genes of *Peromyscus leucopus*. Immunogenetics 32: 371-379.

Crew, M. D., M. E. Filipowsky, M. S. Neshat, G. S. Smith & R. L. Walford, 1991. Transmembrane domain length variation in the evolution of major histocompatibility complex class I genes. Proc. Natl. Acad. Sci. USA 88: 4666-4670.

David-Watine, B., A. Israel & P. Kourilsky, 1990. The regulation and expression of MHC class I genes. Immunol. Today 11: 286-292.

Deverson, E. V., I. R. Glow, J. V. Coadwell, J. J. Monaco, G. W. Butcher & J. C. Howard, 1990. MHC class II region encoding proteins related to the multidrug resistance family of transmembrane transporters. Nature 348: 738-741.

Diethelm, U. & G. Schuler, 1991. Prognosis in ankylosing spondylitis. Schweiz. Rund. Med. Prx. 80: 584-587.

Doherty, P. C. & R. M. Zinkernagel, 1975. Enhanced immunological surveillance in mice heterozygous at the H-2 gene complex. Nature 256: 50-54.

Finch, C. E., 1990. Longevity, Senescence, and the Genome. The University of Chicago Press, Chicago.

Flaherty, L., 1988. Major histocompatibility complex polymorphism: a nonimmune theory. Hum. Immunol. 21: 3-13.

Flajnik, M. F. & L. D. Pasquier, 1990. The major histocompatibility complex of frogs. Immunol. Rev. 113: 47-63.

Ford, S. P., N. K. Schwartz, M. F. Rothschild, A. J. Conley & C. M. Warner, 1988. Influence of SLA haplotypes on preimplantation embryonic cell number in miniature pigs. J. Reprod. Fertil. 48: 99-104.

Glynne, R., S. H. Powis, S. Beck, A. Kelly, L.-A. Kerr & J. Trowsdale, 1991. A proteasome-related gene between the two ABC transporter loci in the class II region of the human MHC. Nature 353: 357-360.

Goodenow, R. S., J. M. Vogel & R. L. Linsk, 1985. Histocompatibility antigens on murine tumors. Science 230: 777.

Grossberger, D. & P. Parham, 1992. Reptilian class I major histocompatibility complex genes reveal conserved elements in class I structure. Immunogenetics 36: 166-174.

Hall, K. Y., K. Bergmann & R. Walford, 1981. DNA-repair, H-2 and aging in NZB and CBA mice. Tissue Antigens 17: 104-110 1981.

Hanson, I. B. & J. Trowsdale, 1991. Colinearity of novel genes in the class II regions of the MHC in mouse and humans. Immunogenetics 34: 5-11.

Hashimoto, K., T. Nakanishi & Y. Kurosawa, 1990. Isolation of carp genes encoding major histocompatibility complex antigens. Proc. Natl. Acad. Sci. 87: 6863-6867.

Henderson, R. A., H. Michel, K. Sakaguchi, J. Shabanowitz, E. Appella, D. F. Hunt & V. H. Engelhard, 1992. HLA-A2.1-associated peptides from a mutant cell line: a second pathway of antigen presentation. Science 255: 1264.

Hill, A. V. S., C. E. M. Allsopp, D. Kwiatkowski, N. M. Anstey, P. Twumasi, P. A. Rowe, S. Bennett, D. Brewster, A. J. McMichael & B. M. Greenwood, 1991. Common West African HLA antigens are associated with protection from severe malaria. Nature 352: 595.

Honda, H. & A. Rostami, 1989. Expression of major histocompatibility complex class I antigens in rat muscle cultures: the possible developmental role in myogenesis. Proc. Natl. Acad. Sci. 86: 7007.

Howard, J., 1991. Disease and evolution. Nature 352: 565-567.

Howard, J., 1992. Fast forward in the MHC. Nature 357: 284-285.

Hughes, A. L., 1991. Two models of evolution of the class I MHC, pp. 95-101 in Molecular Evolution of the Major Histocompatibility Complex, edited by J. Klein & D. Klein. Springer-Verlag, Berlin.

Hughes, A. L. & M. Nei, 1988. Pattern of nucleotide substitution at major histocompatibility complex class I loci reveals overdominant selection. Nature 335: 167.

Ito, K., L. Van Kaer, M. Bonneville, S. Hsu, D. B. Murphy & S. Tonegawa, 1990. Recognition of the product of a novel MHC TL region gene (27[b]) by a mouse gamma-delta T cell receptor. Cell 62: 549-561.

Jameson, S. C., C. Rada, R. Lorenzi, A. G. Diamond, G. W. Butcher & J. C. Howard, 1990. Cloning, expression, and evolution of rat classical and nonclassical class I genes. Transplant. Proc. 22: 2510.

Janick-Buckner, D. & C. M. Warner, 1990. An analysis of class I and II major histocompatibility antigen expression on C57BL/6 lymphocytes during aging, pp. 413-428 in Genetic Effects on Aging II, edited by D. E. Harrison. Telford Press, Inc., Caldwell, N.J.

Janick-Buckner, D., C. J. Briggs, T. E. Meyer, N. Harvey & C. M. Warner, 1991. Major histocompatibility complex antigen expression on lymphocytes from aging strain A mice. Growth, Develop. & Aging 55: 53.

Johnson, T. E., 1988. Thoughts on the selection of longer living rodents. Growth, Develop. Aging 52: 207-209.

Johnson, T. E., D. B. Friedman, N. Foltz, P. A. Fitzpatrick & J. E. Shoemaker, 1990. Genetic variants and mutations of Caenorhabditis elegans provide tools for dissecting the aging processes, pp. 101-127 in Genetic Effects on Aging II, edited by D. E. Harrison. Telford Press, Inc., Caldwell, N.J.

Joly, E., L. Mucke & M. B. A. Oldstone, 1991. Viral persistence in neurons explained by lack of major histocompatibility class I expression. Science 253: 1283-1285.

Kaufman, J., K. Skjoedt & J. Salomonnsen, 1990. The MHC molecules of nonmammalian vertebrates. Immunol. Rev. 113: 83-117.

Kelly, A., S. H. Powis, L. A. Kerr, I. Mockridge, T. Elliot, J. Bastin, B. Uchanska-Ziegler, A. Ziegler, J. Trowsdale & A. Townsend. 1992. Assembly and function of the two ABC transporter proteins encoded in the human major histocompatibility complex. Nature 355: 641.

Kindt, T. J. & D. S. Singer, 1987. Class I major histocompatibility complex genes in vertebrate species: what is the common denominator? Immunol. Res. 6: 57-65.

Kirkwood, T. B. L., 1977. Evolution of ageing. Nature 270: 301-304.

Kirkwood, T. B. L. & M. R. Rose, 1991. Evolution of senescence: late survival sacrificed for reproduction. Phil. Trans. R. Soc. London. B 332: 15-24.

Kirisits, M. J., H. W. Kunz, A. L. Cortesse Hasett & T. J. Gill III, 1992. Genomic DNA sequence and organization of a TL-like gene in the grc-G/C region of the rat. Immunogenetics 35: 365-377.

Klein, J., 1986. Natural history of the major histocompatibility complex. Wiley, N.Y.

Klein, J., 1987. Origin of major histocompatibility complex polymorphism: the trans-species hypothesis. Human Immunology 19: 155.

Klein, J. & F. Figueroa, 1986. Evolution of the major histocompatibility complex. Crit. Rev. Immunol. 6: 295.

Koizumi, A., R. L. Walford & L. Hasegawa, 1987. Differences

in H-2 recombinant mice in the β-napthoflavone inducibility of the mixed function monooxygenase, P_1-450. Immunogenetics 26: 169-173.

Koller, B. H., D. E. Geraghty, R. DeMars, L. Duvick, S. S. Rich & H. T. Orr, 1989. Chromosomal organization of the human major histocompatibility complex class I gene family. J. Exp. Med. 169: 469-480.

Koller, B. H., P. Marrack, J. W. Kappler & O. Smithies, 1990. Normal development of mice deficient in β_2m, MHC class I proteins, and CD8+ T cells. Science 248: 1227.

Kuhner, M. K., D. A. Lawlor, P. D. Ennis & P. Parham, 1991. Gene conversion in the evolution of human and chimpanzee MHC class I loci. Tissue Antigens 38: 152-164.

Kunz, H. W., T. J. Gill, B. D. Dixon, F. H. Taylor & D. L. Greiner, 1980. Growth and reproduction complex in the rat. J. Exp. Med. 152: 1506-1508.

LaFuse, W. & M. Edidin, 1980. Influence of the mouse major histocompatibility complex, H-2, on liver adenylate cyclase activity and on glucagon binding to liver cell membranes. Biochem. 19: 49.

Lawlor, D. A., J. Zemmour, P. D. Ennis & P. Parham, 1990. Evolution of class I MHC genes and proteins: from natural selection to thymic selection. Annu. Rev. Immunol. 8: 23-63.

Lerner, S. P., C. P. Anderson, R. L. Walford & C. E. Finch, 1988. Genotypic influences on reproductive aging of inbred female mice: effects of H-2 and non-H-2 alleles. Biol. Reprod. 38: 1035-1043.

Lerner, S. P. & C. E. Finch, 1991. The major histocompatibility complex and reproductive functions. Endocrine Reviews 12: 78-90.

Lundberg, A. S. & H. O. McDevitt, 1992. Evolution of major histocompatibility complex class II allelic diversity: direct descent in mice and humans. Proc. Natl. Acad. Sci. 89: 6545-6549.

Meyer, T. E., M. J. Armstrong & C. M. Warner, 1989. Effects of H-2 haplotype and gender on the lifespan of A and C57BL/6 mice and their F1, F2, and backcross offspring. Growth, Devel. & Aging 53: 175-183.

Monaco, J. J., S. Cho & M. Attaya, 1990. Transport protein genes in the murine MHC: possible implications for antigen processing. Science 250: 1723-1726.

Nathenson, S. G., J. Geliebter, G. M. Pfaffenbach & R. A. Zeff, 1986. Murine major histocompatibility complex class I mutants: molecular analysis and structure-function implications. Ann. Rev. Immunol. 4: 471.

O'Neill, H. C. & I. F. C. McKenzie, 1980. Quantitative variation in H-2 antigen expression. Immunogenetics 3: 226-239.

O'Neill, A. E., K. Reid, J. C. Garberi, M. Karl & L. Flaherty, 1986. Extensive deletions in the Q region of the mouse major histocompatibility complex. Immunogenetics 24: 368.

Parham, P., D. A. Lawlor, C. E. Lomen & P. D. Ennis, 1989. Diversity and diversification of HLA-A,B,C alleles. J. Immunol. 142: 3937.

Pease, L. R., R. M. Horton, J. K. Pullen & Z. Cai, 1991. Structure and diversity of class I antigen presenting molecules in the mouse. Crit. Rev. Immunol. 11: 1.

Potts, W. K., C. J. Manning & E. K. Wakeland, 1991. Mating patterns in seminatural populations of mice influenced by MHC genotype. Nature 352: 619-621.

Powis, S. J., E. V. Deverson, W. J. Coadwell, A. Ciruela, N. S. Huskisson, H. Smith, G. W. Butcher & J. C. Howard, 1992. Effect of polymorphism of an MHC-linked transporter on the peptides assembled in a class I molecule. Nature 357: 211.

Rada, C., R. Lorenzi, S. J. Powis, J. van den Bogaerde, P. Parham & J. C. Howard, 1990. Concerted evolution of class I genes in the major histocompatibility complex of murine rodents. Proc. Natl. Acad. Sci, USA 87: 2167.

Roberston, M., 1991. Proteasomes in the pathway. Nature 353: 300-301.

Rose, M. R. & B. Charlesworth, 1980. A test of evolutionary theories of senescence. Nature 287: 141-142.

Rose, M. R., 1984. Laboratory evolution of postponed senescence in *Drosophilia melanogaster*. Evolution 38: 1004-1010.

Rose, M. R., 1988. Response to 'Thoughts on the selection of longer-lived rodents'. Growth, Develop. & Aging 52: 209-211.

Rose, M. R., 1990. Evolutionary genetics of aging in *Drosophilia*, pp. 41-56 in Genetic Effects on Aging II, edited by D. E. Harrison. Telford Press, Inc., Caldwell, N.J.

Sidman, C. L., E. A. Luther, J. D. Marshall, K.-A., Nguyen, D. C. Roopenian & S. M. Worthen, 1987. Increased expression of major histocompatibility complex antigens on lymphocytes from aged mice. Proc. Natl. Acad. Sci. 84: 7624.

Singer, D. S., R. Ehrlich, L. Satz, W. Frels, J. Bluestone, R. Hodes & S. Rudikoff, 1987. Structure and expression of class I MHC genes in miniature swine. Vet. Immunol. Immunopath. 17: 211-221.

Smith, G. S. & R. L. Walford, 1977. Influence of the main histocompatibility complex on ageing in mice. Nature 270: 727-729.

Solano, A. R., M. L. Sanchez, M. L. Sardanons, L. Dada & E. J. Podesta, 1988. Leutenizing hormone triggers a molecular association between its receptor and the major histocompatibility complex class I antigen to produce cell activation. Endocrin. 122: 2080.

Spies, T., G. Blanck, M. Bresnahan, J. Sands & J. L. Strominger, 1989. A new cluster of genes within the human major histocompatibility complex. Nature 243: 214.

Spies, T., M. Bresnahan, S. Bahram, D. Arnold, G. Blanck, E. Mellins, D. Pious & R. DeMars, 1990. A gene in the human histocompatibility complex class II region controlling the class I antigen presentation pathway. Nature 348: 744-747.

Stroynowski, I., 1990. Molecules related to class-I major histocompatibility complex antigens. Annu. Rev. Immunol. 8: 501-530.

Tanaka, K., T. Yoshioka, C. Bieberich & G. Jay, 1988. Role of the major histocompatibility complex class I antigens in tumor growth and metastasis. Ann. Rev. Immunol. 6: 359.

Tarleton, R. L., B. H. Koller, A. Latour & M. Postan, 1992. Susceptibility of β_2-microglobulin deficient mice to Trypanosoma cruzi infection. Nature 356: 338-340.

Tiwari, S. L. & P. I. Terasaki, 1985. HLA and Disease Associations. Springer-Verlag, N.Y.

Trowsdale, J., I. Hanson, I. Mockridge, S. Beck, A. Townsend & A. Kelly, 1991. Sequences encoded in the class II region of the MHC related to the 'ABC' superfamily of transporters. Nature 348: 742-744.

Turnley, A. M., G. Morahan, H. Okano, O. Bernard, K. Miko-shiba, J. Allison, P. F. Bartlett & J. F. A. P. Miller, 1991. Dysmyelination in transgenic mice resulting from expression of class I histocompatibility molecules in oligodendrocytes. Nature 353: 566-569.

Uehara, H., 1991. Mouse Oct-3 maps between the tcl12 embryonic lethal gene and the Qa gene in the H-2 complex. Immunogenetics 34: 266-269.

Verland, S., M. Simonsen, S. Gammeltoft, H. Allen, R. A. Flavell & L. Olsson, 1989. Specific molecular interaction between the insulin receptor and a D product of the MHC class I. J. Immunol. 143: 945.

Vincek, V., F. Figueroa, T. J. Gill III, A. L. Cortess Hassett & J. Klein, 1990. Mapping in the mouse of the region homologous to the rat growth and reproduction complex (grc). Immunogenetics 32: 293-295.

Walford, R. L., 1990. The major histocompatibility complex and aging in mammals, pp. 31-41 in Molecular Biology of Aging, edited by C. Finch & T. E. Johnson. Alan R. Liss, Inc.

Wang, C.-R., A. Livingstone, G. W. Butcher, E. Hermel, J. C. Howard & K. Fischer Lindahl, 1991. Antigen presentation by neoclassical MHC class I gene products in murine rodents, pp. 441-462 in Molecular Evolution of the Major Histocompatibility Complex, edited by J. Klein & D. Klein. Springer-Verlag, Berlin.

Warner, C. M., M. S. Brownell & M. A. Ewoldsen, 1988. Why aren't embryos immunologically rejected by their mothers? Biol. Reprod. 38: 17.

Warner, C. M., M. S. Brownell & M. F. Rothschild, 1991. Analysis of litter size and weight in mice differing in Ped gene phenotype and the Q region of the H-2 complex. J. Reprod. Immunol. 19: 303-313.

Watkins, D. I., Z. W. Chen, A. L. Hughes, M. G. Evans, T. F. Tedder & N. L. Letvin, 1990. Evolution of the MHC class I genes of a New World primate from ancestral homologues of human non-classical genes. Nature 346: 60-63.

Watkins, D. I., S. N. McAdam, X. Liu, C. R. Strang, E. L. Milford, C. G. Levine, T. L. Garber, A. L. Dogon, C. I. Lord, S. H. Ghim, G. M. Troup, A. L. Hughes & N. L. Letvin, 1992. New recombinant HLA-B alleles in a tribe of South American Amerindians indicate rapid evolution of MHC class I loci. Nature 357: 329-333.

Watson, A. L. M., R. S. Gelman, R. M. Williams & E. J. Yunis, 1990. Murine chromosomal regions influencing lifespan, pp. 473-488 in Genetic Effects on Aging II, edited by D. E. Harrison. Telford Press, Inc., Caldwell, N.J.

Wines, D. R., J. M. Brady, E. M. Southard & R. J. MacDonald, 1991. Evolution of the rat kallikrein gene family: gene conversion leads to functional diversity. J. Mol. Evol. 32: 476-492.

Yamazaki, K., G. K. Beauchamp, Y. Imai, J. Bard, S. P. Phelan, L. Thomas & E. A. Boyse, 1990. Odortypes determined by the major histocompatibility complex in germfree mice. Proc. Natl. Acad. Sci. 87: 8413-8416.

Yewdell, J. W. & J. R. Bennink, 1990. The binary logic of antigen processing and presentation to T cells. Cell 62: 203.

Yunis, E. J., A. L. M. Watson, R. S. Gelman, S. J. Sylvia, R. Bronson & M. E. Dorf, 1984. Traits that influence longevity in mice. Genetics 108: 999.

Zijlstra, M., M. Bix, N. E. Simister, J. M. Loring & R. Jaenisch, 1990. β_2-microglobulin deficient mice lack CD4$^-$8$^+$ cytolytic T cells. Nature 344: 742.

M.R. Rose and C.E. Finch (eds.), Genetics and Evolution of Aging, 270–284, 1994.
© 1994 *Kluwer Academic Publishers. Printed in the Netherlands.*

Genetic influences on glucose neurotoxicity, aging, and diabetes: a possible role for glucose hysteresis

Charles V. Mobbs
Fishberg Center for Neurobiology, Mt. Sinai School of Medicine and Molecular Medicine and Diagnostic Laboratory, Bronx VAMC, Gustave Levy Pl., Box 1065, New York, NY 10129, USA

Received and accepted 23 August 1993

Abstract

Glucose may drive some age-correlated impairments and may mediate some effects of dietary restriction on senescence. The hypothesis that cumulative deleterious effects of glucose may impair hypothalamic neurons during aging, leading to hyperinsulinemia and other age-correlated pathologies, is examined in the context of genetic influences. Susceptibility to toxic effects of gold-thio-glucose (GTG) is correlated with longevity across several mouse strains. GTG and chronic hyperglycemia induce specific impairments in the ventromedial hypothalamus similar to impairments which occur during aging. GTG and a high-calorie diet both induce chronic hyperinsulinemia, leading initially to hypoglycemia, followed by the development of insulin resistance and hyperglycemia. Aging in humans and rodents appears to entail a similar pattern of hyperinsulinemia followed by insulin resistance. In humans, genetic susceptibility to high-calorie diet-induced impairments in glucose metabolism is extremely common in many indigenous populations, possibly due to the selection of the 'thrifty genotype'. It is suggested that the 'thrifty genotype' may entail enhanced sensitivity to the neurotoxic effects of glucose, and may represent an example of antagonistic pleiotropy in human evolution. These data are consistent with the hypothesis that genetic susceptibility of hypothalamic neurons to the cumulative toxic effects of glucose (glucose neurohumoral hysteresis) may correlate with genetic influences on longevity.

Introduction

Blood-borne substances, including glucose, may drive some age-correlated impairments. Neurohumoral hysteresis (Mobbs, 1989; Mobbs, 1990) refers to the phenomenon by which cumulative exposure to normal levels of blood-borne substances during aging is thought to gradually impair neurons which regulate those substances. Glucose neurohumoral hysteresis refers specifically to the possible cumulative deleterious effects of normal glucose levels during aging on neurons in the ventromedial hypothalamus (VMH), which in turn cause pancreatic hyperactivity, in some genotypes leading to insulin resistance, pancreatic impairments and diabetes (Fig. 1). This hypothetical mechanism is suggested by four observations: (i) a subset of VMH neurons which regulate glucose homeostasis is spe-

cifically sensitive to damage by elevated glucose and glucose derivatives; (ii) these neurons are particularly vulnerable to impairments and loss with age; (iii) damage to these glucose-sensitive VMH neurons causes pancreatic hyperactivity followed by insulin resistance and hyperglycemia; (iv) pancreatic hyperactivity, often followed by insulin resistance, pancreatic impairments, and hyperglycemia, increases with age. These observations suggest the theory of glucose neurohumoral hysteresis: that normal (post-prandial) hyperglycemia causes residual VMH impairments, leading to pancreatic hyperactivity and insulin resistance, in some genotypes leading to higher post-prandial glucose, leading to more VMH impairments, etc. (Fig. 1). One implication of this theory is that dietary restriction may delay some age-correlated pathologies by reducing cumulative exposure to glucose and attenu-

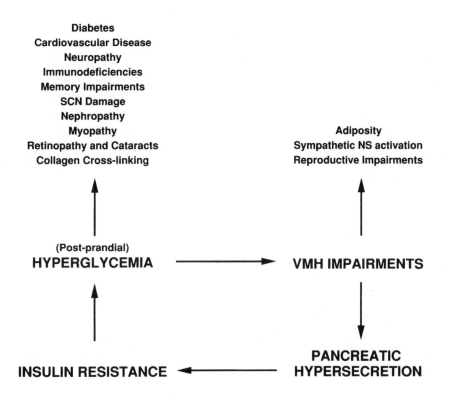

Fig. 1. Graphical depiction of the essential elements of glucose neurohumoral hysteresis. For details see text.

ating the process of glucose hysteresis (*ibid.*).

Glucose hysteresis describes a time-dependent mechanism by which an environmental influence, dietary intake, may cause the acquisition of impairments with age, and thus contribute to the process of senescence. Clearly, however, senescence entails an intimate interaction between temporally acquired (environmental) impairments and genetic susceptibility to those impairments: the effects of environment on phenotype increase with age. If glucose hysteresis is a major mechanism in senescence (of mammals, at least), one would predict a genetic relationship between longevity and susceptibility to glucose-induced pathologies. That is, all other factors held constant, genotypes with the least susceptibility to glucose hysteresis should live the longest. I review here evidence that in rodents and humans the genetic susceptibility to glucose toxicity correlates with genetic influences on longevity, especially genetic susceptibility to mature-onset (non-insulin-dependent) diabetes (NIDDM).

Link between genetic susceptibility to hypothalamic damage by glucose derivatives and longevity

The theory of glucose hysteresis was first suggested by two observations. First, we and others had obtained considerable evidence that the cumulative exposure to estrogen leads to age-correlated impairments in the neuroendocrine system controlling female rodent reproductive function (*ibid.*). Many of these studies entailed demonstrating that elevated levels of estrogen would produce impairments in young female rodents similar to those ordinarily exhibited only in older female rodents (Mobbs *et al.*, 1984; Mobbs & Finch, 1992). However, since such studies usually entailed either non-physiological levels or non-cycling physiological levels of estrogen, it was possible that it was the non-physiological presentation of the hormone that caused the impairments, rather than the cumulative exposure to the hormone. However, we and others also showed that removing the ovaries when the animals were young delayed the development of some neuroendocrine impairments (Mobbs, Gee & Finch,

1984; Mobbs *et al.*, 1985; Nelson *et al.*, 1987; Mobbs & Finch, 1992). Similar studies in the regulation of the rat adrenocortical system suggested that elevating corticosterone accelerated age-correlated impairments in the neuroendocrine system regulating corticosterone, and reducing corticosterone delayed these impairments (Landfield, Waymire & Lynch, 1978; Landfield, Baskin & Pitler, 1981; Sapolsky, Krey & McEwen, 1986). Although none of these studies have been absolutely definitive, evidence continues to grow that if elevated levels of a hormone cause impairments, even physiological levels of that hormone may cause cumulative impairments during aging, although at a slower rate.

The second observation, which led particularly to the neurohumoral theory of glucose hysteresis (a more general form, based on gene memory, is discussed below), was that neurons in the VMH (a region of the mediobasal hypothalamus containing the ventromedial nucleus, the arcuate nucleus, and the cell-poor area between them) appear to be particularly sensitive to the deleterious effects of elevated glucose and glucose derivatives. For example, chronic hyperglycemia resulting from alloxan-induced pancreatic damage specifically reduces the size of nuclei in neurons in the VMH, but does not affect nuclei in neurons in the dorsomedial nucleus and periventricular nucleus (Akmayev & Rabkina, 1976). The effect of chronic hyperglycemia on VMH nuclei is not due to acute responses to elevated glucose, since the well-fed state is normally associated with *larger* ventromedial nuclei than the fasted (relatively hypoglycemic) state (Pfaff, 1972). In another study, hyperglycemia resulting from streptozotocin-induced pancreatic damage caused many neuropathologies (fragmented endoplasmic reticulum, loss of organelles, irregular nuclei) in the arcuate nucleus (Bestetti & Rossi, 1980). In the genetically diabetic db/db mouse, the number of degenerating neurons progressively increases, and neuronal density decreases, in the ventromedial and arcuate nuclei compared with controls (Garris, West & Coleman, 1985). In the genetically diabetic Chinese hamster, the area, number of neurons, and neuronal density of the ventromedial nucleus are all significantly reduced compared with nondiabetic controls (Garris *et al.*, 1982). In fact, diabetes is associated with an acceleration of the normal age-correlated loss of neurons in this nu-

cleus (Smith-West & Garris, 1983). Hyperglycemia in db/db mice is associated with a progressive loss of estradiol receptors in the VMH (Garris & Coleman, 1984; Garris, Coleman & Morgan, 1985). Streptozotocin-induced hyperglycemia also causes a progressive loss of estradiol receptors in the VMH, as well as a loss of lordosis behavior (Ahdieh, Hamilton & Wade, 1983). This study also suggested that the loss of estrogen receptors after several weeks of hyperglycemia was not simply due to an acute effect of hyperglycemia, since daily injection of insulin prevented the loss of receptors and behavior, but restoration of lower blood sugar by insulin injection after several weeks of hyperglycemia did not reverse the impairments (Ahdieh, Hamilton & Wade, 1983).

The sensitivity of VMH neurons to glucose toxicity was provocative because these neurons are also particularly vulnerable (compared, for example, to other hypothalamic neurons) to aging. Sabel and Stein (1981) reported a progressive age-correlated loss of neurons from the rat ventromedial nucleus of 30-month-old rats compared with the number in 3-month-old rats; in contrast, neurons in the nearby lateral amygdala were not significantly reduced (Sabel & Stein, 1981). Similarly, in perhaps the most extensive analysis of hypothalamic neuron number during aging published to date, Sartin and Lamperti (1985) reported an age-correlated decrease in neuron number in both the ventromedial nucleus and arcuate nucleus counted in 3-, 12-, and 24-month-old male rats. In this study neurons in the VMH decreased 30% by 24 months, whereas six other hypothalamic nuclei, including the medial preoptic area, exhibited no age-correlated neuronal loss (Sartin & Lamperti, 1985). Hsu and Peng (1978) reported a 50% loss in neurons in the arcuate nucleus of female rats by 24 months compared with the number in rats 4 months old, although they did not replicate this loss in male rats (Peng & Hsu, 1982). Smith-West and Garris (1983) reported a 20% reduction in the ventromedial nuclear area from 3 months to 15 months, as well as a reduction in nuclear density, in the Chinese hamster.

A particularly striking example of the sensitivity of neurons in the VMH to glucose toxicity is their remarkable vulnerability (at least in mice) to the toxic effects of gold-thio-glucose (GTG) (Liebelt & Perry, 1967). This simple derivative of glucose, when injected i.p. at appropriate doses in mice, causes a specific lesion in the VMH (*ibid.*). Three days after minimum effective doses of GTG, this lesion extends from the cell-poor area between the arcuate nucleus and the ventromedial nucleus, to the ventrolateral aspect of the VMH. Extensive analysis has demonstrated very few other toxic effects of GTG when injected at doses sufficient to lesion these VMH neurons. The lesion does not occur in insulin-deficient animals, and can be blocked by a number of glucose derivatives. It is therefore generally assumed that GTG destroys specific neurons in the VMH which are sensitive to glucose and/or insulin. We therefore speculated that the neurons sensitive to GTG are the same ones which are sensitive to the toxic effects of hyperglycemia and possibly to aging as well. Thus we began to examine the possibility that GTG-lesioned mice might prove useful as a model for accelerated aging of the neuroendocrine regulation of glucose homeostasis.

In examining published dose-response curves of GTG potency, it appeared that there might be a correlation between longevity and genetic susceptibility to VMH damage or death induced by GTG. Figure 2 depicts, for several mouse strains, the relationship between longevity and susceptibility to GTG toxicity. Mean lifespan was obtained from Henninger and Dorey (1982), reflecting the cumulative experience of 'several years' before 1982 at Jackson Laboratories. Although mean lifespan is more dependent on environment than maximum lifespan, and can only serve as a general indication of the genetic determinants of longevity, this parameter was chosen for the present analysis as the best standardized source of information on longevity for the several strains in which GTG toxicity had been determined by a detailed dose-response curve (Liebelt *et al.*, 1960). GTG sensitivity in this study was determined by injecting i.p. about 20 mice from each strain with the following doses of GTG: 0, 0.1, 0.2, 0.3, 0.4, 0.6, 0.8, 1.0, 1.2, 1.4, and 1.5 mg/gm

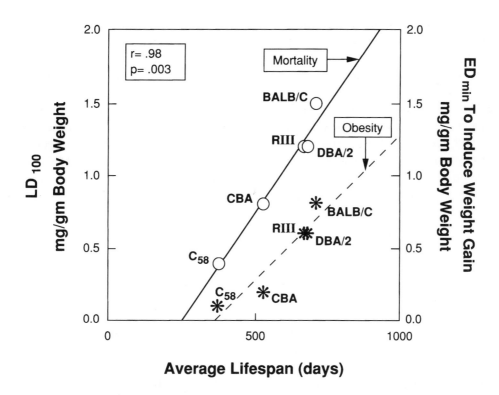

Fig. 2. Correlation between mean longevity and susceptibility to gold-thioglucose (GTG) toxicity for several strains of mice. Mean longevity data were obtained from Henninger and Dorey (1982). Susceptibility was expressed as minimum dose of GTG (within 0.1 mg/gm body weight) required to induce significant gain in weight, or 100% mortality, 60 days after injection.

body weight. Sixty days after injection, each group was assessed for body weight gain and mortality. As a measure of susceptibility to GTG toxicity in each strain, I have used the minimum dose of GTG required to induce a significant gain in body weight (generally correlated with VMH damage) 60 days after injection, and the minimum dose required to cause 100% mortality. As demonstrated in Fig. 2, there is a surprising correlation, at least in the strains investigated by Liebelt *et al.*, between the mean lifespan and susceptibility to GTG toxicity, measured either by induction of obesity or lethality. These data suggest the possibility that the genetically determined sensitivity of hypothalamic neurons to the toxic effects of glucose could influence longevity.

Liebelt *et al.* also found that the CBA strain was, except for the C_{58} strain, the most sensitive to GTG and exhibited the most reliable response. In general we (Bergen & Mobbs, unpublished) have obtained similar results. Liebelt *et al.* further found that when CBA mice are crossed with several other strains, the resulting F1 hybrids all exhibit sensitivity to GTG similar to the sensitivity of CBA mice. Liebelt *et al.* interpreted these data as indicating that 'genetic factors involved in goldthioglucose obesity, and derived from the CBA parent, are expressed as the dominant influence in F1 hybrids'. This suggests the interesting possibility that F1 hybrids derived from CBA mice might also exhibit the shorter longevity of CBA mice, inherited as a dominant trait, and that the longevity trait would segregate with the susceptibility to GTG toxicity. Such a result would facilitate identification of major genetic loci controlling longevity in CBA mice, and suggest that a relevant phenotype is hypothalamic susceptibility to (gold-thio-) glucose toxicity.

Similarity of gold-thioglucose-induced VMH lesions to age-correlated VMH impairments

To further explore GTG-lesioned mice as a model for studying normal age-correlated processes, we began to characterize VMH lesions induced by GTG (Bergen *et al.*, 1992). Oxytocin receptors (as a marker of neurons in the ventrolateral VMN) and neuropeptide Y (NPY) mRNA in the arcuate nucleus were assessed in male CBA mice given GTG (0.5 mg/ gm bw, i.p.) three months before sacrifice. Oxytocin receptors were assessed using standard

I^{125}-ornithine vasotocin receptor autoradiography. NPY mRNA was assessed by *in situ* hybridization, entailing single-stranded NPY cDNA probes. Half of each group of mice were fasted for 72 h, then sacrificed at 1900 h, one hour after lights went out. The other half of each group were fasted for 8 h, then injected i.p. with glucose (2 mg/gm body weight) at 1845 h and sacrificed at 1900 h. Brains were fresh-frozen in dry ice, then cut into 10 micron sections for analysis by *in situ* hybridization. Single-stranded neuropeptide Y cDNA probes were labelled with tritium by asymetric PCR (using a plasmid generously supplied by S. Sabol). Sections were first exposed to film for 3 weeks, then dipped in emulsion and exposed for another 3 weeks. Six weeks after the GTG injection, cresyl violet stain revealed a loss of cells in the VMH. Oxytocin receptor binding was readily demonstrable in the VMH of control mice, although not influenced by fasting, and was completely abolished by the GTG injection. NPY mRNA in the arcuate nucleus was lower in fed (glucose-injected) GTG-lesioned mice compared to fed controls. In addition, the induction of NPY mRNA by a 72-h fast was lower in GTG-treated mice compared with controls.

We found evidence for similar impairments in the VMH of aging rats (Mobbs & Kleopoulos, 1992). Six-, 12-, and 18-month-old rats were fasted for 72 h, or fasted for 8 h and given access to 10 ml sweetened milk 45 min before sacrifice. Oxytocin receptor binding, measured as described above, was not influenced by fasting, but decreased progressively in the ventrolateral VMN of 6-, 12-, and 18-month-old fed and fasted male Sprague-Dawley rats. In addition, NPY mRNA in the arcuate nucleus of fed rats decreased progressively with age. In 6-month-old rats NPY was significantly induced by a 72-h fast. However, the induction of NPY mRNA decreased progressively with age, and even in 12-month-old rats the induction was not significant. We have also found that growth-hormone releasing hormone (GHRH) mRNA in the arcuate nucleus decreases progressively with age (Kleopoulos & Mobbs, 1993). These data suggest that during aging in rats VMH impairments occur which are similar to those induced by GTG in young mice, and that these impairments could lead to age-correlated pathologies (e.g., the loss of GHRH mRNA could lead to decreased growth hormone during aging).

To clarify the potential functional significance of age-correlated impairments in VMH neurons, we examined in aging male rats functions regulated by VMH neurons (Mobbs, 1990). Rats with VMH lesions exhibit an increased adiposity setpoint which is defended by increased conversion of calories into fat and increased eating, especially during the light phase. To examine if aging rats exhibit similar behavior, which would suggest the presence of functional lesions, 6-, 12-, and 18-month-old male Sprague-Dawley rats (n = 12/age) were fasted for 72 h. For 1 week after fasting, rats were restricted to consuming only as much food as before the fast. From 1 to 2 weeks after fasting, rats were allowed to eat *ad lib*. Rats weighed more with increasing age; older rats lost the same weight during fasting as younger rats. Rats of all ages gained weight even when restricted to pre-fast consumption, and ate more during the *ad lib* period. By two weeks, most rats had regained initial body weights. At all times, the ratio of food eaten during the light to food eaten in the dark (L:D) was greater in older than younger rats. Thus older rats will defend the age-related increase in adiposity in a manner similar to VMH-lesioned rats. One of the most diagnostic indications of VMH lesions is that VMH-lesioned animals are generally described as 'finicky': they will consume *more* highly palatable (high carbohydrate) food, but *less* unpalatable (bitter) food, compared with controls. In a separate study in aging C57Bl/6NXC3H F1 mice, we found that 24-month-old mice consumed more sweetened milk (50% diluted Carnation condensed milk) than 6-month-old mice, both in a well-fed state or after a 24-h fast. Thus as rodents age, they exhibit, even in middle age, functional changes similar to those induced by VMH lesions. In the context of the histological studies described above, these studies suggest that during aging VMH impairments occur which are similar to those induced by GTG and which have progressive functional consequences.

GTG and diet-induced impairments in glucose homeostasis: hyperinsulinemia followed by insulin resistance

We next examined possible physiological consequences of glucose-induced lesions which might influence longevity. It has been hypothesized that the VMH neurons sensitive to GTG possess specific glucoreceptors presumably related to the regulation of glucose homeostasis (Liebelt & Perry, 1967). We therefore examined glucose homeostasis in mice injected with GTG (Bergen & Mobbs, 1992). To examine the metabolic consequences of GTG lesions, male C57XC3H mice were injected i.p. with GTG (0.8 mg/gm body weight in 0.1 M Na Citrate). Blood glucose (after an overnight fast) was assessed 3 days and 6 weeks after GTG injection. Three days after injection, GTG-injected mice had *lower* levels of plasma glucose than vehicle-injected controls but weighed more. This result was consistent with previous studies reporting that rats were slightly hypoglycemic three weeks after electrolytic VMH lesions (Frohman, Goldman & Bernardis, 1972). Six weeks after injection, GTG-treated mice continued to weigh more, but began to exhibit *higher* levels of plasma glucose compared to controls, both after a 48-h fast and after an i.p. injection of glucose. GTG-treated mice also exhibited elevated levels of insulin, either in the fasted state or after an i.p. injection of glucose. Similar results were obtained in two other strains of mice (CBA/N and C57Bl/6J). GTG-induced elevations of plasma insulin are consistent with other studies reporting that pancreatic hypersecretion of both insulin and glucagon occurs within minutes of an electrolytic VMH lesion in rats (Berthoud & Jeanrenaud, 1979; Rohner-Jeanrenaud & Jeanrenaud, 1980; Rohner-Jeanrenaud & Jenrenaud, 1984), well before hyperphagia or insulin resistance develops (Jeanrenaud, Halimi & Van de Werve, 1985). In addition, VMH lesions cause hypersecretion of pancreatic islet polypeptide (amylin) (Tokuyama *et al.*, 1991). We interpret these data as suggesting that an early effect of the GTG lesion is pancreatic hypersecretion. Three days after the lesion, the resulting hyperinsulinemia causes a decrease in blood glucose. However, by 6 weeks after the lesion, pancreatic hypersecretion leads to insulin resistance and hence to an elevation of plasma glucose. These data are consistent with the hypothesis that specific, insulin-dependent lesions in glucose-sensitive neurons of the VMH can lead to impairments of glucose metabolism, entailing pancreatic hypersecretion and hypoglycemia, followed by hyperglycemia.

Since GTG lesions are specific but not physiological, we developed an alternative, nutritional model for glucose-induced impairments in glucose homeostasis. C57Bl/6J mice were given a high-calorie diet (based on Surwit *et al.*, 1988; 1990) which caused a higher cumulative exposure to blood glucose than normal mouse chow. Mice on this diet for 2 weeks exhibited *lower* levels of blood glucose, after a 24-hour fast, than controls on normal mouse chow. Such a result is consistent with reports that this diet causes a persistent increase in insulin levels (Surwit *et al.*, 1990) and that a diet high in sucrose also causes a persistent increase in insulin levels and a temporary 'improvement' in glucose tolerance (Kergoat, Bailbe & Porthe, 1987). Thus, although in the fed state mice on this diet are exposed to elevated levels of glucose, in the fasted state these mice are relatively hypoglycemic, compared with controls. However, we found (in confirmation of the results of Surwit *et al.*) that mice on this diet for 3 months began to exhibit higher levels of blood glucose, both in the fasted state and in response to glucose challenge, compared with controls. Thus mice on a simple carbohydrate diet exhibit a similar progression to GTG-injected mice: hyperinsulinemia entailing early hypoglycemia followed by hyperglycemia (presumably due to the development of insulin resistance in response to the pancreatic hypersecretion). Although this progression is slower when induced by diet than GTG, such a result would be expected if the diet is inducing VMH impairments similar to those induced by GTG, but at a much slower (and more physiological) rate.

Genetic analysis of susceptibility to diet-induced insulin resistance suggests that this trait, like susceptibility to GTG toxicity, is determined by a dominant genetic influence. Surwit *et al.* (1990) demonstrated that the high-calorie diet used in the above studies induced insulin resistance in C57Bl/6J mice but not in A/J mice. Of particular interest, fasting plasma insulin levels were similar between the two strains when both were on a diet of mouse chow, but when the two strains were put on the high carbohydrate diet, insulin levels were several times higher in the C57Bl/6J mice than the A/J mice. Thus the C57 mice seem to be particularly sensitive to diet-induced hyperinsulinemia. When C57Bl/6J mice were crossed with A/J mice, the diet-induced insulin resistance in the F1 hybrids was similar to the insulin resistance induced in the C57 parent strain. In contrast the F1 hybrids did not exhibit

diet-induced hyperglycemia exhibited by the C57 parental strain. These data suggested that the genes controlling diet-induced insulin resistance are inherited in a dominant fashion, whereas the genes controlling diet-induced hyperglycemia (an extreme manifestation of insulin resistance, possibly entailing late-stage pancreatic impairments as well) are inherited in a recessive fashion. Analysis of recombinant inbred strains derived from crosses of these strains indicated that the number of genes controlling susceptibility to diet-induced insulin resistance was 'relatively few' (Surwit *et al.*, 1990). Using these same inbred recombinant lines, it was possible to clearly dissociate the genetic factors controlling insulin resistance from those controlling hyperglycemia. Such analyses raise the possibility of examining a potential genetic relationship between diet-induced impairments in glucose homeostasis (hyperinsulinemia, insulin resistance, or hyperglycemia), and longevity of these strains.

The studies described above suggest that GTG-induced hypothalamic impairments and diet-induced (glucose-induced?) impairments entail hyperinsulinemia, followed initially by lower blood glucose, then by the development of (presumably compensatory) insulin resistance, followed eventually by elevated blood glucose. If glucose-induced VMH damage occurs gradually during aging, as hypothesized above, then one would predict a similar progression of events during aging: an age-correlated increase in pancreatic secretion, initially in the presence of hypoglycemia (i.e., an 'improvement' of glucose metabolism), followed at later ages by insulin resistance and, eventually in some individuals, by hyperglycemia (i.e., an impairment in glucose metabolism). Discussed below are studies on age-correlated changes in glucose metabolism generally consistent with this prediction.

Age-correlated changes in glucose metabolism: hyperinsulinemia followed by insulin resistance

Impairments in glucose regulation in humans, rodents, and other species is progressively impaired with age even in individuals without diabetes (Brancho-Romero & Reaven, 1977; Chlouverakis, Jarret & Keen, 1967; Davidson, 1979; DeFronzo, 1979; Reaven & Reaven, 1985; Sandberg *et al.*, 1973; Welborn, Stenhouse & Johnstone, 1969). Although it is clear that in humans and rodents aging is generally associated with elevated plasma levels of pancreatic hormones (see below), at least at some age, in humans and rats it is difficult to determine unequivocally if hyperinsulinemia precedes (and contributes to) the development of insulin resistance or if hyperinsulinemia is a compensation for age-correlated development of insulin resistance. The reason it is difficult to dissociate the increase in insulin levels and insulin resistance generally is that both increase incrementally during aging in humans and rats. Such an incremental increase in both insulin and insulin resistance is a plausible outcome from the gradual impairments which occur in VMH neurons, but clearly cannot prove causality.

We have found that using oral glucose tolerance tests in early middle-aged rats, hyperinsulinemia can be detected in the presence of hypoglycemia; only later does hyperglycemia develop (Mobbs, 1991). Six-, 12-, and 18-month-old male Sprague-Dawley rats were trained to drink 50% Carnation condensed milk. Just after lights went out, the rats were given access to 10 ml of milk, and sacrificed 40 min later. Plasma was taken for assessment of glucose and insulin. We found that 12-month-old rats had lower plasma glucose and higher levels of insulin than 6-month-old rats, generally consistent with Brancho-Romero and Reaven (1977). However, plasma glucose was higher in 18-month-old rats than in 6 month-old rats, and insulin levels were not different. The increase in insulin in middle age, even in hypoglycemic animals, followed by a drop in insulin concomitant with the development of hyperglycemia in older animals, is similar to the progression exhibited in the development of mature-onset diabetes (DeFronzo, Bondadonna & Ferrannini, 1992; Zimmet, 1992).

Since we had shown that mice undergo characteristic progressive changes in glucose homeostasis after GTG lesions and during exposure to a high carbohydrate diet, we examined the possibility that during aging mice might exhibit a similar progression (hyperinsulinemia in the presence of hypoglycemia, followed only later by the development of insulin resistance) which would provide clearer evidence for hyperinsulinemia preceding insulin resistance than can be obtained in humans and rats. We examined (Mobbs & Bergen, 1992) plasma glucose levels in 6-month and 24-month old male

C57XC3H mice after fasting, stress (5 min restraint at 15, 30, and 45 min, sample taken at 45 min), glucose injection (15 min after 2 mg/gm bw, i.p.), or oral glucose intake (40 min after 1 ml 6% glucose). Plasma glucose was obtained by tail bleeding and analyzed with a glucose meter. Blood glucose was lower in 24-month-old mice than in 6-month-old mice, under all of the experimental conditions examined. Examining mice at several ages between 2 months and 24 months, we found an age-correlated monotonic decrease in fasting levels of plasma glucose, with significant decreases between 2, 6, 12, and 24 months. Despite the age-correlated decrease in glucose, plasma insulin was greater in 24-month-old mice than in 6-month-old mice, in both the fasted state and 15 min after i.p. glucose. However, the ratio of insulin to glucose was greater in old mice, suggesting that although the older mice were hypoglycemic, some insulin resistance had already developed. Similar results were obtained in C57Bl/6N and CBA mice. It is possible that the increase in plasma insulin is due in part to decreased clearance of insulin with age (Minaker et al., 1982). However, in vitro pancreatic secretion and islet volume increase with age (Bonnevie-Nielsen, Skovgaard & Lernmark, 1983; Jeffrey et al., 1986), concomitant with the development of hypoglycemia (Leiter et al., 1988). These data support the hypothesis that pancreatic secretion increases with age (at least in mice) before the development of insulin resistance. It is plausible that pancreatic hypersecretion in early middle age could actually cause insulin resistance later, since hypersecretion of insulin (Davidson & Casanello-Ertl, 1979; Rizza et al., 1985; Soman & DeFronzo, 1980), glucagon (Baron et al., 1987; Del Prato et al., 1987), and amylin (Leighton & Cooper, 1988) can all cause insulin resistance and hyperglycemia. The drop in insulin in older individuals could plausibly be due to a well-documented 'burn-out' phenomenon, in which pancreatic hyperactivity can lead to pancreatic exhaustion (i.e., insensitive beta-cells) correlated with an increase in blood glucose (Lazaris, Goldberg & Kozlov, 1985).

Humans exhibit similar impairments in glucose metabolism during aging. The prevalence of diabetes in the United States at ages 20-44, 45-54, 55-64, and 65-74 is 2%, 8.5%, 13.4%, and 18.7% respectively (Harris et al., 1987). The development of mature-onset diabetes entails hypersecretion of insulin before any abnormalities of glucose tolerance, followed by a further increase of insulin when impaired glucose tolerance develops, followed by reduction of insulin to the normal range after diabetes develops (Saad *et al.*, 1989; DeFronzo *et al.*, 1992). Age-correlated impairments in glucose metabolisms, and mature-onset diabetes, are characterized by increased resistance to insulin (Rowe *et al.*, 1983; Chen *et al.*, 1985; Meneilly *et al.*, 1987). Plasma insulin levels increase during aging even in humans with normal glucose tolerance (Chlouverakis *et al.*, 1967; Davidson, 1979; DeFronzo, 1979; Sandberg *et al.*, 1973; Welborn *et al.*, 1969). In early middle-aged non-diabetic humans, as in rats and mice, hyperinsulinemia can be detected in the presence of lower levels of blood glucose (Chlouverakis, Jarrett & Keen, 1967). Furthermore, during aging humans become more susceptible to hypoglycemia caused by sulfonylurea-induced insulin secretion (Ferner & Neil, 1988). Plasma levels of pancreatic glucagon also increase during aging in humans (Hayashi, 1980). Another indication of pancreatic hyperactivity during aging in humans is that mature-onset diabetes and aging are associated with the deposition of amylin (Westermark *et al.*, 1987). The cause of pancreatic hypersecretion in aging humans is not certain, but in monkeys (Hamilton & Brobeck, 1963) and possibly in humans (Buzzi *et al.*, 1987), VMH lesions may cause obesity and diabetes associated with hyperinsulinemia. Interestingly, GTG may lead to the development of hypoglycemia in humans (Yao *et al.*, 1992), consistent with our observations on the early effects of GTG on blood glucose in mice. Thus the pattern of age-correlated impairments in glucose metabolism in humans is similar to the pattern exhibited by aging rodents, and also to mice with GTG-induced VMH lesions or exposed to diabetogenic diets.

As discussed above, we had shown that GTG and a high-calorie diet cause hyperinsulinemia followed, at least initially, by hypoglycemia. Thus these models of accelerated glucose-induced hypothalamic damage cause an acceleration in the normal age-correlated increase in plasma insulin, consistent with the hypothesis that even normal levels of glucose may cause age-correlated impairments in the VMH, leading to hyperinsulinemia and hypoglycemia. A more persuasive test of this hypothesis, however, would be to examine if reducing exposure to glucose during aging would delay the development of age-correlated changes in glucose metabolism. We therefore examined (Mobbs, Kleopoulos & Bergen, 1993) if dietary restriction, which attenuates many age-correlated impairments (Yu, Masoro & McMahan, 1985) and decreases integrated exposure to plasma glucose (Masoro, Katz & McMahan, 1989), would delay the early and robust age-correlated decreases in plasma glucose. In this study, 2-month-old mice were given *ad lib* diets or given access to food only every other day. After one week on this regimen, blood glucose was determined (from tail vein) in all mice after a 24-hour fast. At this point, the fasting level of glucose was slightly lower in the restricted mice. Three months after beginning the regime, however, fasting blood glucose levels of the *ad lib* mice had dropped significantly, compared to the levels they had been 3 months earlier and compared to the levels exhibited by mice on a restricted diet for 3 months. In contrast, fasting glucose levels in the restricted mice did not drop during that period. These studies are consistent with the hypothesis that even normal levels of glucose can cause VMH damage, leading to pancreatic hypersecretion during aging followed in rats and humans by the development of insulin resistance and, in genetically predisposed individuals, the development of diabetes.

Genetic influences on diet-induced impairments in glucose metabolism in humans: selection for and against the 'thrifty genotype'

Epidemiological studies in human populations examining the interaction of genetics and environment on the incidence of mature-onset diabetes suggest a genetic influence on diet-induced impairments in humans (Zimmet, 1992). Many (especially 'traditional' or aboriginal) ethnic groups are highly susceptible to mature-onset diabetes. However, these groups did not exhibit significant incidence of diabetes until after conversion from their traditional low-calorie diets to Western high-calorie diets (Zimmet, 1992). Thus these human populations appear to exhibit a particularly strong genetic susceptibility to diet-induced mature-onset diabetes, which, as discussed above, appears to represent a particularly severe form of normal age-correlated impairments in glucose metabolism.

The evolutionary significance of this susceptibility to diet-induced impairments in glucose metabolism, and possible relationship to glucose hysteresis, is suggested by detailed studies of a single population on the South Pacific island of Nauru. The (approximately 5,000) Micronesians on this island exhibit the world's second highest recorded incidence and prevalence of mature-onset diabetes (Dowse *et al.*, 1992), after the Pima Indians of Arizona. As far as can be determined, the incidence of mature-onset diabetes on Nauru was negligible until the 1950s, yet by 1975-1976 the prevalence of diabetes in the population over 45 old was greater than 40% (Dowse *et al.*, 1992), and by 1987 the prevalence in the population over 60 was an extraordinary 60%. Before 1950 this population had been acted on by three major circumstances which selected for resistance to starvation. First, the island is relatively remote, so the ancestors of the present islanders presumably endured a prolonged sea voyage, with an attendant scarcity of food. Second, the island itself has little arable land, and food supplies were precarious. Finally, during the Japanese occupation of the island in World War II, about 25% of the population died of malnutrition. In the early 1950s, however, substantial phosphate deposits were discovered, and income from these mines, distributed over the relatively few native islanders, made the Nauruans among the world's wealthiest people. In less than a generation, the average diet of the Nauru was transformed from one of almost lethal caloric deprivation to one entailing higher caloric intake than most Western diets (Diamond, 1992). This wealth has lead to what would appear to be one of the world's most hospitable environments, yet, remarkably, the population of Nauru has one of the world's shortest human lifespans (Zimmet, Dowse & Finch, 1990).

It is generally thought that the susceptibility of indigenous populations to (presumably) diet-induced diabetes has arisen as a result of natural selection (Neel, 1962). Thus it has been hypothesized that in human populations, such as American Indians or many South Pacific islanders, which were subject to periods of low food availability or famine, genotypes are selected (the 'thrifty genotype') which utilize consumed calories very efficiently (Neel, 1962). However, when these genotypes are exposed to a high-calorie diet for extended periods of time, this very efficiency leads to obesity and mature-onset diabetes, and thus becomes deleterious. Thus under conditions of high caloric intake, the 'thrifty genotype' is selected against. Presumably, in Western ethnic groups the 'thrifty genotype' (and hence the incidence of mature-onset diabetes) is relatively rare, even in the presence of high-calorie diets, because it has been selected against during many generations of relative affluence (Neel, 1962; Dowse *et al.*, 1992; Diamond, 1992).

This 'thrifty genotype' hypothesis is supported by recent studies (Dowse *et al.*, 1992) of the Nauru population which appear to demonstrate that even over the course of 1-2 generations natural selection is reducing the representation of the 'thrifty genotype'. From 1975-1976 until 1987, the age-standardized prevalence of mature-onset diabetes has remained relatively constant and the rate of progression from impaired glucose tolerance to diabetes has increased. However, the overall age-standardized incidence of diabetes and the prevalence of impaired glucose tolerance has decreased significantly, as has the rate of progression from normal glucose tolerance to impaired glucose tolerance or diabetes. Although there were no changes in apparent risk factors for diabetes, diabetic Nauruans had higher mortality and, of particular interest, lower fertility, during that period. In this context it may be pertinent that on average mature-onset diabetes occurs at earlier ages in Nauru than in the West, such that the prevalence of diabetes is much higher dur- ing the ages of reproduction in Nauru than in the West. Thus presumably the deleterious effects of the 'thrifty genotype' in Western populations have been largely delayed to the post-reproductive period, as would be predicted from population genetics.

What is the mechanism by which a highly efficient metabolic phenotype, selected for in a calorie-poor environment, leads to a shortened lifespan in a calorie-rich environment? If glucose hysteresis is a major component leading to impaired glucose tolerance, a strong prediction is that hyperinsulinemia should precede the onset of hyperglycemia. The Nauru (Zimmet, 1992), as well as the Pima Indians and other ethnic groups (Zimmet, Dowse & Bennet, 1991), exhibit clear evidence of such a progression. It is well-established in the Nauru and many other groups that in the progression from normal glucose tolerance to (mature-onset) impaired glucose tolerance, insulin increases, but in the transition from impaired glucose tolerance to mature-onset diabetes, insulin levels fall back to roughly normal levels (Zimmet, 1992; DeFronzo *et al.*, 1992). In our studies rats exhibit a similar U-shaped curve during aging. Of particular interest, prospective longitudinal studies in the Nauru show that of individuals under the age of 29 with normal glucose tolerance, those who later developed impaired glucose tolerance had significantly higher glucose-induced plasma insulin, and those who went on to develop diabetes had even higher glucose-induced insulin levels when they were young. Coupled with the observation that in general the Nauru are hyperinsulinemic compared with Western populations (Zimmet, Dowse & Bennet, 1991), it appears that a major phenotype of the 'thrifty genotype' is a neuroendocrine hypersensitivity to glucose-induced insulin release. This phenotype naturally entails a more efficient use of ingested calories, a selective advantage in a calorie-poor environment. It is possible, though speculative, that such a phenotype, which is more susceptible to the acute effects of glucose to induced insulin secretion, would also be more susceptible to the cumulative toxic effects of glucose (i.e., glucose hysteresis). Since studies with GTG in mice demonstrate dramatic differences in the genetic susceptibility of VMH neurons to damage by this glucose derivative (correlated with lifespan), and differences in susceptibility to glucose-induced VMH damage would lead to differences in insulin secretion which correlate with the development of diabetes, it is tempting to speculate that the 'thrifty genotype' in humans (and its relationship to longevity) may entail particular susceptibility to VMH damage by glucose.

The above analysis suggests a relationship between glucose hysteresis and the life-prolonging effects of dietary restriction, and suggests a possible selection strategy by which it may be possible to enrich an outbred population (of mammals, at least) for a longer-lived genotype. If the 'thrifty genotype' is deleterious under circumstances of a calorie-enriched environment, it seems plausible to consider the predominant genotype of populations sensitive to the life-prolonging effects of dietary restriction as still too 'thrifty' relative to the ambient caloric intake. To select for a less 'thrifty' genotype (hence one which would react to the previous 'normal' levels of caloric intake as if they were 'restricted'), one could put an outbred population on a calorie-enriched diet. Such a diet would be expected to generally shorten lifespan, but some individuals would presumably be less 'thrifty', and therefore would live longer and possibly could be selected for by their longer reproductive lifespans. If after repeated selection the population on the high-calorie diet has a lifespan similar to the pre-selected group, a diet which had been 'normal' for the unselected population would effectively constitute a restricted diet for the new selected population, so would be expected to extend lifespan beyond the lifespan of the original population on the 'normal' diet. Most likely, the restricted diet which increased the lifespan of the original population would be lethal to the new, longer-lived population, since the new population is less efficient. Nevertheless this strategy may make it possible to produce a new population with a longer lifespan, at least on the 'normal' diet than the unselected population. Such a result is the opposite from that obtained in fruit flies, in which selection for resistance to starvation (as opposed to calorie-induced mortality) led to longer-lived populations (Hutchinson, Shaw & Rose, 1991). Thus it is possible that the 'thrifty genotype' has played a different role in the evolution of longevity in insects as opposed to mammals.

Pathological implications: 'normal' aging vs. diseases of aging

The physiological complications of glucose hysteresis are so common in human populations that this process can plausibly be considered as a component of 'normal' aging, as opposed to simply a component of a specific disease, mature-onset diabetes. Mature-onset diabetes is so frequent in aging human populations (in the population over 60 years of age, ranging from about 20% in the U.S. to about 60% in Nauru) that even this extreme form of pathology is plausibly a component of normal aging in humans at least. Even when diabetics are removed from the population, plasma glucose in humans increases after middle age (Zimmet & Whitehouse, 1979), concomitant with developing insulin resistance and hyperinsulinemia. Hyperinsulinemia is so common that a rise in insulin levels may be an almost universal feature in middle-aged humans (Mobbs, 1990), although insulin levels may decrease later in life. The pathological consequences of pancreatic hypersecretion are far more general than simply predisposition to mature-onset diabetes (Mobbs, 1990; Fig. 1). In particular, hyperinsulinemia is a major risk factor for a constellation of associated pathologies, referred to as 'Syndrome X', which includes cardiovascular disease, hypertension, and obesity (Reaven, 1988; see also Modan *et al.*, 1985; Zavaroni *et al.*, 1989), and some cancers (Yam, 1992). Similarly, pathological consequences of VMH damage extend far beyond a predisposition to diabetes (Mobbs, 1990; Fig. 1). For example, GTG-induced VMH lesions impair fasting-induced suppression of specific, nutritionally-regulated activities of the sympathetic nervous system (Young & Landsberg, 1980), and aging in humans is accompanied by specific increases in sympathetic tone (Rowe & Troen, 1980). Of particular interest is the relationship between glucose hysteresis and the age-correlated increase in adiposity, and the several age-correlated pathologies for which obesity is a risk factor. One interpretation of the relationship between these factors is that glucose hysteresis, by causing VMH damage and hyperinsulinemia, predisposes to both increased adiposity as well as other pathologies. Thus glucose hysteresis plausibly contributes to several of the most clinically relevant age-correlated impairments in humans. In addition, to the extent that glucose hysteresis plays an important role in those age-correlated impairments (including mortality) which can be attenuated by dietary restriction, glucose hysteresis must plausibly be considered as a component of normal aging.

Molecular mechanism of glucose hysteresis

The mechanism by which the cumulative exposure of specific neuroendocrine cells to glucose may be deleterious is unknown, although the specific susceptibility of VMH neurons and pancreatic cells to glucose toxicity is presumably related to their role in sensing glucose. At least 3 distinct mechanisms are plausible. Peripheral neuropathy is perhaps the most common complication of diabetes, and the best-established mechanism for diabetic neuropathy is the aldose reductase pathway (Dyck, 1990). It seems plausible that the cumulative exposure of glucose may gradually damage neurons in the VMH through a similar mechanism. It has also been proposed that glucose leads to cumulative deleterious effects during aging through gradual covalent modification of macromolecules (Cerami, 1985), and VMH neurons and beta cells may be particularly sensitive to these effects. In addition, it is possible that some neurotoxic effects are due to the early induction of hypoglycemia due to glucose-induced hyperinsulinemia exhibited during early middle age.

Of particular interest, however, is the possibility that the cumulative toxic effects of glucose may be analogous to the estrogen-induced VMH impairments (estrogen hysteresis) contributing to senescence of the reproductive system (Mobbs, 1989; Mobbs, 1990). Both glucose and estrogen hysteresis could be caused by a common molecular phenomenon, called 'gene memory' or 'transcriptional hysteresis', in which estrogen (Tam, Hache & Deeley, 1986) and glucose (Roy *et al.*, 1990) induce metastable changes in chromatin state, so that genes induced or repressed by a hormone often may not fully return to their previously uninduced or active state after removal of the hormone. Thus over many exposures to hormone (i.e., during aging) genes may tend to stay in a chronically induced or repressed state. A salient prediction of such a mechanism is that genes should exhibit persistent effects of its regulatory hormone, and/or exhibit hyper-responsiveness, during aging, leading to physiological impairments. We have recently obtained data on the reproductive senescence of female rats consistent with such a prediction (Kleopoulos & Mobbs, 1992). We have specifically suggested that acute effects of estrogen on phospholipase C may mediate facilitatory effects of estrogen on reproductive function, but persistent and cumulative effects could lead to reproductive impairments in aging female rodents (Mobbs *et al.*, 1991). Interestingly, a similar relationship between glucose-induced impairments and phospholipase C activity has been suggested (Zawalich, 1990). Since we have recently found (Mobbs, Kleopoulos & Funabashi, 1993) that glucose induces jun-b and c-fos in specific neurons in the VMH, at least some of which also express the pancreatic form of glucokinase (thought to comprise the main glucose sensor in the pancreas), the glucose-sensitive neurons expressing these gene products may be susceptible to glucose toxicity by a mechanism analogous to glucose toxicity in the pancreas.

In conclusion, a cumulative toxic effect of glucose on VMH neurons (and possibly on pancreatic and other cells) may be a risk factor in the development of age-correlated impairments of glucose metabolism and other age-correlated impairments. It is likely that effects of glucose on VMH neurons is only one of several possible mechanisms by which glucose influences age-correlated impairments, and the role of genetic susceptibility to glucose-induced damage in determining longevity remains to be established. However, the theory of glucose hysteresis provides a framework within which these issues may be addressed in detail.

Acknowledgements

I would like to acknowledge Hugo Bergen, Toshiya Funabashi, and Steven Kleopoulos, who contributed essentially to many of the experiments reported here, Michael Kaplitt for valuable discussions, and Dr. John Rowe and Paul Glenn for their support. These studies were supported by the American Federation for Aging Research, the American Diabetes Association, and the Glenn Foundation for Medical Research, of which CVM was a Fellow.

References

Ahdieh, H.B., J. Hamilton & G.N. Wade, 1983. Copulatory behavior and hypothalamic estrogen and progestin receptors in chronically insulin-deficient female rats. Physio. Behav. 31:219-223.

Akmayev, I.G. & A.E. Rabkina, 1976. CNS-pancreas system. The hypothalamic response to insulin deficiency. Endokrinologie 68:211-220.

Baron, A.D., L. Schaeffer, P. Shragg & O.S. Kolterman, 1987. Role of hyperglucagonemia in maintenance of increased rates of hepatic glucose output in type II diabetes. Diabetes 36:274-284.

Bergen, H. & C.V. Mobbs, 1992. Hypothalamic lesion by gold-thio-glucose (GTG) leads to hyperglycemia in mice. Endocrine Soc. Abs., 74th Ann. Meet. Prog. Abs., p.91

Bergen, H., S.P. Kleopoulos, J. Pfaus & C.V. Mobbs, 1992. Effect of gold-thio-glucose (GTG) on hypothalamic oxytocin receptors and neuropeptide Y (NPY) mRNA. Soc. Neurosci. Abs., Vol. 18: p. 1485.

Berthoud, H.R. & B. Jeanrenaud, 1979. Acute hyperinsulinemia & its reversal by vagotomy following lesions of the ventromedial hypothalamus in anesthetized rats. Endocrinology 105:146-151.

Bestetti, G. & G.L.Rossi, 1980. Hypothalamic lesions in rats with long-term streptozotocin-induced diabetes mellitus. Acta Neuropathol 52:119-127.

Bonnevie-Nielsen, T., L.T. Skovgaard & A. Lernmark, 1983. Beta-cell function relative to islet volume and hormone content in the isolated perfused mouse pancreas. Endocrinology 112: 1049-1056.

Brancho-Romero, E. & G.M. Reaven, 1977. Effect on age and weight on plasma glucose and insulin responses in the rat. J. Am. Geriatr. Soc. 7:299-302.

Buzzi, S., G. Buzzi, A. Buzzi & C. Baccini, 1987. Hypothalamic syndrome in a woman with three sewing needles in the brain. Lancet 1:1313.

Cerami, A., 1985. Glucose as a mediator of aging. J. Am. Ger. Soc. 33:626-634.

Chen, M., R.N. Bergman, G. Pacini & D. Porte, 1985. Pathogenesis of age-related glucose intolerance in man: Insulin resistance and decreased beta-cell function. J. Clin. Endo. Metab. 60:13-20.

Chlouverakis, C., R.J. Jarrett & H. Keen, 1967. Glucose tolerance, age, and circulating insulin. Lancet 1:806-809.

Davidson, M.D., 1979. The effect of aging on carbohydrate metabolism. A review of the English literature and a practical approach to the diagnosis of diabetes mellitus in the elderly. Metabolism 28:688-705.

Davidson, M.D. & D. Casanello-Ertl, 1979. Insulin antagonism on cultured rat myoblasts secondary to chronic exposure to insulin. Horm. Metab. Res. 11:207-209.

DeFronzo, R.A., 1979. Glucose tolerance and aging. Evidence for tissue insensitivity to insulin. Diabetes 28:1095-1101.

DeFronzo, R.A., R.C. Bonadonna & E. Ferrannini, 1992. Pathogenesis of NIDDM: A balanced overview. Diabetes Care 15:318-368.

Del Prato, S., P. Castellino, D.C. Simonson & R.A. DeFronzo, 1987. Hyperglucagonemia and insulin-mediated glucose metabolism. J. Clin. Invest. 79:547-556.

Diamond, J.M., 1992. Diabetes running wild. Nature 357:362-363.

Dowse, G.K., P.Z. Zimmet, C.F. Finch & V.R. Collins, 1992. Decline in incidence of epidemic glucose intolerance in Nauruans: Implications for the 'Thrifty Genotype'. Am. J. Epidemiology 133:1093-1104.

Dyck, P.J., 1990. Resolvable problems in diabetic neuropathy. J. NIH Res. 2:57-62.

Ferner, R.E. & H.A.W. Neil, 1988. Sulfonylureas and hypoglycemia. Brit. Med. J. 296:949-950.

Frohman, L.A., J.R. Goldman & L.L. Bernardis, 1972. Studies of insulin sensitivity in vivo in weanling rats with hypothalamic obesity. Metabolism 21:1133-1139.

Garris, D.R., A.R. Diani, C. Smith & G.C. Gerritsen, 1982. Depopulation of the ventromedial hypothalamic nucleus in the diabetic Chinese hamster. Acta Neuropathol. 56:63-66.

Garris, D.R. & D.L. Coleman, 1984. Diabetes-associated changes in estradiol accumulation in the aging C57BL/KsJ mouse brain. Neurosci. Lett. 49:285-290.

Garris, D.R., L.R. West & D.L. Coleman, 1985. Morphometric analysis of medial basal hypothalamic neuronal degeneration in diabetes (db/db) mutant C57BL/KsJ mice: Relation to age and hyperglycemia. Dev. Brain Res. 20:161-168.

Garris, D.R., D.L. Coleman & C.R. Morgan, 1985. Age- and diabetes-related changes in tissue glucose uptake and estradiol accumulation in the C57BL/KsJ mouse. Diabetes 34:47-52.

Hamilton, C.L. & J.R. Brobeck, 1963. Diabetes mellitus in hyperphagic monkeys. Endocrinology 73:512-515.

Harris, M.I., W.C. Hadden, W.C. Knowler & P.H. Bennett, 1987. Prevalence of diabetes & impaired glucose tolerance and plasma glucose levels in U.S. population aged 20-74. Diabetes 36:523-524.

Hayashi, K., 1980. Glucose tolerance in the elderly with special reference to insulin and glucagon responses. Wakayama Med. Rep. 23:29-39.

Henninger, H.J. & J.J. Dorey, 1982. Handbook on genetically standardized JAX mice. The Jackson Laboratory, Bar Harbor, ME.

Hutchinson, E.W., A.J., Shaw & M.R. Rose, 1991. Quantitative genetics of postponed aging in Drosophila melanogaster. II. Analysis of selected lines. Genetics 127:729-737.

Hsu, H.K. & M.T. Peng, 1978. Hypothalamic neuron number in old female rats. Gerontology 24:434-440.

Jeanrenaud, B., S. Halimi & G. Van de Werve, 1985. Neuroendocrine disorders seen as triggers of the triad: obesity-insulin-resistance-abnormal glucose tolerance. Diabetes Metab. Rev. 1:261-291.

Jeffrey, I.J.M., C. Gordon, A.P. Yates & H. Fox, 1985. Obesity, hyperinsulinemia and hyperplasia of the pancreatic islets in aging (C3H/HeJXC57Bl/6J) F1 hybrid mice. Hormone Metab. Res. 18:210-212.

Kergoat, M., D. Bailbe & B. Portha, 1987. Effect of high sucrose diet on insulin secretion and insulin action: a study in the normal rat. Diabetologia 30:252-258.

Kleopoulos, S.P., L. Krey & C.V. Mobbs, 1992. Regulation of hypothalamic oxytocin receptors and lordosis reflex during reproductive senescence of female Fisher rats. Soc. Neurosci. Abs., Vol. 18:p.1486.

Kleopoulos, S.P. & C.V. Mobbs, 1993. Hypothalamic growth-hormone releasing hormone mRNA levels, and response to fasting, decrease with age in male rats. Soc. Neurosci. Abs., Vol 19 (in press).

Landfield, P., J. Waymire & G. Lynch, 1978. Hippocampal aging and adrenocorticoids: A quantitative correlation. Science 202:1098-1102.

Landfield, P.W., R.K. Baskin & T.A. Pitler, 1981. Brain aging correlates: retardation by hormonal-pharmacological treatments. Science 214:581-584.

Lazaris, J.A., R.S. Goldberg & M.P. Kozlov, 1985. Studies on diabetes mellitus after ventromedial hypothalamic lesions in adult and aged rats. Endocrinol. Exp. 19:67-76.

Liebelt, R.A., K. Sekiba, A.G. Liebelt & J.H. Perry, 1960. Genetic susceptibility to goldthioglucose-induced obesity in mice. Proc. Soc. Expt. Biol. Med. 104:689-694.

Liebelt, R.A. & J.H. Perry, 1967. Action of gold thioglucose on the central nervous system. Handbook of Physiology Vol. 1 (Section 6), C.F. Code, Ed., pp. 271-285. Waverly Press, Baltimore.

Leighton, B. & G.J.S. Cooper, 1988. Pancreatic amylin and calcitonin gene-related peptide cause resistance to insulin in skeletal muscle in vitro. Nature 335:632-635.

Leiter, E.H., F. Premdas, D.E. Harrison & L.G. Lipson, 1988. Aging and glucose homeostasis in C57BL/6J male mice. FASEB J. 2:2807-2811.

Masoro, E.J., M.S. Katz & C.A. McMahan, 1989. Evidence for the glycation hypothesis of aging from the food-restricted rodent model. J. Gerontol. 44:B20-B22.

Meneilly, G.S., K.L. Minaker, D. Elahi & J.W. Rowe, 1987. Insulin action in aging man: Evidence for tissue-specific differences at low physiological insulin levels. J. Gerontol. 42:196-201.

Minaker, K.L., J.W. Rowe, R. Tonino & J.A. Pallotta, 1982. Influence of age on clearance of insulin in man. Diabetes 31:851-855.

Mobbs. C.V., K. Flurkey, D.M. Gee, K. Yamamoto, Y.N. Sinha & C.E. Finch, 1984. Estradiol-induced adult anovulatory syndrome in female C57BL/6J mice: Age-like neuroendocrine, but not ovarian, impairments. Biology of Reproduction 30:556-563.

Mobbs, C.V., D.M. Gee & C.E. Finch, 1984. Reproductive senescence in female C57BL/6J mice: Ovarian impairments and neuroendocrine impairments that are partially reversible and delayable by ovariectomy. Endocrinology 115:1653-1662.

Mobbs, C.V., D. Cheyney, Y.N. Sinha & C.E. Finch, 1985. Age-correlated and ovary-dependent changes in relationships between plasma estradiol and luteinizing hormone, prolactin & growth hormone in female C57BL/6J mice. Endocrinology 116:813-820.

Mobbs, C.V., 1989. Neurohumoral hysteresis as a mechanism for senescence; Comparative aspects. In: Scanes, C.G. & Schriebman, M.P. (Eds.) Development, Maturation & Senescence of the Neuroendocrine System. Academic Press, pp. 223-252.

Mobbs, C.V., 1990. Neurotoxic effects of estrogen, glucose, and glucocorticoids: Neurohumoral hysteresis and its pathological consequences during aging. Reviews of Biological Research on Aging, Vol. 4, pp. 201-228.

Mobbs, C.V., M. Kaplitt, L.-M. Kow & D.W. Pfaff, 1991. PLC-alpha as a common mediator of estrogen and other hormones. Molecular & Cellular Endocrinology. 80:C187-C191.

Mobbs, C.V. & C.E. Finch, 1992. Estrogen-induced impairments as a mechanism in reproductive senescence of female C57Bl/6J mice. J. Gerontol. 47:B48-B51.

Mobbs, C.V. & S.P. Kleopoulos, 1992. Regulation of hypothalamic neuropeptide Y, (NPY) by fasting during aging in male rats. Soc. Neurosci. Abs., Vol. 18: p. 1485.

Mobbs, C.V. & H. Bergen, 1992. Glucose metabolism during aging in mice: Hypoglycemia and hyperinsulinemia. Endocrine Soc.Abs. 74th Ann. Meet. Prog. Abs., p.370

Mobbs, C.V., S.P. Kleopoulos & H. Bergen, 1993. Hypoglycemia precedes hyperglycemia during aging: effect of dietary enhancement and dietary restriction. Endocrine Soc. Abs., 75th Ann. Meet. Prog. Abs., p. 367.

Mobbs, C.V., S.P. Kleopoulos & T. Funabashi, 1993. A glucokinase/ AP-1 glucose transduction mechanism in the ventromedial hypothalamic satiety center. Soc. Neurosci. Abs., Vol. 19 (in press).

Modan, M., H. Helkin, H. Almog, A. Lusky, A. Eshkol, M. Shefi, A. Shitrit & Z. Fuchs, 1985. Hyperinsulinemia. A link between hypertension, obesity, and glucose intolerance. J. Clin. Invest. 75:809-817.

Neel, J.V., 1962. Diabetes mellitus: a thrify genotype rendered detrimental by 'progress'? Am. J. Hum. Gen. 14:353- 362.

Nelson, J., M.D. Bergman, K. Karelus & L.S. Felicio, 1987. Aging of the hypothalamic-pituitary-ovarian axis: Hormonal influences and cellular mechanisms. J. Steroid Biochem. 27:699-705.

Peng, M.T. & H.K. Hsu, 1982. No neuron loss from hypothalamic nuclei of male rats in old age. Gerontology 28:19-22.

Pfaff, D.W., 1972. Histological differences between ventromedial hypothalamic neurons of well-fed and underfed rats. Nature 223:77-79.

Reaven, G.M. & E.P. Reaven, 1985. Age, glucose intolerance, and non-insulin-dependent diabetes mellitus. J. Am. Geriatr. Soc. 33:286-290.

Reaven, G.M., 1988. Role of insulin resistance in human disease. Diabetes 37:1595-1607.

Rizza, R.A., L.J. Mandarino, J. Genest, B.A. Baker & J.E. Gerich, 1985. Production of insulin resistance by hyperinsulinemia in man. Diabetologia 28:70-75.

Rohner-Jeanrenaud, F. & B. Jeanrenaud, 1980. Consequences of ventromedial hypothalamic lesions upon insulin and glucagon secretion by subsequently isolated perfused pancreases in the rat. J. Clin. Invest. 65:902-910.

Rohner-Jeanrenaud, F. & B. Jeanrenaud, 1984. Oversecretion of glucagon by pancreases of hypothalamic-lesioned rats: A re-evaluation of a controversial topic. Diabetologia 27:535-539.

Rowe, J.W., K.L. Minaker, J. A., Pallota & J.S. Fliers, 1983. Characterization of the insulin resistance of aging. J. Clin. Invest. 71:1581-1589.

Rowe, J.W. & B.R.Troen, 1980. Sympathetic nervous system and aging in man. Endocrine Rev. 1:167-178.

Roy, S., R. Sala, E. Cagliero & M. Lorenzi, 1990. Overexpression of fibronectin induced by diabetes or high glucose: Phenomenon with a memory. Proc. Nat. Acad. Sci. 87:404-408.

Saad, M.F., D.J. Pettitt, D.M. Mott, W.C. Knowler, R.G. Nelson & P.H. Bennett, 1989. Sequential changes in serum insulin concentration during development of non-insulin-dependent diabetes. Lancet (17 June 1989):1356-1358.

Sabel, B.A. & D.G. Stein, 1981. Extensive loss of subcortical neurons in the aging rat brain. Exp. Neurol. 73:507-516.

Sandberg, H., N. Yoshimine, S. Maeda, D. Symons & J. Zavodnick, 1973. Effects of an oral glucose load on serum immunoreactive insulin, free fatty acid, growth hormone, and blood sugar levels in young and elderly subjects. J. Am. Geriatr. Soc. 10:433-438.

Sartin, J.L & A.A. Lamperti, 1985. Neuron numbers in hypothalamic nuclei of young, middle-aged and aged male rats. Experientia 41:109-111.

Sapolsky, R.M., L. Krey & B.S. McEwen, 1986. The neuroendocrinology of stress and aging: The glucocorticoid cascade hypothesis. Endocr. Rev. 7:284-301.

Smith-West, C. & D.R. Garris, 1983. Diabetes-associated hypothalamic neuronal depopulation in the aging Chinese hamster. Dev. Brain Res. 9:385-389.

Soman, V.R. & R.A. De Fronzo, 1980. Direct evidence for downregulation of insulin receptors by physiologic hyperinsulinemia in man. Diabetes 29:159-163.

Surwit, R.S., C.M. Kuhn, C. Cochrane, J.A. McCubbin & M.N. Feinglos, 1988. Diet-induced Type II diabetes in C57Bl/6J mice. Diabetes 37:1163-1167.

Surwit, R.S., M.F. Seldin, C.M. Kuhn, C. Cochrane & M.N. Feinglos, 1990. Control of expression of insulin resistance and hyperglycemia by different genetic factors in diabetic C57BL/6J mice. Diabetes 40:82-87.

Tam, S.P., J.G. Hache & R.G. Deeley, 1986. Estrogen memory effect in human hepatocytes during repeated cell division without hormone. Science 234:1234-1237.

Tokuyama, Y, A. Kanatsuka, H. Ohsawa, T. Yamaguchi, H. Makino, S. Yoshida, H. Nagase & S. Inoue, 1991. Hypersecretion of islet amyloid polypeptide from pancreatic islets of ventromedial hypothalamic-lesioned rats & obese Zucker rats. Endocrinology 128:2739-2744.

Welborn, T.A., N.S. Stenhouse & C.G. Johnstone, 1969. Factors determining serum insulin response in a population sample. Diabetologia 5:263-266.

Westermark, P., C. Wernstedt, D.W. Wilander, T.D. O'Brien & K.H. Johnson, 1987. Amyloid fibrils in human insulinoma and islet of Langerhans of the diabetic cat are derived from a neuropeptide-like protein also present in normal islet cells. Proc. Natl. Acad. USA 84:3881-3885.

Yam, D., 1992. Insulin-cancer relationships: Possible dietary implication. Medical Hypotheses 38:111-117.

Yao, K., Y. Uchigata, H. Kyono, H. Yokoyam, Y. Eguchi, H. Fukushima, K. Yamauchi & Y. Hirata, 1992. Human insulin-specific immunoglobulin-G antibody and hypoglycemic attacks after the injection of gold-thioglucose. J. Endo. Invest. 15:43-48.

Young, J.B. & L. Landberg, 1980. Impaired suppression of sympathetic activity during fasting in goldthioglucose-treated mouse. J. Clin. Invest. 65:1086-1094.

Yu, B.P., E.J. Masoro & C.A. McMahan, 1985. Nutritional influences on aging Fisher 344 rats. I. Physical, metabolic, and longevity characteristics. J. Gerontol. 40:657-670.

Zavaroni, I., E. Bonora, M. Pagleria, E. Dell'Aglio, L. Luchetti,G. Buonanno, A. Bonah, M. Bergonzeni, L. Gnudi, M. Passen & G. Reaven, 1989. Risk factors for coronary artery disease in healthy persons with hyperinsulinemia & normal glucose tolerance. New Engl. J. Med. 320:702-706.

Zawalich, W., 1990. Multiple effects of increases inphosphoinositide hydrolysis on islets and their relationship to changing patterns of insulin secretion. Diabetes Res. 12:101-111.

Zimmet, P. & S. Whitehouse, 1979. The effect of age on glucose tolerance. Diabetes 28:617-628.

Zimmet, P., G. Dowse & P. Bennet, 1991. Hyperinsulinemia is a predictor of non-insulin-dependent diabetes mellitus. Diabetes Metab. 17:101-108.

Zimmet, P.Z., 1992. Challenges in diabetes epidemiology- From West to the rest. Diabetes Care 15:232-252.

Zimmet, P.Z., G.K., Dowse & C.F. Finch, 1990. The epidemiology and natural history of NIDDM- lessons from the South Pacific. Diabetes Metab. Rev. 6:91-124.

M.R. Rose and C.E. Finch (eds.), Genetics and Evolution of Aging, 285–293, 1994.
© 1994 *Kluwer Academic Publishers. Printed in the Netherlands.*

Genetic heterogeneity of gene defects responsible for familial Alzheimer disease

Rudolph Tanzi, Sandra Gaston, Ashley Bush, Donna Romano, Warren Pettingell, Jeffrey Peppercorn, Marc Paradis, Sarada Gurubhagavatula, Barbara Jenkins & Wilma Wasco
The Laboratory of Genetics and Aging, Neuroscience Center, Department of Neurology, Massachusetts General Hospital, Harvard Medical School, Charlestown, MA 02129, USA

Received and accepted 22 June 1993

Abstract

Inherited Alzheimer's disease is a genetically heterogeneous disorder that involves gene defects on at least five chromosomal loci. Three of these loci have been found by genetic linkage studies to reside on chromosomes 21, 19, and 14. On chromosomes 21, the gene encoding the precursor protein of Alzheimer-associated amyloid (APP) has been shown to contain several mutations in exons 16 and 17 which account for roughly 2-3% of familial Alzheimer's disease (FAD). The other loci include what appears to be a susceptibility gene on chromosome 19 associated with late-onset (> 65 years) FAD, and a major early-onset FAD gene defect on the long arm of chromosome 14. In other early- and late-onset FAD kindreds, the gene defects involved do not appear to be linked to any of these three loci, indicating the existence of additional and as of yet unlocalized FAD genes. This review provides a historical perspective of the search for FAD gene defects and summarizes the progress made in world-wide attempts to isolate and characterize the genes responsible for this disorder.

Introduction

Alzheimer's disease (AD) is a devastating neurodegenerative disorder that is characterized by dramatic personality changes and global cognitive decline (Terry & Katzman, 1983). It currently affects approximately four million Americans, taking more than 100,000 lives each year. Since mean survival from the time of diagnosis ranges from three to twenty years (average of eight years), a tremendous burden is placed on family members during the course of the disease. In fact, it is estimated that for patients living at home, the cost of care averages $ 18,000 annually. In the U.S., AD patients fill more than 50% of all nursing home beds at an annual cost of $ 36,000 for each patient. The overall cost to society is presently more than $ 90 billion per year. Presently, no effective therapy exists for this disorder and it is expected that by next century more than 14 million Americans will be affected (Alzheimer's Association Statistical Data on Alzheimer's Disease). Clearly, one of the major challenges to those studying the aging process will be deciphering the etiological events leading to AD.

The genetics of Alzheimer's disease

The basic etiology of AD remains unknown, although advanced age and a positive family history of dementia (clustering of AD cases in kindreds) appear to represent prominent risk factors. Confirmed diagnosis of AD is only possible by autopsy or biopsy and depends on observing higher than expected amounts of beta-amyloid ($\beta A4$) plaques (senile plaques) and neurofibrillary tangles (NFT) in the brain (Glenner & Wong, 1984; Terry & Katzman, 1983). A major difficulty in developing rational therapies for AD stems from the lack of information regarding its etiology. The identification of particular environmental agents causing human neurodegenerative diseases is an arduous task. However, in some cases of AD the primary cause lies not in the environment, but in the genome of the patient. In this review, we will concentrate on genetic studies of AD including the potential role(s) played by the genes encoding the amyloid βA4 precursor protein (APP) family, and a novel locus on chromosome 14.

The evidence for familial forms of Alzheimer's disease (FAD) derives from family-, survey-, and life table-based analyses (reviewed in St George-Hyslop et al., 1989). FAD has been clearly demonstrated to be a genetically heterogeneous disorder, involving multiple genetic loci on at least three chromosomes (chromosomes 14, 19, and 21). The actual proportion of AD that is considered to be inherited is debatable, since it is often difficult to assess whether the disorder is familial or sporadic in most kindreds where familial clustering of the disorder is observed. In many pedigrees, particularly in those with late onset (>65 years) AD, at-risk, presymptomatic family members may die of other age-related illnesses (e.g. heart disease) before showing symptoms of dementia, making it difficult if not impossible to determine if these cases of AD have a genetic component. Consequently, estimates of the proportion of AD that is inherited range from 10% to nearly 100%. Overall estimates of life-time risk of developing AD in first-degree relatives of probands with AD suggest that approximately 50% of AD is inherited (Farrer et al., 1991). Meanwhile, a relatively low concordance rate of 40% in monozygous twins (Breitner & Murphy, 1992) implicates non-genetic factors in the expression of AD.

The role of the amyloid β protein precursor gene

Perhaps the greatest clues to the etiology of AD have been derived from studies of the neuropathological lesions associated with AD, and particularly the amyloid-containing senile plaques. The cores of senile plaques are made up primarily of βA4, a 39-43 amino acid peptide (Glenner & Wong, 1984) derived from a much larger precursor protein, APP. The APP gene produces multiple mRNA transcripts (Kang et al., 1987: Goldgaber et al., 1987; Tanzi et al., 1987; Robakis et al., 1987; Ponte et al., 1988; Tanzi et al., 1988; Kitaguchi et al., 1988; De Sauvage & Octave, 1989; Jacobsen et al., 1991), the majority of which contain an alternatively-

spliced exon ecoding a Kunitz protease inhibitor (KPI) domain. The form(s) of APP that actually give rise to the βA4 in amyloid deposits remain unknown. The APP gene is expressed ubiquitously throughout the body and brain in a differential pattern (Tanzi et al., 1987, 1988). The predominant form of APP RNA in brain is APP695 (lacking the KPI domain) and is abundantly synthesized by large neurons such as cortical pyramidal cells (Bahmanyar et al., 1987; Palmert et al., 1987; Goedert, 1987; Higgins et al., 1987; Lewis et al., 1987; Neve, Finch & Dawes, 1988; Tanzi & Hyman, 1991; Tanzi/Hyman & Wenniger, 1993; Hyman, Wenniger & Tanzi, 1993).

The APP gene resides on chromosomes 21 and was mapped in 1987 to the same vicinity as that of a locus for early-onset FAD (Kang et al., 1987; Goldgaber et al., 1987; Tanzi et al., 1987; Robakis et al., 1987, St George-Hyslop et al., 1987). When APP was tested for genetic linkage to FAD in the same four early-onset FAD pedigrees that were used to show linkage of the disorder to DNA markers on chromosome 21, at least one obligate crossover event was detected in each pedigree in at least one affected individual displaying an early age of onset, thereby diminishing the possibility that the recombinants were due to the occurrence of sporadic AD. These results indicated that APP was not tightly linked to FAD in these families (Tanzi et al., 1987b). A further assessment of the potential genetic role of APP in FAD was prompted by the findings that FAD is a genetically heterogeneous disorder, thus the APP gene could still represent the gene defect in some pedigrees (Schellenberg et al., 1988; St George-Hyslop et al., 1990), and that a mutation in the βA4 region of APP segregates with hereditary cerebral hemorrhage with amyloidosis-Dutch type (HCHWA-D, Levy et al., 1990). Patients in these families generally die of hemorrhages in their fourth to fifth decade due to the accumulation of βA4 deposits in cerebral arteries.

In light of these reports, Goate and colleagues sequenced exon 17 of APP in patients from a chromosome 21-linked FAD pedigree that exhibited no apparent crossovers with APP. A missense mutation causing an amino acid substitution (V->I) at codon APP717 was found in affected individuals in two separate pedigrees (Fig. 1, Goate et al., 1991). Meanwhile, the same mutation is absent in 250 unrelated, normal individuals from the same popu-

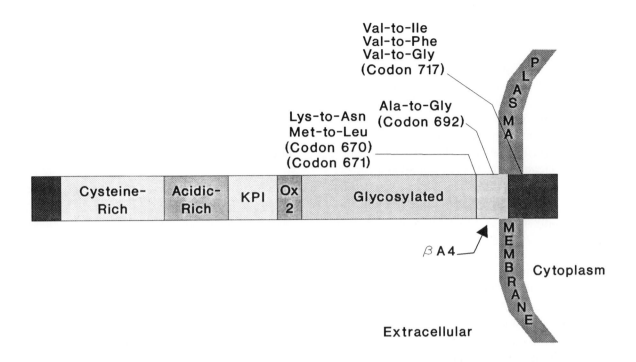

Fig. 1. The APP molecule (as it would be predicted to reside within the plasma membrane) and the sites of FAD-associated mutations found in the portions encoded by exons 16 and 17. A signal peptide, cysteine-rich region, acidic region, and roughly two-thirds of the βA4 domain are contained in the extracellular portion of APP. The rest of βA4 is contained within the transmembrane domain followed by the cytoplasmic portion. Two extracellular domains encoded by alternatively-spliced exons include the Kunitz protease inhibitor (KPI) and the OX-2-like region (II). All of these domains are significantly conserved in the APLP protein family. The APP717 (V->I) mutation was first described by Goate *et al.* (1991). Later, four additional FAD-associated mutations were discovered and include two more in codon 717 (V->F; Murrell *et al.*, 1991, and V->G; Chartier-Harlin *et al.*, 1991), one at codon 692 (A->G; Hendriks *et al.*, 1992), and a double missense substitution in a Swedish patient at codons 670 and 671 (K->N and M->L, respectively; Mullan *et al.*, 1992).

lation as the positive FAD pedigree (Goate *et al.*, 1991; other unpublished findings), implying that this change is not simply a rare polymorphism in APP but could actually represent a gene defect. The discovery of the APP717(V->I) mutation spurred a worldwide search for the presence of this and other APP mutations. Subsequently, four additional FAD-associated mutations in APP were found (see Fig. 1) including two more changes in codon 717 (V->F; Murrell *et al.*, 1991, and V->G; Chartier-Harlin *et al.*, 1991a), one at codon 692 (A->G; Hendriks *et al.*, 1992), and a double missense substitution in a Swedish patient at codons 670 and 671 (K->N and M->L, respectively; Mullan *et al.*, 1992).

Despite the intriguing finding of various APP mutations that appear linked to FAD, overall, the APP gene appears to underlie a very small proportion of inherited AD. In our laboratory, we have

examined a large set of FAD kindreds for the presence of recombinants with and mutations in the APP gene. One of the large, early-onset (<65 years) FAD pedigrees, FAD4, which supplied strongly suggestive evidence for linkage to chromosome 21 (St George-Hyslop *et al.*, 1987) reveals an apparent recombinant with APP, predicting that the gene defect in pedigree FAD4 is not due to a mutation in APP. When all nineteen exons of APP were sequenced in affected individuals from each branch of this pedigree, no mutations were found (Tanzi *et al.*, 1992).

We also found no mutations in the exons encoding the βA4 region of APP (16 and 17) in 20 early-onset and ten late-onset FAD pedigrees (Tanzi *et al.*, 1992). When combined with data obtained in other laboratories, over 170 FAD kindreds have now failed to reveal any mutations in exons 16 and 17 of APP (Chartier-Harlin *et al.*,

1991b; Schellenberg *et al.*, 1991b; Crawford *et al.*, 1991; Tanzi *et al.*, 1992; Naruse *et al.*, 1991; other unpublished observations). Collectively, of approximately 180 FAD pedigrees sequenced for mutations in exons 16 and 17, only eleven, all early-onset, or under 6%, have revealed mutations. Meanwhile, the overall number of FAD pedigrees found to lack missense substitutions specifically at APP codon 717 is now over 300 (Chartier-Harlin *et al.*, 1991b; Goate *et al.*, 1991; Kamino *et al.*, 1992; Van Duijn *et al.*, 1991; Crawford *et al.*, 1991; Tanzi *et al.*, 1992; other unpublished observations). The absence of any amino acid alteration in, or genetic linkage to the APP gene in most FAD kindreds suggests that overall only 2-3% of FAD is associated with mutations in APP (Tanzi *et al.*, 1992; Kamino *et al.*, 1992). Moreover, these data indicate that one or more FAD loci must be located elsewhere in the genome, and that these other loci account for the vast majority (>97%) of FAD.

The mechanism by which the known APP mutations lead to AD neuropathogenesis (assuming they are etiologic of the disease) is not yet clear. The substituted amino acids may disrupt membrane integrity, or affect the anchoring of processing of APP. It has recently been shown that βA4 is generated in a soluble form by cultured cells (Haass *et al.*, 1992; Seubert *et al.*, 1992; Shoji *et al.*, 1992). This finding prompted a study by Citron *et al.* (1992) in which the effects of the Swedish APP mutations (codons 670 and 671) were assessed with regard to generation of βA4. This mutation was chosen because it occurs at the N-terminus of the βA4 domain (see Fig. 1) and may affect the rate of cleavage at this site. Interestingly, approximately sevenfold higher levels of soluble βA4 were produced by the transfected cells expressing a transgene for the mutant APP versus the wild-type construct. These results have not yet been corroborated for the other APP mutations including the most common form at APP717 (V->I). Thus, it remains unclear as to whether this is the fundamental and universal mechanism of the known Alzheimer-associated mutations.

An alternative mechanism for the APP717 mutations involves the disruption of a putative regulatory stem-loop structure (Tanzi & Hyman, 1991b) residing at position 40 of the βA4 domain. This stem-loop resembles the so-called iron-responsive elements (IRE) in the ferritin and transferrin receptor RNAs, where they regulate mRNA translation and stability in response to iron concentration. All three mutations at APP717 destabilize the stem of the IRE-like structure in APP, perhaps, resulting in altered mRNA translation or stability, ultimately leading to accelerated amyloid formation.

The search for other FAD genes

Genetic linkage analysis is an extremely powerful technique for localizing disease genes so that they can be subsequently isolated. However, the effectiveness of linkage analysis is greatly limited by the nature of the disorder under study. In early attempts at genetic linkage analysis in FAD, it was anticipated that it would be quite difficult, given the late onset of the disorder and consequent limited informativeness of most pedigrees and the low heterozygosity of most restriction fragment length polymorphism (RFLP) markers. Analysis was initially concentrated on chromosome 21, given the similar neuropathology in middle-aged patients with Down syndrome (trisomy 21). However, the recent finding that missense substitutions in the APP gene on chromosome 21 gene account for only 2-3% of inherited AD (Tanzi *et al.*, 1992; Kamino *et al.*, 1992) has prompted a more comprehensive search of the human genome for additional FAD loci.

Methods for scanning the human genome have dramatically improved in the past few years for two major reasons. First, the FAD pedigrees which we, and others, have been banking since 1982 have increased in number and in genetic informativeness as additional family members have aged and become affected. In fact, we have now collected several FAD pedigrees capable of individually yielding significant positive lod scores for linkage. Second, a new class of highly informative genetic markers known as simple sequence repeats (SSR) has been developed. These markers include common dinucleotide repeats such as $(dGdT)_n$, which frequently vary in length and are present at roughly every 100,000 bp throughout the human genome (Weber & May, 1989).

SSR are typed by polymerase chain reaction (PCR) amplification of a simple sequence repeat using flanking unique primers. The amplified allelic products are then viewed directly on polyacrylamide gels. SSR markers frequently possess heterozygosity values of greater than 0.7 and include as many as 15 alleles or more. Over the past two years, SSR have all but replaced the more labor-intensive and expensive, yet less informative RFLP markers. Many individual laboratories around the world are currently generating SSR markers of greater than 70% heterozygosity and 5-10 cM spacing across each chromosome of the human, mouse and rat genomes. These 'index' markers will ultimately form a standard battery, available to all via their PCR primer sequences, for detecting linkage of disease genes in any region of the genome. Screening of the human genome has revealed two additional FAD loci, one associated primarily with late-onset (>65 years) FAD on chromosome 19 (Pericak-Vance et al., 1991), and a major early onset FAD gene on chromosome 14 (Schellenberg et al., 1992); St George-Hyslop et al., 1992; Van Broeckhoven et al., 1992; Mullan et al., 1992).

Chromosome 19 and the evolutionarily conserved APP gene family

Chromosome 19 has been reported to harbor a late onset FAD gene defect (Pericak-Vance et al., 1991). Interestingly, the genetic linkage data suggesting the existence of an FAD locus on chromosome 19 are significant only when the affected members and not the asymptomatic, at-risk individuals are included in the analysis. In other words, an affected pedigree member (APM) analysis is necessary to obtain positive linkage on this autosome. The inclusion of at-risk individuals excludes linkage with chromosome 19. The most plausible explanation for this effect is that the putative chromosome 19 locus is not a 'causative' gene defect (100% penetrance) but, instead, represents a 'susceptibility' or predisposing locus with incomplete penetrance. In this case, the inheritance of the defective gene may not be sufficient to trigger the disease phenotype, but does so only in combination with other genetic factors or environmental influences.

The majority of AD cases occur in individuals over the age of 65 years and most FAD pedigrees being studied are represented by small nuclear kindreds with late-onset disease. Due to the large variability in the age of onset in such pedigrees, familial versus familial clustering of sporadic disease often cannot be clearly established. These observations could be explained by incomplete penetrance, and raise the possibility that other FAD loci may actually involve susceptibility as opposed to causative gene defects.

The identity of the putative chromosome 19 FAD defect is not known. However, we have recently isolated and described an APP-like protein, APLP, mapped the gene encoding this protein to the long arm of chromosome 19, and have proposed that this gene represents a candidate for the chromosome 19 late onset FAD locus (Wasco et al., 1992; Wasco et al., in press). APP appears to be a member of a highly conserved protein family. In 1989, an APP-like protein, APPL, was isolated from Drosophila (Rosen et al., 1989). Subsequently, Luo, Tully and White (1992) demonstrated a functional homology between APPL and human APP when they elegantly demonstrated that transgenes expressing either human APP695 or Drosophila APPL were equally capable of rescuing a fast phototaxis defect in mutant Drosophila lacking the APPL gene. Another APP-like gene was isolated in the form of a partial cDNA from rat testes (Yan et al., 1990).

The human APLP gene on chromosome 19 encodes a novel member of the APP-like gene family which resembles a membrane-associated glycoprotein with a predicted structure that is highly similar to that of APP. APLP is 43% identical and 66% similar to APP, and contains virtually all of the identified domains and motifs that characterize APP (see Fig. 1). These include an N-terminal cysteine-rich region consisting of 12 cysteines, a zinc-binding motif, an acidic-rich domain, an alternatively-transcribed Kunitz protease inhibitor domain, potential N-glycosylation sites, a cytoplasmic clathrin-binding domain, and several phosphorylation sites. In APP, these conserved C-terminal phosphorylation sites appear to play a role in secretion and metabolism of the molecules, as evidenced by the ability of protein kinase C to increase release of APP following cleavage within the extracellular portion of the βA4 domain (Gandy, Czernik & Greengard, 1988). Meanwhile, the conserved clathrin-binding domain plays a role in endocytic trafficking by allowing ligand-independent internalization. Overall, these data suggest that APP, APLP, APPL, and the rat testes APP-like genes are members of a highly conserved gene family. Interestingly, although APP and APLP display very similar hydrophilicity profiles throughout, APLP does not contain the amyloidogenic βA4 domain.

Since the βA4 domain is missing in APLP, this protein could not serve as a substrate for βA4 amyloid formation. However, the overall conservation of amino acid sequence and domain structure within APP, APPL, and APLP suggests that these proteins may very well share common functions and be similarly processed. If APP and APLP were to interact with the same post-translational factors involved with maturation and metabolism, then an alteration in, or the overproduction of APLP might affect the overall processing of APP and the generation of βA4 amyloid.

It is also possible that factors or events leading to the upregulation of APLP might simultaneously result in increased expression of APP, shuttling more APP into the amyloidogenic pathway. The accelerated production of βA4 amyloid in Down Syndrome patients is most likely the result of the extra copy of the APP gene which leads to increased amounts of APP. In Down Syndrome brain, increased levels of APP may overburden intracellular and plasma membranes and/or the metabolic machinery responsible for processing APP thereby leading to amyloid formation. The overproduction of APLP molecules might similarly overburden plasma or intracellular membranes containing APP. In this case, APLP could effectively 'compete' for the same kinases and proteases that interact with, and regulate the metabolism of APP, perhaps ultimately leading to the generation of βA4 amyloid from alternatively processed APP.

The proposed hypothesis that APLP may carry the potential to interfere with APP maturation or metabolism expression requires that these two genes are expressed in the same sets of cells. In collaboration with Dr. Brad Hyman, we have performed *in situ* hybridization studies revealing a virtually identical regional distribution and cellular specificity for APP and APLP in rat brain and in the human hippocampal formation. Intracellularly, both APP and APLP messages are present in the neuronal cell soma as well as dendrites, indicating that both of these genes produce messages that are directly translated at nerve terminals and synapses. Immunohistochemically, both APP and APLP can

be visualized in association with intracellular organelles, including the trans-golgi apparatus (Wasco *et al.*, 1992), where they are most likely undergoing post-translational modification through similar pathways.

In future studies, the APLP gene will be screened for the presence of mutations in late onset FAD patients to test whether it is the putative late onset FAD gene on chromosome 19. We are also attempting to isolate other members of the APP/APLP gene family and have recently isolated a second APP-like gene, APLP2, which is presently being characterized (Wasco *et al.*, 1993).

A major FAD locus on chromosome 14

The increase in the number of FAD pedigrees available for genetic linkage analysis and the emergence of highly informative SSR markers have made it considerably easier to scan the total human genome for additional FAD loci. Using SSR technology, we as well as other laboratories have recently discovered a major FAD gene defect on chromosome 14 in the vicinity of the markers *D14S43* and *D14S53* which map to the region 14q24.3 (Schellenberg *et al.*, 1992, St George-Hyslop *et al.*, 1992; Van Broeckhoven *et al.*, 1992; Mullan *et al.*, 1992). In our data set (Boston and Toronto) the combined FAD pedigrees tested yield a highly significant peak lod score of +23.4 in the vicinity of the above two markers. This is indicative of the existence of a major FAD locus in chromosome 14 which appears to be most tightly linked to the early-onset FAD pedigrees.

Virtually all of the FAD pedigrees being screened in our laboratory have been previously tested for linkage to chromosomes 19 and 21. Of these, all demonstrate some degree of positive genetic linkage with chromosome 14 and six of the largest early-onset FAD kindreds have individually provided significant lod scores (> +3.0) for markers on this autosome. One of these large chromosome 14-linked pedigrees, FAD4, also yields a positive lod score (2.99) with a pericentromeric marker (*D21S52*) on chromosome 21 (Tanzi *et al.*, 1992). The FAD4 pedigree also shows a clear crossover with, and contains no mutations in, the APP gene (Tanzi *et al.*, 1992). Thus this pedigree provides evidence for genetic linkage to both chromosomes 14 and 21, although the peak lod score on chromosome 14 is significantly higher (6.99). A similar situation exists for a pedigree presented in the Van Broeckhoven *et al.* (1992) study of two Belgium FAD kindreds.

Although it is possible that two gene defects (on chromosomes 14 and 21) are responsible for FAD in these pedigrees, alternative explanations also exist. One possibility is that the linkage of the gene defect in FAD4 and the Belgium pedigree to pericentromeric markers on chromosome 21 is simply spurious, a false positive result. Linkage analysis is, after all, based on probability leaving room for such an event. Alternatively, the phenomenon may be explained by non-random sorting of a particular chromosome 21 with the mutant chromosome 14 due to a pericentromeric feature of chromosome 21 (in the region of the genetically linked pericentromeric markers *D21S52* and *D21S13*). Both chromosomes 14 and 21 are acrocentric, and Robertsonian translocations of chromosome 21 involve chromosome 14 more than 95% of the time. It is, therefore, possible that a specific feature of the pericentromeric region of a chromosome 21 segregating in these two pedigrees non-randomly segregates with the chromosome 14 harboring the mutant FAD gene.

The region of chromosome 14 (14q24.3) that shows genetic linkage to FAD includes a number of candidate genes, most notably an HSP70 gene family member and the cFOS oncogene. The APP promotor contains the AP-1 transcriptional element which interacts with FOS-JUN complexes to modulate transcriptional regulation. A breakdown in this regulation could result in overexpression of APP and propagate a situation similar to that which occurs in Down Syndrome patients where increased expression of APP (due to trisomy 21) appears to result in accelerated amyloid formation. With respect to HSP70, this family of proteins function, among other things, as molecular chaperones. HSP70 molecules might bind APP or the $\beta A4$ peptide and serve to prevent amyloid formation. Although both cFOS and HSP70 represent reasonable gene candidates, it more likely that the chromosome 14 FAD locus is a novel gene which may or may not play a direct role in amyloid formation.

Positional cloning strategies employing recombination analysis are currently being used in an attempt to derive flanking markers around the chromosome 14 FAD gene. We are also presently cloning human DNA from the linked region into Yeast Artificial Clones (YACs) which can then be employed to screen for expressed sequences in the linked region. Once obtained, candidate cDNAs can then be characterized, sequenced, and tested for potential mutations associated with FAD. Once isolated, the chromosome 14 FAD gene defect is certain to provide a major piece of the Alzheimer's etiological puzzle.

Acknowledgements

This work was supported by NIH grant NS30428, a grant from the American Health Assistance Foundation, and a Fellowship from the French Foundation (RET).

References

Bahmanyar, S., G. A. Higgins, D. Goldgaber, D. A. Lewis, J. H. Morrison, M. C. Wilson, S. K. Shankar & D. C. Gajdusek, 1987. Localization of amyloid beta-protein messenger RNA in brains from patients with Alzheimer's disease, Science 237: 77-80.

Breitner, J. C. S. & E. A. Murphy, 1992. Twin studies of Alzheimer's disease .2. some predictions under a genetic model. Am. J. Hum. Genet. 44: 628-634.

Chartier-Harlin, M-C., F. Crawford, H. Houlden, A. Warren, D. Hughes, L. Fidani, A. Goate, M. Rossor, P. Roques, J. Hardy & M. Mullan, 1991a. Early-onset Alzheimer's disease caused by mutations at codon 717 of the β-amyloid precursor protein gene. Nature 353: 884-846.

Chartier-Harlin, M-C., F. Crawford, K. Hamand, M. Mullan, A. Goate, J. Hardy, H. Backhovens, J. J. Martin & C. van Broeckhoven, 1991b. Screening for the β-amyloid precursor protein mutation (APP717: Val→Ile) in extended pedigrees with early onset Alzheimer's disease. Neurosci Lett 129: 135-135.

Crawford, F., J. Hardy, M. Mullan, A. Goate, D. Hughes, L. Fidani, P. Roques, M. Rossor & M-C. Chartier-Harlin, 1991. Sequencing of exons 16 and 17 of the β-amyloid precursor protein gene in 14 families with early onset Alzheimer's disease fails to reveal mutations in the β-amyloid sequence. Neurosci. Lett. 133: 1-3.

Citron, M., T. Oltersdorf, C. Haass, L. McConlogue, A.Y. Hung, P. Seubert, C. Vigo-Pelfrey, I. Leiberburg & D. J. Selkoe, 1992. Mutation in the β-amyloid precursor protein in familial Alzheimer's disease increase β-protein production. Nature 360: 672-674.

De Sauvage, F. & J-N. Octave, 1989. A novel mRNA of the A4 amyloid precursor gene coding for a possibly secreted protein, Science 245: 651-653.

Farrer, L. A., R. H. Myers, L. Connor, A. Cupples & J. H. Growdon, 1991. Segregation analysis reveals evidence of a major gene for Alzheimer's disease. Am. J. Hum. Genet. 48: 1026-1033.

Gandy, S., A. J. Czernik & P. Greengard, 1988. Phosphorylation of Alzheimer's disease amyloid precursor peptide by protein kinase C and calcium/calmodulin-dependent protein kinase II. Proc. Natl. Acad. Sci. (U.S.A.) 85: 6218-6221.

Glenner G. G. & C. W. Wong, 1984. Alzheimer's disease: initial report of the purification and characterization of a novel cerebrovascular amyloid protein. Biochem. Biophys. Res. Commun. 120: 885-890.

Goate, A. M., M. C. Chartier-Harlin, M. C. Mullan, J. Brown, F. Crawford, L. Fidani, A. Guiffra, A. Haynes, N. Irving, L. James, R. Mant, P. Newton, K. Rooke, P. Roques, C. Talbot, M. Pericak-Vance, A. Roses, R. Williamson, M. Rossor, M. Owen & J. Hardy, 1991. Segregation of a missense mutation in the amyloid precursor protein gene with familial Alzheimer's disease. Nature 349: 704-706.

Goedert, M., 1987. Neuronal localization of amyloid beta protein precursor mRNA in normal human brain and Alzheimer's disease, EMBO J. 6: 3627-3632.

Goldgaber D., J. I. Lerman, O. W. McBride, U. Saffiotti & D. C. Gajdusek, 1987. Characterization and chromosomal localization of a cDNA encoding brain amyloid of fibril protein. Science 235: 877-880.

Haass, C., M. G. Schlossmacher, A. Y. Hung, C. Vigo-Pelfrey, A. Mellon, B. L. Ostaszewski, I. Leiberburg, E. H. Koo, D. Schenk, D. B. Teplow & D. J. Selkoe, 1992. Amyloid β-peptide is produced by cultured cells during normal metabolism. Nature 359: 322-325.

Hendriks, L., C. M. van Duijn, P. Cras, M. Cruts, W. van Hul, F. van Harskamp, A. Warren, M. G. McInnis, S. E. Antonarakis, J-J. Martin, A. Hofman & C. van Broeckhoven, 1992. Presenile dementia and cerebral haemorrhage linked to a mutation at codon 692 of the β-amyloid precursor protein gene. Nature Genetics 1: 218-221.

Higgins, G. A., D. A. Lewis, S. Bahmanyar, D. Goldgaber, D. C. Gajdusek, W. G. Young, J. H. Morrison & M. C. Wilson, 1987. Differential regulation of amyloid beta-protein mRNA expression within hippocampal neuronal subpopulations in Alzheimer's disease, Proc. Nat. Acad. Sci. (U.S.A.) 85: 1297-1301.

Hyman, B. T., R. E. Tanzi, K. Marzloff, R. Barbour & D. Schenk, 1992. Kunitz protease inhibitor-containing amyloid β-protein precursor immunoreactivity in Alzheimer's disease, J. Neuropathol. Exp. Neurol. 51: 76-83.

Hyman, B. T., J. J. Wenniger & R. E. Tanzi, 1993. Nonisotopic in situ hybridization of amyloid beta protein precursor in Alzheimer's disease: Expression in neurofibrillary tangle bearing neurons and in the microenvironment surrounding senile plaques, Mol. Brain Res. 18: 253-258.

Jacobsen, S. J., H. A. Muenkel, A. J. Blume & M. P. Vitek, 1991. A Novel species-specific RNA to alternatively spliced amyloid precursor protein mRNAs, Neurobiology of Aging 12: 575-583.

Kamino, K., H. T. Orr, H. Payami, E. M. Wijsman, M. E. Alonso, S. M. Pulst, L. Anderson, S. O'dahl, E. Nemens, J. A. White, A. D. Sadovnick, M. J. Ball, J. Kaye, A. Warren, M. McInnis, S. E. Antonorakis, J. R. Korenberg, V. Sharma, W. Kukull, E. Larson, L. L. Heston, G. M. Martin, T. D. Bird & G. D. Schellenberg, 1992. Linkage and mutational analysis of familial Alzheimer's disease kindreds for the APP gene region. Am. J. Hum. Genet. 51: 998-1014.

Kang, J., H. G. Lemaire, A. Unterbeck, J. Salbaum, L. Masters, K. H. Grzeschik, G. Multhaup, K. Beyreuther & B. Mueller-Hill, 1987. The precursor of Alzheimer's disease amyloid A4 protein resembles a cell-surface receptor. Nature 325: 733-736.

Kitaguchi, N., Y. Takahashi, Y. Tokushima, S. Shiojiri & H. Ito, 1988. Novel precursor of Alzheimer's disease shows protease inhibitory activity. Nature 331: 530-532.

Levy, E., M. D. Carman, I. J. Fernandez-Madrid, M. D. Power, I. Lieberburg, G. Sjoerd, S. G. van Duinen, G. Bots, W. Luyendijk & B. Frangione, 1990. Mutation of the Alzheimer's disease amyloid gene in hereditary cerebral hemorrhage, Dutch type. Science 248: 1124-1126.

Lewis, D. A., G. A. Higgins, W. G. Young, D. Goldgaber, D. C. Gajdusek, M. C. Wilson & J. H. Morrison, 1987. Distribution of the precursor of amyloid-beta-protein messenger RNA in human cerebral cortex: Relationship to neurofibrillary tangles and neuritic plaques. Proc. Natl. Acad. Sci. (U.S.A.) 85: 1691-1695.

Luo, L., T. Tully & K. White, 1992. Human amyloid precursor protein ameliorates behavioral deficit of flies deleted for Appl gene. Neuron 4: 595-605.

Mullan, M., F. Crawford, K. Axelman, H. Houlden, L. Lilius, W. Winblad & L. Lannfelt, 1992. A pathogenic mutation for probable Alzheimer's disease in the N-terminus of β-amyloid. Nature Genetics 1: 345-347.

Murrell, J., M. Farlow, B. Ghetti & M. Benson, 1991. A mutation in the amyloid precursor protein associated with hereditary Alzheimer's disease. Science 254: 97-99.

Naruse, S., S. Igarashi, H. Kobayashi, K. Aoki, I. Inuzuki, K. Kaneko, T. Shimizu, K. Ihara, T. Kojima, T. Miyatake & T. Tsuji, 1991. Mis-sense mutation Val→Ile in exon 17 of amyloid precursor protein gene in Japanese familial Alzheimer's disease. Lancet 337: 978-979.

Neve, R. L., E. A. Finch & L. R. Dawes, 1988. Expression of the Alzheimer amyloid precursor gene transcripts in the human brain. Neuron 1: 669-677.

Palmert, M. R., T. E. Golde, M. L. Cohen & D. M. Kovacs, R. E. Tanzi, J. F. Gusella, M. F. Usiak, L. H. Younkin & S. G. Younkin, 1988. Amyloid protein precursor messenger RNAs: differential expression in Alzheimer's disease. Science 242: 1080-1084.

Pericak-Vance, M. A., J. L. Bebout, P. C. Gaskell, L. H. Yamaoka, W-Y. Hung, M. J. Alberts, A. P. Walker, R. J. Bartlett, C. A. Haynes, K. A. Welsh, N. L. Earl, A. Heyman, C. M. Clark & A. D. Roses, 1991. Linkage studies in familial Alzheimer's disease: evidence for chromosome 19 linkage. Am. J. Hum. Genet. 48: 1034-1050.

Ponte, P., P. Gonzalez-DeWhitt, J. Schilling, J. Miller, D. Hsu, B. Greenberg, K. Davis, W. Wallace, I. Lieberburg, F. Fuller & B. Cordell, 1988. A new A4 amyloid mRNA contains a domain homologous to serine proteinase inhibitors. Nature 331: 525-527.

Robakis, N. K., N. Ramakrishna, G. Wolfe & H. M. Wisniewski, 1987. Molecular cloning and characterization of a cDNA encoding the cerebrovascular and the neuritic plaque amyloid peptides. Proc. Nat. Acad. Sci. (U.S.A.) 84: 4190-4194.

Rosen, D. R., L. Martin-Morris, L. Luo & K. White, 1989. A Drosophila gene encoding a protein resembling the human β-amyloid protein precursor. Proc. Natl. Acad. Sci. (USA)

86: 2478-2482.

Schellenberg, G. D., T. D. Bird, E. M. Wijsman, D. K. Moore, M. Boehnke, E. M. Bryant, T. H. Lampe, D. Nochlin, S. M. Sumi, S. S. Deeb, K. Beyreuther & G. M. Martin, 1992. Absence of linkage of chromosome 21q21 markers to familial Alzheimer's disease. Science 241: 1507-1510.

Schellenberg, G. D., T. D. Bird, E. M. Wijsman, H. T. Orr, L. Anderson, E. Nemens, J. A. White, L. Bonnycastle, J. L. Weber, M. E. Alonso, H. Potter, L. L. Heston & J. Martin, 1992. Genetic linkage evidence for a familial Alzheimer's disease locus on chromosome 14. Science 258: 668-671.

St George-Hyslop, P. H., R. E. Tanzi, R. J. Polinsky, J. L. Haines, L. Nee, P. C. Watkins, R. H. Myers, R. G. Feldman, D. Pollen, D. Drachman, J. Growdon, A. Bruni, J-F. Foncin, D. Salmon, P. Frommelt, L. Amaducci, S. Sorbi, S. Piacentini, G. D. Stewart, W. J. Hobbs, P. M. Conneally & J. F. Gusella, 1987. The genetic defect causing familial Alzheimer's disease maps on chromosome 21. Science 235: 885-889.

St George-Hyslop, P. H., R. D. Myers, J. L. Haines, L. A. Farrar, R. E. Tanzi, K. Abe, M. F. James, P. M. Conneally, R. J. Polinsky & J. F. Gusella, 1989. Familial Alzheimer's disease: Progress and problems. Neurobiology of Aging 10: 417-425.

St George-Hyslop, P. H., J. L. Haines, L. A. Farrer, R. Polinsky, C. van Broeckhoven, A. Goate, D. R. Crapper-McLachlan, H. Orr, A. C. Bruni, S. Sorbi, I. Rainero, J. F. Foncin, D. Pollen, J. M. Cantu, R. Tupler, N. Voskresenskaya, R. Mayeux, J. Growdon, L. Nee, H. Backhovens, J. J. Martin, M. Rossor, M. J. Owen, M. Mullan, M. E. Percy, H. Karlinsky, S. Rich, L. Heston, M. Montes, M. Mortilla, N. Nacmias, G. Vaula, J. F. Gusella, J. A. Hardy and the FAD collaborative group, 1990. Genetic linkage studies suggest that Alzheimer's disease is not a single homogeneous entity. Nature 347: 194-197.

St George-Hyslop, P. H., J. Haines, E. Rogaev, M. Mortilla, G. Vaula, M. Pericak-Vance, J-F. Foncin, M. Montesi, A. Bruni, S. Sorbi, I. Rainero, L. Pinessi, D. Pollen, R. Polinsky, L. Nee, J. Kennedy, F. Macciardi, E. Rogaeva, Y. Liang, N. Alexandrova, W. Lukiw, K. Schlumpf, R. Tanzi, T. Tsuda, L. Farrer, J-M. Cantu, R. Duara, L. Amaducci, L. Bergamini, J. Gusella, A. Roses & D. Crapper-McLachlan, 1992. Genetic evidence for a novel familial Alzheimer's disease locus on chromosome 14. Nature Genetics 2: 330-334.

Shoji, M., T. Golde, J. Ghiso, T. T. Cheung, S. Estus, L. M. Shaffer, X-D Cai, D. M. McKay, R. Tintner, B. Frangione & S. G. Younkin, 1992. Production of Alzheimer amyloid β-protein by normal proteolytic processing. Science 258: 126-129.

Seubert, P., C. Vigo-Pelfrey, F. Esch, M. Lee, H. Dovey, J. Whaley, C. Swindlehurst, R. McCormack, R. Wolfert, D. Selkoe, I. Leiberburg & D. Schenk, 1992. Isolation and quantification of soluble Alzheimer's β-peptide from biological fluids. Nature 359: 325-327.

Tanzi, R. E., J. F. Gusella, P. C. Watkins, G. A. Bruns, P. St George-Hyslop, M.L. Vankeuren, S. P. Patterson, D. M. Kurnit & R. L. Neve, 1987. Amyloid beta protein gene: cDNA, mRNA distribution, and genetic linkage near the Alzheimer locus. Science 235: 880-884.

Tanzi, R. E., P. H. St George-Hyslop, J. L. Haines, R. J. Polinsky, L. Nee, J-F Foncin, R. L. Neve & J. F. Gusella, 1987b. The genetic defect in familial Alzheimer's disease is not tightly linked to the amyloid β-protein gene. Nature 329: 156-157.

Tanzi, R. E., A. I. McClatchey, E. D. Lamperti, L. Villa-Komaroff, J. F. Gusella & R. L. Neve, 1988. Protease inhibitor domain encoded by an amyloid protein precursor mRNA associated with Alzheimer's disease. Nature 331: 528-530.

Tanzi, R. E. & B. T. Hyman, 1991a. Studies of amyloid β-protein precursor expression in Alzheimer's disease, Annals NY Acad. Sci. 640: 149-154.

Tanzi, R. E. & B. T. Hyman, 1991b. Alzheimer's mutation Nature 350: 564.

Tanzi, R. E., G. Vaula, D. M. Romano, M. Mortilla, T. L. Huang, R. G. Tupler, W. Wasco, B. T. Hyman, J. L. Haines, B. J. Jenkins, M. Kalaitsidaki, A. C. Warren, M. G. McInnis, S. E. Antonarakis, H. Karlinsky, M. E. Percy, L. Connor, J. Growdon, D. R. Crapper-McLachlan, J. F. Gusella & P. H. St George-Hyslop, 1992. Assessment of amyloid β protein precursor gene mutations in a large set of familial and sporadic Alzheimer's disease cases. Am. J. Hum. Genet. 51: 273-282.

Tanzi, R. E., J. T. Wennigers & B. T. Hyman, 1993. Regional distribution and cellular specificity of APP alternative transcripts are unaltered in Alzheimer's disease and control hippocampul formation. Molecular Brain Research 18: 246-252.

Terry, R. D. & R. Katzman, 1983. Senile dementia of the Alzheimer type. Ann. Neurol. 14: 497-506.

Van Broeckhoven, C., H. Backhovens, M. Cruts, G. de Winter, M. Bruyland, P. Cras & J-J. Martin, 1992. Mapping of a gene predisposing to early-onset Alzheimer's disease to chromosome 14q24.3. Nature Genetics 2:334-339.

Van Duijn, C. M., L. Hendriks, M. Cruts, J. A. Hardy, A. Hofman & C. van Broeckhoven, 1991. Amyloid precursor protein gene imitation in early-onset Alzheimer's disease. Lancet 337: 978.

Wasco, W., K. Bupp, M. Magendantz, J. Gusella, R. E. Tanzi & F. Solomon, 1992. Identification of a mouse brain cDNA that encodes a protein related to the Alzheimer-associated amyloid β precursor protein. Proc. Natl. Acad. Sci. (USA) 87: 2405-2408.

Wasco, W., J. D. Brook, J. F. Gusella, D. E. Housman & R. E. Tanzi, 1993. The amyloid precursor-like protein gene maps to the long arm of chromosome 19. Genomics 15: 237-239.

Wasco, W., S. Gurubhagavatula, M. D. Paradis, D. M. Romano, S. S. Sisodia, B. T. Hyman, R. L. Neve & R. E. Tanzi, 1993. Isolation and characterization of APLP2 encoding a homologue of the Alzheimer's associated amyloid β protein precursor. Nature Genetics 5: 95-99.

Weber, J. L. & P. E. May, 1989. Abundant class of human DNA polymorphisms which can be typed by the polymerase chain reaction. Am. J. Hum. Genet. 44: 388-396.

Yan, Y. C., Y. Bai, L. Wang, S. Miao & S. S. Koide, 1990. Characterization of cDNA encoding a human sperm membrane protein related to A4 amyloid protein. Proc. Natl. Acad. Sci. USA 87: 2405-2408.

M. R. Rose and C. E. Finch (eds.), Genetics and Evolution of Aging, 294–306, 1994.
© 1994 *Kluwer Academic Publishers. Printed in the Netherlands.*

Abiotrophic gene action in *Homo sapiens*: potential mechanisms and significance for the pathobiology of aging

George M. Martin
Departments of Pathology and Genetics, University of Washington, Seattle, WA 98195, USA

Received and accepted 22 June 1993

Key words: abiotrophy, genetic polymorphisms, dominant mutations in man, recessive mutations in man, senescent phenotype

Abstract

A subset of genetic loci of *Homo sapiens* are reviewed that: 1) have the potential for allelic variation (either mutation or polymorphism) such that degenerative and/or proliferative phenotypic aberrations may be of relatively late onset ('abiotrophic'); 2) have phenotypic features which overlap, to some extent, with those of important age-related disorders of man (many of which are systematically tabulated in this review); 3) have had significant characterization at the biochemical genetic level. The ascertainment bias of physicians to discover strong phenotypic effects ('non-leaky' mutations) obscures the fact that, for many such instances, there exist numerous other alleles of lesser effects, including those whose gene actions probably escape the force of natural selection.

The patterns of 'normal' aging in *Homo sapiens* are quite variable and, hence, difficult to define. It seems likely that the 'wild-type' alleles of a number of loci will also be found to have antagonistic pleiotropic effects that contribute to the syndromology of senescence in our species.

Introduction

Our species is by far the best studied mammal and, from many points of view (notably its pathophysiology), the best understood organism. Until recent years, the major gap in our knowledge has been its formal genetics. Because of the many advances in cytogenetics, somatic cell genetics and, especially, in molecular genetics, the situation is rapidly changing. Thus, there are increasing opportunities to couple the enormous amount of detailed information concerning phenotypic variation with variations in the constitutional genotype.

The dominant theme of current thinking on the evolutionary biology of aging is that senescent phenotypes are nonadaptive nature-nurture interactions that have escaped the force of natural selection. The genetic component consists of a potentially large number of loci falling into either of two broad classes of constitutional mutations or polymorphisms. My interpretation of the literature is that one class consists of variants with little effect upon reproductive fitness, either because of the intrinsic nature of the gene action or because of the evolution of suppressor genes. The second class includes wild-type and variant genes characterized by obvious pleiotropic features, having been selected because of enhanced reproductive fitness, but with ultimate deleterious effects playing out during the postreproductive life span; this type of gene action has been referred to as negative or antagonistic pleiotropy (Rose, 1991). Efforts to explore the latter type of mutation in man are outlined in the contribution to this volume by Dr. Roger L. Albin. Given the present difficulties in documenting small effects upon reproductive fitness in man, we here take a more global approach to the delineation of genetic loci of potential relevance to the pathobiology of man. The proposition that we wish to explore is that numerous abiotrophic mutations of man, *including* those which impair reproductive fitness, can delineate sets of genetic loci which influence senescence. Thus, for each locus that a medical geneticist has been able to define because of its pronounced effects or senescence, there may exist numerous alternative alleles with similar, but more subtle effects. Wild-type alleles at such loci could also be implicated as playing important roles in aging, either as 'progeroid' or 'antiprogeroid' genes.

The approach will be to present a biochemical genetic classification, using only examples for which there is significant information regarding the underlying biochemical genetic basis. This 'ground rule' precludes discussion of some very important examples of abiotrophic gene action in man, including the prototypic example of Huntington's chorea. It also precludes discussion of one the author's major research interests, Werner's syndrome (Progeria of the Adult). Except for the recent successes in mapping this autosomal recessive mutation to 8p12, however (Goto *et al.*, 1992; Schellenberg *et al.*, 1992a), the subject has been recently reviewed (Martin, 1989, 1991).

These are still 'early times' in human biochemical genetics, and so the number of examples for which we have some detailed information about gene structure and gene action is limited. The picture is likely to change dramatically in the next ten or twenty years. Eventually, we may be able to discover the biochemical genetic bases of individual variations of late-life functional decrements that are independent of disease. Examples would include declines in the efficiencies of temperature homeostasis and of various enzyme inductions.

I am grateful to two seminal references, which readers should consult for further documentation: Scriver *et al.* (1989) and McKusick (1992). The latter is available 'on line' via Internet or Telnet (phone 301-955-7058 for information). This data base is continually updated.

Some definitions

Like most of my colleagues in mammalian gerontology, I shall use the terms aging and senescence interchangeably, referring to those mostly nonadaptive changes in structure and function that gradually unfold after the attainment of sexual maturity and the young adult phenotype. A partial list of pathological features that most physicians and pa-

Table 1. Examples of pathological processes associated with the senescent phenotype of *Homo sapiens*.

Body system	Phenotype
Cardio-vascular	atherosclerosis, arteriosclerosis, medial calcinosis, basement membrane thickening of capillaries; hypertension; increased susceptibility to thromboembolism; myocardial lipofuscinosis; myocardial hypertrophy with interstitial fibrosis, valvular fibrosis and valvular calcification
Central nervous	β-amyloid depositions; lipofuscin depositions; neuritic plaques; neurofibrillary tangles; gliosis; regional neuronal loss, leading to disorders such as Parkinson's disease and Alzheimer's disease
Peripheral nervous	segmental demyelination with decreased nerve conduction velocity
Special senses	loss of visual accommodation; ocular cataracts; senile macular degeneration; loss of high frequency auditory acuity; loss of olfactory acuity
Respiratory	chronic obstructive pulmonary disease; interstitial fibrosis; decreased vital capacity
Renal	glomerulosclerosis and loss of nephron units; interstitial fibrosis
Male reproductive	decreased spermatogenesis; hyalinization of seminiferous tubules; benign prostatic hypertrophy; adenocarcinoma of the prostate
Female reproductive	depletion of ovarian primordial follicles; ovarian stromal cell hyperplasia; endometrial atrophy and hyperplasia; endometrial carcinoma; carcinoma of breast and ovary; 'fibroid' tumors of the myometrium; vaginal atrophy
Musculo-skeletal	skeletal muscle atrophy and interstitial fibrosis; osteoporosis; osteoarthritis
Hemato-poietic	anemia; chronic lymphocytic leukemia; chronic myelogenous leukemia; myelofibrosis; myoclonal gammopathies and multiple myeloma; polycythemia vera
Endocrine	interstitial fibrosis of thyroid; hypercortisolemia; asynchrony of growth hormone-release hormone release; amyloid depositions in beta cells of pancreatic islets; non-insulin-dependent diabetes mellitus

Table 1. Continued.

Gastro-intestinal	colonic polyps and adenocarcinoma; diverticulosis of colon; gastric antral atrophy; fatty infiltration and brown atrophy of pancreas; adenocarcinoma of pancreas; cholelithiasis; periodontal disease
Integu-mentary	epidermal atrophy; pigmentary alterations; basophilic alteration of collagen; senile elastosis; basal cell and squamous cell carcinomas; regional atrophy and hypertrophy of adipocytes; thinning and greying of hair of scalp and body

thologists associate with the senescent phenotype is given in Table 1.

The terms 'abiotrophy' (noun) and 'abiotrophic' (adjective) apparently had their origins in the writings of a turn-of-the-century London neurologist, Sir William Richard Gowers (1902). In his perceptive address to his neurological colleagues at National Hospital for the Paralysed and Epileptic, Queen-Square, Bloomsbury, delivered on February 21, 1902, Professor Gowers clearly appreciated the possibility that unusual inheritance could explain some late-onset types of degenerative pathology, including what we now know to be X-linked forms of muscular dystrophy and Friedriech's disease (an excellent example, now known to be autosomal recessive with both myocardial and neural degenerative changes). The concept and terminology was therefore subsequently employed principally by neurologists concerned with late-onset degenerative disorders of the central nervous system. A medical dictionary (Friel, 1985) defines abiotrophy (*a* = negative + Gr. *bios* = life + *trophe* = nutrition) as the 'progressive loss of vitality of certain tissues or organs leading to disorders or loss of function; applied especially to degenerative hereditary diseases of late onset, e.g., Huntington's chorea.' I define abiotrophic mutations as those which have relatively delayed expression (and certainly without apparent phenotypes at birth), in which the gene action results in either *degenerative* types of pathology (the affected cell tissues having previously been functioning sufficiently well to preclude a clinical diagnosis), or *proliferative* types of pathology, or both. The latter could include any sorts of abnormalities in proliferative homeostasis, including atrophies, hyperplasias, benign neoplasias and malignant neoplasias (Martin, 1979).

The biochemical genetic basis of abiotrophic mutations of Homo sapiens

Disorders of carbohydrate metabolism

Non-insulin-dependent diabetes mellitus

Sometimes referred to as Type II diabetes mellitus, it is differentiated from the much rarer juvenile forms of diabetes which are insulin-dependent and whose pathogenesis is considered to be immune-mediated. The key metabolic abnormality is hyperglycemia, the result of overproduction of glucose by the liver coupled with inadequate utilization by peripheral tissues. Numerous slowly developing degenerative lesions eventually become apparent, including retinopathy, coronary artery atherosclerosis and myocardial infarction, peripheral vascular insufficiency (sometimes with ensuing gangrene), glomerulosclerosis with renal insufficiency, and peripheral neuropathy. The precise pathogenetic pathways leading to these serious and often fatal late complications are yet to be defined, but a major contributor may be the glycation of basement membrane proteins of the microvasculature. The beta cell of the islets of Langerhans in the pancreas is dysfunctional, possibly as a result of more than one mechanism. Genetic studies have clearly demonstrated multiple contributory loci. While many forms of the disease are likely to have a polygenic basis, recent research has revealed two different monogenic loci which may be sufficient to result in unusually early onsets of the disorder.

In some autosomal dominant pedigrees, nonsense mutations (Katagiri *et al.*, 1992) and missense mutations (Stoffel *et al.*, 1992) have been demonstrated in the structural gene for glucokinase (GCK), following linkage studies implicating GCK as a candidate gene (Froguel *et al.*, 1992; Hattersley *et al.*, 1992). These observations were made possible by the discovery of a polymorphic dinucleotide repeat sequence some 10 kb 3' to the GCK gene (Matsutani *et al.*, 1992). This pattern of discovery (elucidation of new, highly informative polymorphic markers; establishment of linkage; recognition of a plausible candidate gene; identification of specific mutations co-segregating in particular pedigrees) is a research paradigm that is likely to become increasingly productive. Glucokinase (EC 2.7.1.1) is a functionally unique member of the hexokinase family of enzymes. It catalyzes the phosphorylation of glucose at the sixth carbon position of glucose, the first step in glucose metabolism. Its structural gene maps to the short arm of chromosome 7.

Linkage analysis has revealed a second locus on the long arm of chromosome 20 (probably 20q13) (Bell *et al.*, 1991) contributing to this form of diabetes. The responsible gene mutation remains to be identified. This mutation results in an unusually mild form of the disease, despite its relatively early onset. A number of affected individuals have been followed from the third decade of life, when the diagnosis may sometimes be made, until the sixth or seventh decade, at which time less than half had developed retinopathy.

The major significance of the above studies for gerontology, specifically the work on glucokinase, is that they provide definitive evidence that the clinical type of diabetes mellitus so prevalent in older human subjects, including all of its major devastating degenerative long-term complications, may clearly be caused by a *primary*, *constitutional* defect in carbohydrate metabolism.

Amino acid metabolism

Gyrate atrophy of the choroid and retina
In addition to a variable age of onset of the choroidal and retinal degenerations, homozygotes for this autosomal recessive disorder develop ocular cataracts whose morphologies (posterior subcapsular) overlap with those of 'senile' cataracts. Early symptoms include reduction of night vision, but declines in visual acuity may not be manifested in some patients until the sixth or seventh decades. About 10% of patients also suffer mild proximal muscle weakness. Mitochondria in liver, iris and skeletal muscle have been described as having abnormal morphologies. Scalp and body hair may be sparse.

A deficiency of the mitochondrial matrix enzyme, ornithine delta-aminotransferase, has been documented in skin fibroblast cultures, peripheral blood lymphocytes, skeletal muscle, hair roots and liver. This enzyme catalyzes the reversible interconversion of ornithine and alpha-ketoglutarate to glutamate semialdehyde and glutamate. Enzyme deficiency leads to an accumulation, in various fluids and tissues (including plasma, cerebrospinal fluid, urine and the aqueous humor of the eye), of the amino acid L-ornithine and certain of its metabolites. The precise pathogenetic sequence leading from enzyme deficiency to the degenerative processes is not known. One interesting hypothesis is that the high ornithine concentrations inhibit glycine transamidinase, resulting in reduced creatine synthesis and, thus, a reduction in the pools of creatine and creatine phosphate. These compounds are thought to be of crucial importance in such processes as the energy transduction of vision. There are high concentrations of creatine kinase and creatine phosphate in both rod and cone photoreceptor cells.

The disease gyrate atrophy of the choroid and retina and the structural gene for ornithine-delta-aminotransferase has been mapped to chromosome 10 (10q26) (reviewed by Ramesh, Gusella, & Shih, 1991). Some 24 different allelic mutations have so far been identified at the ornithine aminotransferase (OAT) locus that are responsible for the disease in different families (Brody *et al.*, 1992). While 20/24 of these alleles produced normal amounts of normally sized mRNA transcripts, only 2/24 had normal amounts of enzyme protein, as measured immunologically.

The significance of the above studies for gerontology is that they implicate a biochemical disorder in mitochondrial function as a cause for several phenotypes that overlap with what can be seen in aging human subjects. Moreover, they illustrate that such degenerative pathologies may be derived from defects other than oxygen-mediated free radicals, currently the most popular molecular theory of aging. The curious specificity of pathology remains to be explained, however. While it would seem quite doubtful that the specific enzyme deficiency in this rare recessive disorder occurs in any significant number of older human subjects as they age, the research points to a broad biochemical domain of interest to gerontologists.

Homocystinuria
This autosomal recessive disorder is usually caused by a deficiency of cystathionine β-synthase, leading to the accumulations of both homocysteine and methionine in the plasma. Various compounds dis-

tal to the enzymatic block are synthesized at lower rates in these patients. The reason for invoking this inborn error of metabolism in the context of gerontology is that, among the many phenotypic manifestations, one observes osteoporosis, atherosclerosis and thromboembolism. Less frequent complications include cataracts and retinal degeneration. The skin typically has an atrophic appearance and there is electrophysiological evidence of a myopathy. Of these phenotypes, the most significant has to do with the cardiovascular components, which involve cerebral, carotid, coronary and peripheral arteries. There is now compelling evidence that increased levels of plasma homocysteine constitute an independent risk factor for coronary artery disease, the major cause of death among aging cohorts of human subjects in western societies (Clark *et al.*, 1991; Genest *et al.*, 1990). A most intriguing possibility is that, among the population of hyperhomocystinuric adults who are at increased risk for vascular diseases, there are individuals who are heterozygous for cystathionine-β-synthase mutations. In Ireland, the frequency of enzyme deficient newborns is exceptionally high (10 per 573,206, or about 1 per 57,000 (S. Cahalane, cited in Scriver *et al.*, 1989). Assuming Hardy-Weinberg equilibrium, that would give a frequency of heterozygotes of about 0.8-0.9%, a non-trivial contribution to the morbidity load of that population.

Despite a considerable amount of research, the pathogenetic sequences leading from mutation to phenotype are poorly understood. A gene coding for the cystathionine β-synthase has been mapped to the long arm of chromosome 21 (21q22.3) (Munke *et al.*, 1988). The gene for cystathionine β-synthase maps to the subtelomeric region on human chromosome 21q and to proximal mouse chromosome 17. Only a single mutant allele has so far been identified, a G to A transition resulting in the substitution of serine for glycine at residue 307 (Gu *et al.*, 1991). There are a variety of phenotypes among individual patients, however. It is thus possible that many different mutant alleles will be discovered in various populations. It is also possible that another structural gene locus will be discovered, as there is evidence in the rat that the functional protein is a tetramer consisting of two different subunits.

Metabolism of purines and pyrimidines

Exertional myopathy associated with adenosine monophosphate deaminase-1 deficiency

AMP deaminases are widely distributed in mammalian cells, but there appear to be tissue-specific isoforms, probably the result of both differential expression of two structural genes (which, for the human, have both been mapped to 1p21-p13), as well as to alternative splicing of the mRNA of at least one of these genes (Sabina *et al.*, 1990). Heritable deficiencies of the enzyme that is relatively specific for skeletal muscle can result in a remarkable range of ages of phenotypic expressions – from age 1 to age 70! The clinical picture is one of myopathy induced by some degree of exertion, giving a picture of easy fatiguability. The enzyme clearly plays an important role in the energy metabolism of skeletal muscle, but the precise mechanisms are not understood. Astonishingly, a single mutant allele with a double mutation (including a nonsense mutation resulting in a truncated protein) has been found in 12% of Caucasians and 19% of African-Americans, but in none of 106 Japanese subjects (Morisaki *et al.*, 1992).

There have been no comprehensive anatomic, histochemical, biochemical and physiological investigations of the skeletal muscle alterations in aging cohorts of subjects homozygous and heterozygous for ADA-1 deficiency, as compared to aging subjects with wild-type enzyme.

Disorders of lipid metabolism

Atherosclerosis

It is now quite evident that allelic variations at many loci are major determinants of the age of onset and the rate of progression of this most important abiotrophic disorder of man. In Western societies, it is by far the major cause of death, including those who survive to the tenth decade (Table 2). Quantitative measures of the extent of involvement of coronary arteries have shown approximately linear increases of the extent of raised, lipid-rich fibrotic plaques in the coronary arteries of some 19 different population-ethnic groups from around the world, with significant differences in the rates of increase among these various groups (Eggen & Solberg, 1968). Figure 1 gives a composite statistical analysis of all the data, which, unfor-

Table 2. A selected list of causes of death for 90- to 94-year-old males and females (all races) dying in the United States in 1980. (Adapted from Vital Statistics of the United States, Vol. II, Part B, 1985).

Causes of death	Total number of deaths
All causes	111,471
Major cardiovascular[a]	78,657
Malignant[b]	9,082
Pneumonia	6,359
Accidents	2,004
Chronic obstructive pulmonary disease[c]	1,467
Diabetes mellitus	1,347
Nephritis, nephrosis, renal failure	1,204
Septicemia	574
Suicide	101
Chronic liver disease and cirrhosis	92

[a] Includes entry nos. 390-448 from the Ninth Revision of the International Classification of Disease, 1975
[b] Includes lymphatic and hematopoietic malignant neoplasms
[c] Includes deaths from related conditions

tunately, were limited to the age range of 20-60. It is beyond the scope of this overview to present a comprehensive review of all of the implicated genetic loci. Interested readers should consult several excellent recent book chapters and monographs (Motulsky & Brunzell, 1992; Bearn, 1992; Nora, Berg, & Nora, 1991). We shall instead illustrate the importance of constitutional genetic variations with two examples. The LDH receptor mutations of Familial Hypercholesterolemia were the first to be explored in depth. This research was recognized by a Nobel Prize for Drs. Joseph Goldstein and Michael Brown. An example of an area of more recent research is the Lp(a) locus, which appears to be emerging as one of the most polymorphic loci in man, its various alleles serving as independent risk factors that are relatively insensitive to environmental variables.

Familial hypercholesterolemia: Mutations at the low density lipoprotein (LDL) receptor locus
Lipoprotein molecules consist of globular aggregates of lipids surrounded by a shell of protein, called apolipoprotein. Numerous mutations have

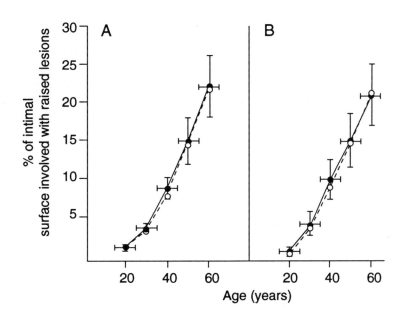

Fig. 1. Extent of gross lesions of coronary artery atherosclerosis in human males as a function of age by decades. Solid circles are unweighted means of the means (± 95% confidence limits) for 19 different location-race groups involving 9031 subjects dying from causes other than coronary heart diseases, diabetes or hypertension. Open circles are weighted means, obtained by normalizing the sample sizes of the various location-race groups. Panel A: data from the three major coronary arteries. Panel B: data from the anterior descending branch of the left coronary artery. (After Martin, Ogburn & Sprague, 1975)

been characterized among subsets of apolipoprotein molecules that are responsible for unusual susceptibilities to atherosclerosis primarily, it is thought, because of alterations in the metabolism of cholesterol and various lipid moeities. The first such locus to be studied in molecular detail was the locus coding for the cell receptor for one such class of lipoprotein molecules characterized by its centrifugal behavior (low density lipoprotein, LDL). The most recent and authoritative review of the molecular genetics of these rather common 'autosomal dominant' mutations is that of Hobbs *et al.* (1990). Like virtually all 'dominant' disorders of man (Huntington's disease being a rare exception), carriers with a mutation in only a single allele typically have significant phenotypic manifestations. Homozygotes or compound heterozygotes have qualitatively similar but much more severe phenotypes, with earlier onset and earlier complications. Homozygotes have exceedingly high levels of total serum cholesterol, typically ranging from 650 to 1000 mg/dl. (The mean for 'clinically normal' white American males, ages 20-29 years is 174 mg/dl; only 5% of this population will have total serum cholesterol levels greater than 231 mg/dl.) Heterozygotes (ages $\geq$ age 20) for a mutation at the LDL locus have values of 368 ± 78 mg/dl. The degree of hypercholesterolemia correlates with the extent of atherosclerotic disease. The principal complication is coronary artery thrombosis with myocardial infarction. For homozygotes, this may occur in childhood or adolescence. For heterozygotes, there are no clinical manifestations of coronary artery disease until over age 50 for half the males and until over age 60 for half the females. The latter are thought to be protected by their premenopausal levels of estrogens.

In a period of just five years following the characterization of the wild-type LDL receptor gene in 1985, more than 150 different mutant alleles have been described! This is a dramatic indication of the accelerating pace of research in human genetics. These mutations are among the most prevalent monogenic causes of morbidity and mortality in many human populations. The mutations have been classified in five groups. Class 1 mutations are null alleles, in which there is no detectible protein product. The protein products of Class 2 alleles are defective in transporting LDL cholesterol between the endoplasmic reticulum and the Golgi apparatus.

While the products of Class 3 mutations are successfully transported to the cell surface, they are defective, to varying degrees, in the efficiency with which they bind LDL. The LDL receptors coded for by Class 4 mutations, while wild-type in terms of transport and binding functions, are defective in their ability to cluster in clathrin-coated pits, and are referred to as internalization mutations. Finally, the products of Class 5 mutations are wild-type in transport, binding and internalization functions, but do not efficiently release their ligands within endosomes and, consequently, there is a deficiency of recycling to the cell membrane. The frequencies of these different types of mutations vary substantially among populations, presumably primarily related to particular founder effects. For example, among South African Jews, almost all of whom emigrated from Lithuania, 1/67 carry a single type of mutant allele. For most populations, the summed frequencies of all types of mutations are currently thought to be of the order of 1/500. In terms of absolute numbers, this makes a substantial contribution to the phenotype of senescing human subjects.

Lp(a) polymorphism

In addition to its lipid component and its apolipoprotein (apo B-100), Lp(a) (lipoprotein a) has a unique structural feature, a glycoprotein called apoprotein(a) or apo(a) that is linked to apoprotein B-100 via disulfide bonds. Its cDNA, which maps to 6q27 (the gene symbol is LPA) is highly homologous to plasminogen, the gene for which is closely linked to the apo(a) gene at 6q26-q27. Both genes code for a series of repeated 'kringle' domains. Certain kringle regions of proteases, like the serine protease region of plasminogen, are important in the regulation of coagulation, including functions such as the binding to fibrin. The regions in the homologous structures of the Lp(a) molecule are altered, however, in that the arginine residues critical to binding are deleted from all but a single kringle region (kringle 4) and all have sites capable of N glycosylation. While the exact mechanisms whereby varying types and concentrations of Lp(a) influence atherosclerosis (in particular, coronary artery ischemic disease) are unknown, some workers believe they act at the stage of thrombosis, which is often the immediate triggering event for a myocardial infarction at the site of an advanced atheromatous plaque. Rath and Pauling (1990) have

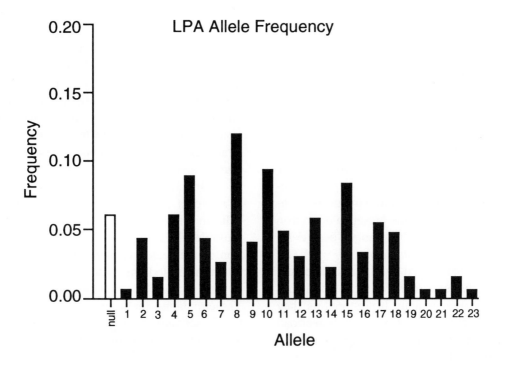

Fig. 2. Graphic representation of LPA allele frequencies. The frequencies of all 24 alleles are as follows: Null = .061; 1 = .007; 2 = .043; 3 = .014; 4 = .061; 5 = .089; 6 = .043; 7 = .025; 8 = .118; 9 = .039; 10 = .093; 11 = .046; 12 = .029; 13 = .057; 14 = .021; 15 = .082; 16 = .032; 17 = .054; 18 = .046; 19 = .014; 20 = .004; 22 = .014; and 23 = .004. (from Kamboh, Ferell & Kottke, 1991)

noted that mammalian species that have lost the ability to synthesize ascorbate (such as humans, guinea pigs and the Indian fruit-eating bat) have gained the LPA gene. It is of interest that an associated correlation may be increased susceptibility to atherogenesis for the group of animals expressing LPA. For example, while mice do not have LPA genes, they are very resistant to the development of atherosclerotic types of lesions, even when fed lipid-rich experimental diets. Transgenic mice expressing a cDNA for human LPA, however, such that they have levels of Lp(a) equivalent to the median level found in human populations ($\sim$ 10 mg/dl), develop such lesions (Lawn *et al.*, 1992).

The LPA gene of humans has at least 24 different alleles, most of which occur at rather high frequencies (Kamboh, Ferell, & Kottke, 1991) (Fig. 2). Thus, it is one of the most informative polymorphic loci in man. Most of us are compound heterozygotes at this locus. This polymorphism, attributable mainly to variable numbers of tandem repeats of kringle 4, has the effect of producing an enormous range of variation in the blood levels of Lp(a), with a range of about a thousand fold! There are good

empirical data indicating that, the higher the concentration, the greater the risk of coronary artery disease. Thus, almost a third of patients with premature coronary heart disease have levels of Lp(a) above the 95th percentile of the general population. This appears to be independent of other known risk factors, such as smoking and hypertension. The risks from having polymorphic forms of LPA *and* mutations at the LDL receptor locus are additive.

Metabolism of elemental micronutrients

Hemochromatosis

The responsible gene mutation for what is considered to be the principal hereditary form of this disease of excessive iron storage has not yet been cloned, although it is thought to map to 6p21.3, close to the HL-A complex (Gruen *et al.*, 1992). This subject is dealt with elsewhere in this volume by Albin, who suggested that this mutation might be considered as an example of negative pleiotropic gene action of relevance to the biology of aging (Albin, 1988). The fully expressed phenotype in patients homozygous for mutation includes features such as muscle weakness, lethargy, diabetes mellitus, joint pain, hypogonadism (loss of libido and potency), exertional dyspnea and varying degrees of cardiac failure, alterations of skin pigmentation, thinning of hair and osteoporosis. The major pathology is cirrhosis of the liver, with a complicating primary hepatocarcinoma being a major cause of death; these entities are unusual aspects of senescence in most populations. Nevertheless, the principal pathophysiological outcome, an increased body load of iron, fits well with the concept of free-radical-mediated macromolecular damage as a major mechanism of aging. In the presence of an iron overload, lipid peroxidation has indeed been demonstrated, both *in vivo* and *in vitro*.

An important unanswered question is the extent to which heterozygous carriers for the major hemochromatosis gene accumulate iron stores as they age, and the degree to which such iron overload might contribute to the senescent phenotype of such individuals, particularly among males, in whom phenotypic expression is more likely. In some populations, there have been claims that the prevalence of heterozygotes may be as high as 10%. Given the evidence of a progressive rise in the levels of serum ferritin in heterozygotic carriers (Meyer *et al.*, 1988), such subjects may indeed be at risk, depending, of course, upon the iron content of diet and other environmental and genetic variables, yet to be described.

Sodium retention via mutations in angiotensinogen
Age-related increases in both systolic and diastolic blood pressure ('essential' hypertension) have been well-documented in cross-sectional studies in a variety of populations (reviewed by McNeil & Silagy, 1991). Longitudinal studies, however, indicate that there are interesting individual variations among people, some exhibiting little evidence for blood pressure elevations during aging (Kannel & Gordon, 1978). Moreover, hypertension appears to be rare in at least 20 different remote tribal groups (reviewed by Burke & Motulsky, 1992). Genetic and dietary influences on the metabolism of sodium have been thought to be important modulators of hypertension, since increased salt retention can be related to increased peripheral vascular resistance, the underlying hemodynamic abnormality in hypertension. A metabolic pathway of major significance to sodium homeostasis and the regulation of blood pressure is the renin-angiotensin cascade (Drazu & Pratt, 1991). Renin is an enzyme released from the kidneys in response to a variety of regulatory signals, including the load of sodium sensed by the distal tubular epithelium, the renal perfusion pressure, vasopressin, and a substance originating in the atrial muscles of the heart that enhances the excretion of sodium (atrial natriuretic factor). Its substrate is the pro-hormone angiotensinogen, a tetradecapeptide synthesized in the liver. The product of this reaction, angiotensin I, serves as a substrate for a protease 'converting enzyme' that results in angiotensin II, a potent octapeptide with multiple physiological actions, including stimulation of the secretion of aldosterone, renal tubular retention of sodium, vasoconstriction, and effects on central nervous system sympathetic outflow. The circulating renin-angiotensin-aldosterone system is activated during sodium depletion and volume contraction. Given the importance of this system for homeostasis of blood pressure, one might anticipate that allelic variations at genes controlling these several steps could contribute to certain forms of essential hypertension. A recent three-pronged approach to this question, employing a genetic linkage (sib-pair) analysis, a molecular characterization of alleles, and a population-based association study of adult hypertensives vs. controls has indeed verified this prediction for the case of the angiotensinogen gene (Jeunemaitre *et al.*, 1992). There is substantial polymorphism at that locus, certain alleles being implicated as constitutional susceptibility factors for hypertension. One speculative interpretation of that research is that these polymorphisms may have developed under selective pressure to permit an increased pool of tissue angiotensinogen. This may be one effect of certain alleles associated with hypertension. It is conceivable that significant increases in the concentration and/or the compartmentalization of that substrate might have conferred some evolutionary advantage in terms of more rapid control of blood pressure and ancillary sympathetic functions. This polymorphic system might thus prove to be an example of antagonistic pleiotropy, since those carrying such alleles appear to be at greater risk of developing hypertension.

Disorders of protease and protease inhibitor metabolism

Antithrombin deficiency

Antithrombin is a serine protease inhibitor acting within the intrinsic coagulation pathway. It inhibits thrombin, factor IXa and factor Xa. Some 1-2% of patients with venous thromboembolic disease have a deficiency of antithrombin due to an autosomal dominant mutation. The prevalence of such mutations among geriatric subjects with thromboembolic disease is unknown, but is unlikely to be less than 1-2%. Mutations have been described that involve different functional domains of antithrombin, including alterations of a heparin-binding site. Deletions have also been described.

By age 50, about 85% of subjects carrying an autosomal mutation will have experienced at least one thrombotic or embolic event, most of them having had a number of such episodes. We need much more information concerning the potential role of minor allelic variations in determining the probability of an individual developing thromboembolism, particularly after certain precipitating factors, such as trauma, infection, prolonged bed rest and various surgical procedures.

Alpha-1-antitrypsin deficiency

This glycoprotein (a1AT) inhibits a number of serine proteases. Particular attention has been given to its ability to inhibit the elastase present in neutrophiles, a principal component of the cellular inflammatory reaction to a number of different types of tissue injury.

The first observations concerning a relationship between heritable variations in the levels of a1AT and a late onset human degenerative disorder was with pulmonary emphysema, a condition that, together with chronic bronchitis, is responsible for chronic obstructive pulmonary disease. It is now clear that a major environmental contributor to this condition is cigarette smoke, one reaction to which is an infiltration of neutrophiles into the bronchopulmonary tree. The unopposed action of neutrophile elastase leads to tissue destruction and emphysema. The extent to which the very common finding of emphysema in elderly subjects ('senile emphysema') is morphologically comparable is controversial. Some authors regard this process as a 'hyperinflation' of alveoli, without significant associated tissue destruction. In the author's experience, it is difficult to rule out prior tissue destruction in senile emphysema. Moreover, there is also concomitant fibrosis and chronic bronchitis. The extent to which this is related to exogenous injury vs. some contribution of endogenous age-related alteration is unknown. In any case, it is clear that heritable variations in the efficiency of protection from elastase-mediated injury, from whatever cause, could be of major significance in the rates at which these phenotypes reach some clinically significant threshold (Perlmutter & Pierce, 1989).

The structural gene for a1AT (abbreviated PI or AAT) is on the long arm of chromosome 14, at 14q32.1. More than 75 allelic variants have so far been described, a large proportion of which are characterized by a deficiency of the gene product (Brantly, Nukiwa, & Crystal, 1988).

In addition to pulmonary emphysema, a1AT deficiencies have been associated with cirrhosis of the liver, both in children and in adults (Cox, 1989). Thus, there is a rich genetic substrate for determining the outcome of nature-nurture interactions for these forms of emphysema and cirrhosis.

Familial Alzheimer's Disease (FAD)

This is the subject of a separate chapter in this volume by Rudy Tanzi. The reader should consult that chapter for a fuller set of references. Despite the great importance of this field, it will therefore be dealt with briefly, in the context of the present biochemical genetic classification of abiotrophic disorders.

There are several reasons for the working hypothesis that altered interactions between certain yet-to-be-identified proteases and protease inhibitors are essential components in the pathogenesis of both familial and sporadic forms of the disease. 1) The only specific gene defect so far established at the molecular level involves the structural gene for the β-amyloid precursor protein (reviewed by Fidani & Goate, 1992). Two of the three major isoforms derived from the differentially spliced messenger of that gene include domains coding for a serine protease inhibitor (reviewed by Beyreuther et al., 1992). 2) Highly insoluble aggregates of β-amyloid, the monomeric unit of which is a proteolytic derivative of the β-amyloid precursor protein, accumulate in the blood vessels and neuritic plaques of patients with familial and sporadic forms of the disorder (Beyreuther *et al.*, 1992). They may also accumulate at synapses (Probst *et al.*, 1991; Adams, 1992). 3) At least some mutant forms of the protein appear to result in the increased production of β-amyloid (Citron *et al.*, 1992; Cai, Golde & Younkin, 1993). 4) The gene for the β-amyloid precursor protein is localized on the long arm of chromosome 21, in the region associated with phenotypic expression of the Down syndrome (trisomy 21); those patients develop the lesions of Alzheimer's disease, including deposits of β-amyloid, prematurely (Beyreuther *et al.*, 1992). 5) There is evidence that altered forms of processing of the precursor protein may be of major importance in neurotoxicity (Fukuchi *et al.*, 1992a). 6) The toxicity of these altered products of proteolytic degradation may be specific for neuronal cells (Fukuchi *et al.*, 1992b; Fukuchi, Sopher, & Martin, 1993). 7) The comparative studies of the rates of accumulation of β-amyloid in the brains of mammalian species of contrasting maximum life span potentials are consistent with the view that the process is coupled to some intrinsic process of biological aging rather than merely to chronological time (Selkoe *et al.*, 1987). (Rodent species, however, appear to be exceptionally resistant to the spontaneous development of deposits of β-amyloid in old age.)

One must keep open the possibility, however, that in some forms of Alzheimer's disease, β-amyloid is indeed an epiphenomenon. There are a limited number of patterns of reaction to injury within the central nervous system. The characteristic histological findings may have a degree of non-specificity, resulting from several biochemically distinct pathogenetic mechanisms. A key test of this possibility will come when one determines the nature of the gene products associated with the several genetically distinctive autosomal dominant forms of FAD (Schellenberg *et al.*, 1992b; St. George-Hyslop *et al.*, 1992; Van Broeckhoven *et al.*, 1992; Mullan *et al.*, 1992).

Conclusions

We have taken only a very small sample of the hundreds, or perhaps thousands (Martin, 1978) of genetic polymorphisms and mutations that manifest themselves during adult life via degenerative and/or proliferative aberrations. These genetic variants mimic, to some extent, what one might see in aging among large populations of individuals. Apparently 'normal' individuals in such populations exhibit striking variations in patterns of aging. It is difficult in fact to define 'normal' aging in our species.

Without question, many of the genetic variants described above reduce reproductive fitness and thus might be considered less appropriate models than, for example, the autosomal dominant mutations of familial Alzheimer's disease. One must keep in mind, however, the strong ascertainment bias of the medical profession. It is reasonable to expect that, for each allele with an effect sufficiently great to bring the patient to the attention of a physician, there exist many alleles with more subtle modulating effects upon the maintenance of macromolecular integrity and the maintenance of proliferative homeostasis. This has been clearly established for a number of the loci discussed above. Moreover, it is quite likely that many apparently wild-type alleles act over time to produce deleterious by-products or epiphenomena that escape the force of natural selection. Such gene action is at the heart of what we characterize as aging. Although beyond the scope of this review, it will also be important to consider how environmental factors can modulate such gene actions.

Acknowledgements

The author thanks his close colleagues Thomas D. Bird and Arno G. Motulsky for assistance in discovering the origins of the term 'abiotrophy'. He also thanks Ms. Janice Garr for her continued assistance with preparing manuscripts for publication. Support for research related to issues dealt with in this review have come from research and training grants from the National Institute on Aging (AG00057, AG01751, AG08303, AG05136, and AG08985).

References

Adams, I. M., 1992. Structural plasticity of synapses in Alzheimer's disease. Mol. Neurobiol. 5: 411-419.

Albin, R. L., 1988. The pleiotropic gene theory of senescence. Supportive evidence from human genetic disease. Ethol. Sociobiol. 9: 371-382.

Bearn, A. G., editor-in-chief, 1992. Genetics of Coronary Heart Disease, Institute of Medical Genetics, University of Oslo, Norway.

Bell, G. I., K. Xiang, M. V. Newman, S. H. Wu, L. G. Wright, S. S. Fajans, R. S. Spielman & N. J. Cox, 1991. Gene for non-insulin-dependent diabetes mellitus (maturity onset diabetes of the young subtype) is linked to DNA polymorphism on human chromosome 20q. Proc. Natl. Acad. Sci. USA 88: 1484-1488.

Beyreuther, K., T. Dyrks, C. Hilbich, U. Monning, G. Konig, G. Multhaup, P. Pollwein & C. L. Masters, 1992. Amyloid precursor protein (APP) and beta A4 amyloid in Alzheimer's disease and Down syndrome. Prog. Clin. Biol. Res. 379: 159-182.

Brantly, M., T. Nukiwa & R. G. Crystal, 1988. Molecular basis of alpha-1-antitrypsin deficiency. Am. J. Med. 84: 13-31.

Brody, L. C., G. A. Mitchell, C. Obie, J. Michaud, G. Steel, G. Fontaine, M. F. Robert, I. Sipila, M. Kaiser-Kupfer & D. Valle, 1992. Ornithine delta-aminotransferase mutations in gyrate atrophy. Allelic heterogeneity and functional consequences. J. Biol. Chem. 267: 3302-3307.

Burke, W. & A. G. Motulsky, 1992. Hypertension, pp. 170-191 in The Genetic Basis of Common Diseases. edited by R. A. King, J. I. Rotter & A. G. Motulsky. Oxford Press, N.Y.

Cai, X.-D., T. E. Golde & S. G. Younkin, 1993. Release of excess amyloid β protein from a mutant amyloid β protein precursor. Science 259: 514-516.

Citron, M., T. Oltersdorf, C. Haass, L. McConlogue, A. Y. Hung, P. Seubert, C. Vigo-Pelfrey, I. Lieberburg & D. J. Selkoe, 1992. Mutation of the β-amyloid precursor protein in familial Alzheimer's disease increases β-protein production. Nature 360: 672-674.

Clark, R., L. Daly, K. Robinson, E. Naughten, S. Cahalane, B. Fowler & I. Graham, 1991. Hyperhomocystinuria: an independent risk factor for vascular disease. N. Engl. J. Med. 324: 1149-1155.

Cox, D. W., 1989. al-antitrypsin deficiency, pp. 2409-2437 in The Metabolic Basis of Inherited Disease, Volume II, 6th edition, edited by C. R. Scriver, A. L. Beaudet, W. S. Sly, & D. Valle. McGraw Hill, N.Y.

Drazu, V. J. & R. E. Pratt, 1991. Renin-angiotensin system: biology, physiology and pharmacology, pp. 1817-1849 in The Heart and Cardiovascular System: Scientific Foundations, volume 2, 2nd edition, edited by H. A. Fozzard, R. B. Jennings, E. Haber, A. M. Katz & H. E. Morgan. Raven Press, N.Y.

Eggen, D. A. & L. A. Solberg, 1968. Variation of artherosclerosis with age. Lab. Invest. 18: 571-579.

Fidani, L. & A. Goate, 1992. Mutations in APP and their role in beta-amyloid. Prog. Clin. Biol. Res. 379: 195-214.

Friel, J. P. editor, 1985. Dorland's Illustrated Medical Dictionary, 26th edition. W. B. Saunders, Philadelphia.

Froguel, P., M. Vaxillaire, F. Sun, G. Velho, H. Zouali, M. O. Butel, S. Lesage, N. Vionnet, K. Clement, F. Fougerousse, Y. Tanizawa, J. Weissenbach, J. S. Beckmann, G. M. Lanthrop, Ph. Passa, M. A. Permutt & D. Cohen, 1992. Close linkage of glucokinase locus on chromosome 7p to early-onset non-insulin-dependent diabetes mellitus. Nature 356: 162-164.

Fukuchi, K., K. Kamino, S. S. Deeb, A. C. Smith, T. Dang & G.

M. Martin, 1992a. Overexpression of amyloid precursor protein alters its normal processing and is associated with neurotoxicity. Biochem. Biophys. Res. Commun. 182: 165-173.

Fukuchi, K., K. Kamino, S. S. Deeb, C. E. Furlong, J. A. Sundstrom, A. C. Smith & G. M. Martin, 1992b. Expression of a carboxy-terminal region of the β-amyloid precursor protein in a heterogeneous culture of neuroblastoma cells: evidence for altered processing and selective neurotoxicity. Mol. Brain Res. 16: 37-46.

Fukuchi, K., B. Sopher & G. M. Martin, 1993. Neurotoxicity of β-amyloid. Nature 361: 122.

Genest Jr., J. J., J. R. McNamara, D. N. Salen, P. W. Wilson, E. J. Schaefer & M. R. Malinow, 1990. Plasma homocyst(e)ine levels in men with premature coronary artery disease. J. Am. Coll. Cardiol. 16: 1114-1119.

Goto, M., M. Rubenstein, J. Weber, K. Woods & D. Drayna, 1992. Genetic linkage of Werner's syndrome to five markers on chromosome 8. Nature 355: 735-738.

Gowers, W. R., 1902. Lancet Abiotrophy. Lancet 1: 1003-1007.

Gruen, J. R., V. L. Goei, K. M. Summers, A. Capossela, L. Powell, J. Halliday, H. Zoghbi, H. Shukla & S. M. Weissman, 1992. Physical and genetic mapping of the telomeric major histocompatibility complex region in man and relevance to the primary hemochromatosis gene (HFE). Genomics 14: 232-240.

Gu, Z., V. Ramesh, V. Kozich, M. S. Korson, J. P. Kraus & V. E. Shih, 1991. Identification of a molecular genetic defect in homocystinuria due to cystathionine β-synthase deficiency (abstract). Am. J. Hum. Genet. 49: 406.

Hattersley, A. T., R. C. Turner, M. A. Permutt, P. Patel, Y. Tanizawa, K. C. Chiu, S. O'Rahilly, P. J. Watkins & J. S. Wainscoat, 1992. Linkage of type 2 diabetes to the glucokinase gene. Lancet 339: 1307-1310.

Hobbs, H. H., D. W. Russell, M. S. Brown & J. L. Goldstein, 1990. The LDL receptor locus in Familial Hypercholesterolemia: Mutational analysis of a membrane protein. Ann. Rev. Genet. 24: 133-170.

Jeunemaitre, X., F. Soubrier, Y. V. Kotelevtsev, R. P. Lifton, C. S. Williams, A. Charru, S. C. Hunt, P. N. Hopkins, R. R. Williams, J.-M. Lalouel & P. Corvol, 1992. Molecular basis of human hypertension: role of angiotensinogen. Cell 71: 169-180.

Kamboh, M. I., R. E. Ferell & B. A. Kottke, 1991. Expressed hypervariable polymorphism of apolipoprotein (a). Am. J. Hum. Genet. 49: 1063-1074.

Kannel, W. B. & T. Gordon, 1978. Evaluation of cardiovascular risk in the elderly. The Framingham Study. Bull. N.Y. Acad. Med. 54: 573-591.

Katagiri, H., T. Asano, H. Ishihara, K. Inukai, M. Anai, J. Miyazaki, K. Tsukuda, M. Kikuchi, Y. Yazaki & Y. Oka, 1992. Nonsense mutation of glucokinase gene in late-onset non-insulin-dependent diabetes mellitus. Lancet 340: 1316-1317.

Lawn, R. M., D. P. Wade, R. E. Hammer, G. Chiesa, J. G. Verstuft & E. M. Rubin, 1992. Atherogenesis in transgenic mice expressing human apolipoprotein(a). Nature 360: 670-672.

Martin, G. M., 1978. Genetic syndromes in man with potential relevance to the pathobiology of aging. Birth Defects: Original Article Series 14: 5-38.

306

Martin, G. M., 1979. Proliferative homeostasis and its age-related aberrations. Mech. Ageing Dev. 9: 385-391.

Martin, G. M., 1989. Genetic modulation of the senescent phenotype in Homo sapiens. Genome 31: 390-397.

Martin, G. M., 1991. Genetic and environmental modulations of chromosomal stability: their roles in aging and oncogenesis. Ann. N.Y. Acad. Sci. 621: 401-417.

Martin, G., C. Ogburn & C. Sprague, 1975. Senescence and vascular disease. Adv. Exp. Med. Biol. 61: 163-193.

Matsutani, A., R. Janssen, H. Donis-Keller & M. A. Permutt, 1992. A polymorphic (CA)n repeat element maps the human glucokinase gene (GCK) to chromosome 7p. Genomics 12: 319-325.

McKusick, V. A., 1992. Mendelian Inheritance in Man, Tenth edition. The Johns Hopkins University Press, Baltimore.

McNeil, J. J. & C. A. Silagy, 1991. Hypertension in the elderly: epidemiology and pathophysiology. Cardiovasc. Drugs Ther. 4 (suppl 6): 1197-1201.

Meyer, T. E., R. D. Baynes, T. H. Bothwell, T. Jenkins, D. Ballaot, P. L. Jooste, A. Green, E. Du-Toit & P. Jacobs, 1988. Phenotypic expression of the HLA-linked iron-loading gene in the Afrikaner population of the western Cape. S. Afr. Med. J. 73: 269-274.

Morisaki, T., M. Gross, H. Morisaki, D. Pongratz, N. Zollner & E. W. Holmes, 1992. Molecular basis of AMP deaminase deficiency in skeletal muscle. Proc. Natl. Acad. Sci. USA 89: 6457-6461.

Motulsky, A. G. & J. D. Brunzell, 1992. The genetics of coronary atherosclerosis, pp. 150-169 in The Genetic Basis of Common Diseases, edited by R. A. King, J. I. Rotter & A. G. Motulsky. Oxford University Press, N.Y.

Mullan, M., H. Houlden, M. Windelspecht, L. Fidani, C. Lombardi, P. Diaz, M. Rossor, R. Crook, J. Hardy, K. Duff & F. Crawford, 1992. A locus for familial early-onset Alzheimer's disease on the long arm of chromosome 14, proximal to the alpha1-antichymotrypsin gene. Nature Genet. 2: 340-342.

Munke, M., J. P. Kraus, T. Ohura & U. Francke, 1988. The gene for cystathionine β-synthase (CBS) maps to the subtelomeric region on human chromosome 21q and to proximal mouse chromosome 17. Am. J. Hum. Genet. 42: 550-559.

Nora, J. J., K. Berg & A. H. Nora, 1991. Cardiovascular Diseases: Genetics, Epidemiology and Prevention, Oxford Monographs on Medical Genetics No 22, pp. 3-40. Oxford University Press, N.Y.

Perlmutter, D. H. & J. A. Pierce, 1989. The a1-antitrypsin gene and emphysema. Am. J. Physiol. 257: L147-L162.

Probst, A., D. Langui, S. Ipsen, N. Robakis & J. Ulrich, 1991. Deposition of β/A4 protein along neuronal plasma membranes in diffuse senile plaques. Acta Neuropathol. 83: 21-29.

Ramesh, V., J. F. Gusella & V. E. Shih, 1991. Molecular pathology of gyrate atrophy of the choroid and retina due to ornithine aminotransferase deficiency. Mol. Biol. Med. 8: 81-93.

Rath, M. & L. Pauling, 1990. Hypothesis: Lipoprotein(a) is a surrogate for ascorbate. Proc. Natl. Acad. Sci. USA 87: 6204-6207.

Rose, M. R., 1991. Evolutionary Biology of Aging. Oxford University Press, NY.

Sabina, R. L., T. Morisaki, P. Clarke, R. Eddy, T. B. Shows, C. C. Morton & E. W. Holmes, 1990. Characterization of the human and rat myoadenylate deaminase genes. J. Biol. Chem. 265: 9423-9433.

Schellenberg, G. D., G. M. Martin, E. M. Wijsman, J. Nakura, T. Miki & T. Ogihara, 1992a. Homozygosity mapping and Werner's syndrome. Lancet 339: 1002.

Schellenberg, G. D., T. D. Bird, E. M. Wijsman, H. T. Orr, L. Anderson, E. Nemens, J. A. White, L. Bonnycastle, J. L. Weber, M. E. Alonso, H. Potter, L. L. Heston & G. M. Martin, 1992b. Genetic linkage evidence for a familial Alzheimer's disease locus on chromosome 14. Science 258: 668-671.

Scriver, C. R., A. L. Beaudet, W. S. Sly & D. Valle, editors, 1992. The Metabolic Basis of Inherited Disease, Sixth Edition. McGraw-Hill, NY.

Selkoe, D. J., D. S. Bell, M. B. Podlisny, D. L. Price & L. C. Cork, 1987. Conservation of brain amyloid proteins in aged mammals and humans with Alzheimer's disease. Science 235: 873-877.

St George-Hyslop, P., J. Haines, E. Rogaev, M. Mortilla, G. Vaula, M. Pericak-Vance, J-F. Foncin, M. Montesi, A. Bruni, S. Sorbi, I. Rainero, L. Pinessi, D. Pollen, R. Polinsky, L. Nee, J. Kennedy, F. Macciardi, E. Rogaeva, Y. Liang, N. Alexandrova, W. Lukiw, K. Schlumpf, R. Tanzi, T. Tsuda, L. Farrer, J-M. Cantu, R. Duara, L. Amaducci, L. Bergamini, J. Gusella, A. Roses & D. Crapper McLachlan, 1992. Genetic evidence for a novel familial Alzheimer's disease locus on chromosome 14. Nature Genet. 2: 330-334.

Stoffel, M., P. Froguel, J. Takeda, H. Zouali, N. Vionnet, S. Nishi, I. T. Weber, R. W. Harrison, S. J. Pilkis, S. Lesage, M. Vaxillaire, G. Velho, F. Sun, F. Iris, P. Passa, D. Cohen & G. I. Bell, 1992. Human glucokinase gene: isolation, characterization, and identification of two missense mutations linked to early-onset non-insulin-dependent (type 2) diabetes mellitus. Proc. Natl. Acad. Sci. USA 89: 7698-7702.

Van Broeckhoven, C., H. Backhovens, M. Cruts, G. De Winter, M. Bruyland, P. Cras & J.-J. Martin, 1992. Mapping of a gene predisposing to early-onset Alzheimer's disease to chromosome 14q24.3. Nature Genet. 2: 335-339.

M.R. Rose and C.E. Finch (eds.), Genetics and Evolution of Aging, 307–314, 1994.
© 1994 *Kluwer Academic Publishers. Printed in the Netherlands.*

Antagonistic pleiotropy, mutation accumulation, and human genetic disease

Roger L. Albin
Department of Neurology, University of Michigan, Ann Arbor, MI 48109, USA

Received and accepted 22 June 1993

Key words: aging, Huntington's disease, hemochromatosis, myotonic dystrophy, Alzheimer's disease

Abstract

The antagonistic pleiotropy theory of senescence is the most convincing theoretical explanation of the existence of aging. As yet, no locus or allele has been identified in a wild population with the features predicted by the pleiotropic theory. Human genetic diseases offer the opportunity to identify potentially pleiotropic alleles/loci. Four human genetic diseases – Huntington's disease, idiopathic hemochromatosis, myotonic dystrophy, and Alzheimer's disease – may exhibit pleiotropic effects and further study of these diseases might result in the identification of pleiotropic genes causing aging. Inability to find an early life selective benefit associated with these disease-causing alleles would favor the major alternative genetic explanation for aging, the mutation accumulation theory.

Introduction

The antagonistic pleiotropy theory of senescence is the most rigorously formulated explanation of the existence of aging. Based on an idea of Medawar (1952), this theory has been systematically articulated by Williams (1957) and Hamilton (1966), and independently formulated by Wallace (1967). The antagonistic pleiotropy theory posits that some individual loci/alleles have different effects on fitness at different ages. If such a locus/allele had a positive effect on fitness at a relatively early age and a deleterious effect on fitness in older animals, the positive effect on fitness would outweigh the negative effect because of their relative positioning in life history. Such pleiotropic alleles would be favored to spread throughout populations and diminish fitness in older animals. Some corollaries of the pleiotropic theory seem to fit generally accepted facts about aging. For example, the pleiotropic theory predicts that there will be no single proximate mechanism of aging. Further, the pleiotropic theory predicts that the mechanisms of aging vary both between species and among organisms within a single species.

While theoretically compelling, it has been hard to accumulate direct evidence supporting the pleiotropic theory. No such pleiotropic allele/locus has been identified in a wild population. This is not surprising as identification of a pleiotropic allele/locus would require a daunting combination of appropriate genetic, demographic, and life history data. A considerable body of experimental data derived from laboratory populations supports the pleiotropic theory (Sokal, 1970; Mertz, 1975; Rose & Charlesworth, 1980, 1981a, 1981b; Rose, 1984; Luckinbill, 1984; Luckinbill & Clare, 1985; Clare & Luckinbill, 1985; Rose & Graves, 1989; Hutchinson, Shaw & Rose, 1991; Hutchinson & Rose, 1991).

The major alternative genetic theory of aging, the mutation accumulation theory, is also based on a suggestion of Medawar (1952) and has similar dependence on the concept that negative effects of an allele/locus on overall fitness wane if the deleterious effect of the allele/locus occurs relatively late in life. Because of the weak selective forces acting against alleles/loci with deleterious effects late in life, these alleles/loci would tend to accumulate within the genome by a process akin to drift and ultimately come to provide a limit on life span. This idea differs from antagonistic pleiotropy in that no selective benefit is attached to these alleles/loci in earlier phases of life. It is possible that antagonistic pleiotropy and mutation accumulation could operate simultaneously to cause aging within a given population/species.

Study of human populations and their diseases may provide candidate alleles/loci for the ascertainment of pleiotropic genes or for finding evidence of mutation accumulation. A good deal is known about human demography and human diseases, and many human diseases have heritable components. For example, common adult-onset diseases such as breast cancer, atherosclerosis, and adult onset diabetes mellitus (AODM) have significant genetic components. These common disorders may not, however, prove suitable in the search for pleiotropic genes. For example, the high prevalence of atherosclerosis and AODM in contemporary Western societies is probably a result of the great changes in lifestyle that have resulted from economic/social modernization. Consequently, the negative effect of these diseases on the fitness of older humans may be an artifact of modern living conditions and not a true reflection of pleiotropic effects under historic conditions.

Another factor that limits the usefulness of some common human diseases in the search for candidate pleiotropic genes or mutation accumulation effects is the fact that modern medical therapies may have altered the natural history of some of these disorders, reducing their deleterious effects on fitness in later life. Finally, the genetics of these disorders is complex and probably involves multiple loci.

The search for candidate pleiotropic genes among human diseases should probably focus on rarer adult onset human diseases with less complicated genetics, distinctive phenotypes, and prior suggestions of some selective advantage. An ideal candidate for a human pleiotropic gene would fulfill the following criteria:

1) Adult onset, preferably past the peak reproductive years.

2) Simple genetics, preferably dominant inheritance with a high level of penetrance to maximize the demographic and life history information obtainable from pedigrees.

3) Phenotype independent or largely independent of environmental factors.

4) No effect of modern medical therapy on the natural history of the disease.

5) Evidence that the allele confers a selective advantage at an age earlier than the onset of disability secondary to the disease caused by the abnormal allele.

6) The existence of populations/pedigrees allowing the accumulation of sufficient demographic data to obtain an estimate of fitness.

While no human genetic disease fulfills all these criteria, at least four human genetic diseases fulfill some of these criteria and may be candidates for the ascertainment of pleiotropic effects.

This strategy is also useful in the search for mutation accumulation effects. Reasonable evidence that an adult onset genetic disease, especially a relatively common one, does not increase fitness relatively early in life would support Medawar's concept of mutation accumulation. The mutation accumulation theory would be supported even more strongly if several adult onset genetic diseases were examined and found to have no positive effect on fitness at a relatively early age.

While the strategy of using adult onset human genetic diseases in the search for pleiotropic or mutation accumulation effects is attractive, there are some specific limitations of working with human genetic diseases. Evaluating the fitness effects of alleles requires considering the effects of both the heterozygous and homozygous states. The phenotypic effect of homozygosity is not known for most human dominant diseases (though see the discussion of Huntington's disease below), and these diseases are the most attractive for pedigree tracing and accumulation of demographic data. Homozygotes of dominant human genetic disorders are quite rare and this may simplify analysis. For recessive disorders, evaluation of the fitness effects of heterozygosity is crucial.

Appropriate estimation of fitness is also crucial. While fitness should properly be measured as the number of progeny that survive to adulthood, those few studies that have attempted to look at the selective benefit of human disease alleles have tended to look at simple fecundity (see discussion of Huntington's disease below) or simply postulated some beneficial effect without measuring actual fitness.

Huntington's disease. Huntington's disease (HD) is a completely penetrant autosomal dominant neurodegenerative disease. HD is a true dominant disorder with homozygotes having the same phenotype as heterozygotes (Wexler *et al.*, 1987). The phenotype is characterized by progressive involuntary movements, impairment of coordination, personality changes, and dementia (Folstein, 1989). Onset is usually in the third and fourth decades of life and death usually occurs 15-20 years after diagnosis. The prevalence within the USA is estimated at 30-45/million (Folstein, 1989; Conneally, 1984). There is no therapy for HD.

The locus responsible for HD has not yet been identified but has been localized to the short arm of chromosome 4. Analysis of widely dispersed pedigrees has revealed that the HD phenotype is invariably associated with the chromosome 4 locus. The spontaneous mutation rate for HD is very low and it has been suggested that the frequency of HD is maintained by some selective advantage conferred by the HD allele (Folstein, 1989; Hayden, 1981).

Clinicians who work with HD families are invariably impressed with the large size of sibships in affected families. This impression has given rise to speculation that the HD allele increases fitness by increasing fecundity (Hayden, 1981). Several studies have attempted to address this issue by comparing the fecundity of HD victims with control populations. Non-affected siblings have been used as controls with the result that HD victims are generally found to produce more offspring than their normal sibs (Hayden, 1981). Some studies, however, report that when unaffected sibs are compared to normal individuals, their fecundity has been found to be depressed (Reed & Neel, 1959; Shokeir, 1975). Six studies have attempted to compare the fecundity of HD victims to that of the general population (Reed & Neel, 1959; Shokeir, 1975; Marx, 1973; Stevens, 1975; Wallace & Parker, 1973; Walker *et al.*, 1983). One study showed depressed fecundity (Reed & Neel, 1959), one showed the same level of fecundity as the general population (Wallace & Parker, 1973), and four showed elevated fecundity as compared to the general population (Shokeir, 1975; Marx, 1973; Stevens, 1975; Walker *et al.*, 1983). The latter four studies all indicated a significant advantage in fitness over the general population. Methodological differences and sampling bias may explain some of the difference in results between the study showing depressed fecundity and those showing increased

fecundity. The study showing depression of fertility in HD victims ascertained HD cases by examining the records of large state-operated mental hospitals (Reed & Neel, 1959). The authors of this study estimated that some of the depression of fecundity was due to premature termination of reproduction due to hospitalization. This study was conducted during the 1950s when permanent and relatively early hospitalization for disorders such as HD was a common practice. On the other hand, two studies reporting increased fecundity in HD were based on efforts to ascertain all HD victims within the study regions and were not so dependent on the records of institutionalized patients (Shokeir, 1975; Walker *et al.*, 1983).

Two studies have attempted to longitudinally assess the possible contribution of the HD allele to fitness. Harper *et al.* (1979) were able to estimate the birth rate of HD carriers in south Wales throughout the 20th century and found that the birth rate of HD carriers has remained constant while that of the general population has declined. These results imply a reproductive advantage for the HD allele. Stine and Smith (1990) estimated the coefficient of selection for the HD allele in the Afrikaner population of South Africa by comparing probable prevalence of HD in the founding (and very small) population of Afrikaners with recent estimates of HD prevalence. Their analysis suggested that the HD allele carries a significant negative impact on fitness.

Some data, then, suggest that the HD allele confers a selective advantage and might fulfill the criteria for a pleiotropic allele. The studies reporting a reproductive advantage for HD victims should, however, be interpreted with caution. None of these studies attempted to control for socioeconomic or class variables that might affect fecundity. HD victims tend to come from lower socioeconomic levels (Reed *et al.*, 1958; Mattson, 1974), which could independently alter demographic characteristics. In addition, all the above data provide birth rate information. Reed and Neel (1959) have suggested that children of HD victims tend to have a higher infant mortality rate, which might offset any selective advantage due to increased fecundity.

Testing of the hypothesis that the HD allele is a pleiotropic gene requires the comparison of a suitable HD population with a carefully matched control population. The best candidate population for this type of study would be the well studied Venezuela pedigree. This remarkably large (hundreds of individuals) pedigree has been carefully characterized by an international team of clinicians and geneticists (Young *et al.*, 1986; Penney *et al.*, 1990). Linkage analysis has been done for the entire pedigree and the HD allele carrier status is known for all members of the pedigree. It should be possible to determine the fecundity and progeny survival rates of all members of the pedigree. Development of a control population matched for economic and demographic variables would permit a test of the hypothesis that the HD allele is a pleiotropic senescence gene.

Idiopathic hemochromatosis. Idiopathic hemochromatosis (IH) is an autosomal recessive inherited disorder of iron metabolism (Bothwell, Charlton & Motulsky, 1983). IH is remarkably common. Approximately 10% of the population of North America and Europe are heterozygous for the IH allele, and the prevalence of homozygotes is approximately 0.3%. Penetrance appears to be complete or nearly complete in men and incomplete in women. Most IH victims develop clinical symptoms between the ages of 40 and 60 (Finch & Finch, 1955; Milder *et al.*, 1980). IH victims suffer from progressive deterioration in liver, endocrine pancreatic, cardiac, and pituitary function. The latter can include gonadal failure (Adams, Kertesz & Valberg, 1991; Finch & Finch, 1955; Milder *et al.*, 1980).

The locus responsible for IH is found on chromosome 6 and is closely linked to the HLA complex. The molecular defect is unknown, but the pathophysiology of IH is well characterized and the pathologic consequences of IH result from excessive gastrointestinal absorption of iron. The human capacity to excrete iron is limited and body iron stores are regulated in large part by intestinal iron absorption. In IH, iron absorption is greatly increased. When normal iron storage capacity is saturated, iron is deposited in abnormal locations with resulting organ dysfunction. Because of menstruation, pregnancy, and lactation, women experience greater normal losses of iron than men. The greater demand for iron faced by women probably accounts for the reduced penetrance of IH in women.

The historic prognosis of IH was dismal, with most patients dying within months of diagnosis (Milder *et al.*, 1980). Medical therapy has had a significant impact on the natural history of IH (Milder *et al.*, 1980; Adams, Speechley & Kertesz, 1991). Insulin therapy for diabetes secondary to IH prolongs life expectancy and institution of regular phlebotomy prior to the development of organ impairment can result in normal life expectancy.

The high frequency of the IH allele has led to speculation that it is the result of a selective advantage. The suggested advantage is amelioration of iron deficiency anemia by enhanced intestinal iron absorption (Rotter & Diamond, 1987; Motulsky, 1979). Iron deficiency anemia is common in industrialized countries and very common in developing nations where diets are often poor in bioavailable iron. IH homozygosity would be especially beneficial to women, who have greater demands for iron than men.

If IH could be proven to have a positive selective effect, then the IH allele would satisfy criteria for a pleiotropic gene. In addition, because of its possible differential benefits/deleterious effects in men and women, IH might be an example of a sexually antagonistic gene (Rice, 1992). Identifying IH homozygotes is relatively straightforward. A combination of serum iron studies and histocompatibility antigen typing can presently identify IH homozygotes (Dadone *et al.*, 1982; Borecki *et al.*, 1990). In the future, cloning of the IH locus will permit definitive ascertainment of IH allele heterozygotes and homozygotes. While medical therapy has altered the natural history of IH, it should still be possible to evaluate the potential selective benefits of IH.

One approach would be to survey a population for IH and determine if iron deficiency anemia is less common among IH homozygotes than among a set of matched controls. This approach should be used in a region where iron deficiency is common and will be possible only after the IH locus has been cloned. At present, IH homozygotes are identified by serum iron studies showing evidence of increased iron stores. By definition, this criterion would exclude IH homozygotes with normal or decreased iron stores and introduce significant sampling bias into the data. Another approach would be to compare the actual fitness of IH homozygotes with an appropriate control population. IH has been extensively studied in Utah (Kravitz *et al.*, 1979), where a relatively stable population and the extensive genealogical records maintained by the Church of the Latter Day Saints permit good ascertainment of pedigrees and could allow development of appropriately matched controls.

Estimating the fitness of IH allele heterozygotes would also be crucial but more difficult. IH allele heterozygotes have a mildly increased ability to absorb iron (Cartwright, 1979) and the high prevalence of the IH allele could be maintained by some selective advantage accruing to the heterozygotes. Identification of IH allele heterozygotes for a study of heterozygote fitness requires molecular genotyping and presupposes molecular identification of the IH locus.

Myotonic dystrophy. Myotonic dystrophy (MyD) is the most common form of adult-onset muscular dystrophy with an estimated prevalence of 3-5/100,000 (Brooke, 1986). MyD is an autosomal dominant disorder with variable expression affecting many organs. The effect of homozygosity is unknown. Myotonia (impaired relaxation of muscles following contraction) and progressive skeletal muscle degeneration are hallmarks of MyD. Other common features include cataracts, endocrine disorders including diabetes and gonadal dysfunction, cardiac arrhythmias, and mild mental retardation. Onset of symptoms often occurs in the young adult years, but in many cases the disease may not become symptomatic until considerably later in life. MyD does occur in children and the prognosis of these unfortunate children is particularly poor. Cardiac dysrhythmias and muscular weakness (which often impair swallowing and pulmonary function) predispose patients to early death. In common with a small number of other genetic diseases, MyD exhibits the phenomenon of 'anticipation' in which the phenotype appears earlier and with greater severity in succeeding generations. A typical pedigree might have a grandparent with cataracts and few other detectable abnormalities, an obviously affected parent in his or her thirties, and a severely affected child. There is no therapy for MyD.

Recent studies have identified the affected locus in MyD on chromosome 19 (Aslanidis *et al.*, 1992; Buxton *et al.*, 1992; Harley *et al.*, 1992). This locus encodes a protein with considerable homology to known protein kinases, enzymes that regulate the activity of other proteins. The basis for the defect in MyD is insertion of unstable trinucleotide repeats. The length of these insertions can increase in suc-

ceeding generations and the increasing length correlates with increasing severity of the clinical manifestations of MyD found in succeeding generations. Other human genetic diseases exhibiting anticipation also have unstable trinucleotide insertions.

MyD has a low spontaneous mutation rate and this fact has given rise to speculation that the prevalence of MyD is maintained by some selective advantage. This speculation has been bolstered by recent molecular data. The MyD allele is in very strong linkage disequilibrium with closely linked markers (Harley *et al.*, 1992; Harley *et al.*, 1991). The linkage data suggest that virtually all MyD cases are descended from a single mutation, a surprising fact in view of the significant impairments that usually result from MyD.

It may be possible to assess the fitness of the MyD allele by surveying MyD patients in the Saguenay region of Quebec, where MyD has the highest known prevalence in the world, estimated at 189/100,000 (Mathieu, De Braekeleer & Prevost, 1990). All cases within this region descend from a single couple who settled in Quebec in 1657 and the MyD allele has been passed over 10-14 generations. This extended pedigree has been carefully reconstructed. The biologic fitness of the MyD allele has been determined by comparing the fitness of contemporary MyD patients with an appropriate control population. To obtain a longitudinal evaluation of the fitness of the MyD allele, it might be possible to compare the historic fitness of individuals within the MyD pedigree with a control historical population.

Alzheimer's disease. Alzheimer's disease (AD) is a common neurodegenerative disease and a tantalizing candidate for a pleiotropic senescence gene and for evaluation of mutation accumulation effects. AD is primarily a disease of the elderly and prevalence increases with age. As many as 15% of individuals over age 65 may suffer from AD and the prevalence is much higher among individuals older than 80. The major manifestation of AD is progressive dementia, often accompanied by dysphasia and dyspraxia.

Several lines of evidence suggest a substantial genetic component in AD. Epidemiological studies aimed at revealing environmental factors contributing to AD have not produced convincing results. Rare cases of AD occur at a relatively early age, and in many of these cases there is clear evidence of autosomal dominant inheritance. In typical AD cases, the late onset of disease prevents the ascertainment of inheritance patterns by conventional pedigree tracing and linkage analysis. Recent studies of typical AD, however, have found that there is a high incidence of dementia in first degree relatives of AD patients (Breitner, 1991). Breitner (1991) has recently summarized the literature examining the incidence of dementia in first degree relatives of AD victims and argues convincingly that AD is inherited in an autosomal dominant fashion with age dependent penetrance and maximum cumulative risk of developing AD around age 90.

Assuming that the majority of AD is genetic in etiology, AD would be among the most prevalent of human genetic diseases. It is possible that there may be considerable genetic heterogeneity underlying the AD phenotype. Linkage studies of early onset pedigrees have identified three different loci responsible for AD (Goate *et al.*, 1991; Pericak-Vance *et al.*, 1991; St. George-Hyslop *et al.*, 1987; Tanzi *et al.*, 1987; Van Broeckhoven, 1987). Whether genetic heterogeneity exists in typical AD can only be determined after molecular genetic identification of the responsible gene(s).

If AD is among the most prevalent of human genetic diseases, its high frequency might result from some selective advantage conferred by alleles causing AD. The fact that AD achieves maximum penetrance late in life makes it an intriguing candidate for ascertainment of pleiotropic gene effects. Up to this point, there have been no suggestions that AD-causing alleles confer a selective advantage. The lack of speculation about the possible selective advantages of AD alleles is probably a result of the fact that most clinicians and scientists interested in AD have not thought of it as a genetic disease. One study of the incidence of dementia among first degree relatives of AD victims found also that AD victims tended to have larger numbers of children than controls (Fitch, Becker & Heller, 1988) though this difference was not statistically significant (N. Fitch, personal communication).

AD-causing alleles are also excellent candidates for mutation accumulation effects. If there was no reasonable evidence that AD-causing alleles had selective benefits at any age, then AD-causing alleles would be a prime example of mutation accumulation.

Evaluation of the effect of AD-causing alleles would seem to be relatively straightforward. AD patients could be identified at autopsy or by epidemiological survey and their fitness compared with a matched set of controls. Alternatively, when the genetic defect(s) underlying typical AD are discovered, it should be possible to compare the fitness of those carrying the allele(s) with a control population. Molecular identification of loci associated with AD may also give clues to the possible effects on fitness of alleles causing AD.

Conclusions. While none of the diseases discussed above fulfills the criteria of a pleiotropic gene, there is suggestive evidence of pleiotropic effects, and further study of these disorders might lead to proof of the existence of a pleiotropic gene in a wild population. The use of human disease populations is a plausible way to ascertain the existence of negative pleiotropy or mutation accumulation and has significant advantages over the use of non-human populations where appropriate demographic and genetic data may be hard to obtain. Negative results regarding selective benefits associated with disease-causing alleles would not disprove the pleiotropic theory, but a positive result would provide powerful evidence in favor of the pleiotropic gene theory of senescence. Negative results regarding selective benefits of disease alleles, however, would provide evidence favoring the mutation accumulation theory.

Acknowledgements

Supported by grant NS01300 from the National Institute of Neurological Disease and Stroke. I thank the anonymous reviewers for their helpful criticisms.

References

Adams, P. C., A. E. Kertesz & L. S. Valberg, 1991. Clinical presentation of hemochromatosis: a changing scene. Am. J. Med. 90: 445-449.

Adams, P. C., M. Speechley & A. E. Kertesz, 1991. Long-term survival analysis in hereditary hemochromatosis. Gastroenterology 101: 368-372.

Aslanidis, C., G. Jansen, C. Amemiya, G. Shutler, M. Mahade-van, C. Tsilfidis, C. Chen, J. Alleman, N. G. M.

Wormskamp, M. Vooijs, J. Buxton, K. Johnson, H. J. M. Smeets, G. G. Lennon, A. V. Carrano, R. G. Korneluk, B. Wieringa & P. J. de Jong, 1992. Cloning of the essential myotonic dystrophy region and mapping of putative defect. Nature 355: 548-551.

Borecki, I. B., G. M. Lathrop, G. E. Bonney, J. Yaouang & D. C. Rao, 1990. Combined segregation and linkage analysis of genetic hemochromatosis using affection status, serum iron, and HLA. Am. J. Hum. Genet. 47: 542-550.

Bothwell, T. H., R. W. Charlton & A. G. Motulsky, 1983. Hemochromatosis, pp. 1269-1298 in The Metabolic Basis of Inherited Disease, edited by J. B. Stanbury, J. B. Wyngaarden, D. S. Frederickson, J. L. Goldstein & M. S. Brown. McGraw-Hill, New York.

Breitner, J. C., 1991. Clinical genetics and genetic counseling in Alzheimer Disease. Ann. Intern. Med. 115: 601-606.

Brooke, M. H., 1986. A Clinician's View of Neuromuscular Diseases. Williams & Wilkins, Baltimore, Maryland.

Buxton, J., P. Shelbourne, J. Davies, C. Jones, T. Van Tongeren, G. Aslanidis, P. de Jong, G. Jansen, M. Anvret, B. Riley, R. Williamson & K. Johnson, 1992. Detection of an unstable fragment of DNA specific to individuals with myotonic dystrophy. Nature 355: 547-548.

Cartwright, G. E., 1979. Hereditary hemochromatosis: phenotypic expression of the disease. New England Journal of Medicine 301: 175-179.

Clare, M. J. & L. S. Luckinbill, 1985. The effects of gene-environment interaction on the expression of longevity. Heredity 55: 19-29.

Conneally, P. M., 1984. Huntington's disease: genetics and epidemiology. American Journal of Human Genetics 36: 506-526.

Dadone, M. M., J. P. Kushner, C. Q. Edwards, D. T. Bishop & M. H. Skolnick, 1982. Hereditary hemochromatosis. Analysis of laboratory expression of the disease by genotype in 18 pedigrees. Am. J. Clin. Pathol. 78: 196-207.

Finch, S. C. & C. A. Finch, 1955. Idiopathic hemochromatosis, an iron storage disease. Medicine 34: 381-430.

Fitch, N., R. Becker & A. Heller, 1988. The inheritance of Alzheimer's disease: a new interpretation. Ann. Neurol. 23: 14-19.

Folstein, S. E., 1989. Huntington's Disease, A Disorder of Families. The Johns Hopkins University Press, Baltimore, Maryland.

Goate, A., M. C. Chartier-Harlin, M. Mullan, J. Brown, F. Crawford, L. Fidani et al., 1991. Segregation of a missense mutation in the amyloid precursor protein gene with familial Alzheimer's disease. Nature 349: 704-706.

Hamilton, W. D., 1966. The moulding of senescence by natural selection. Journal of Theoretical Biology 12: 12-45.

Harley, H. G., J. D. Brook, J. Floyd, S. A. Rundle, S. Crow, K. V. Walsh, M. C. Thibault, P. S. Harper & D. J. Shaw, 1991. Detection of linkage disequilibrium between the myotonic dystrophy locus and a new polymorphic DNA marker. Am. J. Hum. Genetics 49: 68-75.

Harley, H. G., J. D. Brook, S. A. Rundle, S. Crow, W. Reardon, A. J. Buckler, P. S. Harper, D. E. Housman & D. J. Shaw, 1992. Expansion of an unstable DNA region and phenotypic variation in myotonic dystrophy. Nature 355: 545-546.

Harper, P. S., D. A. Walker, A. Tyler, R. G. Newcombe & K. Davies, 1979. Huntington's chorea: The basis for long-term prevention. Lancet II: 346-349.

Hayden, M. R., 1981. Huntington's Chorea. Springer-Verlag, Berlin.

Hutchinson, E. W. & M. R. Rose, 1991. Quantitative genetics of postponed aging in Drosophila melanogaster. I. Analysis of outbred populations. Genetics 127: 719-727.

Hutchinson, E. W., A. J. Shaw & M. R. Rose, 1991. Quantitative genetics of postponed aging in Drosophila melanogaster. II. Analysis of selected lines. Genetics 127: 729-737.

Kravitz, K., M. Skolnick, C. Cannings, D. Carmelli, B. Baty, B. Amos, A. Johnson, N. Mendell, C. Edwards & G. Cartwright, 1979. Genetic linkage between hereditary hemochromatosis and HLA. Am. J. Hum. Genet. 31: 601-619.

Luckinbill, L. S. & M. J. Clare, 1985. Selection for life span in Drosophila melanogaster. Heredity 55: 9-18.

Luckinbill, L. S., R. Arking, M. J. Clare, W. C. Cirocco & S. A. Buck, 1984. Selection for delayed senescence in Drosophila melanogaster. Evolution 38: 996-1003.

Marx, R. N., 1973. Huntington's chorea in Minnesota. Advances in Neurology 1: 237-243.

Mathieu, J., M. De Braekeleer & C. Prevost, 1990. Genealogical reconstruction of myotonic dystrophy in the Saguenay-Lac-Saint-Jean area (Quebec, Canada) Neurology 40: 839-842.

Mattson, B., 1974. Huntington's chorea in Sweden. II. Social and clinical data. Acta Psychiatrica Scandinavica Supplementum 255: 221-235.

Medawar, P. B., 1952. An Unsolved Problem in Biology. H. K. Lewis, London.

Mertz, D. B., 1975. Senescent decline in flour beetle strains selected for early adult fitness. Physiological Zoology 48: 1-23.

Milder, M. S., J. D. Cook, S. Stray & C. A. Finch, 1980. Idiopathic hemochromatosis, an interim report. Medicine 59: 34-49.

Motulsky, A. G., 1979. Letter: Genetics of hemochromatosis. New England Journal of Medicine 301: 1291.

Penney, J. B., Jr., A. B. Young, I. Shoulson, S. Starosta-Rubenstein, S. R. Snodgrass, J. Sanchez-Ramos, M. Ramos-Arroyo, F. Gomez, G. Penchaszadeh, J. Alvir et al., 1990. Huntington's disease in Venezuela: 7 years of follow-up on symptomatic and asymptomatic individuals. Mov. Disord. 5: 93-99.

Pericak-Vance, M. A., J. L. Bebout, P. C. Gaskell, L. H. Yamaoka, W. Y. Hung, M. H. Alberts et al., 1991. Linkage studies in familial Alzheimer's disease: evidence for chromosome 19 linkage. Am. J. Human. Genet. 48: 1034-1050.

Reed, T. E., J. H. Chandler, E. M. Hughes & R. T. Davidson, 1958. Huntington's chorea in Michigan. 1. Demography and genetics. Am. J. Hum. Genet. 10: 201-225.

Reed, T. E. & J. V. Neel, 1959. Huntington's chorea in Michigan. 2. Selection and mutation. Am. J. Hum. Genet. 11: 107-136.

Rice, W. R., 1992. Sexually antagonistic genes: Experimental evidence. Science 256: 1436-1438.

Rose, M. R., 1984. Laboratory evolution of postponed senescence in Drosophila melanogaster. Evolution 38: 1004-1010.

Rose, M. R. & B. Charlesworth, 1980. A test of evolutionary theories of senescence. Nature 287: 141-143.

ose, M. R. & B. Charlesworth, 1981a. Genetics of life history in *Drosophila melanogaster* I. Sib analysis of adult females. Genetics 97: 173-186.

Rose, M. R. & B. Charlesworth, 1981b. Genetics of life history in *Drosophila melanogaster* II. Exploratory selection experiments. Genetics 97: 187-196.

Rose, M. R. & J. L. Graves, 1989. What evolutionary biology can do for gerontology. J. Gerontol. 44: B27-29.

Rotter, J. I. & J. M. Diamond, 1987. News and Views: What maintains the frequencies of human genetic diseases? Nature 329: 289-290.

Shokeir, M. H. K., 1975. Investigations on Huntington's disease in the Canadian prairies. II. Fecundity and fitness. Clin. Genet. 11: 349-353.

Sokal, R. R., 1970. Senescence and genetic load: Evidence from tribolium. Science 167: 1733-1734.

St. George-Hyslop, P. H., R. E. Tanzi, R. J. Polinsky, J. L. Haines, L. Nee, P. C. Watkins *et al.*, 1987. The genetic defect causing familial Alzheimer's disease maps on chromosome 21. Science 235: 885-890.

Stevens, D. L., 1981. Quoted in Hayden, M. R. Huntington's Chorea, Springer-Verlag, Berlin.

Stine, O. C. & K. D. Smith, 1990. The estimation of selection coefficients in Afrikaners: Huntington disease, porphyria variegata, and lipoid proteinosis. Am. J. Hum. Genet. 46: 452-458.

Tanzi, R. E., P. H. St. George-Hyslop, J. L. Haines, R. J. Polinsky, L. Nee, J. F. Foncin *et al.*, 1987. The genetic defect in familial Alzheimer's disease is not tightly linked to the amyloid beta-protein gene. Nature 329: 156-157.

Van Broeckhoven, C., A. M. Genthe, A. Vendenberghe, B. Horsthemke, H. Bakchovens, P. Raeymaekers *et al.*, 1987. Failure of familial Alzheimer's disease to segregate with the A4-amyloid gene in several European families. Nature 329: 153-155.

Walker, D. A., P. S. Harper, R. G. Newcombe & K. Davies, 1983. Huntington's chorea in South Wales: mutation, fertility, and genetic fitness. J. Med. Genetics. 20: 12-17.

Wallace, D. C., 1967. The inevitability of growing old. Journal of Chronic Diseases 20: 475-486.

Wallace, D. C. & N. Parker, 1973. Huntington's chorea in Queensland: The most recent story. Advances in Neurology 1: 223-236.

Wexler, N. S. *et al.*, 1987. Homozygotes for Huntington's disease. Nature 326: 194-197.

Williams, G. C., 1957. Pleiotropy, natural selection, and the evolution of senescence. Evolution 11: 398-411.

Young, A. B. *et al.*, 1986. Huntington's disease in Venezuela: Neurologic features and functional decline. Neurology 36: 244-249.